Green
Cities

Green Cities

An A-to-Z Guide

The SAGE Reference Series on
Green Society
Toward a Sustainable Future

NEVIN COHEN, GENERAL EDITOR
The New School

PAUL ROBBINS, SERIES EDITOR
University of Arizona

SAGE | reference

Los Angeles | London | New Delhi
Singapore | Washington DC

Los Angeles | London | New Delhi
Singapore | Washington DC

FOR INFORMATION:

SAGE Publications, Inc.
2455 Teller Road
Thousand Oaks, California 91320
E-mail: order@sagepub.com

SAGE Publications Ltd.
1 Oliver's Yard
55 City Road
London EC1Y 1SP
United Kingdom

SAGE Publications India Pvt. Ltd.
B 1/I 1 Mohan Cooperative Industrial Area
Mathura Road, New Delhi 110 044
India

SAGE Publications Asia-Pacific Pte. Ltd.
33 Pekin Street #02-01
Far East Square
Singapore 048763

Publisher: Rolf A. Janke
Assistant to the Publisher: Michele Thompson
Senior Editor: Jim Brace-Thompson
Production Editors: Kate Schroeder, Tracy Buyan
Reference Systems Manager: Leticia Gutierrez
Reference Systems Coordinator: Laura Notton
Typesetter: C&M Digitals (P) Ltd.
Proofreader: Rae-Ann Goodwin
Indexer: Julie Sherman Grayson
Cover Designer: Gail Buschman
Marketing Manager: Kristi Ward

Golson Media
President and Editor: J. Geoffrey Golson
Author Manager: Ellen Ingber
Editors: Kenneth Heller, Mary Jo Scibetta
Copy Editors: Anne Hicks, Barbara Paris

Printed in the United States of America

Library of Congress Cataloging-in-Publication Data

Green cities : an A-to-Z guide / general editor, Nevin Cohen.

p. cm. — (The Sage references series on green society: toward a sustainable future)
Includes bibliographical references and index.

ISBN 978-1-4129-9682-2 (cloth) — ISBN 978-1-4129-7381-6 (ebk)

1. Urban ecology (Sociology) 2. Sustainable urban development. 3. Urbanization—Environmental aspects. 4. City planning—Environmental aspects. I. Cohen, Nevin.

HT241.G735 2010 307.76—dc22 2011002869

11 12 13 14 15 10 9 8 7 6 5 4 3 2 1

Contents

About the Editors

Green Series Editor: Paul Robbins

Paul Robbins is a professor and the director of the University of Arizona School of Geography and Development. He earned his Ph.D. in Geography in 1996 from Clark University. He is General Editor of the *Encyclopedia of Environment and Society* (2007) and author of several books, including *Environment and Society: A Critical Introduction* (2010), *Lawn People: How Grasses, Weeds, and Chemicals Make Us Who We Are* (2007), and *Political Ecology: A Critical Introduction* (2004).

Robbins's research centers on the relationships between individuals (homeowners, hunters, professional foresters), environmental actors (lawns, elk, mesquite trees), and the institutions that connect them. He and his students seek to explain human environmental practices and knowledge, the influence nonhumans have on human behavior and organization, and the implications these interactions hold for ecosystem health, local community, and social justice. Past projects have examined chemical use in the suburban United States, elk management in Montana, forest product collection in New England, and wolf conservation in India.

Green Cities General Editor: Nevin Cohen

Nevin Cohen is an assistant professor of Environmental Studies at The New School, in New York City, where he teaches courses in urban planning and sustainable food systems. He serves as co-chair of the Tishman Environment and Design Center, The New School's interdisciplinary environmental research and education center, and home to the university's innovative bachelor program in Environmental Studies, which emphasizes urban ecosystems, sustainable design, and public policy. He has a Ph.D. in Urban Planning from Rutgers University, a Masters in City and Regional Planning from the University of California, Berkeley, and a B.A. from Cornell University.

For the past two decades, Dr. Cohen has worked with Fortune 500 companies on corporate sustainability initiatives. Prior to joining the faculty of The New School, he served as managing principal for GreenOrder, Inc., a consulting firm specializing in sustainable business practices, and held senior research positions at Rutgers University's Center for Environmental Communication, Environmental Defense, the World Resources Institute, Tellus Institute, and INFORM, Inc. As a policy analyst and planner in New York City, Dr. Cohen advised local planning boards and real estate developers on green development strategies. He was also responsible for developing landmark municipal recycling, water conservation, and clean fuel laws in New York City as a policy analyst for the City Council and Manhattan borough president.

Introduction

In 2008 humans shifted from a mostly rural to a mostly urban habitat. Cities worldwide are expected to take in all of the growing population over the next several decades, plus additional rural migrants, so that by 2050, according to the United Nations, the number of people living in urban areas is expected to reach 6.4 billion, out of a total population of 9.2 billion. Most of this urban growth will occur in developing regions of Asia, Africa, and Latin America.

As an increasingly urban planet, our major ecological challenges, from water conservation to biodiversity loss to climate change, are now urban problems, making the quest for green cities essential for human survival. Because global financial, telecommunications, and transportation systems link us inextricably together, the decisions made about cities will affect human and nonhuman species everywhere. As more and more people live in cities, the activities and behavior that cities encourage will determine whether we make the planet less or more habitable. How we plan, design, build, and retrofit our cities will affect the world's natural resource consumption and the health and quality of life for an increasing portion of the world's people. This is a particular burden for developing countries, which face the challenges of rapid urban growth and limited economic resources.

The quest for greener, efficient, and more equitable cities is not a new endeavor. As 19th century industrialization created highly polluted, unsanitary, crowded cities teeming with rural migrants who faced brutal living conditions; urban reformers developed schemes to address these problems, including social housing, public parks, and regulatory innovations like zoning. As the relationship between pathogens and diseases became clear, cities embarked on enormous public works programs to provide clean water from distant sources, and sewage systems to remove waste from cities. Nevertheless, despite these sanitary reforms, the dominant thinking through the 20th century was that the scale and density of cities were problems, and that cities needed to be transformed. Utopian schemes like Ebenezer Howard's networks of garden cities, and Le Corbusier's designs to rationalize the space and social structure in cities, captured the imagination of planners looking for an alternative to existing urban forms.

We now recognize that dense cities are significantly more resource-efficient than disbursed settlements, particularly with respect to the energy consumed by transportation and heating and cooling buildings. Cities also offer opportunities for green social and technological innovations, from urban agriculture to distributed energy generation that, if diffused widely, can address global ecological and social problems. However, our cities have substantial existing infrastructure, built when environmental issues were not well understood or incorporated into the design and development process, so most cities will continue to consume resources unsustainably if not redesigned. Moreover, because cities are centers of consumption in a global economy, their ecological footprints extend far beyond their geographic boundaries. The day-to-day activities of city dwellers not only affect the proximate environment, but as producers and consumers urbanites also affect the communities that supply them with goods and services and serve as sinks for their wastes.

Green cities include new communities conceived as sustainable from the ground up, as well as old industrial communities undergoing the process of greening. The former includes new cities like Dongtan, China, where planners and architects have used technologies to generate renewable energy and minimize energy consumption, and urban designs that enable large numbers of people to live densely with a high quality of life. Most existing cities suffer from a legacy of industrial uses, and contain brownfields (contaminated sites), inadequate transit systems, aging water and sewer infrastructure, and vulnerabilities to threats like sea level rise. Understanding how these communities, from New York City to Bogota, have rebuilt outmoded infrastructure, restored urban ecosystems, and instituted innovative environmental policies can serve as blueprints for the greening of other cities.

Cities that are green enable the people who live and work in them to carry out their daily lives in an environmentally sound manner. Many of the negative impacts of urban living occur through everyday activities, such as how residents move about, what they eat, and how they make their homes and workplaces comfortable. These decisions are influenced by urban design and public policies, so green cities attempt to change the landscape so that the activities of residents are more sustainable as they commute (bicycling), feed their families (urban agriculture), or discard their residuals (composting and recycling). Because cities are part of larger bioregions, green cities contribute to sustainability outside of their boundaries through efforts such as green procurement and purchasing policies, watershed protection, and regional open space preservation.

Green cities are also resilient in the face of large-scale risks. Recognizing that it is impossible to predict threats that may arise from natural phenomena (such as sea level rise) and technological failures (such as blackouts), green cities are designed with increased capacity to weather these occurrences without inexorable damage. Increasing resilience requires understanding a city's carrying capacity and building on the ecosystem services provided by wetlands, green roofs, and other forms of natural capital to provide multiple forms of protection against disturbances. It also depends on citizen participation in environmental planning and a commitment to addressing socio-ecological issues such as environmental justice.

This volume, *Green Cities*, in *The SAGE Reference Series on Green Society: Toward a Sustainable Future*, provides an overview of the key concepts that urban planners, policy makers, architects, engineers, and developers use to understand the sustainability dimensions of the urban environment. It identifies the cities that have taken steps to become greener, and discusses the strategies they have used. It reviews the broad concepts associated with green cities. It discusses technologies, infrastructures, and programs that contribute to the greening of cities.

Cities face enormous environmental challenges, and the entries in this volume, from case studies of greener cities to discussions of green urban design, infrastructure, and processes, can help us transform our cities into healthier, sustainable communities where a growing urban population can thrive. As engines of creativity and economic development, green cities can help develop the breakthrough technologies that can reduce our emissions of greenhouse gases and avert the worst consequences of climate change. Cities can both be greener and serve as a force for global greening. We hope that readers will come away from these articles with a greater understanding of the issues facing cities, and the opportunities for our cities to become a force for sustainability.

Nevin Cohen
General Editor

Reader's Guide

City Organizations, Movements, and Planning

Agenda 21
Brownfields
Carrying Capacity
Charette
City Politics
Civic Space
Ecoindustrial Parks
Environmental Impact Assessment
Environmental Planning
Green Communities and Neighborhood
    Planning
Green Design, Construction, and
    Operations
Greenfield Sites
Infrastructure
Intermodal Transportation
Millennium Development Goals
Mitigation
NIMBY
Personal Rapid Transit
Resilience
Sustainability Indicators
Sustainable Development
Transit-Oriented Development
Transport Demand Management

City Profiles

Austin, Texas
Bahía de Caráquez, Ecuador
Bangkok, Thailand
Barcelona, Spain
Beijing, China
Bogotá, Colombia
Chattanooga, Tennessee
Chernobyl, Ukraine
Chicago, Illinois
Copenhagen, Denmark
Curitiba, Brazil
Dongtan, China
Dzerzhinsk, Russia
Hamburg, Germany
Kabwe, Zambia
Kampala, Uganda
La Oroya, Peru
Linfen, China
London, England
Los Angeles, California
Malmö, Sweden
Mexico City, Mexico
New York City, New York
Norilsk, Russia
Portland, Oregon
Reykjavik, Iceland
Rio de Janeiro, Brazil
San Francisco, California
Seattle, Washington
Stockholm, Sweden
Sukinda, India
Sumgayit, Azerbaijan
Sydney, Australia
Tianying, China
Vancouver, Canada
Vapi, India

Green City Challenges

Adaptation, Climate Change
Adaptive Reuse
Air Quality
Biodiversity
Carbon Footprint
Coastal Zone Management
Combined Sewer Overflow
Commuting
Construction and Demolition Waste
Denitrification
Density
Ecological Footprint
Ecosystem Restoration
Embodied Energy
Energy Efficiency
Environmental Justice
Environmental Risk
Food Deserts
Food Security
Garbage
Greywater
Gridlock
Heat Island Effect
Indoor Air Quality
Landfills
Light Pollution
Natural Capital
Nonpoint Source Pollution
Ports
Power Grids
Recycling in Cities
Sea Level Rise
Stormwater Management
Transit
Waste Disposal
Water Conservation
Water Pollution
Water Sources and Delivery
Water Treatment
Watershed Protection
Wetlands

Green City Solutions

Bicycling
Biophilia
Bioregion
Bluebelts
Bus Rapid Transit
Carbon Neutral
Carbon Trading
Carpooling
Cities for Climate Protection
Citizen Participation
Combined Heat and Power
   (Cogeneration)
Community Gardens
Compact Development
   (New Urbanism)
Composting
Congestion Pricing
Conservation Subdivision
Daylighting
Distributed Generation
District Energy
Ecovillages
Green Belt
Green Energy
Green Fleets (Vehicles)
Green Housing
Green Infrastructure
Greening Suburbia
Green Jobs
Green Landscaping
Green Procurement and Purchasing
Green Roofs
Greyfield Development
Habitat Conservation and Restoration
Healthy Cities
Historic Preservation
Infill Development
LEED (Leadership in Energy and
   Environmental Design)
Location-Efficient Mortgage
Masdar Ecocity
Mayors Climate Protection Agreement
Parks, Greenways, and Open Space
Renewable Energy
Smart Growth
Traffic Calming
Universal Design
Urban Agriculture
Urban Forests
Walkability (Pedestrian-Friendly Streets)
Xeriscaping

List of Articles

List of Contributors

Alexis, Gwendolyn Yvonne
Monmouth University

Anderson, Christopher
Auburn University

Aylett, Alexander
University of British Columbia

Ballamingie, Patricia
Carleton University

Baptista, Sandra
The Earth Institute at Columbia University

Bassett, Deborah R.
University of Washington

Bates, Felicia
University of Houston–Downtown

Tom Bauler
Free University of Brussels

Beder, Sharon
University of Wollongong

Birge-Liberman, Phil
Syracuse University

Boslaugh, Sarah
Washington University in St. Louis

Bridgeman, Bruce
University of California, Santa Cruz

Bryson, Jeremy
Syracuse University

Buehler, Ralph
Virginia Tech University

Busà, Alessandro
Center for Metropolitan Studies, Berlin

Byrne, Jason
Griffith University

Can, Liu Li
International Center for Communication Development

Carr, Constance
Humboldt University, Berlin

Carroll, Walter F.
Bridgewater State College

Carswell, Andrew T.
University of Georgia

Chattopadhyay, Aparajita
International Institute for Population Sciences (IIPS), Mumbai

Chiaviello, Anthony R. S.
University of Houston–Downtown

Cidell, Julie
University of Illinois

Corfield, Justin
Geelong Grammar School, Australia

Crot, Laurence
University of Neuchâtel

Dixon, Megan
College of Idaho

Dooley, Michael D.
Knox College

Doshi, Ameet D.
Georgia Perimeter College

Dudley, Michael Quinn
University of Winnipeg

Faass, Josephine S.
*Rutgers, The State University
    of New Jersey*

Fenner, Charles R., Jr.
State University of New York, Canton

Gachechiladze, Maia
*Central European
    University*

Gasper, Rebecca R.
University of Maryland

Golub, Aaron
Arizona State University

Goodier, Chris
Loughborough University

Grover, Velma I.
*United Nations University–Institute
    for Water, Environment and
    Health (UNU–IWEH)*

Guha, Mohua
*International Institute for Population
    Sciences (IIPS), Mumbai*

Hagerman, Chris
Portland State University

Harper, Gavin D. J.
Cardiff University

Harrington, Jonathan
Troy University

Helfer, Jason A.
Knox College

Hurley, Patrick T.
Ursinus College

Johnson, Sherrill
Independent Scholar

Kasper, Debbie V. S.
Sweet Briar College

Keane, Timothy P.
Saint Louis University

Kelley, Ingrid
Energy Center of Wisconsin

Kofoworola, Oyeshola Femi
University of Toronto

Krayenhoff, E. Scott
*University of British
    Columbia*

Lanfair, Jordan K.
Knox College

Lang, Steven
*LaGuardia Community
    College*

Law, Caroline Man Yee
University of Hong Kong

Lee, Megan
University of Georgia

LeVasseur, Todd J.
University of Florida

Lwasa, Shuaib
Makerere University

Lyons, Donald
University of North Texas

Maassen, Anne
University of Durham

Machler, Leonard
Arizona State University

Martins, Rafael D'Almeida
University of Campinas

May, Shannon
University of California, Berkeley

McKendry, Corina
University of California, Santa Cruz

Miller, Caroline Lomax
Massey University

Moos, Markus
University of British Columbia

Mullaney, Emma Gaalaas
Pennsylvania State University

Newell, Josh
University of Southern California

Nieuwenhuis, Paul
Cardiff University

Nuñez, Maura Troester
University of Colorado, Boulder

Ohayon, Jennifer Liss
University of California, Santa Cruz

Opdyke, Matthew R.
Point Park University

Otto, Judith
Framingham State College

Panda, Sudhanshu Sekhar
Gainesville State College

Paolini, Federico
University of Siena

Parker, Jonathan
University of North Texas

Qun, Zeng Li
Beihai College, Beihang University

Ren, Guoqiang
University of Washington

Ren, Zhiqiang
Independent Scholar

Schroth, Stephen T.
Knox College

Schuppert, Fabian
Queen's University Belfast

Scott, Austin Elizabeth
University of Florida

Silva, Carlos Nunes
University of Lisbon

Smith, Julie
Countryside and Community Research Institute

Smith, Stephen
Gainesville State College

Snell, Carolyn
University of York

Spraker, Deanna
Virginia Polytechnic Institute

Staddon, Chad
University of the West of England, Bristol

Stead, Dominic
Delft University of Technology

Stewart, Iain D.
University of British Columbia

Susilo, Yusak Octavius
University of the West of England

Syed, Jawad
University of Kent

Tyman, Shannon
University of Oregon

van Vliet, Bas
Wageningen University

Weaver, Susan H.
Independent Scholar

Weissman, Evan
Syracuse University

White, Kenneth S.
San Jose State University

Wickstrom, Stefanie
Independent Scholar

Woodworth, A. Vernon
Boston Architectural College

Xiujuan, Li
Beihai College, Beihang University

Zimmermann, Petra A.
Ball State University

Green Cities Chronology

c. 1700 B.C.E.: King Minos of Crete has running water for bathing and sewage disposal in his palace at Knossos.

c. 500 B.C.E.: Athens, Greece, establishes what may have been the first city dump in the Western world, accompanied by a ban against throwing garbage into the streets.

1388: In an early attempt to control municipal pollution, the English Parliament prohibits dumping wastes in the public waterways.

1580: Queen Elizabeth creates a green belt around London by prohibiting construction of new buildings within three miles of the city wall.

1657: The Common Council of New Amsterdam (now New York City) establishes public dumps, prohibits dumping refuse in the street, and requires residents to keep the streets in front of their homes clean.

1833: English chemist and meteorologist Luke Howard describes the "urban heat island" effect in *The Climate of London*, noting that the city "partakes much of an artificial warmth, induced by its structure, by a crowded population, and the consumption of great quantities of fuel."

1842: Edwin Chadwick issues a report on the sanitary conditions among England's laboring classes, which argues that environment and disease are intimately linked, and that governments should be concerned with matters such as establishing municipal water supplies and sewage systems to improve public health.

1848: Great Britain's Public Health Act establishes the General Board of Health, which lays the groundwork for modern public health.

1849: The Croton Aqueduct Department forms in New York City to build a comprehensive sewer system for the city. Seventy miles are constructed by 1855, and by the 1890s New York City has 464 miles of sewers, more than any other American or European city except Chicago. Part of the motivation is recurrent cholera epidemics, including one in 1849 that causes 5,071 deaths (cholera is transmitted through fecal matter, hence keeping sewage out of the drinking water supply is crucial). The new sewer system proves effective as the last cholera epidemic in the city is in 1892 and causes fewer than 120 deaths.

1854: John Snow uses geographic analysis to trace the source of a cholera outbreak in London to a particular water pump, bolstering his theory that cholera is transmitted through water (long before the specific organism that causes cholera had been identified) and bolstering support for making the provision of clean drinking water a civic responsibility.

1863: The world's first subway opens in London, England.

1892: The Sierra Club, one of the most influential environmental organizations in the United States, is founded in San Francisco.

1894: New York City approves a referendum for a subway system to be constructed at public expense to reduce traffic congestion in the city streets. The first segment opens in 1902, and the entire system is in operation by 1908 (although it will be greatly expanded in years to come). It is the first in the world to have a four-track system (local and express lanes in both directions).

1894: Detroit's Potato Patches program establishes the first organized community garden program in the United States, promoting self-sufficiency, and providing food for the impoverished during an economic crisis.

1900: The Sanitary District of Chicago reverses the flow of the Chicago River, sending sewage and other pollutants into the Sanitary and Ship Canal, significantly improving the water quality in Lake Michigan.

1928: Large-scale use of geothermal power is initiated when a borehole is sunk in Reykjavik, Iceland. This system now provides hot water for 80 percent of the households in Reykjavik, as well as domestic heating.

1930: A thermal inversion in the Meuse Valley of Belgium traps pollutants from nearby industrial plants in the local atmosphere, causing about 60 deaths and many illnesses.

1931: Charleston, South Carolina, passes the nation's first zoning ordinance establishing historic districts in order to preserve older neighborhoods and the city's unique sense of place.

1941–1945: Gas rationing in the United States due to World War II encourages carpooling as a patriotic activity.

1952: London, England, experiences a thermal inversion in December that, combined with air pollution from automobiles, factories, and coal-burning furnaces, blankets the city in smog that is believed to have caused about 3,000 excess deaths.

1954: The U.S. Supreme Court rules in *Berman v. Parker* that city governments may use the principle of eminent domain to seize properties in areas considered to be blighted and redevelop them as part of an urban renewal campaign.

1955: The Air Pollution Control Act of 1955 becomes the first federal legislation in the United States concerned with air pollution, although its scope is limited to providing funds for air pollution research.

1961: Jane Jacobs's *The Death and Life of Great American Cities* criticizes urban renewal policies (in particular those championed by Robert Moses in New York City) that destroy neighborhoods, favor automobiles over people, and ignore the benefits of the human-scale, pedestrian-friendly environments provided by many of the world's great historical cities.

1962: Community activists and historical preservationists succeed in preventing the destruction of Ghirardelli Square in San Francisco. This former chocolate factory becomes the first example of successful adaptive reuse, repurposing a historic structure.

1963: The Clean Air Act establishes a federal program within the U.S. Public Health Services and authorizes research into monitoring and controlling air pollution, making it the first federal legislation in the United States to deal with the control of air pollution.

1969: Scottish landscape architect Ian McHarg publishes *Design With Nature,* a landmark work on ecological planning.

1970: The Clean Air Act authorizes the development of U.S. state and federal regulations to limit emissions from both stationary (such as industrial) and mobile sources (such as automobiles).

1970: The New York City Traffic Department coins the term "gridlock" to describe traffic congestion, in particular when an intersection is blocked because cars moving in one direction cannot clear the intersection before the light turns red, blocking traffic attempting to move in the cross direction.

1970s: The term *brownfield* is coined to refer to abandoned industrial and commercial facilities (initially steel plants in the United States) that may be contaminated by hazardous waste such as chemical pollutants.

1971: A fire in the basement of an office building in Binghampton, New York, highlights the dangers of using PCBs (polychlorinated biphenyls) in construction when contaminated soot spreads throughout the building through the air conditioning ducts.

1972: The U.S. Congress passes the Coastal Zone Management Act, initiating a process of sustainable development of coastal zones.

1974: Curitiba, Brazil, develops the first bus rapid transit (BRT) system that uses buses on roads to provide services more typical of rail systems. Innovations for this type of bus system include providing a dedicated travel lane, permanent stations and stops, more efficient methods of boarding and fare collection, and relatively fewer stops placed further apart.

1976: The Tax Reform Act encourages adaptive reuse in the United States by allowing faster tax write-offs for projects that recycle or restore historic properties.

1978: U.S. President Jimmy Carter declares Love Canal, a toxic waste dump in Niagara Falls, New York, a national emergency due to chemical pollution. Over 1,000 families are evacuated at public expense.

1978: New York City begins Operation Green Thumb to encourage community gardening. The city allows residents to use vacant lots for gardens for the nominal fee of $1 per year. By 1991 the city reports that there are over 500 community gardens in the city.

1980s: Staten Island (New York City) establishes the first successful Bluebelt, a system of local septic tanks, rather than an island-wide conventional storm water system.

1981: Seaside, Florida, often cited as the first New Urbanist development, is built on the model of traditional southern towns with narrow streets and alleys, and houses with porches and pitched roofs.

1981: A task force of experts from Colorado State University and the Associated Landscape Contractors of Colorado coin the term "xeriscape," meaning landscaping that is water-efficient and bioregionally appropriate.

1983: The Environmental Protection Agency recommends that the residents of Times Beach, a small town in Eastern Missouri, evacuate due to dioxin contamination. The chemical was a contaminant in oil spread on the roads to control dust in the 1970s.

1984: Los Angeles reduces vehicle traffic by 50 percent in preparation for the Olympics through ride-sharing schemes and other measures, although the plan to reserve specific freeway lanes for cars carrying three or more passengers was abandoned.

1986: Chernobyl, Ukraine (formerly of the Soviet Union), is the site of the worst nuclear disaster in history. The amount of radiation released is estimated to be 100 times that of the atomic bombs dropped in 1945 on Hiroshima and Nagasaki.

1987: The United Church of Christ Commission for Racial Justice issues a study demonstrating that the location of toxic waste sites is more closely related to the race of neighborhood residents than to either income or social class.

1987: The infamous New York City "garbage barge" cruises the Atlantic and Gulf coasts looking for a dump that will accept the over 3,000 tons of commercial refuse it carries. Eventually the barge returns to New York where its cargo is incinerated, but the incident raises awareness of the growing problem of garbage disposal in America's cities.

1989: New York City enacts a broad recycling law (Local Law 19), in large part to reduce air pollution caused by the incineration of garbage. By 1996 the city claims that 41 percent of its refuse is diverted to recycling.

1989–1993: Water metering trials of 53,000 British households estimate that compulsory metering could reduce domestic water use by 11 percent. However, the trial also demonstrates that conservation effects may not last because consumers quickly become used to the presence of the water meter.

1991: Germany adopts stringent measures to reduce the quantity of garbage by requiring that manufacturers, distributors, and retailers collect and recycle all packaging waste from their products.

1992: The Earth Summit held in Rio de Janeiro results in the document "Agenda 21," which calls for national governments to adapt strategies for sustainable development and to cooperate with nongovernmental organizations and other countries in implementing them.

1992: Mayor Richard M. Daley establishes a Department of the Environment for the city of Chicago, which undertakes the Brownfields Initiative to redevelop abandoned industrial areas, as well as projects such as tree planting to improve the aesthetic qualities of city streets.

1993: William Kunstler's *The Geography of Nowhere: The Rise and Decline of America's Man-Made Landscape* argues that urban sprawl has created a suburban environment that is ecologically destructive as well as counterproductive to normal human living.

1993: The International Coalition for Local Environmental Initiatives creates the Cities for Climate Protection network to support urban measures to reduce greenhouse gas emissions. It had more than 750 members in 2010.

1995: The Global Ecovillage Network is founded, providing an administrative home for those interested in creating intentional communities that are sustainable, support human development, and minimize human ecological impact.

1996: William Rees and Mathis Wakernagel develop the concept of the "ecological footprint," which signifies all the resources used by a particular population or species in their book *Our Ecological Footprint: Reducing Human Impact on the Earth*.

1999: The Ecuadorian capital city of Bahía de Caráquez embarks on a campaign of rebuilding on principles of ecological sustainability following repeated damage from flooding and mudslides.

1999: The Congress for the New Urbanism estimates that seven percent of regional and superregional malls are grayfields (abandoned or underused commercial properties) with an additional 12 percent of U.S. malls likely to enter greyfield status by 2004.

2000: New York City introduces Green Building Tax Credits, which offer tax breaks to developers whose buildings meet energy efficiency standards.

2001: Bangkok begins enforcing a series of regulations intended to lessen the city's environmental impact and improve the health and welfare of its citizens. These measures include monitoring vehicle emissions, enforcing the use of covering sheets on trucks and construction sites, and providing free vehicle inspections and tune-ups.

2001: The U.S. Green Building Council founds the Green Building Certification Institute to certify Leadership in Energy and Environmental Design (LEED) professionals who are qualified to evaluate the sustainability of buildings.

2003: Great Britain institutes traffic congestion pricing, charging a daily rate for drivers to enter the central zone of London. The program is successful: Studies indicate that traffic in the affected area fell by 21 percent 2002–2006.

2003: Zipcars, the world's largest car sharing program, introduces hybrids to its Seattle fleet.

2003: A massive blackout in Ontario, Canada, and the eastern and midwestern United States affects an estimated 55 million people, and points out the vulnerability of the electrical power grid.

2003: Kansas City, Missouri, adopts a walkability plan intended to reduce automobile use and encourage walking. It includes plans to ensure safe routes for local children to walk to school, reversing decades of suburban development that assumed that any trip beyond the immediate neighborhood would be made by automobile.

2004: A study by SMARTRAQ, a land use research project conducted by Georgia Tech University, finds that residents who live in parts of Atlanta, Georgia, characterized by urban sprawl drive 40 percent more on weekends and 30 percent more during the week than people who live in areas of the city more conducive to walking.

2005: The E.U. Emission Trading Scheme, a carbon-trading scheme involving 25 of the then-27 European Countries, officially begins.

2005: The Brookings Institute issues a report stating that the average partial carbon footprint of people living in metropolitan areas was 14 percent lower than that of the U.S. population at large. The reasons cited include less automobile use, compact housing, and mixed-development neighborhoods.

2005: The U.S. Conference of Mayors endorses the U.S. Mayors Climate Protection Agreement, which commits mayors to working to reduce greenhouse emissions in their cities. As of 2009, 969 mayors representing nearly 85 million people signed the agreement.

2007: Barcelona begins a bike-sharing program in which members pay an annual fee plus a small charge for each half-hour of bicycle use. The program begins with 30,000 subscribers and 400 bicycles, and has grown to 186,000 subscribers and 6,000 bicycles.

2007: New York City passes Local Law 86, more commonly known as the Green Buildings Act, which requires that new municipal buildings and renovations and additions to existing buildings meet LEED sustainability standards.

2007: San Francisco, California, bans polystyrene foam (Styrofoam) containers for take-out food, requiring that containers used for that purpose be compostable or recyclable.

2008: Gasoline prices go over $4 per gallon in the United States, encouraging the use of public transportation, as well as alternatives such as biking and walking.

2008: SustainLane chooses Portland, Oregon, for the second year in a row as the greenest of the green cities. Fifty U.S. cities were evaluated on 16 criteria related to urban sustainability.

2009: San Francisco, California, passes the most stringent recycling and composting ordinance in the United States. Several other cities have mandatory recycling, but San Francisco is the first to require composting as well.

2009: Sydney, Australia, holds a one-hour citywide blackout in March to raise awareness of global warning and encourage conservation.

2009: The European Commission names Stockholm, Sweden, first European Green Capital.

Sarah Boslaugh
Washington University in St. Louis

Adaptation, Climate Change

Anthropogenic (human-caused) climate change will seriously test our ability to plan and manage cities. Taken-for-granted urban institutions such as housing, industries, schools, hospitals, and transport systems will all likely be affected. Although the scale and intensity of impacts will vary, anticipated problems will have pernicious consequences, especially for the poor and politically disenfranchised. This article considers climate change adaptation concepts, issues, and responses. Beginning by briefly defining mitigation, adaptation, vulnerability, and resilience, this article identifies key players, provides examples of how some cities are already addressing threats, and discusses limitations to current adaptation practices.

With rising temperatures, heat waves will become more frequent and severe. Droughts, like this one in Louisiana, will threaten food and water supplies.

Source: iStockphoto

Key Concepts

Four key concepts are central to how the world's cities are adapting to climate change. Early climate change responses hinged on the notion of reducing or offsetting human-caused greenhouse gases—an idea referred to as *mitigation*. However, some degree of climate change is now inevitable. The differential effect of climate change on natural and social systems is a function of three interrelated factors: sensitivity to biophysical change, the duration of their exposure to altered conditions (e.g., higher temperature, less rainfall), and the capacity to adapt to new conditions. Intense long-term exposure, extreme sensitivity, and low adaptive capacity will heighten vulnerability to harm, but some cities will be able to quickly "bounce back" from perturbations without suffering high death rates and/or long-term damage to infrastructure, life-sustaining systems, and social institutions. This capacity to withstand harm is termed *resilience*. Climate change *adaptation* seeks to forecast the likely impacts and then to figure out how best to modify cities to cope with these expected changes and their environmental, social, and financial consequences.

Likely Urban Effects of Climate Change

The Intergovernmental Panel on Climate Change has predicted that global warming will bring numerous biophysical changes to the world's cities, with knock-on environmental, social, political, and economic effects. For instance, with rising temperatures, heat waves in many places will become more frequent and severe, and urban "heat island" effects will kill many more people. Increased higher-intensity storm events and more rainfall will cause widespread flooding, with concomitant property and infrastructure damage. Frequent droughts will threaten food and water supplies. Increased insect-borne diseases will incapacitate many urban residents. Rising sea levels will devastate some low-lying coastal cities. Urban biodiversity will also likely decline markedly. However, these impacts will disproportionately affect cities in developing countries because of their predominantly warm climates, nature of existing built environments, scale and intensity of expected impacts, high costs of adaptation, and size of the affected populations. Large numbers of people will be displaced as climate refugees.

Numerous urban development practices will exacerbate these impacts, making adaptation more difficult. Unregulated clearing of vegetation increases stormwater run-off, worsening flooding and heightening landslide risk. Development on floodplains and coastlines increases vulnerability to flooding and coastal erosion, raising insurance premiums. Squatter settlements may be especially vulnerable because of unregulated construction methods, inadequate sanitation, their inappropriate location, and substandard infrastructure—flooding, landslides, fire, and epidemic disease will all take their toll. In addition, very high densities within many Asian, South American, and African megacities will expose more people to "heat island" effects. The challenge for urban professionals (e.g., planners, engineers, and healthcare practitioners, etc.) is how best to adapt cities to these and other expected changes.

Adaptation Measures

Adaptation measures typically rely on new technologies, institutional reform, citizen education, and capacity building. They can be summarized as forecasting impacts, monitoring current conditions, risk assessment and education, mapping vulnerable places, disaster preparedness, liability assessment, promoting behavioral change (e.g., water and energy efficiency or decreased automobile use), building adaptive capacity, and retrofitting built environments (e.g., installing building insulation, photovoltaics, and green roofs, or planting more street trees).

Cities worldwide have begun to respond in various ways, ranging from inexpensive, local measures (e.g., distributing free compact fluorescent lightbulbs to households) to major new regional infrastructure projects (e.g., solar power stations, wind farms, new dams, water recycling projects, and coastal seawalls). For example, Shanghai, China, has a heat wave monitoring and emergency response program; Gold Coast, Australia, has built a desalination plant; Durban, South Africa, is working to protect biodiversity;

Venice, Italy, is building sea walls; and citizens in Mumbai, India, are pushing for renewable energy laws. Other cities are modifying their building codes and town planning regulations to require new development to use green building technologies, to strengthen buildings against storm damage, and to identify areas where future development may be prohibited. Many cities are also expanding their conservation reserves and attempting to join fragmented habitat patches through "wildlife corridors" to limit biodiversity loss.

Although adaptive technologies will play a role, they are not a panacea; they are oftentimes expensive and insufficiently tested and cannot redress institutional problems. For instance, significant shortcomings of much adaptation work to date have been insufficient citizen involvement in decision making and inadequate efforts to build adaptive capacity among traditionally marginalized and vulnerable populations—an issue of equity and justice.

Challenges of Fairness and Equity

Governments from the national to local levels are involved in climate change adaptation, as are supranational government organizations (e.g., the United Nations) and nongovernment organizations (e.g., ICLEI–Local Governments for Sustainability). Community groups, charities, nonprofits, and businesses (real estate, finance, insurance, etc.) are also playing roles, but the capacity for climate change adaptation is limited by unstable and corrupt systems of governance, a lack of political will, ineffective leadership, corporate greed, citizen apathy, unhealthy populations, and limited access to resources. Communities especially vulnerable to climate change impacts include low-income groups, refugees, homeless, some migrants, socioeconomically marginalized groups, and the like. Some commentators have begun to recognize this problem, labeling it an environmental justice/climate justice concern warranting immediate attention. The challenge is to ensure that adaptive measures are rapidly implemented without undermining participatory planning processes or marginalizing vulnerable groups.

See Also: Biodiversity; Carbon Footprint; Carbon Trading; Cities for Climate Protection; Density; Energy Efficiency; Environmental Justice; Environmental Planning; Green Communities and Neighborhood Planning; Green Infrastructure; Heat Island Effect; Mitigation; Parks, Greenways, and Open Space; Renewable Energy; Resilience; Smart Growth; Sustainable Development; Urban Forests.

Further Readings

Adger, W. N. "Scales of Governance and Environmental Justice for Adaptation and Mitigation of Climate Change." *Journal of International Development*, 13 (2001).

Bicknell, J., et al. *Adapting Cities to Climate Change: Understanding and Addressing the Development Challenges*. London: Earthscan, 2009.

Davoudi, S., et al. *Planning for Climate Change: Strategies for Mitigation and Adaptation for Spatial Planners*. London: Earthscan, 2009.

Füssel, H. "Adaptation Planning for Climate Change: Concepts, Assessment Approaches, and Key Lessons." *Sustainability Science*, 2 (2007).

Satterthwaite, D., et al. *Adapting to Climate Change in Urban Areas: The Possibilities and Constraints in Low- and Middle-Income Nations*. London: International Institute for Environment and Development, 2007.

Schelesinger, M. E., et al. *Human-Induced Climate Change: An Interdisciplinary Assessment.* Cambridge, MA: Cambridge University Press, 2007.

Westra, L. *Environmental Justice and the Rights of Ecological Refugees.* London: Earthscan, 2009.

Wilby, R. L. and G. L. W. Perry. "Climate Change, Biodiversity and the Urban Environment: A Critical Review Based on London, UK." *Progress in Physical Geography*, 30 (2006).

Jason Byrne
Griffith University

ADAPTIVE REUSE

Adaptive reuse is a way of preserving existing buildings through a change in the purpose of the building from what was originally intended, usually after the property has matured within the property life cycle. One fundamental characteristic of adaptive reuse is the conversion of a previously underused building to one that has an economically viable use. This change is notably different from merely preserving the structure for museum-like purposes.

Although people have always found new uses for older structures, the term *adaptive reuse* did not come into existence until the 21st century. The recycling of buildings has since become an important tool for the historic preservation community, which has been intent on protecting historically significant buildings from demolition. Historic preservation advocates attempt to maximize the hidden value of real property and to provide for the reemployment of the structure. However, it is instructive to note that there still exist several hundred million square feet of idle retail space, further exacerbating both the need and the opportunity for adaptive reuse to occur.

Reasons for Implementing Adaptive Reuse

Public and private entities have significant purposes for engaging in adaptive reuse projects. Curbing urban sprawl by reducing the number of new structures on the outskirts of communities is one of the most convincing reasons for implementing adaptive reuse strategies. This type of plan is particularly attractive to communities in which available land is scarce. In implementing such an approach, cities reinforce the importance of existing structures that are linked to the area's cultural heritage. Moreover, cities make a point of showcasing these reused buildings as key components of greater redevelopment efforts.

Older unused properties have prohibitive social costs. Many times, older buildings may have little or no value in their current condition. Subsequently, these buildings may become abandoned and run down, attracting vandals, homeless, arsonists, and drug dealers. This results in decreased property values, tax revenues, and services provided to a metropolitan area's residents. In the long term, this discourages investment in a community. By reinvesting in existing buildings, cities have the potential to revitalize blighted areas while creating construction jobs for inner-city residents, as well as an improved visual environment.

The environmental impacts of advocating adaptive reuse can be evidenced in the reduction of construction and demolition debris that results from the destruction of older buildings and the construction of new ones. Overall, waste generated through construction and demolition activities accounts for roughly half of all landfill waste in U.S. landfills, according to the U.S. Environmental Protection Agency. In addition, it has been estimated that rehabilitation costs associated with adaptive reuse are between 15 and 20 percent less than comparable new construction, even when problems such as asbestos exist. Adaptive reuse projects also have been found to have an economic multiplier effect within communities, in that more jobs are created by retrofitting projects of a certain dollar amount than by similarly priced new construction projects. Extending the useful life of existing buildings instead of constructing new ones minimizes the use of new building materials, the energy required to produce and transport those materials, and the energy required for construction.

Adaptive reuse has become a more popular option in recent years thanks to shifts in both public and private incentives. Developers recognize the prohibitive costs and associated difficulties in securing building permits in several areas, thus making traditional development options quite costly. They are further induced by a variety of tax incentives that spur developers toward adaptive reuse, as opposed to more traditional development options. One of the first federal initiatives toward this form of historic preservation was the 1976 Tax Reform Act, which permitted faster tax write-offs for restoration or recycling of properties of historic value. The 1981 Economic Recovery Tax Act went even further by permitting a tax deduction of up to 25 percent of the value for restoration of such buildings. The Historic Preservation Tax Credit gives tax breaks toward the rehabilitation of income-producing properties. The tax credit can even be leveraged with the low-income housing tax credit to provide reuse opportunities for low- to moderate-income clients. Developers can also use a variety of state and local incentive programs that aid them in financing adaptive reuse projects. Owners of older buildings may decide to provide adaptive reuse in an attempt to protect the asset or maximize its resale value, especially when the use of the building has become obsolete. Public agencies that face an eroding tax base may also be encouraged to find reuse for their own underused assets to stimulate revitalization and economic growth.

Examples of Adaptive Reuse

For adaptive reuse to occur, there are two criteria: an existing building and a person or organization willing to reuse the structure for a purpose for which it was not originally intended. Historic preservationists and sociologists also assert that the character and spirit of a building cannot be compromised for an adaptive reuse project to be successful. Although the possible options available for the reuse of existing buildings are innumerable, some of the more visible examples of adaptive reuse are industrial buildings that have been converted to museums, art studios, live–work units, offices, residential units, schools, and retail.

Ghirardelli Square in San Francisco is considered the first successful adaptive reuse of an industrial complex. Once a chocolate factory, the city block housing the Ghirardelli factory was saved by historic preservationists and community activists in 1962, amid fears that the existing structures would be demolished. Since that time, a number of other adaptive reuse projects have been undertaken, with many occurring in the northeast, where a number of centuries-old buildings have been preserved by citizens concerned about the historic character of the neighborhoods and the structural integrity of the existing built environment. Still, as industrial shifts occur and current building uses inevitably become obsolete, opportunities for adaptive reuse will manifest themselves.

Other Pertinent Issues Regarding Adaptive Reuse

For an adaptive reuse project to successfully be implemented, a series of actions generally ensue as part of a sometimes lengthy process. First, a pro forma analysis allows the owner, developer, and various stakeholders to determine whether the proposed adaptive reuse project has economic feasibility. Assuming that it does offer the potential for a satisfactory internal rate of return, an initial assessment and evaluation must then be conducted. Different eras of construction have particular architectural types and building materials that may give the redevelopment team an understanding of the overall building's integrity. Pre–Civil War buildings, for example, differ from those built in later periods in terms of foundation, load-bearing capacity, masonry materials, window and door patterns, interior finish and details, roof systems, and heating, ventilation, and air conditioning systems. In some cases, sensitive tests may need to be conducted by professionals on wall reinforcement, concrete strength, wood decay, and moisture content. Some proponents of adaptive reuse even advocate the additions of sustainable components on redevelopment, such as the inclusion of local, renewable, and reused building materials.

Although many adaptive reuse projects receive widespread praise on their completion, there are several institutional barriers that serve to slow the progress of completion of adaptive reuse projects. Building codes, often implemented with the intent of incorporating advancements in building and construction technologies, do not implicitly endorse the preservation of the existing characteristics of some older buildings. For example, features like steep stairs and open stairwells do not comply with requirements set forth by the Americans with Disabilities Act and create obstacles for developers intent on preserving the integrity of the structure. Developers may need to work with local government officials to determine whether the new use of the refurbished property conforms to zoning regulations. Nevertheless, there is evidence that municipal governments show some flexibility with adaptive reuse developers on both zoning and impact fees. From a long-term perspective, there has been some criticism that local governments have oftentimes been more reactive than proactive in converting existing buildings for residential purposes. For many cities, there is no coordinated policy of identifying the contribution that obsolete buildings can make in easing the present shortage of affordable housing that exists in many municipalities. Adaptive reuse can help alleviate some of the burdens apparent in the disequilibrium between obsolete office space and the demand for residential accommodation close to the urban center.

Environmental, safety, and cost concerns also pervade when performing adaptive re-use projects. Old industrial buildings are often brownfields, contaminated with toxic chemicals that need to be remediated before construction. Many boiler systems, as well as insulation areas, run the risk of containing asbestos, which usually requires replacement or remediation by a specialized contractor. Old wiring and outdated plumbing systems may also need to be replaced—another expensive task. Energy-efficient improvements may also need to be implemented, which may reduce operating costs over the renewed life of the building. Some changes made for the sake of energy efficiency, such as the replacement of older windows with storm windows and/or double-glazed windows, may detract from the original character of the building, however. Because of the uncertainties involved in redeveloping an existing building, the projected costs of an adaptive reuse project have much higher variance than those associated with new construction. For operators of properties converted through adaptive reuse, the daily maintenance costs may also be more costly than for newer buildings.

There are valuation concerns for those who either reside in adaptive reuse properties or are considering renovating a property for adaptive reuse. Historic properties are notably undervalued when appraised against newer properties that are nearby and used for the same purpose—the perception being that the value premium of a historic designation never actually materializes. In addition, the character and integrity evident within the structure during its previous use must be preserved for appraisers to consider full value in an adaptive reuse project. There is some evidence, however, that properties located near an adaptive reuse property experience an increase in value over time, thus strengthening the case for such a project to materialize. There are also environmental risks involved with conversion and redevelopment opportunities. These risks include such things as underground storage tanks in converted gasoline stations and other contaminants evident in industrial settings. As adaptive reuse properties continue to gain prominence among consumers, concerns about being able to capture value within such projects will likely dissipate over time.

See Also: Brownfields; Green Design, Construction, and Operations; Historic Preservation; Sustainable Development.

Further Readings

Austin, Richard L. *Adaptive Reuse: Issues and Case Studies in Building Preservation.* New York: Van Nostrand Reinhold, 1988.

Bullen, Peter A. "Adaptive Reuse and Sustainability of Commercial Buildings." *Facilities,* 25 (2007).

Burchell, Robert W. *The Adaptive Reuse Handbook.* New Brunswick, NJ: Rutgers University Center for Urban Policy Research, 1981.

Campbell, Jan. "Is Your Building a Candidate for Adaptive Reuse?" *Journal of Property Management,* 61/1 (1996).

Cantell, Sophie F. "The Adaptive Reuse of Historical Industrial Buildings: Regulation Barriers, Best Practices and Case Studies." Master's Thesis, Virginia Polytechnic Institute and State University, 2005.

Diamonstein, Barbaralee. *New Uses, Old Places: Remaking America.* New York: Crown, 1986.

Dickinson, James. "Adaptive Reuse: Towards a Sociology of the Built Environment." Paper presented at the annual meeting of the American Sociological Association, San Francisco, California, August 14, 2004.

Gause, Jo Allen. *New Uses for Obsolete Buildings.* Washington, D.C.: Urban Land Institute, 1996.

Heath, Tim "Adaptive Reuse of Offices for Residential Use: The Experiences of London and Toronto." *Cities,* 18/3 (2001).

McLaughlin, Sara Beth. *Large-Scale Adaptive Reuse: An Alternate to Big-Box Sprawl.* Master's Thesis, University of Pennsylvania, 2008.

Rabun, J. Stanley and Richard Kelso. *Building Evaluation for Adaptive Reuse and Preservation.* Hoboken, NJ: John Wiley & Sons, 2009.

Rombouts, Christine. "Promoting Preservation: Financing Incentives Help Brokers Restore and Reuse Historic Buildings." http://www.ciremagazine.com/article.php?article_id=208 (Accessed April 2009).

Andrew T. Carswell
University of Georgia

AGENDA 21

Agenda 21 is a blueprint for sustainable development into the 21st century. It sets out environmental strategies for the management of coasts, oceans, and water; the monitoring and reduction of chemical waste; the eradication of radioactive waste; and the conservation of natural vegetation and soils. It builds on plans for the development of sustainable farming and other socioeconomic proposals including measures to improve healthcare, to reduce poverty, and to develop fair and environmentally friendly trade policies.

The basis for Agenda 21 was agreed on during the Earth Summit in Rio de Janeiro in 1992 and was signed by 179 heads of state and government. The document reflects a global consensus and political commitment at the highest level on development and environment cooperation to deal with the pressing problems of today and the need to prepare for the challenges of the future.

The aim is to promote global sustainable development at a more fundamental level than traditional aid programs, based on common needs and interests and on collective responsibility. Agenda 21 is an action plan at all scales, requiring local governments to develop their own Local Agenda 21 to spread understanding of, and action for, sustainable development.

It is a comprehensive document, with 40 chapters in four sections, dealing with the following:

1. Social and economic dimensions, including combating poverty, changing consumption patterns, changing population and demographic dynamics, promoting health, promoting sustainable settlement patterns, and integrating environment and development into decision making.

2. Conservation and management of resources for development, including atmospheric protection, combating deforestation, protecting fragile environments, conservation of biological diversity (biodiversity), and control of pollution. It also takes into consideration agricultural practices, biotechnology, toxic chemicals, solid waste, and sewage.

3. Strengthening the role of major groups such as women, indigenous peoples, children, and youth, and actors such as nongovernmental organizations, local authorities, workers, farmers, business entrepreneurs, and industry.

4. Means of implementation through finance, technology transfer, education, research and science, capacity building, international negotiation, legal and institutional frameworks, and information.

Agenda 21 makes clear that population, consumption, and technology are the primary driving forces of environmental change. The document not only encompasses issues such as poverty, excessive consumption patterns, health, education, and cities and urban centers but also provides options for conserving forests and diversity of species and avoiding degradation of the land, air, and water.

Agenda 21 calls on governments to adopt national strategies for sustainable development. It recognizes that sustainable development requires national strategies, plans, and policies. The efforts of nations need to be linked by international cooperation through such organizations as the United Nations. The broadest public participation and the active involvement of the nongovernmental organizations and other groups should also be encouraged. These include nongovernment organizations and the public at large. Agenda 21 puts most of the responsibility for leading change on national governments but says they need to work in a broad series of partnerships ranging from international organizations to citizens' groups.

In fact, rather than just governments, different social actors have a role to play in these collective efforts including business, industry, trade unions, scientists, teachers, indigenous people, women, youth, and children. As Agenda 21 says, only a global partnership will ensure that all nations will have a safer and more prosperous future.

A major theme of Agenda 21 is the need to eradicate poverty by giving poor people more access to the resources they need to live in a sustainable way. The world is confronted with poverty, hunger, disease, illiteracy, and the continuing deterioration of ecosystems on which humanity depends for its well being. By adopting Agenda 21, industrialized countries recognized that they have a greater responsibility toward the environment than poor nations, which produce relatively less pollution. The richer nations also promised more funding to support other nations to develop in ways that have lower environmental impacts. Beyond funding, nations need assistance in building the capacity to plan and carry out sustainable development decisions. This requires the transfer of information and skills.

The pathway for sustainable development should reverse both poverty and environmental destruction. Agenda 21 provides principles regarding what needs to be done to reduce wasteful and inefficient consumption patterns in some parts of the world, while encouraging increased but sustainable development in others. It offers a repertoire of policies and programs to achieve a sustainable balance between consumption, population, and the Earth's life-supporting capacity. It also provides insights into some technologies and techniques that need to be developed to provide for human needs while carefully managing natural resources.

According to Agenda 21, the only way to ensure a prosperous future is to deal with environment and development issues together in a balanced manner. The implementation of Agenda 21 is intended to involve action at international, national, regional, and local levels. Many national and state governments have legislated or advised that local authorities take steps to implement the plan locally, as recommended in chapter 28 of the document.

Such programs are often known as Local Agenda 21 or LA21. During the Rio Summit it was agreed that local councils would produce their own plan—a Local Agenda 21. It contains a direct call to all local governments to create their own action plans for sustainable development. These Local Agenda 21 action plans translate the principles and mandates of Agenda 21 into concrete service strategies for each local community.

Local Agenda 21 is a local-government-led, community-wide, participatory effort intended to establish a comprehensive action strategy for environmental protection, economic prosperity, and community well being in the local jurisdiction or area. This requires the integration of planning and action across economic, social, and environmental spheres and involves consulting with the community, acknowledging that communities have the local knowledge needed to make sensible decisions for their future. Therefore, key elements for Local Agenda 21 are full community participation, assessment of current conditions, target setting for achieving specific goals, monitoring, and reporting.

Discussion and action on Agenda 21 spread rapidly to the policy realm, and a series of other summits involving the global community have addressed the theme over time. Different world conferences took the discussions forward, including the world conferences for Human Rights, Women, Population, Climate and Global Warming, and Food.

As a result of this series of summits and conferences, awareness has increased and people have gradually been realizing the importance of the environment for sustainable development and increasing their concern about social and cultural issues for sustainable development in the community.

See Also: Adaptation, Climate Change; Cities for Climate Protection; Sustainable Development; Water Pollution.

Further Readings

International Council for Local Environmental Initiatives. *The Local Agenda 21 Planning Guide: An Introduction to Sustainable Development Planning.* Toronto, Canada: International Council for Local Environmental Initiatives, 1996. http://www.idrc.ca/en/ev-9322-201-1-DO_TOPIC.html#begining (Accessed September 2009).
Sitarz, Daniel. *Sustainable America: America's Environment in the 21st Century—The U.S. Agenda 21.* Flagstaff, AZ: Earth Press, 1998.
United Nations Environment Programme. "Agenda 21." http://www.unep.org/Documents .Multilingual/Default.asp?DocumentID=52 (Accessed April 2008).

Rafael D'Almeida Martins
University of Campinas

AIR QUALITY

Photochemical smog, most commonly associated with Los Angeles, is caused by the interaction of solar radiation with pollutants like nitrogen oxides, peroxyacytyl nitrate, and volatile organic compounds.

Source: iStockphoto

The standard composition of a well-mixed atmosphere is approximately 78 percent nitrogen, 21 percent oxygen, 0.93 percent argon, and 0.038 percent carbon dioxide, plus water vapor (spatially and temporally variable) and numerous trace constituents. Because of the negative impacts of poor air quality, such as acid precipitation or compromised human health, an understanding of both normal and disturbed atmospheric conditions is necessary. Air quality is an assessment of the state of the atmosphere that uses a combination of various characteristics describing the ambient atmosphere, including concentration of pollutants, visibility, turbidity, and thermal pollution.

Air quality problems are often common to urbanized regions, where people engage in industrial activities, vehicular traffic is concentrated, and other conditions that deteriorate the ambient atmosphere exist. Efforts to mitigate compromised air quality range from international treaties and protocols to local regulations and ordinances.

Pollutants that compromise the "clean" characteristics of the air may come from nonhuman or human (anthropogenic) sources. Nonhuman sources of pollutants include volcanic eruptions, which send ash and sulfurous gases into the atmosphere, diminishing local visibility and possibly contributing to cooling on a global scale. Lightning-caused forest fires create airborne particulates, or soot, and anaerobic decomposition in wetlands is a source of methane. Because there is little that can be done with respect to natural sources of air pollutants, efforts to restore and maintain air quality usually focus on reducing anthropogenic sources of pollution. Anthropogenic sources of air pollutants include particulate matter caused by combustion engines, chemicals released by industrial and agricultural processes, and the use of consumer products.

Components of Air Quality

Increasing the concentrations of pollutants compromises the integrity of the ambient air and causes human health problems. Sulfur dioxide, for example, affects the respiratory and circulatory systems of humans, with particularly sensitive populations (e.g., the elderly, asthmatics, children) suffering the most. Carbon monoxide can impair the transport of oxygen to the brain and other organs. Toxic chemicals in the air can cause a range of adverse health problems.

Another component of air quality is visibility, or the ability to view objects at a distance. Visibility is described as the farthest horizontal distance at which the unaided eye can view an object. Atmospheric substances can impair visibility, reducing the visual quality of the air. Particulates may reflect or scatter solar radiation, which decreases viewing ability. Smog, largely produced by coal combustion, as well as by industrial and vehicular emissions that interact with incoming sunlight, reduces visibility. Haze, another cause for diminished visibility, is also caused by air pollution.

Turbidity refers to a reduction in the cloud-free part of the atmosphere's ability to transmit light. It is a function of both atmospheric depth (measured in the vertical) and the components in the air. In the troposphere, turbidity is typically caused by anthropogenic activities and can be described by Beer's Law. Turbidity is both spatially and temporally variable. An urban megalopolis, for example, typically exhibits greater turbidity than a proximate rural locale, owing to the greater presence of aerosols, tropospheric ozone, and other factors. Turbidity is not to be confused with global dimming—a phenomenon describing the overall reduced amount of solar radiation striking the Earth's surface.

Thermal pollution refers to a reduction in air quality stemming from an increase in average ambient temperatures. The most well-known example of this is the heat island effect, in which urban areas have warmer temperatures than proximate rural surroundings. Implicated in the heat island effect are urban building geometries (which impede the transport of heat), surface materials with high heat capacities, waste heat generated by heating and cooling, and a concentration of vehicular traffic.

The Clean Air Act

In the United States, the Clean Air Act established the role of the Environmental Protection Agency in the stewardship of the atmosphere. Passed in 1970 and last amended in 1990, the act set standards (the National Ambient Air Quality Standards) for certain pollutants deemed harmful or health-compromising; these are carbon monoxide, sulfur dioxide, nitrogen dioxide, ozone, and particulate matter. The pollutants are summarized by the Environmental Protection Agency in the Air Quality Index (AQI), with a value of 100 corresponding to the levels established by the National Ambient Air Quality Standards. Lower AQI values signify cleaner air, and values above 100 indicate the possibility of harm to sensitive populations. The danger increases as the AQI rises, and the population potentially affected by the deteriorated air also rises. National air quality forecasts are issued daily for cities across the United States.

From a human health perspective, tropospheric ozone and particulates may be some of the most dangerous pollutants that reduce air quality. Ozone irritates the lungs and is especially problematic for asthmatics, the elderly, and young children. However, high concentrations also endanger the general population. Ozone warnings are issued by the National Weather Service when levels become dangerous. Particulates—pollutants in solid or liquid phase—also damage the respiratory system. Coarsely sized particulates (between a diameter of 2.5 and 10 μm) accumulate in the lungs. Vehicular traffic (on paved and unpaved roads) adds these particles to the atmosphere. Fine particulates (less than 2.5 μm) may lodge deeply within the lungs, creating a high health risk; the origins of these particulates are largely fossil fuel combustion and biomass (including wood) burning.

Inversions, meteorological phenomena in which air temperature increases with height, can render polluted urban air more harmful than normal. By having a layer of warmer air atop cooler air, the dispersion of harmful atmospheric substances is inhibited, and air does not rise away from the near-surface environment. Inversions trap cooler air, along with pollutants, closer to the surface. This produces a pollution dome that rests above the city. Inversions may be the result of high pressure systems, which may stagnate and create conditions in which the winds are too light to transport pollutants from a city.

Some cities may suffer from the deterioration of air quality partly as a consequence of the local physical geography. The smog problem in Los Angeles is exacerbated by the San Bernardino Mountains, which impede the flow of air out of the city. In addition, cool Pacific sea breezes contribute to a local inversion by forcing the warmer, less-dense air aloft. Typically, air (and pollutants within it) is dispersed by winds. However, mountains serve as obstacles that inhibit this. Mexico City, an urban area with serious pollution problems, suffers from its location in a basin, which sets up a circulation pattern that traps air pollution.

Smog and the City

Smog (the word stems from the terms *smoke* and *fog*) is typically an urban phenomenon. Smog can be classified as sulfurous or photochemical. Sulfurous smog, which appeared after the Industrial Revolution took hold, forms when sulfur dioxide reacts with water vapor; it is a by-product of coal combustion. London, in particular, was severely affected in the first half of the 1900s, when homes used coal to generate heat, the by-products of which mixed with fog. Several cities have suffered from "killer" smog events, including

London (1952) and Donora, Pennsylvania (1948). Today, sulfurous smog stems largely from electrical power generation. Emissions of sulfur dioxide and nitrogen oxides can form acidic precipitation, which can harm aquatic and terrestrial ecosystems.

Photochemical smog, or ground-level ozone, arises from the interaction of solar radiation with pollutants, and thus is most common on sunny days. Interactions with sunlight lead to greater amounts of tropospheric ozone, a ground-level pollutant and greenhouse gas. This type of smog is commonly associated with Los Angeles. The pollutants include nitrogen oxides, peroxyacytyl nitrate, and volatile organic compounds. Photochemical smog is a more contemporary air quality problem than sulfurous smog. It results from the emissions from vehicular traffic, industrial activity, and chemical solvents. Interactions with sunlight lead to greater amounts of tropospheric ozone, a ground-level pollutant and greenhouse gas. Among the negative effects of photochemical smog are stresses on respiratory systems, damage to plant cells, and damage to rubber and paint.

Cities and urban areas have often been the sites of severe reductions in air quality. Concern about the pollution in Beijing spurred concerns by athletes and coaches before the 2008 Summer Olympics. Unfiltered exhaust from cars and motorbikes created a layer of smog that blanketed Tehran, Iran, before pollution reduction measures were enacted. Therefore, some local or other governmental agencies have developed measures, such as regulations, to combat air quality deterioration. For example, many German cities have established low-emissions zones. In addition, many local governments have established air quality agencies to combat the problem. However, some critics point out that regulations emphasize control of the primary pollutants (direct emissions) while underestimating the amount of secondary pollutants (products of primary pollutants) in the atmosphere.

Although cities and urbanized regions generate much of the pollution that compromises the quality of air, atmospheric winds transport that pollution to locations far from the source. Rural areas downwind of cities receive pollution from urban sources. Pollutants from industrial activity in the United Kingdom and (West) Germany led to acid precipitation over Sweden, acidifying soils and damaging Swedish lakes (and the fish within). Maintaining air quality, then, is a transboundary issue.

See Also: Cities for Climate Protection; Heat Island Effect; Indoor Air Quality; Los Angeles, California.

Further Readings

AirNow.gov. "Air Quality Index (AQI)—A Guide to Air Quality and Your Health." http://airnow.gov/index.cfm?action=static.aqi (Accessed March 2009).

Godish, Thad. *Air Quality,* 4th Ed. Boca Raton, FL: Lewis, 2004.

Kidd, J. S. and Renee A. Kidd. *Air Pollution Problems and Solutions.* New York: Chelsea House, 2006.

U.S. Environmental Protection Agency. "Air Pollution Emissions Overview." http://www.epa.gov/air/oaqps/emissns.html (Accessed March 2009).

Petra A. Zimmermann
Ball State University

AUSTIN, TEXAS

Austin, the capital of Texas, is a city of almost 800,000 people, located in the central hill country of the state and home to a diverse group of citizens. It has the reputation of being a major American center of progressive culture and politics. Many of the unique attributes of Austin contribute to its reputation as a "green city" as well.

Austin, the capital of Texas, is a major center of progressive culture and politics. Austin's citizens have been involved in environmental movements since the 1970s.

Source: iStockphoto

For example, Austin is home to the University of Texas, which alone adds a school-year population of over 50,000 students to the city. Austin also has a large and influential music scene, which causes a mass migration of hopeful rock stars. The combination of young high-tech workers, university students and professors, and a large artistic community makes Austin the most liberally minded city in the largely conservative state of Texas. Thus, it is no surprise that Austin would be a national vanguard of green living.

Austinites are known for their desire to "Keep Austin Weird." This is a slogan that is plastered on bumper stickers and billboards throughout the city. It refers to the Austinites' desire to promote small business and maintain the local charm of the "little blue island" in the middle of the politically conservative red Texas sea. This commitment to the city manifests itself in the high level of community involvement with environmental issues. Beginning in the 1970s, citizens became concerned with the state of their environment. During the 1970s and 1980s, many city ordinances were passed to protect Austin's rivers and lakes. One particular focus that continues to this day is the preservation and maintenance of Barton Springs, a green and blue oasis almost adjacent to downtown Austin.

The government of the City of Austin has been integral in pursuing a sustainable, green city. Because of the liberal tendencies of many Austin citizens, the municipal government is expected to be active in promoting, and even mandating, green living. The city supports green living by promoting the use of renewable energy, enforcing green building codes, supporting efficient public and private transportation, advocating for watersheds, having a proactive stance toward climate change, and emphasizing green gardening. Over the 2009–2012 time span, the City of Austin has planned to complete the transition to renewable-energy power for all municipal facilities. Austin produces more wind power and biodiesel fuel than any city in the United States.

The city has mandated that by 2015, every new home in Austin have an optional solar power system in the building plans, which would result in rendering the house energy neutral. The city is already participating in sustainable building, having recently rebuilt the Austin City Hall with mostly recycled materials.

Austin also is widely known for its cost-effective metro bus system. Many university students use public transportation, rather than drive to school. The city planned to unveil its downtown rail system in mid-2009. Many areas of Austin are carefully designed for pedestrian traffic, and the city is renowned for being bicycle friendly. Bikes can be rented from any public library or from one of the many bike rental shops in the city. The Lance Armstrong Bicycle Highway is under construction, intended to add a number of bicycling paths throughout much of the greater Austin area.

In the downtown area, creative transportation is taking the place of traditional taxis: Human-powered pedicabs carry passengers from restaurants to clubs, emitting zero greenhouse gases. The city government has also established the Watershed Protection Development Review, a development process dedicated to reducing flooding and pollution while protecting metropolitan watersheds. The review's website provides up-to-date information that publishes all the city's decisions regarding every area watershed.

Climate change is another issue that the city government has addressed, passing measures to ensure that all municipal facilities and transportation will transition to carbon-neutral operation by 2020. Carbon caps are slated to be put in place to control all utility emissions in the city.

Another city department is Grow Green, which focuses on gardening solutions that conserve water in an extremely dry ecotone. Grow Green promotes the concept of xeriscaping—a desert-oriented gardening design and practice for dry climates that focuses on water conservation—along with gardening practices that minimize chemical use. The department provides information for home gardeners, along with certified training for gardening and lawn maintenance professionals.

Business and Citizen Greening

Many businesses that focus on ecofriendly living have chosen to locate in Austin. The organic food giant Whole Foods was launched from downtown Austin. In 2008, Office Depot opened its first "green store" in the city. A prototype for certifying green building design and construction, Leadership in Energy and Environmental Design (LEED) operates from Austin to promote future business construction across the United States. Another substantial development in Austin's green business community is Dell's Children's Medical Center, the first LEED "Platinum Certified" hospital. The hospital conserves resources and is a sustainable site.

The motto "Keep Austin Weird" has given rise to many local groups that focus on environmental needs specific to the Austin area. One of the largest civic environmental groups in Texas is Austin's Environment Texas. This group advocates water conservation and solar power, providing an array of resources for citizens to learn about environmental issues. Environment Texas operates a website with accessible links to local, state, and national representatives, assisting citizens in lobbying on environmental issues. Keep Austin Beautiful is a nonprofit group that seeks to educate citizens and involve them in waste reduction and resource conservation. Ten thousand annual volunteers keep that organization functioning.

Active community involvement is probably the major key to Austin's success as a green city. The Sustainable Food Center is a community-based organization that advocates locally grown organic food. Founded in 1993, long before locally grown food became popular, the group hosts a farmers market to provide affordable, nutritious food to various groups and private citizens in Austin. Many other groups exist to support specific local areas of environmental concern, such as the city's Greenbelt recreation area and Barton Springs. The plethora of nonprofit environmental groups helps Austin citizens participate more actively in environmental policies by creating easy opportunities for the average citizen to become involved.

The combination of an active city government and active citizens' advocacy groups produces a city purposefully pursuing green living. As many of the city government's new, environmentally friendly ordinances go into effect, Austin will most likely emerge in the vanguard of green cities, leading the nation toward environmentally conscious and sustainable living.

See Also: Environmental Planning; Watershed Protection; Xeriscaping.

Further Readings

City of Austin. "Green City Austin." http://www.ci.austin.tx.us/greencityfest/default.htm (Accessed April 2009).

Reichman, Trevor. "Green City Guide." http://planetgreen.discovery.com/travel-outdoors (Accessed April 2009).

Wassenich, Red, et al. *Keep Austin Weird: A Guide to the Odd Side of Town.* Atglen, PA: Schiffer, 2007.

Anthony R. S. Chiaviello
University of Houston–Downtown

B

Bahía de Caráquez, Ecuador

Since 1999, following repeated devastation from flooding and mudslides, the Ecuadorian coastal city of Bahía de Caráquez has committed itself by law to rebuilding according to standards of ecological sustainability. The city, renowned for its beaches and neotropical forest, was essentially rebuilt from the ground up, with an emphasis on ecotourism and upscale hotels, and has since become one of Ecuador's most popular resorts for upper-class tourists. Through a partnership with the San Francisco–based Planet Drum Foundation, Bahía has applied an ecological strategy known as "bioregionalism," as it continues to develop both urban infrastructure and local ecosystems.

Bahía is a small city of great regional political and economic importance located on a sandy peninsula at the mouth of the Río Chone estuary on the country's western coast. In addition to hosting thousands of tourists from Ecuador and abroad each year, it is the county municipal center, transportation hub for both highway and ferry traffic, and core manufacturing and shopping area for the region. The shifting sands and shallow channel of the Río Chone have historically limited Bahía's usefulness as a port. As a result, the city has been spared the environmental harms that often accompany commercial shipping and heavy industry. Ecuadorian dry neotropical forests, such as those that surround Bahía, are areas of high biodiversity and endemism and are considered a top priority by conservationists. Average precipitation in the region fluctuates significantly: Rainfall in an El Niño year can be 200 times that of a dry year.

Bahía featured noteworthy ecologically minded efforts before its formal commitment to become an ecocity. It was protected by several natural resource management laws: In 1989 area beaches were set aside as a National Reserve Area, and in 1990 the municipality established the Frigate Bird Islands Mangrove Bird Sanctuary. Pioneering local projects include a recycling facility for both organic and inorganic matter in the Bahía market center, a participatory organic farm and environmentalist school in Río Muchacho, an agroecology project at Encarnación Organic Farm, and an "ecological" housing subdivision for the homeless. Alternative tourism opportunities include bird watching in the mangroves of nearby Isla del Corazón (Heart Island); participatory local farm tours; an archaeological museum in Bahía; the nearby archaeological site of Chirije; tours to nearby forests, caves, mangroves, and wetlands; a local workshop to produce handmade recycled paper, and an Environmental Interpretation Center in which visitors to the area are taught about local ecology and conservation efforts.

Dealing With Disaster

On August 4, 1998, a magnitude 7.1 (Mw) earthquake hit the coast of Ecuador 10 kilometers north of Bahía, reducing 200 buildings to rubble. The earthquake came after a devastating six months for the city in which relentless El Niño rains flooded the streets and melted the surrounding hillsides; between December 1997 and August 1998, mudslides crushed houses, buried roads, destroyed bridges, and rendered homeless some 5,000 of the city's 25,000 residents.

Faced with the challenging decisions of recovery, county mayor Fernando Cassis Martinez passed a law declaring Bahía de Caráquez a "Ciudad Ecologica" (Ecological City). The declaration, issued on February 23, 1999, identified a list of priorities for the city to move forward according to principles of sustainability, which it defined as shared environmental responsibility and long-term goals. The list included the establishment of an Environmental Affairs Municipal Department, environmental education programs for government employees, and a citizens environmental awareness campaign, as well as a move to designate the dry neotropical forest in the Bahía urban area a biodiversity preserve.

Mayor Martinez invited Peter Berg, founder of the Planet Drum Foundation, to advise the city in the design and implementation of a new sustainable development plan. Planet Drum was founded in San Francisco in 1973 around the concept of bioregionalism. A bioregion is defined as a distinct area of land, often defined by a watershed, with interconnected plant and animal communities and natural systems. Planet Drum seeks to teach individuals to live within the unique natural confines of their bioregion, believing that education is the key to sustainable living practices.

Steps Toward Recovery

The foundation opened a field office in Bahía in 2000 and has maintained a significant presence in the city ever since. In its first project to rehabilitate the local environment, Planet Drum partnered with a recently founded local nonprofit, EcoBahía Centro de Educación Ambiental (EcoBahía Environmental Learning Center), in two small-scale mangrove reforestation efforts designed to mitigate local pollution going from houses into the bay. EcoBahía represents a wide range of the local community—its current president also teaches agriculture at nearby Colegio Técnico San Vicente, and it currently has a membership near 100, including those who are working class, professionals, business leaders, homeless, activists, and students.

Planet Drum and EcoBahía began another revegetation project in 2001 on the hillsides of the María Auxiliadora barrio, which had been entirely swept away, down to clay subsoil, by the 1998 mudslides. Volunteers planted six acres of barren land with native tree species to stem future erosion and possibly provide a corridor for local wildlife. Today, barrio residents continue to maintain the area, known as La Bosque de las Ruinas (The Forest Among the Ruins), as an urban park with educational and tourism potential.

As Planet Drum remains involved in the redevelopment of Bahía, revegetation projects seem to be the most successful efforts to date: approximately 50 areas have been replanted with 300 to 500 seedlings each. In addition to providing habitat for native plants and animals, these trees will help to stabilize the slopes towering over Bahía and prevent future mudslides. Additional revegetation projects have been geared toward food crops, fruits, and cattle fodder.

Action and Education

In tandem with revegetation efforts, Planet Drum opened a Bioregional Education Program in 2005, in which local schoolchildren meet twice a week for classes and field trips about the local environment and restoration efforts. Both the revegetation and education initiatives are implemented by volunteers, who are typically college age and from the United States, recruited and housed by Planet Drum for stays of one to three months. Though Planet Drum states that its goal is ultimately to have all initiatives directed and sustained by local residents and institutions, its efforts currently rely heavily on foreign volunteers. Planet Drum maintains an office and a field projects manager in Bahía who coordinates volunteers and technical assistance.

The city of Bahía has recently expanded its recycling program to include a neighborhood composting project, and the city now recycles a majority of the organic waste from household kitchens and the large central market for compost.

The EcoCity declaration provided impetus to reform several local industries, including a shrimp farm, Camaronera Bahía. Ecuador's shrimp industry has long been targeted by social and environmental justice movements for its destructive impact on threatened mangrove ecosystems and local livelihoods. The renovated shrimp farm, now EcoCamaronera Bahía, became the world's first certified-organic shrimp production system. A partner company, GreenAqua Ltd., produces and distributes patented inputs for the shrimp industry including food, natural growth stimulators, preservatives, antibiotics, and vaccines derived primarily from local minerals and plants. Naturland, a German organic-certification company, has certified the breeding labs, packaging sites, and food providers to ensure a clean chain of production.

Despite these innovative production methods, independent research suggests that EcoCamaronera Bahía may continue to marginalize local residents, preventing access to the coast and sources of drinking water. Certification criteria are concerned primarily with technical production factors, instead of broader social and ecological sustainability, and were not available in Spanish-language publications. The farm has also come under criticism for yielding few opportunities for local employment and for exploiting former mangrove swamps, designated by Ecuadorian law as public land, for private profit.

The city has drafted plans for green development beyond existing projects, ranging from renewable energy systems to artificial wetlands for sewage filtration. However, a struggling national economy yields little funding support; Bahía's development has relied primarily on international fundraising and private donations, which have dried up since the 1999 declaration. In addition, green projects thus far have not featured job creation as a top priority, which has hindered local recruitment.

See Also: Biodiversity; Bogotá, Colombia; Infrastructure; Sustainable Development.

Further Readings

Berg, Peter. "Conservation, Preservation and Restoration in Ecuador." *Earth Island Journal,* 16/1 (2000).

Gertz, Emily. "On the Edge of the Future: What Are Some of the World's Poorest Cities Getting Right, and What Can They Teach Us?" *Momentum Magazine* (May–August 2009).

Gobierno Cantonal de Sucre (Sucre County Government). "Bahía de Caráquez—Ecociudad" ("Caráquez Bay—Ecocity"). http://www.bahiadecaraquez.com/ecociudad.htm (Accessed October 2009).

Planet Drum Foundation. "Eco Ecuador Works: Collaboration for a Long-Term Sustainable City." http://www.planetdrum.org/eco_ecuador.htm (Accessed October 2009).

Yepes, Carolina LeMarie. *Certifying Destruction: Integral Analysis of Organic Certification of Shrimp Aquaculture Industry in Ecuador, Executive Summary.* Quito, Ecuador: Corporación Coordinadora Nacional para la Defensa del Ecosistema Manglar, 2007.

Emma Gaalaas Mullaney
Pennsylvania State University

Bangkok, Thailand

The city of Bangkok has served as the capital of Thailand for more than 240 years. Situated on the Cho Phraya River delta, the city experienced rapid growth from the 1960s through the 1990s, its population increasing from 2.1 million to 5.9 million during those 30 years. By 2000, its official population had reached 6.3 million, although the actual number of residents was much higher. With growth came a multitude of environmental problems. To its credit, Bangkok has taken these problems head-on and for its efforts has garnered recognition as one of the world's greenest cities. In 2007, Bangkok was included on the list of Fifteen Greenest Cities in the World by Grist, a well-regarded environmental nonprofit organization. The following year it was awarded the title "Clean and Green Land" by the Association of Southeast Asian Nations (ASEAN).

Planning to Be Green

Bangkok's commitment to being an environmentally sustainable city dates back to 1992, when the Bangkok Metropolitan Administration (BMA) undertook a series of integrated systems studies to assess the full range of environmental issues. BMA paid particular attention to developing the social processes that would ensure the effort's viability over time. Though the Asian economic crisis that struck in July of 1997 sapped the city's financial resources, little momentum was lost. Bangkok's Agenda 21 (BA21) was adopted in 1998, setting in place a 10-point plan that would guide city development for the next 20 years.

Patterned on the United Nations' model, BA21 proposed conventional goals—improvement of the transportation network, investment in urban greening, and land use reform to provide a higher quality of life for residents—but it also promised to provide "good governance . . . to meet the challenges of the future." Even as it established a broad agenda,

Bangkok took care to translate it to the local level to ensure broad public commitment to and investment in attaining the goals.

While the city has made laudable progress in becoming a sustainable city, the combined pressures of growth and increased affluence keep moving the endpoint. Heightened awareness of the environmental threats associated with global warming has added a new layer of urgency. In response, the BMA prepared its Action Plan on Global Warming Mitigation 2007–2012 (APGWM) and adopted Bangkok 2020: Sustainable Metropolis Plan to succeed BA21.

The Challenges

The first environmental challenge taken under BA21 was a clean air campaign. The concentration of industrial activity in Bangkok coupled with the ever-increasing traffic congestion cast a filthy pall over the city. A multi-pronged approach was taken. The road network was expanded in an effort to decrease congestion. Improvements to the mass transit system were planned but were stalled for years by the financial crisis. Other programs were initiated, including phasing out two-stroke motorcycle engines, which produce particularly noxious tailpipe emissions. The taxi fleet was converted from running on conventional gasoline to cleaner-burning liquid petroleum gas. Perhaps most significantly, automobile emission standards were adopted that were even more strict than those set by the European Community. In combination, these actions reduced small particulate pollution by 47 percent over the course of a decade.

Despite this success, air quality is a continuing problem for Bangkok. As in every major city, increased private vehicle use is a major culprit. Transportation accounts for 50 percent of the city's CO_2 emissions. While the APGWM called for roadway improvements to reduce congestion, it also proposed congestion pricing for central Bangkok and improvements to the transit, bus, and bicycle systems to reduce reliance on private cars.

The greening effort in some respects benefited from the economic downturn. Though the downtown core of Bangkok is densely populated, as late as 2002, nearly 40 percent of the land in the greater city was still undeveloped. This was partly because large tracts of land had been made inaccessible by the network of roads hurriedly built to accommodate the growth of the boom decades, and partly because the 1997 financial collapse had stopped development dead in its tracks. The city took advantage of the situation to develop a system of 24 metropolitan parks and nearly 3,000 smaller parks and gardens that now account for a major portion of the city's green infrastructure. APGWM expanded the urban greening effort from parks into neighborhoods by calling for 85 million trees to be planted in and around the city by 2012.

Despite the progress made in addressing air pollution and urban greening, in nearly every aspect, water poses a problem for Bangkok. The city sits a mere six feet above sea level and is subject to intermittent flooding. Often called the "Venice of the East," it faces the same problems of subsidence and seawater intrusion into groundwater wells as the Italian city. The canals, or *klongs*, that brought a Venetian comparison, are polluted by illegally discharged untreated effluent and run off from solid waste. The public sewer system, which was started in the 1990s, is still incomplete. Most buildings have private septic systems.

Bangkok draws more than 90 percent of its drinking water from the Chao Phraya and Mae Klong rivers, both of which are badly polluted by untreated wastewater. The remaining

9 percent of drinking water comes from wells, many of which are contaminated by urban and agricultural runoff. Groundwater reserves are being overdrawn with two consequences: Several aquifers suffer from saltwater intrusion, and areas of Bangkok are sinking. The subsidence problem was most critical in the early 1980s when the subsidence rate was 12 centimeters per year. Groundwater pricing policies and the expansion of the tap water supply have helped slow the rates of groundwater overdraft and subsidence, the latter to 2.5 centimeters per year by 2001. Despite the rate decrease, the area affected by the problem is expanding. Furthermore, because of the soft clay composition of the Bangkok plain, subsidence would still continue for a number of years even if overdrafting were to cease overnight.

Exacerbating Bangkok's water problems, climate change has caused river flow rates to be highly variable. According to a report published by the Organisation for Economic Co-operation and Development in 2008, nearly 1 million Bangkok residents are currently vulnerable to flooding, attributable to the effects of climate change. Experts believe that a dyke system is essential. The APGWM does not address flooding, but it does call for increasing the city's wastewater treatment facilities and for a reduction in household wastewater.

Solid waste poses another seemingly intractable problem. The city generates about 9,400 tons each day, but the waste management system can only handle about 7,500 tons. The excess waste is stacked on vacant lots and thrown in the canals. Previous resource recovery and recycling programs proved inadequate. To a limited extent, scavengers have stepped into the void. With a projected daily generation in 2015 of 16,000 tons, alternatives to landfills are being pursued, and the city is considering organizing scavenger groups.

The Outlook

The City of Bangkok has always taken a proactive approach to environmental problems. Programs like "The Bangkok Big Switch," aimed at significantly reducing energy usage, will continue to engage the public. Never inclined to rest on past accomplishments, Bangkok seems well aware that being green is a process, not a condition.

See Also: Agenda 21; Air Quality; Congestion Pricing; Sea Level Rise; Waste Disposal; Water Pollution.

Further Readings

Bangkok Metropolitan Administration. Action Plan on Global Warming Mitigation 2007–2012. http://www.baq2008.org/system/files/BMA+Plan.pdf (Accessed December 2009).

The Cities Alliance. "Livable Cities: The Benefits of Urban Environmental Planning." Washington, D.C.: The Cities Alliance, 2007. http://www.unep.org/urban_environment/PDFs/LiveableCities.pdf (Accessed December 2009).

Polprasert, Chongrak. "Water Environment Issues of Bangkok City, Thailand: Options for Sustainable Management." *ScienceAsia*, 33/1 (2007). http://researchers.in.th/file/chorchat/PDF13.pdf (Accessed December 2009).

Susan H. Weaver
Independent Scholar

BARCELONA, SPAIN

Barcelona is a Mediterranean city that has experienced a major socioeconomic urban revival in the run-up to the 1992 Summer Olympics and has, according to some sources, the highest quality of urban life in Europe, mainly because of its mild climate, cultural infrastructure, and availability of public spaces, such as parks and beaches. As a signatory to a number of international agreements on climate change, Barcelona is committed to becoming a sustainable city; however, it currently faces issues in waste disposal, water, and energy supply that pose serious obstacles to this ambition.

Barcelona's Profile

Barcelona, a Mediterranean city, is the regional capital of the autonomous community of Catalonia. With approximately 1.6 million inhabitants, it is Spain's second-largest city. It should not be confused with the Province of Barcelona (a larger entity with over 5 million inhabitants) or the Barcelona Metropolitan Area (approximately 3 million inhabitants), which is the sixth-largest city-region in Europe, accounting for roughly 12.5 percent of total Spanish gross domestic product. Historically, the City of Barcelona has been a major industrial center, with both heavy and light industries; however, over the last decades the importance of finance, banking, and tourism has increased. The city's 10 administrative districts have some powers in the area of planning and infrastructure, but most power lies with the Barcelona city council (Ajuntament de Barcelona), which has jurisdiction in the fields of urban planning, transport, municipal taxation, policing, environment, culture, and health—though it should be noted that many competencies are shared with the regional government (Generalitat de Catalunya) and the central Spanish state (and the latter two take legal precedence).

Climate Change

The city council is signatory to a series of international agreements to reduce greenhouse gas emissions, and it is a member of the Klimabündnis, Energie-Cites, ICLEI, and the C40. The Heidelberg Declaration (1994) committed the city to reduce carbon (equivalent) emissions by 20 percent by 2005 with respect to 1987 levels (a target that was not met), and as a member of Klimabündnis, carbon (equivalent) should be reduced by 27 percent by 2010 with respect to 1997 levels. Barcelona also created its own Agenda 21, an attempt to involve a broad range of stakeholders in the city. In 2006, Barcelona emitted 5.3 million tons of carbon dioxide, a 10.78 percent increase since 1999; however, Barcelona's per capita emissions (about 3 tons carbon dioxide equivalent in 2002) are low compared with those of other cities (5–10 tons in northern European cities). This is attributed to a benign climate, a low-carbon energy mix (mainly nuclear), a compact city model, and increased use of natural gas. This low baseline is in part the reason that commitments to reduce emissions are not being met (other reasons identified being negative financial returns on the required investments to further reduce emissions). Emissions by sector show that almost a third are from waste (landfill and incineration), 25 percent from transport, 19 percent from the domestic sector, 8 percent from commercial activities, and 15 percent from industry. Across the sectors, over 73 percent of emissions are from the city's energy use.

Energy

In 2006, Barcelona's energy consumption accounted for 15.5 percent of Catalonia's energy consumption and 3.1 percent of the Spanish state's. The city's energy mix is predominantly nuclear (49 percent), followed by fossil fuels (23 percent natural gas, 18 percent gasoline, 4 percent liquefied natural gas), and only about 5 percent from renewable sources (mostly hydroelectric). Energy consumption has been increasing consistently, reflecting deindustrialization and the growth of the service sector (and its electricity-intensive cooling demands), as well as an increase in the use of natural gas (up by almost 40 percent) for domestic heating and a growing demand for transport fuels (driven by the "ring road effect").

The Barcelona Energy Improvement Plan is the main instrument to manage energy in the city. The plan provides a diagnosis of the status quo, a scenario-based analysis, and an action plan. It concluded that the high dependence on external energy supply means that Barcelona's emission reduction's potential is through the promotion of renewable energy in the city and energy efficiency. Policies include awareness-raising campaigns (such as "Barcelona saves energy" in the 1990s) and demonstration projects, such as solar thermal installations in public swimming pools and solar photovoltaic systems in schools, community centers, and other public buildings. Barcelona is the first European city to have introduced a "solar thermal ordinance" (in 2000), a bylaw that requires new buildings and refurbishments above a specified size to supply 60 percent of warm water from solar thermal energy, resulting in savings of almost 2 tonnes of carbon dioxide equivalent emissions. Cities across Spain have followed Barcelona's example, and in 2007, Barcelona won the "ManagEnergy" Local Energy Action Award for its commitment to sustainability.

Transport and Air Quality

Transport accounts for 33 percent of energy used in the city, 25 percent of carbon equivalent emissions, and 88 percent of other harmful emissions (nitrogen oxide, sulfur dioxide, carbon monoxide, and particles of 10 μm or less). Private and commercial transport account for 91.1 percent of energy consumed in transportation activities and over 95 percent of transport-related carbon dioxide emissions (constituting just over half of all journeys). Public services consume only 8.9 percent of the total energy dedicated to transport, while accounting for 48 percent of trips (the underground is the most efficient, consuming less than 4 percent of transport-related energy for over a fifth of trips).

Concentrations of particulate matter and nitrous oxide (both the result of road traffic) regularly exceed European Union health standards. The regional government is implementing a plan that targets air quality in Catalonia's worst-affected areas, such as the Barcelona Metropolitan Area, aiming to reduce air pollution to comply with the current standards legislated by the European Union by 2010. In 2007, the city council introduced "Bicing," a public hire bicycle network integrated into the larger public transport system across the city; it is estimated that during the first months of implementation, 960 tons of carbon dioxide were avoided.

Water

Water policy is one of the priority policy areas of Barcelona, mainly because of a series of dry winters over the past decades that have left Barcelona's water reservoirs far below desired levels. Coupled with Barcelona's trend toward a more diffuse city model (expected to increase water demand), this means that water deficit will become an even more serious problem over the next decades. So far, water policy has focused on demand-side management strategies, in particular by raising awareness about water scarcity and attempting to create a "new water culture." Water quality is another issue—rivers of the region are some of the most polluted and degraded in western Europe, mainly as a result of water discharged from wastewater plants. Since the European directive on urban wastewater (1991), a comprehensive program of water treatment plants has been implemented, and the situation has improved dramatically.

Waste Management

Selective waste collection is carried out by the city council according to the directives established in the regional and metropolitan program on waste management. About a third of waste was collected in this way in 2007. Waste processing is handled by the Metropolitan Environment Entity, which oversees several energy recovery plants, composting plants, landfill sites, and ecoparks. The latter two receive by far the most waste (32.2 and 41.7 percent, respectively), and although ecoparks contribute (minimally) to the city's renewable energy generation (mainly biogas), the landfill sites constitute over a quarter of the city's greenhouse gas emissions. Waste policy focuses on increasing energy recovery and reducing emissions, and a number of demand-side measures are in place, such as awareness-raising campaigns, encouraging the use of metropolitan waste depots by offering tax incentives, and reducing waste by promoting secondhand markets.

See Also: Air Quality; Waste Disposal; Water Conservation.

Further Readings

Ajuntament de Barcelona. "El Medi Ambient de Barcelona: Informe Anual 2007." http://w10.bcn.es/APPS/stnbcneta/es/html/base.jsp?seccion=a_5.jsp&menu=1&submenu=4 (Accessed April 2009).

Ajuntament de Barcelona. "Pla de Millora Energética de Barcelona." 2002. http://www.barcelonaenergia.cat/cas/actuaciones/pmeb.htm (Accessed April 2009).

Área Metropolitana de Barcelona. Entidad del Medi Ambient (2007). "Metropolitan Environmental Data 2007." http://www.amb.es/web/emma/dades/dades07/residus (Accessed April 2009).

Marshall, Tim. *Transforming Barcelona: The Renewal of a European Metropolis.* New York: Routledge, 2004.

Sauri, David, et al. "Changing Conceptions of Sustainability in Barcelona's Public Parks. *The Geographical Review* (February 4, 2009).

Anne Maassen
University of Durham

BEIJING, CHINA

Beijing is one of the world's truly impressive cities, with a 3,000-year history and 15.38 million people. Contrasts between ancient and modern pervade this city, which is surrounded by blue sky, grassland, and clean rivers. Beijing, the capital of the Peoples Republic of China, lies in the north of the North China Plain, adjoining the Inner Mongolian Highland. Covering 16,808 square kilometers in area, Beijing is the political, economic, and cultural center of China. Five rivers run through the city. Beijing has a long history, and it has been China's capital for more than seven centuries.

With the increasing number of cars in Beijing, traffic conditions are worsening, especially during the morning rush hour.

Source: iStockphoto

All these beautiful scenes depend on the changes in Beijing's environment. At the beginning of the 1990s, the primary aim for China was to boost economic growth and reduce poverty, and the rapidly growing economy in China made Beijing a city bustling, flourishing, and brimming with commercial chaos, but the city's extensive growth abated large amounts of resources and caused many environmental problems. Beijing was faced with a dual task; namely, developing its national economy and protecting its ecological environment. Because of its large population and scarce natural resources, Beijing was confronted by many problems with respect to land use, air quality, traffic jams, and ecological conservation. Similar to all the large cities in the world, Beijing had suffered from very rapid development. The major challenges were:

- *Atmospheric pollution:* Because of rapid income growth and falling car prices, the rapid growth in car ownership by families in Beijing inevitably caused air pollution, even if the newer cars were more environmentally friendly. Most of Beijing's air pollution is caused by cars. There were also some factories in and around the city that emitted carbon monoxide, nitrogen oxide, and other pollutants such as particulates and carbon dioxides into the air.
- *Traffic jams:* Because of the increasing number of cars, the traffic conditions in Beijing at that time were becoming worse, especially during rush hour. People wasted a lot of time waiting for the bus and sitting in traffic on their way home.
- *Sand storms:* Partly because of its geographical location, Beijing residents often saw strong winds and sandy weather in spring. Anyone who had spent time in Beijing was familiar with the dust storms that blotted out the sky, filled the eyes with sand, and sometimes decreased visibility to only 50 meters. The dust particles were a serious health hazard as well.
- *Water quality deterioration:* As a result of the excessive pumping of underground water and industrial discharges that were not under control, the Beijing water system was polluted.

Industries such as agriculture, brewing, textiles, and chemicals were heavy consumers of water, and many other manufacturing activities also required large amounts of water in production processes. Water shortages blocked the Beijing economy.

Now, following the joint efforts of the residents and the government, the environment in Beijing has been greatly improved. With the development of the city has come the popular idea that Beijing's environmental issues are important. With the recent economic takeoff, Beijing residents are generally well-off in terms of living standards and are living a more affluent life. At the same time, more and more Beijing residents are interested in environmental issues and are aware of their responsibility for environmental protection. According to the survey of the Beijing Environmental Communication and Education Center and the Beijing Institute of Social-Psychological Studies, Beijing residents are paying more attention to their living conditions, including air quality, sanitation, drinking water, afforestation, trash disposal, and noise.

In recent years, Beijing has made great efforts to improve the environment. The government also has made important arrangements for environmental protection. Ecological consideration was elevated to a strategic position matched by economic, political, cultural, and social development in this new era.

- The atmospheric environmental quality in Beijing has been greatly improved compared with past years. Beijing now uses less coal and is making efforts to use natural gas and other cleaner energy sources. The city consumed 300 million cubic meters of natural gas in 1998, and the amount has grown over 10 times to 4.7 billion cubic meters in 2007. Many factories implement a strategy of low resource consumption and pollution, and some of them have been relocated. For example, Beijing Capital Iron and Steel Group built a new steel plant in Hebei Province. Beijing has eliminated old buses and taxis to reduce air pollution, and the new buses and taxis meet the National Phase III Emission Standard and may face stricter standards in 2009. Many electrically powered buses also were put into use. The authorities are working on a plan to provide natural gas for over-quota use. Now, the average density of sulfur dioxide and nitrogen dioxide in the atmosphere meets the national standard. The only area that does not meet the national standard is inhaled particulates.
- The government took appropriate actions in response to Beijing's traffic conditions. More roads were laid to relieve the congestion in the streets to some extent and to accelerate the flow of traffic, and more underground parking was built. The government also added more public transportation lines. More people can go to work by bus or subway than by bike or car, thus making the roads less crowded. The provisional traffic management measures were entered into force in 2008 and have been effectively carried out, achieving good effects. With all these efforts, the atmospheric environmental quality in Beijing has been greatly improved in comparison with previous years, and the traffic flow in the urban area has decreased, while the average vehicle speed has increased. Beijing residents now live more comfortably and work more efficiently.
- In recent years, more and more people have realized the importance of forestation and grassland. The government invests a lot of money to make the city greener. Grassland and forest repair projects, such as the Three-North Forest Belt project in the north of Beijing, have helped to reduce sandstorms in the city. As a result, there were no sandstorms reported in Beijing in 2003.
- In the area of water environment amelioration in Beijing, sewage treatment has become more and more important. Beijing pays attention to sewage and trash disposal and aims

to deal with 90 percent of wastewater and 97 percent of solid waste in its eight districts. In 2007, 9 of the 14 planned large-scale sewage treatment plants had been built, and the five remaining ones were under construction. The authorities have continued to supplement recycled water into the waterways to keep the water clean. Another supplement is rainwater. Every year, Beijing puts some 10 million cubic meters of rainwater into the rivers and lakes. These efforts have caused a remarkable improvement of water quality. The quality of the city's water has remained stable and now also meets national standards, and the emission of major pollutants is on the decline.

- In 2008, Beijing's environmental protection system achieved great successes as evidenced both by the decreased number of environmental emergencies caused by unprecedented natural disasters and by the delivery of satisfactory environmental quality. Progress has been made in such areas as pollution reduction, construction of environmental protective infrastructures, pollution control of major river basins, and capacity building, as well as environmental economic policies.

Beijing has made remarkable progress in both water protection and elevation of green coverage. At the same time, air quality in the city has improved for the last eight consecutive years. In 2008, days with good air quality reached 274, representing a dramatic increase from 100 such days in 1998. In the same year, Beijing honored its pledge one year ahead of schedule to increase green space, as outlined in its bid to hold the Summer Olympics.

Beijing has accomplished some goals of environmental protection, but there is still a long way to go. To deal with the relationship between the trend of environmental pollution and ecological degradation and increasing national income is a vital issue. Thus, Beijing should carry out a strategy of low resource consumption, proper consumption patterns by consumers, and stable and sustained economic growth. The only way is to strictly follow the strategy of sustainable development that has been proven to be effective in maintaining a greener city environment.

See Also: Air Quality; Environmental Planning; Healthy Cities; Recycling in Cities; Sustainable Development.

Further Readings

Beijing Government. "Beijing Info." http://www.ebeijing.gov.cn/BeijingInfo (Accessed June 2009).

FrogsOnLine. "Beijing Environment and Ecology." http://www.frogsonline.com/asia/-chine/-beijing/beijing-environment.shtml (Accessed June 2009).

Ministry of Environmental Protection of the P. R. China. http://english.mep.gov.cn (Accessed June 2009).

United Nations Environment Programme (UNEP). *Beijing 2008 Olympic Games: An Environmental Review*. New York: UNEP, 2007.

Zhiqiang Ren
Independent Scholar

Guoqiang Ren
University of Washington, Seattle

BICYCLING

Since the summer of 2008, when gasoline prices topped $4 per gallon in the United States, several North American cities, including Washington, D.C.; Lexington, Kentucky; and Montréal, Québec, have created bike-sharing programs. Pictured is Montréal, where the BIXI program offers 3,000 bikes for rent at 300 stations.

Source: iStockphoto

Bicycling is a human-powered form of transportation that is relatively quick and efficient when one travels distances less than five miles. Because it produces no greenhouse gas emissions, bicycling is a sustainable and ecologically friendly means of transportation, especially in cities facing congestion, pollution, and limited parking space. In cities throughout the world, where many trips are shorter than five miles, bicycling offers an attractive alternative to automobiles, buses, and light rail in helping cities become greener. Many cities, however, were designed in ways that fail to accommodate bicycling. Other forms of transportation also possess other attractions that dwarf those of bicycling. Because bicycling provides an alternative, environmentally friendly mode of transport to commuters who would otherwise ride public transportation or a private vehicle, however, it remains a viable option for commuters. As a result, numerous cities have explored the benefits of this practical and economical mode of transportation throughout the last 10 years, making bicycling more popular than it has been in decades.

In many developing countries citizens lack access to gas-powered vehicles or public transportation. For these individuals, bicycling is often the primary mode of transport. Utility cycling, which refers to any biking done for reasons other than recreation or exercise, includes transporting an individual to work or running errands such as picking up groceries. Beijing, China, is perhaps the best-known city for utility biking. Estimates of the number of commuter cyclists in China in the 1980s ran as high as 500 million, although this number is believed to have declined as automobiles and buses have become more available.

At this time, a worldwide trend seeks to improve conditions for utility cyclists. To this end, a number of cities have implemented or are working toward implementing one or more of the following programs:

- Establishing citywide bike-sharing programs, sometimes known as Bixis (bicycle + taxi = Bixi)
- Improving existing roads and creating separate paths for bicyclists to enhance safety and increase convenience
- Creating more parking spaces for bicycles
- Educating the population regarding bicycling's benefits

By undertaking such actions, cities not only enhance their roads to help bikers but also create programs that provide easy access to bikes, as well as helping their populations become more environmentally aware.

Bike-sharing programs also have been implemented in numerous cities in the past few years. These cities include Barcelona, Berlin, Copenhagen, Lyon, London, and Stockholm. All bike-sharing programs are intended to promote easy and affordable access to bicycles within a given city. The logistics of any given program, of course, vary by city. Berlin and Munich, for example, have initiated a "call-a-bike" program, through which one uses a bicycle for a particular length of time. Call-a-bike users walk up to a bicycle parked on the street and call a number on their cell phones. The call results in the bicycle's back wheel becoming unlocked so that the user may ride it. Riders pay 0.08 euros per minute for the bike, and when finished, they park the bicycle on any street corner within a certain zone and then telephone the same number to again lock the bicycle.

In March 2007, Barcelona began a bicycle-sharing program called "bicing." Instead of a pay-by-the-minute system like that in Germany, bicing users pay a 24-euro annual fee for membership in the program, as well as a 0.30-euro per half-hour charge for the actual time a bicycle is used. The bicing program, which began with 30,000 subscribers, 400 bicycles, and 15 stations, has grown to include 186,000 subscribers, 6,000 bicycles, and 400 stations. This growth suggests the potential popularity of bike sharing for cities looking to become greener.

Programs similar to bicing and call-a-bike have been initiated in North America since 2008. During the summer of 2008, gasoline prices topped $4 per gallon in the United States. At that time, several cities began considering public bike-rental or bike-sharing programs as a way of mobilizing alternative and more environmentally friendly means of transportation. Washington, D.C.; Lexington, Kentucky; and Montreal, Quebec have created bike-sharing programs since that time. Other cities such as Chicago, New York, Boston, and Minneapolis have all shown interest in implementing some sort of bike-sharing program. With states as well as cities being forced to tighten their budgets as a result of the recent economic downturn, plans for implementing biking programs in most of these cities are still in the preliminary stages. Minneapolis, however, has received a federal grant for $1.75 million that it is using for a bike-sharing program, putting that city slightly ahead of others.

In addition to providing greater access to bicycles, planners have been developing innovations to make bicycling safer and more convenient by carving out special lanes and paths for cycling. For example, in Bogotá, Colombia, the former mayor created some 186 miles of cycle routes through the city to encourage residents to bicycle. In Portland, Oregon, city planners have used special traffic signals and separate paths to improve bicycling safety.

Despite the initial success of many urban bike-sharing programs, there is reason for skepticism regarding their long-term survival. Bike sharing in Paris has slowed, as half of the 15,000 bicycles originally purchased were either stolen or vandalized. Theft of bicycles is a weakness of bike-sharing programs, as it undermines a city's ability to make the operation financially viable. In addition, U.S. bike-sharing programs grapple with the issue of liability when an individual is injured on a bicycle owned by the government. Also, not everyone is as keen on bicycling as its advocates. For example, when San Francisco first implemented its bike-sharing program, as well as putting in 40 miles worth of new bike lanes in 2006, city leaders were surprised to find themselves being sued by antibike activists, which slowed the process of greening the city with biking.

Antibike groups have been mobilizing on venues such as Facebook and online blogs. Incidents of antibike road rage have also been reported. These motorists are often angered by cyclists who don't obey traffic rules and by bike riders who at times exhibit aggressive behavior toward car drivers.

Whether rhetorically or as a matter of policy, taking a probiking stance is highly popular in many areas. Qiu Baoxing, China's deputy minister of construction, has called for a return to the bicycle for Beijing. This call comes as an overload of car traffic has caused severe traffic jams, with the number of cars on the streets increasing daily. Colleges also are embracing bicycles—two University of Denver students were able to raise $50,000 to purchase 600 bicycles for a campus-wide bike-sharing program that debuted in the fall of 2009. Many other college towns and campuses across the United States also are trying to initiate bike-sharing or rental programs.

Biking has great potential for daily use by vast numbers of people, especially those living in cities. Biking reduces greenhouse emissions but also makes a personal statement about the environment. However, a massive global shift in which utility biking is substituted for other transportation options appears unlikely in the near future. In developed countries, with widespread access to cars, buses, trains, and subways, a number of roadblocks remain for utility bikers. These hindrances include bad weather, crowded sidewalks, and issues of motivation. Cities and the individuals who live there must share a common vision for cycling to reach its full potential to green cities.

See Also: Commuting; Gridlock; Healthy Cities; Infrastructure; Transit; Walkability (Pedestrian-Friendly Streets).

Further Readings

Cradock, A. L., et al. (2009). "Factors Associated With Federal Transportation Funding for Local Pedestrian and Bicycling Programming and Facilities." *Journal of Public Health Policy,* 2009 Supplement 1, 30: S38–S72 (2009).
Gilderbloom, J. I., et al. "Amsterdam: Planning and Policy for the Ideal City?" *Local Environment,* 14/6:473–93 (2009).
Wray, J. Harry. *Pedal Power: The Quiet Rise of the Bicycle in American Public Life.* Boulder, CO: Paradigm Publishers, 2008.

Stephen T. Schroth
Jason A. Helfer
Michael D. Dooley
Knox College

BIODIVERSITY

Biodiversity, short for biological diversity, is the variety and variability of life on Earth. It is the outcome of ongoing evolutionary processes that began about 4 billion years ago. The term gained currency in the 1980s as people became more aware of human-driven mass extinction and the many costs to society that are associated with rapidly diminishing biodiversity. The concept encompasses not only genetic and species diversity but also the multiple dimensions, dynamics, and processes of organisms and their ecological systems or ecosystems. Thus, biodiversity science investigates the past and present forms, patterns, functional traits, and interactions of ecological components, including genetic sequences, organisms, populations, communities, ecosystems, and landscapes. The global biodiversity crisis is one of the greatest challenges facing humanity. Solutions to this complex problem require human adaptive responses coordinated with strategies to address the other major sustainability challenges of our time, including global climate change, human population growth, energy demand, food insecurity, and poverty.

Biogeography is the study of the geographical distribution of biodiversity over space and time. The highest concentrations of biodiversity have been found in tropical ecosystems such as tropical rainforests and coral reefs. Habitat heterogeneity has also been associated with high levels of biodiversity. Species richness and species diversity are commonly used biodiversity metrics. Species richness is the number of species per unit of habitat. Species diversity is a measure of both species richness and species abundance for a given unit of habitat. At this time, our global inventory of plant, animal, and microbial species remains far from complete. Despite this uncertainty about the total number of living species that exist worldwide, estimates range in the tens of millions. To date, taxonomists have classified and catalogued fewer than 2 million living species.

Given these limitations of our biodiversity knowledge, there is also a great deal of uncertainty about the current rate of species extinction on Earth. Nevertheless, recent estimates of species loss per year range in the tens of thousands. Species extinctions and other types of biodiversity loss have occurred throughout the Earth's natural history. Evidence suggests that at least five mass extinction episodes occurred long before the emergence and spread of human influence on ecosystems and the environment. Although the biosphere has been transformed by human action for the past tens of thousands of years, the current human-induced mass extinction episode is unprecedented. Direct and indirect human impacts on the biosphere have greatly expanded and accelerated over recent decades.

Cities can contribute to biodiversity loss through direct disturbance of ecosystems, as well as the consumption of products manufactured or grown in a manner that damages ecosystems. As the global human population continues to grow, and urbanization continues, these impacts are expected to magnify. Some of the major threats to biodiversity posed by contemporary human activities and urban development are:

- Habitat alteration and loss resulting from land-use changes (e.g., deforestation, agricultural modernization, and urbanization)
- Over-exploitation and depletion of natural resources (e.g., overfishing)
- Alterations to biomes and hydrology as a result of anthropogenic climate change (e.g., glacial retreat and rising sea levels)
- Introductions of invasive species
- Nutrient loading and pollution of aquatic systems

To support biodiversity management and decision making, more research is needed to identify the conditions and processes that generate and enhance biodiversity. Indeed, some researchers have already been seeking to improve understanding of the human activities that contribute to biodiversity renewal and enrichment. Scientists are developing novel approaches for integrating multiple measures of biodiversity. For example, in addition to determining species richness and species diversity, biodiversity can also be measured by identifying and classifying the functional traits, roles, and relationships that are present or absent in a given ecosystem.

Biodiversity indicators (e.g., water quality and habitat connectivity) and indexes (e.g., the Ecological Footprint and the Living Planet Index) are being developed to monitor human impacts on ecosystems. They are needed to evaluate performance toward meeting international targets such as the goals put forth in the United Nations Convention on Biological Diversity.

Biodiversity forms the central basis for human well-being by providing multiple ecosystem services including soil formation, nutrient cycling, pollination, food and fiber production, water purification, carbon storage, flood control, and cultural, psychological, spiritual, aesthetic, and recreational benefits. Biodiversity dynamics also present some risks to human health such as emerging infectious diseases. Current societal responses to the global biodiversity crisis reflect many different values, motivations, and priorities. Some individuals and organizations prioritize the conservation of so-called biodiversity hotspots of species richness, rarity, endemism, or threat. Others emphasize the importance of maintaining or enhancing both social and ecological resilience. Many are driven by economic motives. Measures that have been implemented or proposed to manage biodiversity and promote healthy ecosystem functioning include:

- Establishing protected areas and other types of restrictions on resource use such as environmental zoning regulations to protect headwaters, wetlands, riparian areas, freshwater lakes, coastal lagoons, estuaries, and coral reefs
- Developing market-based mechanisms and economic incentives for funding biodiversity conservation, such as payments to individuals or groups for ecosystem services (e.g., forest and watershed protection)
- Strengthening institutions and partnerships for biodiversity knowledge sharing and to increase public awareness and engagement on biodiversity issues
- Improving local-level community involvement through participatory approaches to resource management

Cities can be designed and planned to support biodiverse ecosystems through the preservation of open space, the protection of critical habitat, the restoration of degraded ecosystems, and the design of new forms of habitat, such as vegetated (green) roofs.

An overarching goal of global biodiversity management is to sustain and restore ecological integrity. To reach this ambitious goal it is necessary to integrate different types of empirical data, develop geospatial monitoring and modeling tools, and coordinate multiple levels of planning and action. Therefore, it is important to create the institutional and technological infrastructure to facilitate rapid and open exchange of biodiversity information and knowledge. Since the early 1990s, a branch of the information and communications technology sector known as biodiversity informatics has advanced the standardization of digital data collection, storage, archiving, distribution, and analysis in support of ecosystem monitoring and assessments. There are now numerous Internet-based initiatives offering open access to biodiversity databases.

See Also: Bioregion; Conservation Subdivision; Ecosystem Restoration; Habitat Conservation and Restoration.

Further Readings

Avise, John C., et al. "In the Light of Evolution II: Biodiversity and Extinction." *Proceedings of the National Academy of Sciences,* 105 (2008).
Biodiversity Informatics Facility at the American Museum of Natural History's Center for Biodiversity and Conservation. http://biodiversityinformatics.amnh.org (Accessed August 2009).

Center for International Earth Science Information Network. "Species Distribution Grids."
    http://sedac.ciesin.columbia.edu/species (Accessed August 2009).
Encyclopedia of Life. http://www.eol.org (Accessed July 2009).
Global Biodiversity Information Facility. http://www.gbif.org (Accessed August 2009).
Group on Earth Observations Biodiversity Observation Network. http://www
    .earthobservations.org/cop_bi_geobon.shtml (Accessed August 2009).
Kinzig, Ann P., et al., eds. *The Functional Consequences of Biodiversity: Empirical Progress
    and Theoretical Extensions*. Princeton, NJ: Princeton University Press, 2002.
Millennium Ecosystem Assessment. *Ecosystems and Human Well-Being: Biodiversity
    Synthesis*. Washington, D.C.: Island Press, 2005.
Species 2000. http://www.sp2000.org (Accessed July 2009).
United Nations Convention on Biological Diversity. http://www.cbd.int/doc/legal/cbd-un-en
    .pdf (Accessed August 2009).

Sandra Baptista
The Earth Institute at Columbia University

BIOPHILIA

Biophilia is the theory that humans have a genetic predisposition towards natural environments. Over the past three decades, a growing body of scholars have responded to the global decline in biodiversity (resulting from human activities, e.g., land clearing for agriculture, pollution, etc.) by searching for evidence of such a trait. Scholars from diverse disciplines such as psychology, biology, geography, philosophy, planning, and economics have advanced an argument that humans are inherently "ecocentric" or biophilic animals—that is we are drawn to natural environments.

The Meaning(s) of Biophilia

Biophilia is a term that refers to a purportedly instinctive drive that impels humans to favor certain aspects of natural environments. While the term has been attributed to psychologist and philosopher Erich Fromm—who referred to "a psychological affinity for life"—it was renowned entomologist E. O. Wilson who popularized (and slightly modified) the term in his widely cited book of the same name. Wilson has defined biophilia as "an innate tendency to focus on life and lifelike processes," suggesting that from infancy, humans are attracted to living things "like moths to a porch light." Moreover, animals seem to have played a pivotal role in human evolution; interactions with animals appear to have shaped our cognitive capabilities.

Joined by social ecologist Stephen Kellert, Wilson sparked an efflorescence of research into whether there might be a genetic underpinning for human attitudes toward nature. They, and others, have since argued that because humans have evolved within "nature" (here meaning biotic environments), and since the human mind is an evolutionary construct, humans may be genetically driven to value or seek out (some) natural environments.

Biophilia Research

Numerous studies investigating this "biophilia hypothesis" have tended to confirm Wilson and Kellert's assertions. Empirically grounded evidence suggests that humans are intrinsically drawn to at least some types of natural environments, supporting the idea that we may have a genetically inherited predisposition toward life. Many studies have shown, for example, that people with a view of—or access to—natural environments recover faster from illness and surgery, are better able to resist mental illness, are more affable, are more socially adjusted, can manage their life affairs better, can concentrate longer on difficult tasks, recover faster from exposure to stress, and become ill less often than their counterparts who lack access to nature.

Seeking a genetic explanation for these observations, some anthropologists, geographers, biologists, and psychologists, among others, have argued that a preference for the natural environments in which early humans evolved may have become encoded into our genes. Termed the *savannah hypothesis*, this explanation suggests that early humans (hominids) thrived within habitats that were free from predators and that offered them the greatest prospects of finding food and shelter, thus increasing survival rates and concomitantly the chances of reproducing, subsequently conferring a preference for such environments upon their offspring. These preferences are said to have been inherited by modern humans, genetically encoded into our cognitive capabilities, thus enabling us to recognize such environments as "beneficial," and predisposing us toward them.

Contestations and Disputes

However, it appears there are some limitations to the biophilia hypothesis, not the least of which is the fact that humans seem to have a greater propensity for environmental harm rather than protection—evidenced by numerous global environmental problems. First, evidence from paleontologists suggests that early hominids thrived within a range of habitats throughout the long evolution to modern humans—including forests and woodlands—somewhat undermining the savannah hypothesis.

Second, even if humans are predisposed toward some elements of the natural environment as proponents of biophilia assert (e.g., park-like landscapes consisting of calm water, grasslands, and scattered copses of trees with hilly outcrops), this does not mean that humans are predisposed to valuing all life. For instance many people fear animals like spiders and snakes or have an aversion to densely vegetated areas. It may also mean that we value the environment as a resource to be exploited rather than conserved. Third, research from animal geography has shown that human behaviors toward animals are characterized by both antipathy and affection. There has been a long history of human exploitation of plants and animals, including widespread acts of cruelty.

Fourth, such explanations may paradoxically entrench and continue a longstanding philosophical schism that separates humans from nature. By casting urban environments as bad, harmful, or even "unnatural," such explanations unwittingly posit urban areas as being outside nature. But as the dominant habitat of humanity, cities would also seem to confer an adaptive advantage—in cities we can easily access food, shelter, healthcare, and so on, and thus prosper—the burgeoning human population attests to this. And cities themselves are rarely "dead zones," they oftentimes harbor a wide variety of nonhuman organisms, some of which prosper better than their wildland conspecifics (due to an abundance of food and reduced predation).

Finally, genetic explanations for human behavior toward "natural environments" tend to discount the equally important role of learning and culture—researchers have found it difficult to prove that an affinity for natural places is genetically "hardwired" rather than learned, and their findings thus remain inconclusive. Worse still, assertions that humans are instinctively driven to "explore, hunt and garden" may naturalize behaviors such as colonialism, exploitation, and oppression, legitimizing them as simply "human nature."

Application to Built Environment Research and Practice

Some scholars have recently suggested that there is a need to modify built environments to increase the presence of greenery and animals so as to remedy contemporary urban diseases such as obesity, stress, coronary heart disease, anxiety and depression. Increased access to urban greenspace, they argue, will result in healthier urban populations. Other scholars have begun to radically reconceptualize the long-standing dualism between city and nature. Critically interrogating the notion that cities are "dead zones," they point to the myriad urban ecologies that exist within human life-worlds. Their work has profoundly disrupted the binary of wild nature and cultured humanity to reformulate a more nuanced understanding of the role of humans in nature. Cities have been recast as the habitat of humanity and as inherently natural—not artificial entities. This is not to say that urban environments do not harm nonhuman species and their biogeochemical requisites—they do in numerous ways (e.g., chemical pollutants like endocrine disruptors, acidification of waterbodies, widespread erosion, etc.). But any biophilic explanation for human behaviors must also recognize that as the dominant habitat of humanity, cities should be taken seriously as "ecological," not just sociocultural entities.

See Also: Biodiversity; Environmental Planning; Greening Suburbia; Habitat Conservation and Restoration; Masdar Ecocity; Natural Capital; Parks, Greenways and Open Space; Sustainable Development; Urban Forests.

Further Readings

Byrne, J. and J. Wolch. "Urban Habitats/Nature." In *International Encyclopedia of Urban Geography*, N. Thrift and R. Kitchin, eds. Amsterdam: Elsevier, 2009.
Joye, Y. "Architectural Lessons From Environmental Psychology: The Case of Biophilic Architecture." *Review of General Psychology*, 11 (2007).
Kellert, S. R. *Kinship to Mastery: Biophilia in Human Evolution and Development.* Washington, D.C.: Island Press, 1997.
Kellert, S. R. and E. O. Wilson, eds. *The Biophilia Hypothesis.* Washington, D.C.: Island Press, 1993.
Maller, C., M. Townsend, A. Pryor, P. Brown and L. St. Leger. "Healthy Nature Healthy People: 'Contact With Nature' as an Upstream Health Promotion Intervention for Populations." *Health Promotion International*, 21/1 (2005).
Merchant, C. *Radical Ecology: The Search for a More Livable World.* London: Routledge, 1992.
Wilson, E. O. *Biophilia.* Cambridge, MA: Harvard University Press, 1984.

Jason Byrne
Griffith University

BIOREGION

A *bioregion* is a term and concept that has developed especially over the last 30 years within, but not limited to, the fields of geography, environmental politics, and environmental philosophy, as well as within environmental activist circles. This article examines the historical development of the term, what it signifies, how it has shaped bioregional philosophy, its criticisms, and how the concept relates to green cities.

The term *bioregion* is derived from *bios* (life) and *regia/regere* (region/territory) and refers to the bio-, geo-, and chemical-physical processes and components that constitute a specific geographical territory. Although the term itself did not enter into popular and academic discourse until the late 1970s/early 1980s, when it was popularized by the bioregional movement, intimations of its meaning were present decades prior. Aldo Leopold's "Land Ethic" is one such early progenitor, as is the work of the geographer Carl Sauer. Equally, the development and trajectory of the concepts of holism and ecology can be said to have affected the understanding of a bioregion.

Early and continued use of the term refers to watersheds, so that a bioregion is defined by the streams, rivers, and bodies of water that exist within a given geographical— and thus bioregional—locality. Bioregions may also be identified by bounded geographic and region-specific types of soils, grasslands, forest and tree cover (or lack thereof), plant and animal species, seasonal weather patterns, and regional varieties of rocks. However, the most common bioregional identifier still remains watersheds; the other components listed above tend to be associated with existing within a given, defining watershed. In essence, watersheds and the biologic, geographic, and chemical components that are contained therein are seen as a collective and distinct place that has its own natural and physical characteristics and processes that can be identified by various scientific measuring tools. Some are comfortable with expanding the size of a bioregion to include mountain ranges (e.g., the Rockies bioregion), types of soils (e.g., the midwest prairie), weather patterns (e.g., the Arctic, with its continual snow cover), or connected watersheds (e.g., the bioregion of the Mississippi River and its many tributaries).

Various philosophies and political movements have developed around the concept of a bioregion. Some conservationist biologists have developed a concept of "rewilding," based in part on linking bioregions by corridors so that megafauna can continue their historical migratory routes, which are being threatened by habitat loss. These biologists and wildlife advocates attempt to pass legislation that recognizes and protects ecosystem processes that occur on micro- and macro-bioregional scales.

The concept of a bioregion has directly influenced the bioregional movement, also called "bioregionalism." This movement advocates place-based cultural and linguistic diversity that depends on living within the defined limits of bioregions. Bioregionalists advocate "reinhabitation" within bioregions, and they thus attempt to create bioregionally based lifestyle practices, politics, and appropriate technologies and economies of scale that live within, respect, and depend on the limits of a bioregion. Bioregionalists tend to also be active in local bioregional restoration and protection projects. Theirs can also be said to be a philosophical movement, drawing on the works of ecophilosophers, conservation biologists, and practices of indigenous cultures. Bioregionalism is also a political movement, based especially on various renderings of anarchism and green political ideologies that stress decentralization and community autonomy. Bioregionalists tend to advocate a bioregionally based politics in which the lifestyle and political goals adhere to living in harmony within local bioregional processes and limits.

Some have criticized bioregionalists because they do not recognize that humans are a migratory species, they tend to have an unduly optimistic understanding of the human capacity for both living within limits and of trusting that bioregional governance will not fall prey to power struggles and hierarchy, and they suffer from the naturalistic fallacy, or arguing that because nature "is," this means that this is how humans "ought" to also behave. Some criticisms of the concept of a bioregion include pointing out that global biophysical and geophysical flows and mechanisms affect local bioregions, so there is an arbitrariness of drawing certain bioregional lines, and that the concept is overly environmentally deterministic.

The concept of a bioregion has direct ramifications for the construction and maintenance of green cities. If willing to have human actions be guided by the goal of living within the limits and processes of a bioregion, then cities must address issues of population size, resource dependency and extraction, and transportation and design. City governance would most likely have to be radically altered, and a decentering of global trade agreements would most likely have to occur. The production of appropriate-scale technologies, manufactured to fit within the boundaries of a bioregion, would likely have to occur. Methods of food production and distribution would also need to radically change, as would water catchment and purification design and methods of generating power for cities. Last, waste disposal would have to become local and nonpolluting. In effect, a green city would have to be made to sustainably exist within the carrying capacity of its local bioregion and the resources therein.

See Also: Ecovillages; Environmental Planning; Healthy Cities.

Further Readings

Foreman, Dave. *Rewilding North America: A Vision for Conservation in the 21st Century*. Washington, D.C.: Island Press, 2004.

House, Freeman. *Totem Salmon: Life Lessons From Another Species*. Boston, MA: Beacon Press, 1999.

Meredith, Dianne. "The Bioregion as a Communitarian Micro-Region (and Its Limitations)." *Ethics Place and Environment*, 8 (2005).

Sale, Kirkpatrick. *Dwellers in the Land: The Bioregional Vision*. Santa Cruz, CA: New Society, 1991.

Snyder, Gary. *Turtle Island*. Boston, MA: Shambala, 1974, 1993.

Taylor, Bron. "Bioregionalism: An Ethics of Loyalty to Place." *Landscape Journal*, 19 (2000).

Thayer, Robert, Jr. *LifePlace: Bioregional Thought and Practice*. Berkeley, CA: University of California Press, 2003.

Thomashow, Mitchell. *Bringing the Biosphere Home: Learning to Perceive Global Environmental Change*. Cambridge, MA: MIT Press, 2002.

Worster, Donald. *Nature's Economy: A History of Ecological Ideas*, 2nd Ed. Cambridge, MA: Cambridge University Press, 1994.

Todd J. Vasseur
University of Florida

BLUEBELTS

Bluebelts are undeveloped areas retained by cities to provide stormwater catchment and wetland-based wastewater management to supplement or replace conventional constructed sewage and wastewater treatment systems. With the widespread growth of suburban populations, many standard storm- and wastewater management systems require updating and expansion. Bluebelts are another, environmentally friendly way to sanitize water and ensure proper drainage. In many areas, suburban and rural expansion has outpaced the municipality's ability to create infrastructure to handle wastewater and surface runoff drainage. In some cases, the installation of conventional water treatment piping would destroy protected wetlands. Compared with conventional drainage systems, bluebelts are known for being ecologically sound and extremely cost-effective. In addition, because bluebelts are a positive alternative to building newly constructed, chemical-based systems, many towns and cities are turning to creating them.

Bluebelt systems can update water management systems without completely replacing or expanding existing structures. They use an area's natural waterways, combining them with newly constructed water management structures. One of the oldest and perhaps most famous bluebelts is on Staten Island, a borough of New York City. Many benefits are associated with using bluebelts, and they are becoming widely utilized in other areas of the United States, the United Kingdom, and Canada.

Because the Staten Island bluebelt was the first successful one, the systems of other bluebelts are based on it. At the beginning of the design process, the engineers researched existing drainage systems to combine them with the new bluebelt system. Models were created to gauge the ability of the proposed system to contain and drain water from high levels of rainfall. New drainage systems were designed to connect with preexisting natural waterways to facilitate rapid drainage. These new drainage systems were much like other urban storm drain systems in wide use in the New York metropolitan area. The basic structure of the Staten Island bluebelt is a combination of urban storm drains and existing natural aquatic features. Drainage from urban areas is only one dimension of the bluebelt; the most notable feature of the system is the use of multiple "best management practices" (BMPs). According to a study by the New York City Department of Environmental Protection, BMPs are applied to flood control, water quality improvement, and habitat preservation. In the Staten Island bluebelt, BMPs are applied to wetlands, streams, and basins. The wetlands use naturally occurring plants to remove pollutants from water without the use of chemicals, and naturally existing streams carry excess floodwaters away from populated areas. Special care is taken to control the speed at which the water drains away, so as to not contribute to excess erosion. The basins provide flexible water storage that keeps flooding under control during excessive rainfall events.

The Staten Island Bluebelt

In the early 1980s, Staten Island became more accessible to suburbanites, and homes began to spring up on the less-developed southern end of the island. New York City's water management authorities could not build wastewater or drainage systems quickly enough to service the newly developed areas. In these new developments, subdivisions installed septic tanks instead of island-wide systems for water treatment and drainage. This created widespread flooding and severe water-sanitation issues. New York City decided to look into alternate drainage systems because replacing Staten Island's entire water system would be very costly and would damage existing wetlands. The result of their study was the Staten Island bluebelt system. As of 2009, the Staten Island bluebelt consisted of 16 watersheds (areas that drain into a particular body of water), 10,000 acres, and 40 BMP areas.

Evaluation of the Staten Island bluebelt shows that the system has fulfilled its goal of providing a viable alternative to the conventional water management system. The Staten Island bluebelt has demonstrated that it is more ecologically and fiscally responsible than such conventional systems. According to the city's Department of Environmental Protection, the goals of the bluebelt were to

- develop a storm water management system using existing wetlands,
- collect street runoff in storm sewers,
- improve the capacity of hydraulic structures,
- use BMPs to mitigate flooding impacts at the sewer/wetland interface, and
- construct a separate sanitary sewer system.

Other goals that the bluebelt has achieved (according to a 2006 study by David Hsu) include the

- collection of contaminants from runoff,
- introduction of aquatic plant species to absorb pollutants,
- reduction of stormwater velocities,
- recharge of groundwater supplies, and
- protection and improvement of natural areas within the urban fabric.

The construction and operation of the Staten Island bluebelt has been shown to be more fiscally responsible than if a conventional stormwater system had been built. New York City saved an estimated $80 million, and the ecological systems of Staten Island are healthier as a result. Wildlife is thriving, and biodiversity has increased. The program in Staten Island has been so successful that New York City is looking to expand its bluebelt program into the outlying areas of other boroughs, namely Queens and the Bronx. In 2009, the city was in the exploratory stages of such development, raising funds to finance the next project.

Other Bluebelts

In 2004, the Chartered Institution of Water and Environmental Management, based in the United Kingdom, recommended that urban areas in England begin designating bluebelt areas surrounding their cities to increase drainage and flood protection. The Derby metropolitan area is exploring the implementation of a bluebelt system, and London is looking at the creation of such a system to protect the Thames River.

In the United States, bluebelts are becoming a more and more popular way to restore water ecosystems. Natural springs in northern Florida are increasingly in need of protection. Many citizens' groups, such as the Thousand Friends of Florida, are engaging their communities and trying to gain support for bluebelt systems in the state. Concerned citizens in the San Francisco, California, metropolitan area are also in the process of designing a bluebelt plan that could be implemented once development is halted on delicate floodplains adjacent to San Francisco Bay.

There is a continuing movement in Canada to create a "Sea-to-Sea Green Blue Belt," which would encompass 20,000 hectares (about 50,000 acres) near Victoria, British Columbia. That system was 75 percent complete in 2005 and was waiting on additional funding to purchase the rest of the property necessary to finish the project.

Although it may currently be considered something of a futuristic concept, the idea of an alternative to conventional storm drain systems has begun to spread among city planners. One can expect to see bluebelts becoming more prevalent wherever suitable open space exists in close proximity to population centers. The implementation of bluebelts will help provide more and better ecofriendly and cost-efficient water management systems worldwide.

See Also: Environmental Planning; Green Communities and Neighborhood Planning; New York City, New York; Water Treatment.

Further Readings

Gumb, Dana, et al. "The Staten Island Bluebelt: A Case Study in Urban Stormwater Management." http://66.102.1.104/scholar?hl=en&lr=&q=cache:ypXpjye2t_IJ:documents .irevues.inist.fr/handle/2042/22507+Bluebelt (Accessed April 2009).

Hsu, David. "Sustainable New York City." http://www.nyc.gov/html/oec/downloads/pdf/ sustainable (Accessed April 2009).

New York City Department of Environmental Protection. "Staten Island Bluebelt: Seminar Presentation." http://www.epa.gov/ednnrmrl/events/Bluebeltseminar1 (Accessed April 2009).

Anthony R. S. Chiaviello
University of Houston–Downtown

Bogotá, Colombia

Bogotá, Colombia's capital and largest city, has received considerable international interest for its implementation of progressive policies and programs in several areas including urban mobility, public services, citizenry health and quality of life, and ecological protection. Bogotá can serve as a model for how a city of over 7 million inhabitants, and many of the challenges of a rapidly emerging megacity in the developing world, can achieve notable successes in the improvement and promotion of public and nonmotorized forms of transportation, public space reclamation, and measures to restrict private automobile use.

Transportation Solutions

The TransMilenio System, for example, was created in 2000 to give certain buses the exclusive right to specially allocated lanes. This Bus Rapid Transit system, encompassing 84 kilometers of busways and over 300 kilometers of feeder routes, has significantly increased citizen access to public transport. The system is being built in stages, and by 2030 it is scheduled to cover the entire city. Of the approximately one million passengers transported daily, 5–10 percent of TransMilenio's ridership is made up of former auto users.

Another initiative to increase the daily reliance on nonautomobile modes of transportation is the Ciclorutas project. Composed of a network of approximately 300 kilometers of designated bicycle paths and planned and built mainly between 1998 and 2001, Ciclorutas is one of the most extensive of its kind in the world. This bike network has connectivity with strategic occupational, educational, and recreational spots in the city, as well as the TransMilenio.

To further discourage traffic congestion and address environmental pollution, Bogotá has restricted travel by private vehicles during peak hours on specific days of the week, depending on license plate number. In doing so, the city has limited approximately 40 percent of vehicles from entering the city during rush hours.

In addition, Bogotá hosts Program Ciclovía, in which an average of 117 kilometers of designated streets and main avenues are closed to motorized vehicles on Sundays and holidays to encourage cycling and walking activities. Ciclovía found its roots in social movements in 1974, when over 5,000 cyclists protested pollution, traffic congestion, and lack of dedicated roadways for bicycles. Initially opposed by the transportation and business sectors, the program did not receive institutional support until 1982. By 2005, an average of 400,000 residents aged 18–65 years were participating in Ciclovía every Sunday, and the day often attracts over 2 million participants. Bogotá is also credited with having the world's largest car-free weekday event. Officially approved by public referenda in 2000, this municipal government–sanctioned event prevents almost a million private cars from circulating on the first Thursday in February every year.

Green Space, Cleaner Air

Bogotá has also strengthened the city's sense of community by increasing the number of square meters of green area per inhabitant; at this time, a network of over 1,000 parks covers the city. Reclamation of public space has also proceeded through the pedestrianization of urban streets, with the placement of bollards to restrict car parking on sidewalks and the relocation of vendors into special plazas. Bogotá boasts the world's longest corridor of pedestrian streets, with a 17-kilometer stretch of pedestrian and bicycle pathways that connects several communities to retailers, employment, and public services.

The city government's environmental agency, the Departamento Técnico Administrativo del Medio Ambiente, reports that there have been measurable benefits to ambient air quality since restrictions were placed on private vehicles, the implementation of TransMilenio, and the expansion of cycling and pedestrian infrastructure. Although unsafe levels of air pollution remain an issue, and causal relationships between air quality and the implementation of these programs are often difficult to discern, the air quality monitoring program found reduced emissions of up to 40 percent for certain pollutants, such as sulfur oxides, compared with pre-2001 levels.

The success of these programs required a multifaceted and complementary approach to city planning. For example, enacting restrictions on privately owned cars would not be as tenable if improving public transportation infrastructure had not been simultaneously addressed. Bogotá has been aided in implementing these policies in part as a result of a decentralization process that strengthened the city's autonomy in political and administrative affairs. The Constitution of 1991 instituted a succession of reforms that emphasized the need for decentralization, organization of territorial entities, and enhanced participatory governance. In conjunction with the decentralization program, a shift in fiscal allocations occurred, with the city receiving a significant share of the national income and allowances for increasing urban taxes on property, industry, commerce, gasoline, and car–vehicle licensing came into effect.

Although political and functional decentralization established the municipal government as a body with the authority and resources to make decisions on local issues, the earlier democratization of the mayor's office gave it the incentive to do so. Before 1988, Bogotá's mayors were presidentially appointed, but the introduction of free elections that year caused the mayor's office to become more responsive, legally and institutionally. Furthermore, the development and realization of programs and strategies to improve urban sustainability can be attributed to policy continuity that spanned several different political administrations. Although many of the most critical changes were implemented during Mayor Enrique Peñalosa's term (1998–2000), the administrations of Mayor Jaime Castro (1992–1994) and Mayor Atanas Mockus (1995–1997, 2001–2003) similarly demonstrated levels of political commitment to improving the urban environment.

Despite many of these gains in the urban landscape, environmental degradation in Bogotá is still prevalent and includes urban sprawl into fertile peri-urban areas, inadequate waste management and recycling programs, heavily polluted waterways and bottlenecks in water distribution and treatment systems. Bogotá remains socially segregated, with impoverished districts being disproportionately exposed to urban environmental problems. The exponential growth in urban population, largely a result of rural displacement associated with poverty and violence in the countryside, has led to slums and squatter settlements within and around the urban perimeter. The southern half of the city, where many poorer communities are concentrated, is highly industrialized and suffers from a scarcity of potable water, a lack of sewage and drainage systems and proper garbage disposal, and susceptibility to floods and landslides, and it also has a high incidence of environmentally linked health disorders. Community organizing in low-income settlements has led to urban agriculture projects and informal recycling for a wide variety of materials, but significant environmental challenges remain.

Although many urban sustainability and equity issues remain to be addressed, Bogotá is an example of a city that despite political, economic, and social crises, including high levels of unemployment and several decades of civil conflict, has transformed notable aspects of its urban fabric through local social and political action. Successes include significant expansion and upgrades to public transportation infrastructure and cycling and pedestrian networks, as well as reclaimed public space; these changes were largely implemented within a matter of a few years. Before its construction, proposals for a Bus Rapid Transit system were seen as untenable and far beyond the city's (and country's) financial reach. Nevertheless, the political will was present and Mayor Enrique Peñalosa executed the TransMilenio system within his mayoral term. This initiative was based on a pioneering one in Curitiba, Brazil, and the success of these Bus Rapid Transit systems has encouraged their adoption internationally, including in Delhi, Beijing, Los Angeles, Dublin, Paris, and large cities throughout Latin America. Almost all major cities in Colombia have integrated the Ciclovía program into urban planning approaches, in addition to cities in Mexico, Ecuador, Peru, and Brazil. Other countries in Latin America, such as Ecuador and Peru, are replicating the strategy developed in Ciclorutas, based on the experience in Bogotá.

See Also: Bicycling; Bus Rapid Transit; Curitiba, Brazil.

Further Readings

Gilbert, Alan. "Good Urban Governance: Evidence From a Model City?" *Bulletin of Latin American Research,* 25/3 (2006).

Montezuma, Ricardo. "The Transformation of Bogotá, Columbia: Investing in Citizenship and Urban Mobility." *Global Urban Development,* 1/1 (2005).

Wright, Lloyd and Richard Montezuma. "Reclaiming Public Space: The Economic, Environmental, and Social Impacts of Bogotá's Transformation." Copenhagen: Cities for People Conference Walk 21, June 2004.

Jennifer Liss Ohayon
University of California, Santa Cruz

BROWNFIELDS

The problem of brownfields started in the 1970s, partly because of the closure or downsizing of numerous steel plants, like this Bethlehem Steel plant.

Source: iStockphoto

The term *brownfields* refers to abandoned industrial or commercial facilities, usually contaminated by chemicals and/or other hazardous waste. The size of a brownfield can vary from a small vacant city lot to a large abandoned manufacturing facility spreading over several hundred acres. The extent of contamination will also vary depending on what facility was operating on it before abandonment, and can range from minor surface debris to extensive hazardous soil and/or groundwater contamination

The term *brownfield* was first used in the mid to late 1970s, mainly with reference to the steel industry. Although abandoned, contaminated sites existed before the 1970s, the problem of brownfields became more prominent during that decade for two reasons: The closure and downsizing of numerous steel plants in 1970s as the United States and Western European countries experienced a major structural change of their economies away from manufacturing, and the enactment of environmental legislation that held parties liable for the cost of cleanup of contaminated sites, thus increasing the cost of reusing old industrial facilities

According to the U.S. Environmental Protection Agency, brownfields are "abandoned, idled, or under-used industrial and commercial facilities where expansion or redevelopment is complicated by real or perceived environmental contamination." The definition was slightly altered in 2001 by the U.S.

Small Business and Liability Relief and Brownfield Revitalization Act as "real property, the expansion, redevelopment or reuse of which may be complicated by the presence or potential presence of a hazardous substance, pollutant or contaminant." Initially, if a developer bought a piece of contaminated land, the developer had to pay the clean-up costs to meet the environmental standards for human use of that land. The new brownfields legislation in the United States now limits the financial liability of innocent redevelopers (who were not aware of the contamination) of the contaminated property. Policy goals of these legislations and brownfield programs, among other things, include farmland and open space preservation, public health improvement, reuse of blighted urban land, job creation, tax revenue enhancement, and promotion of environmental justice.

Other countries have taken steps to facilitate the safe reuse of brownfield sites. In Canada, a law passed in 2001 makes the redevelopment of brownfield sites attractive to developers. Some EU countries have developed legislation that includes, in certain cases, an innocent landowner disclaimer (in the Netherlands) and provisions for the municipality to take over remediation work. Others support a "polluter pays principle," and some other European countries (such as France) depend on voluntary cleanup of sites.

Recently there have been a lot of innovative redevelopment strategies to clean up these brownfield sites so that the land is available for use to developers again. There are three main reasons why communities and policy makers are interested in brownfield cleanup and reuse: First, it is believed that the cleanup will reduce the harmful effect of the site's soil and water pollution on human health in particular, and the environment in general. Second, brownfield remediation, by making available already developed land for new uses, helps in stopping the urbanization of green area (forests, agricultural land, and rural areas), which can create other congestion and sprawl issues. Finally, they promote economic growth in inner cities and become an important component of sustainable growth.

On one hand the argument is that reusing the abandoned land increases the opportunities for revitalization and economic expansion and also redirects development away from new pristine land. On the other hand there is an ongoing debate about how well the land can be cleaned and what are the health effects for people who work, shop, or reside on former brownfield sites. Because there are not enough environmental data available on potential short- or long-term hazards of brownfields, there is always a concern that the new policies may not adequately protect the people moving onto brownfields from the hazardous effects.

Another issue with brownfield remediation is that it often requires the transportation of hazardous waste from the remediated site to an off-site disposal facility, thereby increasing the risk of human exposure to the contaminants during removal and transportation and disposal. To minimize these risks, some developers are practicing remediation techniques that take into consideration the potential for environmental exposure and address not only environmental factors but also social responsibility aspects (e.g., minimizing risks to the surrounding communities).

Green and sustainable remediation is an expansion of the current remediation practices and employs strategies through all phases of remediation (i.e., site investigation, remedy selection, remedy design, construction, operation, monitoring, and site closure) to include things such as making use of natural resources, using energy efficiently, reducing negative impacts on the environment, minimizing or eliminating pollution at its source, protecting and benefiting the community at large, and reducing waste to the maximum extent possible. The idea is to minimize the environmental footprint while maximizing the overall benefits of cleanup actions.

See Also: Greenfield Sites; Greyfield Development; Infill Development.

Further Readings

Alberini, Anna, et al. "The Role of Liability, Regulation and Economic Incentives in Brownfield Remediation and Redevelopment: Evidence From Surveys of Developers." *Regional Science and Urban Economics*, 35:327–51 (2005).

Reddy, Krishna R., et al. "Potential Technologies for Remediation of Brownfields." *Practice Periodical of Hazardous, Toxic, and Radioactive Waste Management*, 3/2 (April 1999).

Roberts, Allison Jilayne. "Brownfield Remediation in Kingston and Hamilton, Ontario: A Virtuous Cycle of Civil Society Involvement." Kingston, Ontario, Canada: Queen's University, April 2009.

Schoenbaum, Miriam. "Environmental Contamination, Brownfields Policy, and Economic Redevelopment in an Industrial Area of Baltimore, Maryland." *Land Economics*, 78/1:60–71 (February 2002).

"Sustainable Environmental Remediation Fact Sheet." http://www.ert2.org/ERT2Portal/uploads/SER%20Fact%20Sheet%202009-08%20Final.pdf (Accessed October 2009).

Velma I. Grover
United Nations University–Institute for Water,
Environment and Health (UNU–IWEH)

Bus Rapid Transit

BRT systems, like this one in Eindhoven, the Netherlands, are designed to operate at higher frequencies, and generally have fewer stops spaced farther apart, compared to regular city bus service.

Source: S.P. Smiler/Wikipedia

Bus Rapid Transit (BRT) refers to the adaptation of rubber-tired road buses to serve rapid transit purposes within metropolitan areas. In many ways, the kind of transportation service offered by BRT systems is analogous to traditional rail rapid transit systems, such as metros (subways) or light rail, because they are designed to operate at higher frequencies and generally serve fewer stops, spaced farther apart from one another, than bus stops in a regular city bus service. The first comprehensive BRT system was implemented in Curitiba, Brazil, for its cost advantages and the operational flexibility it offered over rail systems. As a result of the success of the system in Curitiba, BRT has been growing in popularity around the globe, in both the developing and developed worlds, where cities are increasingly faced with limited resources for providing transportation services.

Similar to rail rapid transit systems, BRT is designed to transport larger numbers of riders over longer distances than traditional bus systems operating in mixed traffic. BRT systems incorporate design features meant to distinguish them from regular bus service, including some, if not all, of the following elements: the construction of permanent stations and stops that are an improvement over conventional bus stops, enhanced efficiency of passenger boarding and fare collection to speed up service, the use of traffic signal priority and intelligent transportation system technologies, and service that is branded separately from conventional bus service, often using distinctive bus designs and special labeling as a rapid transit service on transit maps. Often, but not always, BRT buses travel in a dedicated right of way that is separated from the city's public road network. This is done either by marking off a separate lane or, in more advanced circumstances, by constructing a unique roadway—often referred to colloquially as a "busway"—that is physically separated from city streets.

The Birth of BRT

BRT was pioneered in Curitiba, Brazil, in 1974. Similar to a metro system, passengers entered buses not at stops but from specially designed stations where fares were paid before entry. Buses traveled on an exclusive right of way in the median of the road and were specially designed to handle larger loads, with multiple wide doors and a floor that was the same height as that of the station. The design considerations of the buses and stations permitted riders to enter and exit buses much more quickly, and the dedicated right of way and later traffic signal prioritization allowed buses to travel considerably faster than they would on a conventional street. Although notable BRT systems such as Ottawa's Transitway, Los Angeles's Orange Line, Las Vegas's Max, Boston's Silver Line, Cleveland's Euclid Corridor, and Eugene's EmX Green Line have opened recently in North America, choosing a bus over a rail-based strategy for urban rapid transit has been especially popular in the developing world, particularly in Latin America, where BRT planning began. The TransMilenio network in Bogotá, Colombia, is widely cited as the most advanced example of successful BRT planning.

BRT's success in developing countries is largely attributable to its low cost compared with rail systems. The first phase of Bogotá's TransMilenio system cost, on average, $5.5 million/kilometer to construct and was determined to be one-twelfth the cost of building a similar-length metro line. Because buses are already part of almost every city's public transportation system, there is much less need to train or hire new operators or maintenance staff, nor to stock spare parts that are required to service a different kind of technology. Finally, because buses can travel on nearly any paved road, BRT systems have a greater degree of flexibility in terms of the kind of routes and locations the system can serve compared with rail systems. This is useful in the event of an accident or blockage (buses can make a detour to avoid a stalled vehicle, whereas trains are confined to tracks); more important, this flexibility is highly responsive to geographic shifts in passenger demand. For example, if a major trip generator (such as an office park or school) opens away from the BRT line, buses can be routed to service these areas in ways that fixed-rail transit cannot. In many cities, notably in Porto Alegre, Brazil, or along the El Monte busway in Los Angeles County, a busway is used by several different public and private bus operators. This allows operators to use the busway as a trunk line, separated from the congestion of private cars and trucks, and then branch out individually into different neighborhoods at the end of the line. In this way, a large area of the city can be served by public transit in a very efficient manner.

Drawbacks

BRT, however, is not without its disadvantages. In many developed parts of the world with established rail transit, buses are seen as a "lower" form of transportation and are often negatively associated with riders of lower socioeconomic backgrounds. In the United States, this phenomenon is known as "rail bias," and it has led many cities to favor light rail systems to lure middle-class riders to public transportation. In the developing world, many cities that have constructed BRT systems have trouble enforcing the private right of way for buses. Often, other city vehicles and even private cars will illegally take advantage of the busway's exclusivity. In some cities, such as Taipei, congestion occurs from the sheer number of registered buses traveling in the busway. This can be mitigated by building passing lanes for buses at strategic points, but this strategy adds to the amount of land required to construct the actual busway. Because buses have a lower passenger capacity than rail vehicles, more vehicles are needed to service the same number of riders. This has somewhat erroneously led to the belief that BRT systems have, by definition, a lower capacity than heavy rail or even light rail systems. By simply adding more buses and eliminating the potential for bus congestion—through the addition of passing lanes and intelligent traffic monitoring systems—this issue can be largely resolved. Indeed, BRT systems regularly meet or exceed typical ridership capacities experienced on the world's best light rail systems.

The most widely successful examples of BRT, such as Bogotá, Curitiba, and Ottawa, owe their achievement to the integration of the transportation system into the larger sphere of urban planning. In Curitiba, a conscious effort was made to zone higher-density, transit-oriented development and plan major trip generators as close to BRT stations as possible to encourage use and reduce the city's reliance on private automobile transport. Curitiba's BRT system was originally conceived in the late 1960s not as a strategy for transportation, but as a strategy for shaping the urban form. In Bogotá, a major network of bicycle paths named Ciclovía was constructed to complement the TransMilenio BRT system with special, indoor bicycle storage facilities at many major BRT stations. The TransMilenio BRT was also funded in part by an increase in the gasoline tax. Together, these initiatives form part of a larger travel demand management approach. Similar to all strategies for building sustainable cities, BRT should be a coordinated effort with other initiatives to improve citizens' quality of life.

See Also: Bogotá, Colombia; Curitiba, Brazil; Transit; Transit-Oriented Development.

Further Readings

Institute for Transportation and Development Policy. *Bus Rapid Transit Planning Guide (June 2007)*. New York, 2007.
Menckhoff, G., ed. "Latin American Experience With Bus Rapid Transit." *Proceedings From: the Institute for Transportation Engineers—Annual Meeting*. Melbourne, Australia, August 2005.
Transportation Cooperative Research Program. *Volume 1: Case Studies in Bus Rapid Transit*. Washington, D.C., 2003.
U.S. General Accounting Office. *Mass Transit: Bus Rapid Transit Shows Promise*. GAO 01-984. Washington, D.C.: U.S. General Accounting Office, 2001.

Leonard Machler
Aaron Golub
Arizona State University

CARBON FOOTPRINT

The carbon footprint is a measure of the total amount of carbon dioxide (CO_2) and other greenhouse gas emissions that are directly or indirectly caused by an activity or that are accumulated over the life span of a product, person, and organization—or even a city, state, or nation. The term *carbon footprinting* refers to a measure by which a company or individual can calculate how much carbon or CO_2 equivalents they have produced and emitted during a project or time period. The full carbon footprint of an organization (including municipal agencies) encompasses a wide range of emissions sources, from direct use of fuels to indirect impacts such as employee travel or emissions from other organizations up and down the supply chain. When calculating an organization's footprint, it is important to try to quantify as full a range of emissions sources as possible to provide a complete picture of the organization's impact. To produce a reliable footprint, it is important to follow a structured process and to classify all the possible sources of emissions thoroughly. A common classification is to group and report on emissions by the level of control that an organization has over them. A standard classification is defined by the Greenhouse Gas Protocol, a widely used standard for corporate emissions' reporting produced by the World Business Council for Sustainable Development and the World Resources Institute. Three main types of emissions exist:

1. Direct emissions that result from activities that the organization controls: The majority of direct emissions result from combustion of fuels that produce CO_2 emissions (e.g., the gas used to provide heating for a building). Some organizations also directly emit other greenhouse gases (e.g., the burning and production of cement).

2. Emissions from the use of electricity: Workplaces generally use electricity for lighting and equipment. Electricity generation comes from a range of sources, including renewables. In the United Kingdom, however, around 75 percent of the nation's electricity is produced through the combustion of fossil fuels such as coal and gas. Although not directly in control of the emissions, by purchasing the electricity the organization is indirectly responsible for the release of CO_2.

3. Indirect emissions from products and services: Each product or service purchased by an organization contributes toward emissions. The way the organization uses products and

services therefore affects its carbon footprint (e.g., a manufacturing company is indirectly responsible for the CO_2 that is emitted in the transport of the raw materials, as well as for emissions from the distribution, use, and disposal of its finished products).

Calculating a carbon footprint incorporating all three types of emissions can therefore be a complex task. In addition, published carbon footprints are rarely comparable for several reasons:

- Despite emerging international standards, not all organizations follow the same methodology to calculate their carbon footprint.
- Some carbon footprints are expressed on a time period basis, such as annually, and some are measured on a unit basis, such as per product produced.
- Carbon footprints are usually calculated to include all greenhouse gases and are expressed in tons of CO_2 equivalent (tCO_2e). Some, however, calculate the carbon footprint to include CO_2 only and express the footprint in tCO_2 (tons of CO_2).

Calculating the Carbon Footprint

There are usually two major reasons for wanting to determine a carbon footprint—to manage the footprint and reduce emissions over time and/or to report the footprint accurately to a third party.

Calculating an organization's carbon footprint can be an effective tool for ongoing energy and environmental management. If this is the main reason for calculating the carbon footprint, it is often enough just to understand and quantify the key emissions sources through a basic process, typically including gas, electricity, and transport, which is relatively quick and straightforward to do. Having quantified these emissions, potential opportunities for carbon reduction can be identified.

Organizations also increasingly want to calculate their carbon footprint in detail for public disclosure; for example, for marketing purposes, to fulfill requests from customers or investors, or to determine what quantity of emissions they need to offset for them to become "carbon neutral." This therefore requires a more robust approach, covering the full range of emissions for which the organization might be responsible.

Calculation of a basic carbon footprint can be a fairly quick exercise for most organizations. A basic footprint is likely to cover direct emissions and emissions from electricity, as these are the simplest to manage, but exclude some of the indirect emissions. Major emissions sources that must be quantified include on-site fuel and electricity usage and the use of transport that you own. Data must be collected from all utility meters, and the distances traveled by the organization's vehicles must be recorded. The fuel, electricity, and transport consumption figures are then converted to CO_2 using standard emissions factors (which in the United Kingdom are available from Defra and the Carbon Trust).

One commonly used methodology is the Greenhouse Gas Protocol produced by the World Resources Institute and the World Business Council for Sustainable Development. This methodology provides detailed guidance on corporate emissions reporting and is available at http://www.ghgprotocol.org. A more recent standard from the International Organization for Standardization, ISO 14064, also provides guidance on corporate carbon footprint calculation and is available at http://www.iso.org. The United Kingdom has introduced a new measure of CO_2 emissions for goods and services to standardize carbon footprint calculations. PAS 2050:2008, the Specification for the Assessment of the Life Cycle Greenhouse Gas Emissions of Goods and Services, produced by BSI British

Standards, is designed to show customers how much CO_2 has been emitted during production, consumption, and disposal of a range of products. The new calculation, available at http://www.bsigroup.com/pas2050, will give the first "life cycle" CO_2 measure that is standardized across industry. The measure is the first standardized calculation of greenhouse gas emissions for goods and services that allows consumers to compare products easily. A variety of carbon footprint calculators exist, many of them online, from organizations such as the following:

- The U.K. government: http://actonco2.direct.gov.uk/index.html
- The U.S. government: www.epa.gov/climatechange/emissions/ind_calculator.html
- The European Union: www.mycarbonfootprint.eu
- The World Wide Web: http://footprint.wwf.org.uk

Carbon Neutral

This term *carbon neutral* is commonly used for something having net zero emissions (e.g., an organization or product). As the organization or product will normally have caused some greenhouse gas emissions, it is usually required to use carbon offsets to achieve neutrality, which are emissions reductions that have been made elsewhere and are then sold to the organization that seeks to reduce its impact. Carbon neutral can be anything from a person to a building to an organization or even a city or state—the Vatican in Rome is quickly moving to become the first "carbon-neutral state," with the installation of solar panels and the planting of a 37-acre forest in Europe, which is hoped to offset up to 80 tons of CO_2 a year.

Greenhouse Gases

Greenhouse gases are gases that contribute to the greenhouse effect when present in the atmosphere. Six greenhouse gases are regulated by the Kyoto Protocol, as they are emitted in significant quantities by human activities and are thought to contribute to climate change—CO_2, methane, nitrous oxide, hydrofluorocarbons, perfluorocarbons, and sulfur hexafluoride. Each gas, however, has a different global warming implication. Therefore, for simplicity, the mass of each gas emitted is commonly translated into a carbon dioxide equivalent (CO_2e) amount so that the total impact from all sources can be defined as a single figure.

Methods of Minimizing the Carbon Footprint

Several methods exist for reducing an individual's, organization's, or city's carbon footprint:

- Planting trees is one of the most common and simplest forms of carbon footprint reduction. Trees absorb CO_2 from the atmosphere and hence can be used to offset carbon emissions.
- Recycling waste materials such as household, industrial, and construction waste can be a valuable method of carbon footprint reduction, as the carbon content of the new materials that would have otherwise been used can be offset.
- Many energy-saving technologies exist that can contribute toward carbon footprint reduction, from cheap and simple measures such as installing low-energy lightbulbs to more expensive measures such as using electric vehicles. Renewable energy generation can also be used for offsetting a carbon footprint, such as wind turbines and solar panels.

Practical Examples of Carbon Reduction

Seattle's 2008 carbon footprint was about 7 percent smaller than it was in 1990, based partly on energy conservation efforts by Seattle's households and businesses, and Seattle City Light's work to "zero out" the climate pollution associated with the development and delivery of electricity. Emissions from residential and commercial energy use (around 20 percent of their footprint) were both down compared to 1990 levels. Efforts continue, with targets of further 7 percent reductions by 2012, and by an ambitious by 80 percent by 2050. Seattle's Climate Protection Initiative was launched in winter 2005, and since then more than 1,000 mayors, from cities in all 50 states and representing 87 million Americans, have signed on to the agreement.

See Also: Carbon Neutral; Carbon Trading; Energy Efficiency; Sustainability Indicators; Sustainable Development.

Further Readings

British Standards Institute (BSI). PAS 2050:2008, *Specification for the Assessment of the Life Cycle Greenhouse Gas Emissions of Goods and Services* (UK). http://www.bsigroup.com/pas2050 (Accessed December 2009).

Brown, M. A., et al. *Shrinking the Carbon Footprint of Metropolitan America* (May 2008), Brookings Institute. http://www.brookings.edu/reports/2008/05_carbon_footprint_sarzynski.aspx (Accessed December 2009).

City of Seattle Office of Sustainability and Environment. "Seattle Climate Protection Initiative Progress Report." http://www.seattle.gov/climate (Accessed December 2009).

Global Footprint Network. http://www.footprintnetwork.org (Accessed December 2009).

Greenhouse Gas Protocol. http://www.ghgprotocol.org (Accessed December 2009).

International Organization for Standardization (ISO), ISO 14064-1:2006. "Greenhouse Gases Part 1: Specification With Guidance at the Organization Level for Quantification and Reporting of Greenhouse Gas Emissions and Removals." http://www.iso.org/iso/catalogue_detail?csnumber=38381 (Accessed December 2009).

Wiedmann, T. and J. Minx. "A Definition of Carbon Footprint." ISA UK Research and Consulting, 2007. http://www.isa-research.co.uk/docs/ISA-UK_Report_07-01_carbon_footprint.pdf (Accessed December 2009).

Chris Goodier
Loughborough University

CARBON NEUTRAL

The term *carbon neutral* describes any process that over its duration and the lifespan of its products results in absorbing the same amount of environmentally damaging carbon compounds (primarily CO_2) as it releases into the atmosphere. Carbon neutral products and processes are designed to reduce emissions of carbon and sequester any residual emissions. Those processes that release net positive amounts of CO_2 to the atmosphere are known as *carbon positive,* and processes that absorb more CO_2 from atmosphere than they produce are known as *carbon negative.* Global climate change is the consequence of uncontrolled emissions of greenhouse gases, including CO_2, methane, and nitrogen oxides.

Biomass (e.g., trees and vegetation) may be carbon neutral if over their life cycle, burning them or converting them to fuel releases the same quantity of carbon into the atmosphere as they absorbed from the atmosphere through photosynthesis. However, if the biomass is grown on land that is deforested, or requires external inputs (e.g., nitrogen fertilizer) that emit carbon in their production, biomass may release more carbon than is sequestered while growing. Biofuel is the product of converting biomass to a liquid fuel. Scientists disagree about whether biofuel production is carbon neutral.

Because few processes associated with modern human activity are intrinsically carbon neutral, since the mid-19th century, population growth and increasing rates of industrialization and urbanization have led to dramatic increases in CO_2 emissions. As a result, emission reduction efforts have largely been concentrated in cities. To the extent that they are empowered to do so, cities have instituted measures to limit their emissions—reducing traffic congestion, promoting energy conservation, preserving or reintroducing urban forests, introducing reduce-reuse-recycle programs. Cities have also acknowledged that urban forms influence CO_2 emissions and they have begun moving away from the suburban development patterns prevalent since the 1950s to more compact forms, which reduce energy demand and dependence on private automobiles.

While many cities have indicated their intentions to become carbon neutral, with so much of the carbon equation out of their control, few have made formal commitments. Six cities that have include Aguascalientes, Mexico; Sydney, Australia; Curitiba, Brazil; Rizhao, China; Copenhagen, Denmark; and Vancouver, Canada. These cities and nine others have formally committed to becoming carbon neutral by joining the United Nations Climate Neutral Network. Ten countries (including New Zealand, Norway, and Pakistan) and 135 companies and organizations are also members. Each member has outlined a plan to achieve carbon neutrality. Of the city plans, some are highly detailed, while others are more conceptual. Most include plans to reduce dependence on fossil fuels by developing alternative energy sources and improving public transportation systems. Other common components are reforestation plans and stricter air quality monitoring regulations.

Individuals and private organizations can also strive for carbon-neutral status. These individuals or organizations work toward reducing and/or avoiding emitting carbon into the atmosphere to offset unavoidable or unknown carbon emissions by them. Individuals or organizations seeking to become carbon neutral can limit energy usage and emissions from transportation, buildings, equipments, and other carbon-positive processes in which they are involved. They may use energy generated from renewable sources either directly, by generating it through the use of solar panels, wind turbines, or other geothermal heating sources, or by obtaining energy from an approved green energy provider. They may use low-carbon alternative fuels such as biofuels for heating or transportation. They may plant trees for afforestation to offset some of their carbon-positive practices. Companies dealing with fossil fuels—the biggest offenders of carbon positivity in the atmosphere—should work to lock up carbon on a long-term basis by pumping carbon compounds such as CO_2 into old oil wells or coal mines. The process is known as *carbon sequestration*, a promising method for achieving carbon neutrality.

A controversial practice called *carbon offsetting* has been used recently to attain carbon-neutral status. In this process, payment is made to others to remove or sequester

100 percent of the CO_2 emitted into the atmosphere by planting trees, funding carbon projects to reduce future carbon emission, buying carbon credits to remove them through carbon trading, and so on. Individuals, companies, organizations, cities, regions, or countries become carbon negative to offset the carbon positive processes to make the Earth carbon neutral.

The steps of commitment, counting and analyzing, action, reduction, offsetting, and evaluation are followed to be carbon neutral. The commitment to be carbon neutral can be undertaken by individuals or organizations, but political leadership expedites it. In the counting and analyzing step, the emissions that need to be eliminated to reach carbon neutral status must be measured. We may not be able to eliminate the entire amount in one go, but it sets the priorities for action to offset some amount of emitted carbon in the atmosphere by the carbon offset measures listed earlier.

Businesses and local administrations can make use of an environmental management system established by the international standard ISO 14001 (developed by the International Organization for Standardization) to offset carbon emissions to the atmosphere. In the reduction step, everyone is encouraged to reduce their energy consumption. In offsetting steps, individuals or organizations are funded to reduce CO_2 in the atmosphere by undertaking measures discussed earlier.

In summary, to become a genuinely carbon-neutral society will take time. However, many countries and communities are committing to carbon neutrality. Some of the countries that have pledged to reduce emissions to zero are Costa Rica, Iceland, Maldives, New Zealand, Norway, and Tuvalu, through 100 percent use of energy generated from renewable energy sources such as solar, wind, hydroelectric, geothermal, and others. The Vatican City committed in 2007 to be the world's first carbon neutral country. With efforts from government, politicians, big businesses, and of course, individuals, we can soon move closer to the ideal situation of being carbon neutral; that is, having no net effect on the amount of carbon in the atmosphere.

See Also: Carbon Footprint; Carbon Trading; Copenhagen, Denmark; Curitiba, Brazil; Sydney, Australia; Vancouver, Canada.

Further Readings

Martin, L. J. "Carbon Neutral—What Does It Mean?" http://www.eejitsguides.com/environment/carbon-neutral.html (Accessed October 2009).

Panda, S. S. "Biomass." In *Encyclopedia of Global Warming and Climate Change*, S. G. Philander, ed. Thousand Oaks, CA: Sage, 2008.

Revkin, A. "Carbon-Neutral Is Hip, but Is It Green?" *The New York Times* (April 29, 2007). http://www.nytimes.com/2007/04/29/weekinreview/29revkin.html?ex=1335499200&en=d9e2407e4f1a20f0&ei=5124 (Accessed October 2009).

United Nations Environment Programme. "Climate Neutral Network." http://www.unep.org/climateneutral/About/tabid/95/Default.aspx (Accessed December 2009).

Sudhanshu Sekhar Panda
Gainesville State College

CARBON TRADING

Carbon trading is a policy measure aimed at reducing greenhouse gas emissions in an economically efficient manner. It enables nations and firms to trade permits to emit specified amounts of greenhouse gases, particularly carbon, and to offset their extra emissions by paying for reductions to greenhouse gas emissions undertaken by others.

The 1997 Kyoto Protocol requires affluent nations to reduce their greenhouse gases up to 8 percent below 1990 levels by 2012. Signatory nations agreed to allow nations to exceed their target reductions using a range of mechanisms:

- Emissions Trading, which allows countries to buy the rights to discharge emissions above their agreed target from countries that reduce their emissions below their agreed targets.
- Joint Implementation (JI), which allows countries to offset their excess emissions by paying for emissions reductions or carbon sinks in other countries which have agreed to the Protocol.
- Clean Development Mechanism (CDM), which allows countries to offset their excess emissions by paying for emissions reductions or carbon sinks in countries that are not signatories to the Protocol; that is, developing nations.

Emissions Trading

The Emissions Trading system under the Kyoto Protocol is a cap and trade system that began in 2008. The cap is a limit set for total amount of greenhouse gases that are allowed to be emitted over a particular period—usually a year—by specific industries in a particular region or nation. The cap chosen is supposed to be within the estimated capacity of the environment to assimilate the pollutant, or at least a step towards achieving that goal. In reality it is a political decision that usually allows far more emissions than the environmental ideal.

Under the Kyoto Protocol, the cap for each nation is calculated to meet the emissions target it agreed to in 1997. If nations are unable to meet their cap by the end of 2012, they will be penalized by having the excess plus a 30 percent penalty included in their cap for the next five-year compliance period. Each nation, in deciding how to meet its targets, may allocate greenhouse gas allowances to companies and allow them to use carbon trading and offsets to meet them.

Carbon trading schemes are normally limited to large firms in a particular industry sector with significant emissions. A participating firm can sell any emission allowances/permits that are surplus to its requirements to another firm that needs extra allowances, or it can save them up for the future when they might be needed.

Carbon trading is based on the idea that it is cheaper for some firms to reduce their carbon emissions than others, and therefore more cost effective to allow the market to decide where emission reductions will be made than for governments to require uniform reductions across an industry. Firms that find it expensive to reduce emissions are able to buy up emission allowances instead. Those that can reduce emissions cheaply can sell their unwanted allowances.

Emissions trading makes sense if only limited pollution reductions are required—that is, if reductions can be limited to what can be done cheaply. However critics of carbon trading argue that because substantial reductions in greenhouse gases are required to prevent significant climate change, more expensive reductions have to be made in the near future, and there is little point in setting up markets that enable firms to avoid making those expensive reductions.

The two main ways of initially allocating allowances are usually referred to as *grandfathering* and *auctioning*. Grandfathering involves gifting allowances to firms on the basis of their past emissions. Alternatively, allowances can be auctioned off.

The EU emissions trading system began in 2005 and used grandfathering to allocate free allowances to some 13,000 companies, including electricity and heat generators, and producers of cement, ceramics, ferrous metal, glass, and paper. Many governments were overly generous in allocating permits to local firms because they feared their local industries would be at a competitive disadvantage if they had to buy extra permits.

A study by Ilex Energy Consulting for WWF examining six EU countries found none of them had set caps that went beyond business as usual. Because allowances were not in great demand, the market opened at 8 Euros per ton and settled around 23 Euros a few months later, far less than necessary to provide an incentive to reduce emissions. Since 2006 the price has generally been below 10 Euros per ton and has even fallen below 1 Euro on occasion.

Individual European countries have also set up trading programs. For example, Denmark has set up a cap and trade program to cover its electricity sector. The Netherlands has also set up a domestic greenhouse gas emissions trading system. Other nations, including Australia and the United States, are now proposing to introduce carbon trading even though the EU scheme has resulted in no net reductions in emissions. The proposed U.S. scheme will grandfather in 43 new coal plants.

Offsets

Offsets under the JI and CDM mechanisms can be generated by tree plantations, providing renewable energy generation projects and energy efficient technologies to developing countries, or closing down old, dirty plants in Eastern Europe. The resulting emissions reductions are intended to be in addition to what would otherwise have occurred. If a polluting facility goes out of business because of financial difficulties, the emissions reductions cannot be claimed as additional, because the reductions would have happened anyway.

Offsets are attractive to corporations because it means that instead of making expensive changes to their own production processes, they can take advantage of the cheap reductions that are available in poor nations. However, there are many questions about how effectively carbon offset schemes reduce greenhouse gases in the long-term. Offset projects tend to favor cheap methods of reducing carbon emissions rather than renewable energy projects in developing countries.

The use of tree plantations as carbon offsets are particularly problematic. First, there is no accepted method for calculating how much carbon is temporarily taken up by growing trees. Such trees may release their carbon early as a result of fires, disease, or illegal logging, but the necessary long-term monitoring is often not carried out. In many situations, plantations are not sustainable because they use up great amounts of water and agrichemicals. These plantations reduce soil fertility, use up valuable water in dry areas, increase erosion and compaction of the soil, and increase the risk of fire.

Even where CDM projects involve renewable energy, there is the question of "additionality." It is up to those claiming carbon credits to explain how they are reducing greenhouse gas emissions and why these reductions would not have occurred without their investment.

This means the carbon offsets can be rather debatable and often would have occurred anyway. Many CDM projects in China are hydroelectric projects, however the Chinese government tends to favor hydroelectric projects for economic reasons, and it is likely these projects would have occurred without CDM funding.

Almost a third of CDM credits are awarded for the destruction of HFC-23, created during the manufacture of refrigerant gas. However, the value of these credits by 2007 ($6 billion) far exceeded what it would cost producers to destroy the gas ($100 million) and also exceeded the value of the refrigerant gases produced.

Moreover, companies gaining credits for carbon reducing investments don't lose credits for carbon increasing investments. Consequently, even though a company's net investment activity adds to global emissions, it can get credits that it can sell to another company, which is then able to discharge even more greenhouse gases. For example, Mitsubishi has four projects in Brazil that will earn it some 13 million carbon credits over 21 years. It is also investing in an oil field project in Brazil that will emit around 58 times the amount of carbon supposedly reduced by the four carbon credit projects.

Avoiding Change

Carbon trading tends to protect carbon intensive industries by allowing them to buy carbon allowances or offsets rather than reduce their greenhouse gases. In this way, trading can reduce the pressure on companies to change production processes and introduce other measures to reduce their emissions.

Oil companies that want to continue expanding their businesses are the very ones that are promoting emissions trading in the knowledge that it will enable them to continue with business as usual. The additional price of oil due to carbon trading is very small compared with the daily fluctuations in the oil market.

Carbon trading similarly allows affluent nations to delay undertaking the changes in national infrastructure necessary to make the transition to a low carbon economy. Government investment in public transport, changes to urban planning, and the development of renewable energy sources are all expensive and likely to offend vested interests that contribute to political party funds.

See Also: Carbon Footprint; Ecological Footprint; Environmental Justice; Green Infrastructure; Renewable Energy.

Further Readings

Bayon, Ricardo, et al. *Voluntary Carbon Markets: An International Business Guide to What They Are and How They Work* (Environmental Markets Insight Series). London: Earthscan Publications, 2009.

Beder, Sharon. *Environmental Principles and Policies*. Sydney: UNSW Press and London: Earthscan Publications, 2006.

Lohmann, Larry, et al. *Carbon Trading: A Critical Conversation on Climate Change, Privatisation and Power*. Uppsala, Sweden: Dag Hammarskjöld Foundation, 2006.

Transnational Institute. "Carbon Trading: A Brief Introduction." http://www.tni.org/primer/carbon-trading (Accessed December 2009).

Sharon Beder
University of Wollongong

CARPOOLING

Carpooling is the process of sharing private automobile trips, usually to and from work, and usually with fellow commuters. The term is also used to refer to parents' sharing obligations for ferrying their own and other people's children to and from school and other activities. Adult carpool members may live in the same neighborhood or town, or may work for the same employer or employers in proximate locations, or both. The pool may alternate all members as drivers, or one or more drivers in the pool may be the only drivers, with the general understanding that drivers are to be compensated for their effort through reduction in the carpool fee arrangements. Carpools are flexible as to how costs are allocated, again based on pre-arranged agreements. Members may be picked up and dropped off at their homes or at specified collection points. Carpooling may be distinguished from vanpooling in that carpools use one or more private automobiles of the participants, whereas vanpools rely on a van made available strictly for that purpose by an employer or non-profit transit agency.

To attract carpoolers, some states dedicate a high-occupancy vehicle (HOV) lane on busy highways.

Source: iStockphoto

Carpooling Benefits and Incentives

The benefits of carpooling are that it saves fuel (one vehicle with multiple occupants rather than multiple single-occupant vehicles) and reduces traffic congestion, thereby reducing commuter costs, reducing greenhouse gas emissions, and improving highway air quality. It also reduces stress and improves personal time management for members, as they can attend to other tasks (reading, napping, listening to music, chatting, or doing paperwork) during the ride. It reduces wear and tear on the member automobiles that are not being used for that day's commute, and may allow members an automobile insurance premium reduction congruent with the reduction in miles driven on their own automobiles.

Effective carpooling depends on a set of common expectations shared by the members of the pool. When guidelines for wait times, notifications for an absence from the commute, and even personal comfort factors (like the interior temperature of the vehicle, the types of music or radio stations preferred, and whether food and beverages may be brought into the vehicle) are discussed and agreed to in advance, the probability of an arrangement that is successful in the long-term is increased.

Carpools may be formed on an ad hoc basis when neighbors or coworkers decide to share rides on a regular basis. However, given the benefits noted above, others, including employers, state transportation agencies, and metropolitan governments, have also taken an interest in promoting and supporting carpooling through various means, both structural and operational. Structural means include construction of high-occupancy vehicle (HOV) lanes. Operational means include designation of existing lanes as HOV lanes restricted to carpools, vanpools, and buses; reductions in tolls or other forms of congestion pricing relief for carpools; technical assistance and financial support for organizing carpools; purchase of fleets of vans for vanpools; and public awareness campaigns to encourage shared commuting.

A common strategy to incentivize carpooling is the establishment of HOV lanes, and many metropolitan areas have such facilities. HOV facilities are restricted to vehicles carrying a minimum number of travelers, usually two, although the original restriction was often for a minimum of three travelers. A 2008 inventory by the Federal Highway Administration lists 345 HOV facilities in the United States, with the largest concentrations around the Twin Cities, San Francisco, Seattle–Tacoma, Los Angeles, and Houston. (It should be noted that each HOV segment is counted as a separate facility; a center city with highway approaches from the four cardinal directions might thus be counted as having four HOV facilities.) A variety of complicated pricing structures, dependent on time of day and number of occupants, are used in many regions to balance traffic flows between the HOV lanes and the main line.

The advent of web technologies has greatly improved commuters' ability to find and join carpools. State departments of transportation (DOTs) and nonprofit agencies funded by them have developed online matching programs that input personal data about origin, destination, and travel time in order to find potential carpool members in a database of potential participants. Such agencies also often work with employer pools to provide additional incentives for participating employers, such as training sessions, positive public relations, and media exposure. In addition, employers may sponsor their own ride-share and carpool programs, and may incentivize such programs with preferred parking spaces for carpools, reduced parking fees (for paid parking), loaner vehicles for running errands, or payroll incentives for participation.

Effectiveness of Carpooling

Some studies show that HOV lanes have been ineffective at reducing congestion because they do not necessarily carry more drivers than regular lanes; they may have family members who would have commuted together anyway; and they tend to shift people out of public transit or higher-density travel (buses) into carpools. Further, abuse of HOV lanes by single-occupant drivers is a perennial concern. For example, according to research by the Washington Post, HOV cheaters in metropolitan Washington, D.C., got 22,532 tickets in 2007. Interviews with the most egregious violators, some of whom had racked up thousands of dollars in traffic tickets, indicated that the tickets were viewed as a cost well worth it when compared with the time costs of sitting in stalled traffic in the metropolitan region.

The effectiveness of carpooling as a strategy to reduce greenhouse gas emissions and traffic congestion is difficult to assess. Data on carpool participation is not collected directly by the states throughout the United States, but must be imputed from data on high-occupancy auto trips. According to decennial survey information compiled by the U.S. Census Bureau, high-occupancy auto trips (defined as two or more occupants) declined from 19.7 percent of all work trips in 1980 to 12.2 percent of all work trips in 2000.

The National Household Traffic Survey (2001) commissioned by the Federal Highway Administration presents more optimistic data: The survey shows that for urban areas, 16.8 percent of work trips are high-occupancy trips, and for nonwork trips, ride-sharing is the norm for more than 50 percent of all trips (for shopping, social activities, recreation, school, or church-going). An analysis of 2008 HOV lane data compiled by the Federal Highway Administration from data submitted by the states indicates that the peak flows in the HOV lanes range from 4.3 percent of total (Capitol Expressway in the Bay Area) to 43.4 percent (in the Washington, D.C., metropolitan area). It should be noted, however, that most states did not submit data on percentage utilization of the HOV lanes.

Despite advertising campaigns, public relations efforts, public capital investment, and financial incentives (and disincentives), the data indicate that carpooling does not constitute a major contribution to reductions in vehicles on roadways, and hence, does not contribute to significant contribution to reductions in greenhouse gases. A computer modeling study of carbon neutrality utilizing IPCC protocols at California State Polytechnic University Pomona suggested that a 32 percent increase in student carpooling would result in only a 1.6 percent reduction in campus greenhouse gas emissions, even though transportation-related GHG emissions represents 55 percent of the total campus GHG emissions. Much of the research on carpooling and HOV modalities indicates that the choice to carpool is highly sensitive to income levels, regional variation, congestion pricing, and the frequency and amount of enforcement penalties.

See Also: Commuting; Congestion Pricing; Gridlock; Transit; Transportation Demand Management.

Further Readings

Pucher, John and John L. Renne. "Socioeconomics of Urban Travel: Evidence From the 2001 NHTS." *Transportation Quarterly*, 57/3:49–77 (Summer 2003).

U.S. Federal Highway Administration. "A Compendium of Existing HOV Lane Facilities in the United States." December 2008. http://ops.fhwa.dot.gov/publications/fhwahop09030 (Accessed December 2009).

Willson, Richard W. and Kyle D. Brown. "Carbon Neutrality at the Local Level: Achievable Goal or Fantasy?" *Journal of the American Planning Association*, 74/4:497–504 (Autumn 2008).

Judith Otto
Framingham State College

Carrying Capacity

Carrying capacity was first described by ecologists in the context of natural ecosystem functioning, but has since been expanded for use in more human-centered applications. Here, the traditional, ecologically focused notion of carrying capacity is described first; then the many parallels between natural and human-made systems, as well as our fundamental dependence on the Earth's ecosystems, are described in detail. Because it has the potential to function as a policy decision–making aide, the differences between the tenets of carrying capacity and those of traditional neoclassical economic theory are also reviewed. The ways in which the concept of carrying capacity can be interpreted for the purpose of gauging the overall sustainability of people's actions and settlement patterns will also be examined. Against the framework created using the concepts described here, particular attention is paid to the sustainability of cities, as determined using purely economic, versus essentially ecological metrics. Although one of many common settlement patterns, cities are of particular interest with regard to sustainability, as they are both the world's economic engines and loci of its most intensive resource use.

Ecological Carrying Capacity

The concept of carrying capacity originated within the field of ecology, where it is often defined in terms of the population of a species that can be maintained by a given habitat indefinitely without causing irreparable harm to the supporting ecosystem.

In the course of its life cycle, each organism present within a given habitat makes demands of the system through the extraction of nutrients and production of wastes; and in so doing, the presence of one species affects other elements of the system. One organism's hunting behaviors, for example, will impact the population levels of prey species. This, in turn, may have repercussions on other predators that rely on that organism for food, as well as on all other elements of the system with which that prey species interacts.

Each element of an ecosystem is linked to the others, either directly or indirectly, and it is because of these myriad interactions and dependencies that ecosystems are frequently described in terms of food "chains" and "webs." Because of their inherent similarities, individuals of the same species make comparable demands on their supporting ecosystems; therefore, there is a population size above which a population's extraction of resources and introduction of wastes will exceed the system's ability to meet these needs. Simply put, this disequilibrium, which may manifest in altered ecosystem functioning or, in extreme instances, in systemic collapse, occurs because carrying capacity has been exceeded.

Although responsive to internal and external stimuli, different ecosystems possess varying degrees of resiliency. For this reason, harvesting of resources from a system may occur indefinitely if it is done in such a way as to allow for regeneration and does not put undue pressure on other related components of the system. Similarly, naturally occurring associations of organisms and their surroundings can absorb some level of pollution without significant deleterious effects. These productive and absorptive capacities are central to predicting carrying capacity because once they are exceeded, functional breakdown can quickly occur as essential elements of the tightly coupled system are compromised.

At its heart, therefore, the notion of carrying capacity relies on an understanding of ecosystems as dynamic and, in some nondeterministic sense, equilibrium-seeking entities.

Anthropocentric Carrying Capacity

As central as notions of sustainable use and waste production are to the natural sciences, they have considerable merit beyond purely ecological contexts. Recognizing that humans are fundamentally similar to all other organisms in our dependence on the natural world for survival, the concept of carrying capacity has been reinterpreted for the purpose of more anthropocentric applications. In this context, the term *carrying capacity* refers to the level of human consumption and waste discharge that can be maintained for an indefinite period without compromising the natural systems on which such activities depend, regardless whether the people in question actually live within the sending and receiving systems.

More an extension than a reinterpretation, this view of carrying capacity includes humans among the organisms that make up the Earth's ecosystem, and as such, provides a baseline against which the overall sustainability of our activities can be determined. Using this concept as a guide, sustainable population levels can be derived by dividing the level of production that can be maintained indefinitely by the per capita demand made on the related systems. Because the Earth's productive and remedial capacities are essentially finite, it is possible to determine how many people the planet can sustain, given our per capita rates of consumption and waste production. Similarly, the inverse of this formula—namely, our natural capital requirements as a function of productive landscape—provides a now familiar measure of sustainability; namely, the ecological footprint.

Regardless of whether we speak of resource extraction or waste mitigation, the ability of natural systems to provide for our needs is often discussed in terms of "ecosystem services." Though often not easily quantifiable in dollar terms, because most are not transacted in traditional markets and therefore lack prices, ecosystem services are nonetheless recognized as inherently valuable. This value may be described as "direct" or "use value." Examples of this kind of value include the benefit derived from the harvesting of species for consumption, or the pleasure associated with observing wildlife in its natural habitat. Conversely, ecosystem services may provide "indirect," or "non-use values," such as the pleasure some people derive from knowing that a species exists, even if they never encounter or utilize it themselves. Clearly, then, the notion of ecosystem services functions to quantify and categorize the behaviors and outputs of the ecosystems on which we are dependent for our survival. Therefore, we are able to speak of carrying capacity in terms of an extraction of resource services at levels that do not exceed an ecosystem's ability to provide equivalent services in perpetuity.

Carrying Capacity Versus Economic Theory

Given that carrying capacity represents an identifiable baseline against which the sustainability of our actions can be ascertained, why do we not rely more heavily on this concept in the course of policy decision making and agenda setting? Undoubtedly, there are a number of explanations; however, our tendency to use traditional economic models for this purpose is probably the most important in this regard.

The discipline of economics departs dramatically from that of ecology in its thinking about the relationship between humankind and the rest of the ecosphere. According to the ecological perspective, our economy is inseparable from the natural systems from which raw materials are extracted and into which waste materials are discarded. Economists, in contrast, view the economy as a system that is essentially distinct and separate from its ecological supports, termed *natural capital*, which—far from being portrayed as fragile and finite—are described as substitutable with other, human-produced forms of capital. Because of this, neoclassical economic theory argues that resource depletion, the resulting scarcity, and the threat of eventual ecosystem collapse can and will be avoided through the normal functioning of the free market. Adherents to this line of thinking argue that when scarcity threatens, prices rise, leading to decreased consumption and the search for substitutes—both processes that act to diminish pressure on threatened resources, thereby preventing their overuse. Both through scarcity-induced behavioral modifications and the assumption that technological innovations can be relied on to provide resource substitutes, economists argue that market expansion can continue indefinitely.

Although this conclusion is clearly at odds with that reached by ecologists, economically justified policy making nonetheless remains the norm.

Gauging Sustainability Using Carrying Capacity

Despite its lack of policy influence, the concept of carrying capacity does provide a useful baseline against which the long-term effects of human activities can be determined. Lifestyles that rely on more resource services than can be replenished by supporting systems violate those systems' carrying capacities and can therefore be described as "unsustainable." In contrast, "sustainable" behaviors are those that operate within these natural bounds, thereby preserving resources in the quantity and quality available to today's populations for use by future generations of man and by nonhuman entities.

The relationship between behaviors and carrying capacities can be applied at the level of the individual, city, nation, or at any other scale. It is, therefore, an appealing methodological framework with which to quantify sustainability—a concept that is often derided as intuitively appealing, yet challenging to articulate in meaningful, actionable terms.

A number of computations based on the notion of carrying capacity have been conducted for all levels and types of human settlements. Many such assessments have taken the form of "ecological footprints"—calculations that seek to quantify the area of the Earth's surface needed to sustain a given population's current levels of consumption and waste production. In the case of a footprint calculation, the Earth's total amount of productive land is divided by the number of individual human beings alive today to yield the "fair Earth share." Once this number is known—which is essentially the amount of the planet's surface that could be dedicated to sustaining each person if all were allocated an equal portion of available resources, it can be compared to current use levels to determine sustainability.

The overriding conclusion of efforts to gauge sustainability in this way has been that modern, developed world lifestyles violate the bounds of carrying capacity to such a degree as to be grossly unsustainable. Unsurprisingly, the United States is routinely identified as the worst offender in this regard—largely because of Americans' technologically dependent, largely urbanized lifestyles and high levels of waste.

However, the United States is the most extreme example of a general trend; namely, the exploitation of global resources by those who can afford to do so. Simply put, wealthy nations can afford to "borrow" the ecosystem services of poorer nations through the exportation of polluting industries and other unwanted activities and the importation of desirable amenities, many of which encourage the destruction of fragile ecosystems to obtain. In this way, the wealthy effectively expand the area and scope of the ecosystems on which they rely, while encouraging the less-wealthy nations in which these systems are located to exploit them for profit—a process that almost by definition occurs at a rate that violates those systems' carrying capacities.

Although it is true that the residents of some nations, typically those described as "Third World," or "underdeveloped," are found to consume less than their fair Earth share, the planetary balance sheet reflects the fact that as a species, humans are living beyond Earth's carrying capacity. Symptoms of the inevitable collapse that results from continued overuse of ecosystem services already abound and include phenomena such as biodiversity loss, desertification, and global climate change.

The processes that lead to this kind of selfish overuse are often described as the "tragedy of the commons." Resources such as fisheries, pastureland, and even our atmosphere are susceptible to the "tragedy" of eventual collapse through overuse because they possess characteristics of "common pool resources." That is to say, many of our most valuable natural assets are difficult to protect from would-be users, and the use of part of the resource base by one individual or group degrades its overall quality or reduces its availability to other users. Under such circumstances, users seeking maximum personal benefit should take as much as possible, because the negative consequences of extraction are shared equally among all, whereas the benefits accrue wholly to the individual. This dynamic is unsustainable because beyond some typically difficult-to-discern threshold, the system's carrying capacity is violated, leading to its eventual collapse. One need only consider the world's fisheries to realize that technologically enhanced harvesting, a desire for profit, and the ability to ship goods to populations around the globe quickly and inexpensively can easily lead to population decimation.

Part and parcel of these global developmental trends are movements toward an increased reliance on technology, more expansive and comprehensive trade networks, and high and increasing rates of urbanization. All of these factors have been shown to contribute to increased rates of resource service usage, and urban residents in particular appear unaware of their connection to the far-away ecosystems on which they depend for daily subsistence, enhancing the likelihood of overuse. In light of the growing proportion of the planet's population living in cities, the link between urban developmental patterns and overall sustainability is of particular interest, and the concept of carrying capacity provides a useful framework in which to consider this issue.

Cities' Carrying Capacities and Overall Sustainability

Cities can be thought of as collective, spatially defined entities, discernable from their surroundings according to particular political or administrative associations, as well as, in many instances, by the characteristics of their built environments. Although typically much richer in human capital than more-rural areas, cities are not self-sustaining for reasons of density and intensities and types of land uses. Economists view cities as centers of intense interaction among individuals and firms, and therefore, as the physical loci from which the global economy is driven. Ecologists have a much different take.

It is city dwellers' continual reliance on these external supports, not their capacity to generate economic activity that is of interest to ecologists because the nature and intensity of these use patterns does much to determine the overall sustainability of such settlements. Urban centers rely on the continual movement of resources into, and waste products out of, their borders. Today, these flows of physical goods and financial capital between locales are effectively global in scope. Although it is difficult to quantify with any certainty the amount of ecosystem services, and therefore of ecologically rich land needed to sustain a city's population, some estimates put it at approximately an order of magnitude larger than the footprint of the city itself. This finding is especially disturbing when one considers that cities interact with their supporting ecosystems in an essentially unidirectional fashion. That is to say, resources are extracted for use by city dwellers and their wastes introduced into receiving areas without any reciprocal contribution to the functions of resource provision or waste treatment within these systems.

A one-way resource flow is not unique to cities, however. In truth, virtually all human settlement types share this characteristic. Perhaps the more meaningful question, therefore, is not "How do cities impact global carrying capacity?" but "How do the impacts of cities on global carrying capacity compare to those of other developmental forms?" To answer this question, many people have drawn comparisons between urban environments and what is often cast as the polar opposite developmental form: sprawling suburbs.

Although some debate exists as to the mitigating effects of factors, such as technological safeguards and lifestyle choices, there is widespread agreement that urban inhabitants' per capita resource consumption (and therefore, their individual impact on global carrying capacity) is lower than that of suburbanites. Observed reductions in resource intensity result from a confluence of factors related to urban form, such as the height, density, and construction of buildings; the availability of public transit; and the reduced distances between destinations.

The notion of carrying capacity dictates that any developmental style could attain long-term sustainability so long as the demands made on external environments remain at or below the levels at which replenishment and repair occur on a continual basis. Where exactly these "bounds of sustainability" lie is difficult to determine; however, city residents appear closer to living within them than are suburbanites. Despite this reality, there is little doubt that the demands of today's cities continually exceed the Earth's capacity to supply resources and assimilate waste products. As such, although urban development is perhaps "preferable" to its lower-density counterpart, it is far from an environmental panacea, a conclusion made all the more significant as the global trend toward increased urbanization continues unabated.

Conclusion

Although originally conceived to describe ecosystem functioning, the concept of carrying capacity provides a useful alternative to neoclassical economic theory for those interested in determining environmental sustainability. Although the true scope of our reliance on natural ecosystems, as well as the exact nature and value of the services they provide, remains elusive, it is clear that on the global scale, we are living beyond the carrying capacity of our planet. Resource-intensive, urbanized lifestyles are particularly to blame in this regard—a finding that is particularly troubling given the increasing rates of technological advancement and urbanization we are pursuing globally.

See Also: Biodiversity; Ecological Footprint; Natural Capital.

Further Readings

Alberti, Marina. "Measuring Urban Sustainability." *Environmental Impact Assessment Review,* 16:381–424 (1996).

Bohringer, Christoph and Patrick E. P. Jochem. "Measuring the Immeasurable—A Survey of Sustainability Indices." *Ecological Economics,* 63 (2007).

Grafton, R. Quentin, et al. "Private Property and Economic Efficiency: A Study of Common Pool Resource." *The Journal of Law and Economics,* 43 (2000).

Norberg, Jon. "Linking Nature's Services to Ecosystems: Some General Ecological Concepts." *Ecological Economics,* 29:183 (1999).

Peterson, Gary, et al. "Ecological Resilience, Biodiversity and Scale." *Ecosystems,* 1/1:6–18 (1998).

Rees, William E. "Ecological Footprints and Appropriated Carrying Capacity: What Urban Economics Leaves Out." *Environmental and Urbanization,* 4/2:121–30 (1992).

Josephine S. Faass
Rutgers, The State University of New Jersey

CHARETTE

A charette is an intensive design methodology used primarily in urban and land use planning. It rose to prominence with the New Urbanism movement in the 1990s but emerged as a recognized practice in the late 1960s. The charette process evolved as a part of a larger reaction against modernist architecture and planning's claims to "know best" how to shape an ideal urban society through the constructed forms of the city. In its idealized form, the charette process eschews hierarchical, objectivist, comprehensive planning and embraces inclusive, situated, communicative planning. Recognizing that architecture and planning are practices that embody ethics within physical spaces, the charette process seeks to democratize the planning and development of towns and cities through the inclusion of affected persons and interest groups in the design and critical review process. The point is not merely to seek approval for design precepts already enumerated but to engage persons' creative contributions in the formation of what is to be the physical foundation of their community. However, architects, planners, and developers sometimes refer to an intensive design studio without significant public engagement as a charette. This ambiguity is a result of both the etymology of the term as well as an ambivalence regarding the role of persons and interest groups in city, town, and real estate development.

En Charrette: Etymology and Foundational Practices

The first use of the anglicized term *charette* comes from the French *en charette*, literally to be "on the pushcart." In the 19th and early 20th centuries, senior students at the École des Beaux Arts in Paris would gather their plans and drawings for the term's design project and load them onto pushcarts, on which they were carried from the various ateliers in which students produced work to be judged by a jury at a predetermined deadline. The phrase *en charrette* came to signify not only the frantic time of loading final presentation works on to the pushcarts and their journey to the jury but also the final weeks of dogged preparation leading up to the last hectic days of production.

Contemporary charettes inherit both the theoretical framing of the Beaux Arts school and its practices of creative production. Beaux Arts architectural training focused on drawing as a means of visualizing architectural form, from production of conceptual sketches to detailed perspective drawings of the finished site; projects and their sites were predominantly urban. In addition to the critical review of the jury, iterative reviews happened throughout the process of design as peers critiqued each other's works. To be *en charrette* signified collaborative work in the studio with junior colleagues to solve a specific design problem in the face of a looming deadline, amid peers who were also producing their own solutions. Working through the night was typical, even for several nights, as designs were rethought and redone until the presentations were finally loaded on the carts. The pivotal period of gathering and reacting to critical judgment was not at the jury, which was a final evaluation, but, rather, the days of intense work in a shared studio space. At its heart, to be *en charrette* was to iteratively and collaboratively produce drawings as a means of visualizing the solution to an urban problem.

In North America, many architectural schools continue to refer to the final weeks of term spent working in studio as charette; many architecture studios also adopt this usage to refer to the hectic days of all-day, all-night production before a deadline. Proponents of this work style advocate that this concentrated, short burst of productivity leads to greater creativity. No longer distracted by daily routines or other assignments, all energy is concentrated on the single task at hand, and inspiration and criticism abound from one's peers, who are also at work. At the same time, detractors of this method argue that it is no more than the profession's form of ritualized hazing that leads to superficial designs stunted by an arbitrarily short time constraint, and that it encourages substance abuse.

A Reaction to Modernist Architecture and Planning

The contemporary usage of charette to indicate an intensive and inclusive planning process emerged in the 1960s in response to what Carlton Monroe Winslow, Jr., termed the anthropological failure of modernist buildings and town plans. Modernist planning's belief that ideal city spaces would create ideal citizens and support democratic governance was being challenged as modernist master plans failed to deliver the vibrant, democratic communities they promised. As a planning methodology, the charette works to shift the locus of the authoritative knowledge from the ego of the architect or planner and distribute it among various persons and groups within the targeted community. Rather than the master plan presenting the planner's vision for the city, a charette-produced plan is the result of the contributions and compromises of many varied persons and groups.

David Walters has written of this shift as one from modernism to postmodernism, from a focus on the product to a focus on the process. This postmodern orientation not only eschews that there is any single "best" idea but also refuses the premise that any one person or perspective can "know best" or has access to objective, omniscient knowledge. The charette thereby works as a centripetal force on planning authority, recognizing that the master plan to build a desirable, vibrant community can only be achieved by drawing on the expertise of transport and energy specialists, retail and commercial experts, architects and planners, anthropologists and community organizers, real estate developers and municipal officials, amid other specialists, and of course, persons in the general population—or the public.

Integral to the charette method are its transparency, collaboration, visualization, and termination. Rather than producing a master plan from intense work in a private studio or office often far from the development site, a charette produces a master plan after four to seven days of design work done in a public location at or near the site. Each day of the charette is typically broken up into discrete workshops focusing intensively on subsets of the overall design problem, such as water, agriculture, energy, or town programming. Quick sketches are drawn from collaborative brainstorming during workshops that are developed into more detailed drawings during evening and overnight studio sessions. Drawings are posted so that they can be viewed and critiqued in the following days' sessions. Workshops for soliciting responses from persons and interest groups are at minimum set aside at the outset and conclusion of the charette, yet all workshops and the studio space remain open to the public throughout the charette. Through integration of elements of the urban system, all participants—professional and personal—become aware of the complexities of the process and products of development; iteration of design and critique over several days, culminating in a final deadline for presentation of a master plan, guides all participants to compromise toward a solution.

In its ideal form, a charette democratizes the urban planning process, giving voice to the desires and concerns of the people who live and work in the urban setting in question. In so doing, it makes it more likely that a project receives both public support and government approval to proceed and that it enjoys commercial success on completion. In practice, a charette can serve to further conflate developers' interests or the narrow interests of a municipal department, planning agency, or interest group in the name of the public interest. This happens when academics, architects, planners, and other professional experts are allowed to represent the public interest, or the public is only included in a cursory way. When there is limited public announcement of the charette, the public is only invited to participate during common working hours, or the experts spend more time lecturing rather than listening, communicative planning is not taking place. In such cases traditional comprehensive planning continues under the guise of a democratic methodology, purporting to settle any questions as to who has the right to set the terms and to profit from urban (real estate) development, as the public was represented through the inclusion of invited interest groups—such as a diverse group of planners, architects, engineers, and academics, all being paid by the same developer.

A New, Integrated Urbanism

The rise of the New Urbanism movement made the charette a well-known planning practice but also altered its focus away from a process based on discourse ethics and back toward a product based on normative values—an ideal form of the city. Although New Urbanists state that they are committed to citizen-based participatory planning, the citizenry is defined as being the New Urbanists themselves, and the ideal city as that derived from prescriptive principles.

Once called neotraditional planning, New Urbanism advocates a return to town and neighborhood development that favors the pedestrian over the automobile; integrates industrial, retail, and residential functions; and uses form-based zoning, or the appearances of physical structures, to regulate town development. The Charter of the New Urbanism asserts 18 principles of how the city, town, neighborhood, street, and building should be. These prescriptive edicts for urban development arise from an attempt to strengthen the public realm and encourage community formation through face-to-face interactions in physically defined spaces. An attack against the architectural forms and behavioral practices of single-family houses with private yards and automobile commutes, New Urbanism mandates mixed-use development, where people live, work, and play in an area that is walkable or is served by public transport.

Although proponents argue that such design leads to more equitable sharing of space and community formation, critics point to the lack of racial and socioeconomic integration and the prevalence of commuting from New Urbanist towns to other locations within the metropolitan region. Although postmodernist in its appropriation of various, so-called traditional architectural styles into a small town pastiche, the spatial ethics of New Urbanism remain modernist: a revisioning of design determinism in which predetermined spatial relationships are taken to stand for human relationships. As a consequence, the discourse ethics of the charette as a communicative planning tool become a method of validating rule-based and value-based zoning regulations and master plans. To achieve a shared sense of place and aesthetic-as-relational coherence, the public is educated via the charette process to support consumption of a new urban form. New Urbanists reclaim the role of "who knows best"—inclusion in the creative design of the public environment is again denied to the general public—and consumption of a product stands in for a community of persons doing things together.

Performing or Producing Consensus

Whether the charette methodology serves as technique for creating livable, sustainable cities depends on whether it is used performatively or productively. The difference depends on whether persons are treated as creative contributors to the final design—as colleagues in the studio with whom iterative designs are drawn as during the days *en charrette* at the École de Beaux Arts—or whether the public is treated instead as part of the final jury, left to either accept or reject the definitions and forms of an ideal city presented to them.

The revival and reincarnation of the charette process in the United States was intended to end the disenfranchisement of the public from the planning and building of the urban spaces that shape the possibilities of their everyday lives. Through speaking and arguing about place-making, the charette seeks to enable persons to establish a relationship to the city and to each other, enabling both consensus and commitment to a community to emerge. The knowledge of architects, engineers, planners, sociologists, and other area experts is not meant to trump persons' perspectives and desires with normative prescriptions; rather, their expert advice is to serve as guidance, as both a means for enactment and a contributing critique. When the community that is enabled by a charette to introduce and question claims of urban and land use planning becomes synonymous with the various experts who directly materially benefit from development (be they planners, architects, developers, municipal officials, consultants, etc.), the charette does not democratize planning. It instead works to solidify the public's disenfranchisement from control over their local environment, taking public acknowledgement for public participation.

For a charette to meet its promise as a tool of communicative planning that embodies a democratizing, discursive ethics, it must fully engage persons who are not planners or architects as having equal rights to decide what makes a community that they want to live in.

See Also: Citizen Participation; Compact Development (New Urbanism); Green Communities and Neighborhood Planning; Healthy Cities; Smart Growth; Sustainable Development; Walkability (Pedestrian-Friendly Streets).

Further Readings

Bartling, Hugh E. "Disney's Celebration: The Promise of New Urbanism, and the Portents of Homogeneity." *The Florida Historical Quarterly,* 81/1:44–67 (Summer 2002).

Chappel, George S. "Paris School Days: How the Student Lives and Works at the Ecole des Beaux Arts III—The Charrette" (February 1911). In *The Origins of Modern Architecture: Selected Essays From "Architectural Record,"* Eric Uhlfelder, ed. Mineola, NY: Courier Dover, 1998.

Condon, Patrick M. *Design Charrettes for Sustainable Communities.* Washington, D.C.: Island Press, 2007.

Fainstein, Susan. "New Directions in Planning Theory." *Urban Affairs Review,* 35:451–78 (2000).

Frug, Jerry. "The Geography of Community." *Stanford Law Review,* 48/5:1047–108 (May 1996).

Kelbaugh, Doug. "The New Urbanism Author." *Journal of Architectural Education,* 51/2:142–44 (November 1997).

Lennertz, Bill and Aarin Lutzenheiser. *The Charrette Handbook: The Essential Guide for Accelerated, Collaborative Community Planning.* Washington, D.C.: American Planning Association, 2006.

Riddick, W. L. *Charrette Processes: A Tool in Urban Planning.* York, PA: George Shumway, 1971.

Smith, Neil. "Which New Urbanism? The Revanchist '90s." *Perspecta*, 30:98–105 (1999).

Upton, Robert. "Planning Praxis: Ethics, Values and Theory." *The Town Planning Review*, 73/3:253–69 (July 2002).

Walters, David R. *Designing Community: Charrettes, Master Plans and Form-Based Codes*. Oxford: Architectural Press, 2007.

Winslow, Carleton Monroe, Jr. "Architectural Education and Behavioral Science." *Journal of Architectural Education*, 18/2:25–27 (September 1963).

Shannon May
University of California, Berkeley

CHATTANOOGA, TENNESSEE

Over the course of 30 years, the citizens of Chattanooga have worked diligently to clear Chattanooga's air and improve her public image: building parks; making their city bicycle-friendly; and replacing diesel-powered buses with electric ones. Their efforts earned recognition in 1996 both from the President's Council on Sustainable Development and at the Second United Nations Conference on Human Settlements–Habitat II, where the city was given an award for Excellence in Improving the Living Environment. Not only has Chattanooga been recognized for the environmental practices it has adopted, but also for "The Chattanooga Process," which was used to build community consensus around sustainability.

Chattanooga intends to establish itself as the nation's go-to place for information on sustainability, and therefore strives to stay on the cutting edge of environmental issues. In February 2009, the city adopted the Chattanooga Climate Action Plan, which addresses global climate change. The city conducted a carbon footprint analysis and laid out a plan to reduce 2012 greenhouse gas (GHG) emissions levels to 7 percent below those of 1990, largely through changes in transportation and power generation. The intent is to reduce 2020 GHG emissions to 20 percent below the 1990 levels.

Major chapters of the Chattanooga Climate Action Plan address energy efficiency (including alternative energy sources, energy conservation, green building, recycling and waste, and sustainable industry), healthy communities (including the built environment and smart growth, food and agriculture, and transportation) and natural resources (including air quality, biodiversity, green infrastructure, urban and regional forests, and water quality and quantity). The recommendations urge comprehensive public education plans, plans for business participation, government policy and purchasing guidelines, and involvement by area schools.

Chattanooga is a signatory party to the U.S. Conference of Mayors Climate Protection Agreement, and a member of the International Council for Local Environmental Initiatives (ICLEI).

See Also: Air Quality; Intermodal Transportation; Landfills; Sustainable Development.

Further Readings

"Chattanooga: A City Worth Watching." Sustainable Communities Network Case Studies. http://www.sustainable.org/casestudies/tennessee/TN_af_chattanooga.html (Accessed October 2009).

"The Chattanooga Riverpark: Transforming a City and Its Economy: An Urban Parks Institute Success Story Chattanooga, Tennessee." http://www.pps.org/parks_plazas_squares/info/design/success_chatanooga (Accessed October 2009).

City of Chattanooga. "The Chattanooga Climate Action Plan." 2007. http://www.chattanooga.gov/Final_CAP_adopted.pdf (Accessed December 2009).

Clark, Alexandra Walker. *The Hidden History of Chattanooga.* Charleston, SC: The History Press, 2008.

Lewicki, Roy J., et al. *Making Sense of Intractable Environmental Conflicts: Frames and Cases.* Washington, D.C.: Island Press, 2003.

Livingood, James Weston. *Chattanooga: An Illustrated History.* Woodland Hills, CA: Windsor, 1981.

McAdoo, William G. *Crowded Years: The Reminiscences of William G. McAdoo.* London: Jonathan Cape, n.d.

Rogge, Mary E. "Toxic Risk, Community Resilience and Social Justice." In *Sustainable Community Development: Studies in Economic, Environmental and Cultural Revitalization*, Marie D. Hoff, ed. Boca Raton, FL: CRC Press, 1998.

Verdeveld, Ruth and Robin Martin. "Recycling Solid Waste in Chattanooga." *The American Biology Teacher,* 35/2:84–87 (February 1973).

Justin Corfield
Geelong Grammar School, Australia

CHERNOBYL, UKRAINE

Chernobyl is a city in northern Ukraine, close to the border with Belarus. It lends its name to the nearby power plant, whose number four reactor was the subject of the worst disaster in civil nuclear history and one of the worst industrial disasters the world has ever seen. It was the only occurrence of a level 7 disaster on the International Nuclear Event Scale. At the time of the incident, the area was part of the Ukrainian Soviet Socialist Republic, which was part of the former Soviet Union.

The reactor was of a design type called RBMK-1000 (*Reaktor Bolshoy Moshchnosti Kanalniy*; Russian: Реактор Большой Мощности Канальный), which uses graphite as a moderator and water as a coolant. The water boils in the core at 290 degrees Celsius/ 554 degrees Fahrenheit. The design of the reactor leads it to be unstable at low power levels because of the design

The Chernobyl nuclear disaster claimed two immediate victims and resulted in the evacuation of 135,000 people, including 50,000 from the nearby town of Pripyat. This abandoned classroom is in Pripyat.

Source: iStockphoto

of its control rods and because the design has a positive void coefficient. The incident released 100 times the amount of radiation that was released from the atomic bombs that devastated Hiroshima and Nagasaki in 1945.

Cause of the Disaster

On the night that the accident occurred, the operating crew was ordered to perform tests to ascertain whether, in the event of a loss of external power, the spinning turbines could produce sufficient power to keep the coolant pumps running until the emergency diesel generator started up.

The safety systems of the reactor were switched off deliberately to prevent any interruption to the tests. The reactors' design was such that in the event of a power failure, the reactor would SCRAM (control rods would be inserted into the reactor, which in turn would shut down). However, even when the control rods were fully inserted, the reactor still required cooling, as the fission products from reaction would continue to produce heat as they decayed—and this heat could in turn lead to damage of the reactor's core.

Aside from "human error" and "defective technology," another area where the RBMK reactor came under heavy criticism is in the design of the interface between "people" and "machines."

The design of the RBMK reactor has also come under heavy criticism. One notable feature of the reactor is that it has an unusually large "void reactivity coefficient"; this is a parameter that defines how the reactor will respond to the formation of steam bubbles in the water coolant. In many designs, as steam bubbles form, the reactor will produce less energy—this is because water is used to slow down fast neutrons, which are less likely to split atoms of uranium than slower neutrons. As water turns to steam, fewer neutrons are slowed, splitting fewer uranium atoms and releasing less energy. In the RBMK reactor design, graphite was used as a moderator to slow down neutrons, and light water, which absorbs neutrons, to cool the core. The bubbles of steam will not absorb neutrons as readily as water; therefore, as bubbles of steam form, the energy production and reactor temperature increases.

In the Chernobyl reactor, explosive pressure resulted from the coolant water rapidly boiling and turning to steam.

One of the design requirements of the RBMK reactor is that the turbines continue to spin as the reactor powers down.

In the aftermath of the Chernobyl disaster, the remaining RBMK units in operation were mildly reengineered to improve their safety. A number of modifications were made, which also had the effect of reducing the void coefficient from +4.5β to +0.7β, which decreased the chance of a similar incident occurring again:

- The SCRAM time of the reactors was reduced from 18 to 12 seconds, and a "fast SCRAM" system was installed.
- The control rods were redesigned.
- Emergency safety systems were sealed off from unauthorized access.
- The number of manual control rods was increased from 30 to 45.
- Eighty additional "absorbers" were installed within the reactor core to retard low-power operation.
- The reactors were refueled with uranium enriched at 2.4 percent instead of 2 percent. This has the effect of increasing the fuel burn-up rate and increasing neutron absorption.

Timeline of the Disaster

- December 31, 1983, Viktor Bryukhanov, director of the Chernobyl Nuclear Station, signs an acceptance document that the annual works had been completed successfully.
- March 1984, Chernobyl Reactor Four is brought into commercial operation.
- April 25, 1986, Chernobyl Reactor Four is prepared for its safety test, and the reactor's output is slowly reduced to 50 percent of capacity in preparation for the test. As another nearby power station unexpectedly goes offline, the Kiev Grid-Controller requests that further reductions in power be postponed to keep supplying power to the city of Kiev.
- April 25, 1986, 11:04 P.M. The transient fault condition has passed, and the Kiev Grid-Controller gives permission for the test to resume. By this time, the reactor's day shift had departed some time ago, the evening shift is preparing to leave, and the night shift is ready to take over operation of the reactor.
- April 26, 1986, 1:23:04 A.M. The experiment begins. Steam to the turbines is shut off, with the reactor crew unaware of the unstable reactor condition. As the turbine flows, the water flow to the reactor decreases, allowing voids in the cooling water to form as a result of the massive temperature rise within the reactor.
- April 26, 1986, 1.23:40 A.M. A SCRAM of the reactor is initiated with the pressing of button AZ-5 (Rapid Emergency Defense 5). The control rods of the reactor are automatically inserted; however, this takes place at slow speed. (The rods take 20 seconds to travel the full 7-meter RBMK core height. The tip of the control rods is made of graphite, which displaces the moderating coolant and increases the reaction rate.)
- April 26, 1986, 1:23:45 A.M. The reactor output rises rapidly to 530 MW; control rods break as a result of the massive spike of energy and enormous heat being generated within the reactor core.
- April 26, 1986, 1:23:47 A.M. The reactor is now producing 30 GW of thermal power, which is more than the reactor's operational output by a factor of 10. Steam pressure rises to the point at which fuel channels within the reactor are destroyed; large-diameter cooling pipes rupture and fuel rods begin to melt.
- April 24, 1986, 1:24 A.M. A massive steam explosion takes place, blowing the lid of the reactor (known as the biological containment shield), which weighs 2,000 tons. Material is ejected from the reactor, fuel channels are ruptured, and two to three seconds later a hydrogen explosion ensues as a result of the reaction of the hot graphite in the reactor with steam from the cooling water. Burning lumps of material shoot into the air.

Aftermath of the Disaster

The disaster resulted in two immediate deaths as the result of a steam explosion within the plant, and 237 people suffering from acute radiation sickness, of whom 31 died within three months of exposure. The disaster also resulted in the evacuation of 135,000 people, including 50,000 from the town of Pripyat. Because it is impossible to prove the origin of cancers that result in a person's death, statistics detailing Chernobyl's long-term death toll are very hotly debated.

It is not known with any degree of certainty how much radioactive material is left within the core of Chernobyl Reactor Four; however, estimates vary that between 3.8 and 20 percent of the 200 tons of fuel within the reactor were released, resulting in the release of 50–250 Ci of radiation. The issue is a contentious one, with the Ukrainian government agency Interinform contending that 95 percent of the fuel remains within the reactor in one form or another, and some experts, notably Konstantin Checherov of the Kurchatov Institute in Moscow and his German colleague Sebastian Pflugbeil, director of the Society for Radiation Protection in Berlin, contend that "most" of the fuel has been released into the environment.

The effects of the disaster were not contained within the borders of Ukraine; a cloud of radiation was produced that traveled as far as the Turkish Thrace and southern coast of the Black Sea, as far north as Finland, and to the west as far as France, the United Kingdom, and the Isle of Man. On April 27, 1986, following the disaster, workers at the Forsmark NPP in Sweden, 680 miles away from the accident, found particles of radioactive dust on their clothing. When the Swedish workers could find no source of the radiation from their own power plant, the international community became aware that the problem could be elsewhere, in the then–Western Soviet Union.

The authorities encouraged people in Ukraine to participate in May Day celebrations, despite knowing that some of the areas in which parades were taking place were lethally contaminated with radiation.

The effects of the disaster persist to the present day; as recently as April 2009, the U.K. Food Standards Agency reported that 3,000 sheep as far away as Scotland are subject to restrictions because they are contaminated by the aftermath of Chernobyl and are above the safety limit of 1,000 Becquerels of radioactivity per kilogram. These sheep are contaminated as the result of Cesium-137, which was released into the atmosphere, brought down to Earth by rain, and persists in the peat and grass of upland areas.

Cleanup and Recovery

The initial sarcophagus, hastily built in the aftermath of the disaster to allow the other reactors of the plant to continue operating, does not afford adequate protection for the reactor and its contents and does not provide sufficient isolation of the 200 tons of radioactive material believed to be left within the reactor, despite major maintenance work to the structure in 1998 and 1999.

The current sarcophagus is supported by two enormous beams that support the ceiling of the structure. They rest on the west wall of the reactor building, which because of damage sustained during the blast is structurally unsound. Actions have been taken to alleviate this problem by extending an adjacent steel structure known as the "DSSS" to support 50 percent of the roof load.

The structure was never designed to endure a full lifetime of the most harmful radioactivity contained within the plant. A breach of the sarcophagus or partial collapse could cause large amounts of radioactive debris to again enter the atmosphere.

The three remaining reactors also still contain fuel in their cooling ponds, as well as at an intermediate storage facility.

By 2011, it is anticipated that a new containment structure will be built nearby and then moved into position on rails. It will encase Chernobyl Reactor Four and the containment structure built in 1986. The new structure, named the "New Safe Confinement," will contain the debris within the plant. The estimate (as of 2006) for the cost of this endeavor is $1.2 billion. The design incorporates an arched structure, which will be the biggest movable structure ever built, with a span of 270 meters, a length of 150 meters, and a height of 100 meters.

Chernobyl Death Toll

The number of deaths arising from the Chernobyl accident is hotly disputed. As the accident was the largest of its kind in the civil nuclear industry, there are very few precedents. Estimates for the damage caused to the human body as a result of radiation are largely taken from the limited available data from the impact of the two nuclear bombs that were dropped on Japan at the end of World War II.

- A study by the International Atomic Energy Agency, the UN Development Programme, and the World Health Organization claims that the Chernobyl disaster claimed "less than 50 lives." There were 4,000 cases of thyroid cancer in young children and adolescents.
- The Chernobyl League, a group representing those who worked in the relief operations, says that 30 people were killed immediately, 15,000 relief workers were killed as a result of exposure to radiation, and 50,000 were left disabled or ill as a result of the disaster. Five million people were exposed to radiation from the disaster and 52,000 were forced to flee the area around Chernobyl.
- Greenpeace claims that the Chernobyl death toll has been grossly underestimated, with 60,000 additional deaths in Russia as a result of the accident and estimates for Chernobyl and Belarus around 140,000 deaths, in addition to a quarter of a million cancer cases and 100,000 fatal cancers arising as a result of radiation from the disaster. In a report that involved 52 respected scientists and information that had not been previously published in English, the report challenges the U.K. International Atomic Energy Agency Chernobyl Forum account. They state that the International Atomic Energy Agency report is "a gross simplification" of the breadth of human suffering.
- Ian Farlie and David Sumner, two independent radiation scientists, posit that the UN report thoroughly underestimated the death toll from cancers, as it only counted cancer deaths from the most contaminated parts of the countries that were adjacent to the disaster. The omission of the rest of the world, they believe, is a glaring error, especially as the radioactive cloud that ensued in the aftermath of the disaster was known to spread so widely beyond the borders of the countries in the immediate vicinity.

See Also: Environmental Justice; Environmental Planning; Environmental Risk.

Further Readings

Alexivich, Svetlana. *Voices From Chernobyl: The Oral History of a Nuclear Disaster.* New York: Picador. 2006.
Chernobyl.Info. http://www.chernobyl.info (Accessed October 2009).
Marples, David. R. *Chernobyl & Nuclear Power in the U.S.S.R.* New York: St. Martin's, 1986.
Medvedev, Zhores A. *The Legacy of Chernobyl.* New York: W. W. Norton, 1990.
Mould, R. F. *Chernobyl Record: The Definitive History of the Chernobyl Catastrophe.* Bristol: Institute of Physics, 2000.

Gavin D. J. Harper
Cardiff University

CHICAGO, ILLINOIS

Chicago is attempting to transform itself from a blue-collar industrial city into an exemplar of green urbanism. With a history of large-scale environmental change as a result of urban activity, the city has established several programs within the last decade aimed at reducing its environmental footprint by considering the built environment more carefully. Although criticisms exist of the focus on environmental and economic, rather than social, sustainability, Chicago has gone beyond mere aesthetic improvements in its attempts to change the relationship between humans and the environment within its urban context.

The History of Chicago and Its Environment

Chicago has a long history of altering the natural environment on a grand scale. Similar to any other city, its networks of trade stretched out into distant hinterlands, with the city's growth dependent on the transformation of forests into lumber and prairies into grain. Within the city limits, however, three major projects around the turn of the 20th century shaped Chicago's special relationship to its natural environment. First, the original lakefront was filled in with debris from the 1871 Great Chicago Fire and later extended with landfill to form Grant Park, popularly known as "Chicago's front yard." Although this open space has been shared with rail yards and train tracks, it has also remained a distinctive feature of Chicago's skyline. Second, after decades of dealing with refuse being dumped in the Chicago River and then flowing into Lake Michigan, the city's source of drinking water, the Sanitary District of Chicago engineered a reversal of the river in 1900. This sent pollutants into the Sanitary and Ship Canal southward to the Des Plaines River and into the Illinois River, significantly improving the water quality of Lake Michigan and the Chicago River (though reducing water quality downstream). Finally, Daniel Burnham's design of the 1893 World's Columbian Exposition and his 1909 Plan of Chicago brought the City Beautiful movement to the Midwest. Although Burnham's plan was only partially implemented, many of the parks and boulevards he designed were put in place and remain important landmarks today. In addition, Burnham's spirit of large-scale visioning continues to influence planning in Chicago today, especially during the centennial of his original plan.

During the middle of the 20th century, Chicago was not particularly known for being green. Although it was one of the first U.S. cities to develop a regional transportation plan, its heavy industry and rapid suburbanization led to significant air pollution, and social unrest in the 1960s and 1970s was of greater concern than the natural environment. It was not until Mayor Richard M. Daley's tenure that the city's environmental leadership began and a Department of Environment (DOE) was established in 1992. One of the DOE's first projects was the Brownfields Initiative, meant to clean up and redevelop old industrial sites to enhance economic and environmental sustainability. Another early project was the planting of hundreds of thousands of trees along city streets, along with placing concrete planters in the median strips of over 60 miles of city streets. On the one hand, this approach was criticized for focusing on aesthetics rather than on the harder problems of pollution or unemployment; on the other hand, studies have shown that having greenery nearby not only improves property values but also lowers crime and improves public health.

These early initiatives were followed with a more substantial plan in 2001 of sourcing 20 percent of the city's power from clean sources within the state of Illinois by 2006. The goal was not only to reduce emissions from coal-burning power plants but also to create a market for alternative energy within the state, thus spurring economic development. Although this goal was not met, the city did make considerable progress and has been successful in spurring some alternative-energy start-up companies. A more visible project at this same time was the installation of a green roof on City Hall. As is the case with many of Chicago's recent innovations, this one was spurred by a trip abroad by Mayor Daley, where he saw green roofs in Germany and decided to implement them back home (also influenced in part by the 1995 heat wave that claimed hundreds of lives). The roof of City Hall is on average 30 percent cooler than surrounding, traditional urban roofs, with an even greater temperature differential in the summer.

Being Green in Chicago

Today, there are a series of programs under the Department of Environment that are aimed at meeting Mayor Daley's goal of making Chicago the greenest city in the country. The city wants to lead by example, as with the green roof on City Hall and the development of the Chicago Standard, which specifies green building requirements for all municipal buildings. For the most part, these policies are not mandates, although they do steer developers in a certain direction. Green buildings, green roofs, green infrastructure, and green jobs are all explicitly addressed through city policies, some of which are unique to Chicago and most of which have been implemented not only in municipal projects but in the private and nonprofit sectors as well.

LEED (Leadership in Energy and Environmental Design) certification, the de facto designation for green buildings from the U.S. Green Building Council, is required for any buildings funded by the city of Chicago, whether public or private. Beyond this mandate, green building is encouraged through the city's Green Permit program. Established in 2005, this program shortens the permitting process to 30 days for developers who pledge to meet sustainability guidelines and to 15 days if they aim for a higher level of certification. The guidelines include meeting a certain level of LEED certification and installing an additional item such as a green roof or on-site alternative energy generation. Builders who fail to achieve the promised level of LEED certification after project completion are required to pay a fine. This program has worked particularly well in terms of encouraging green urbanism beyond city government. As of April 2009, the city of Chicago has more LEED-certified green buildings than any other city in the United States (64) and more registered and under construction than any city except New York (283).

In addition to green buildings, the city has also become a leader in green roofs, with over 300 such buildings covering over 3 million square feet. Any building that receives financial assistance from the city must include a green roof, and even when it is not required, having a green roof can help in getting permits faster. The special landscaping needs of a green roof have created a niche market and encouraged landscaping businesses to develop in this direction.

Greening in Chicago is not limited to buildings but includes infrastructure as well. For example, the city Department of Transportation uses recycled materials such as asphalt and concrete that incorporates fly ash, and Chicago was one of the first major cities to replace traffic lights with high-efficiency light-emitting diode lights. The Green Alley project has the goal of replacing the city's 1,900 miles of alleyways (the most in the country, totaling about 3,500 acres of impermeable surface) with permeable pavement to reduce stormwater runoff and flooding and increase albedo or solar reflectivity. The pavement technology is currently only usable on low-traffic roadways such as alleys, but in the future it may be applied to all city streets. As of the end of 2008, 80 alleys had been replaced with at least partially permeable pavement.

The city's green goals have always included economic as well as environmental components. For example, the DOE has instituted the Greencorps Chicago program to train unemployed residents and former prisoners for green jobs, including landscaping, computer recycling, and weatherizing homes. Around 50 people per year take advantage of this initiative to learn job skills and gain assistance finding employment. The city has also been somewhat successful at attracting green businesses, including two solar companies.

Chicago's greening has not come without criticism. For example, the *Chicago Tribune* found that the city's endeavor to make 20 percent of its electricity from renewable,

Illinois-based sources has only extended to purchasing carbon credits, most of which can be traced to a preexisting wood-burning power plant in North Carolina and thus are not actually offsetting the city's carbon emissions. The city played a role in developing the Chicago Climate Exchange, which was meant to extend Chicago's historical role as a financial center to a carbon-trading market, but it has come under fire for selling dubious offsets and for focusing on making money rather than on reducing global warming. Nevertheless, Chicago has done a remarkable job of promoting environmentally responsible practices in all arenas, from individual households to city government to private companies, and in working to transform the city's image from industrially blighted to clean and green.

See Also: Brownfields; Cities for Climate Protection; City Politics; Green Infrastructure; Green Jobs; Green Roofs; LEED (Leadership in Energy and Environmental Design); Portland, Oregon.

Further Readings

Chamberlain, Lisa. "Mayor Daley's Green Crusade." *Metropolis*, July 2004.
Cronon, William. *Nature's Metropolis: Chicago and the Great West.* New York: W. W. Norton, 1991.
Department of the Environment, City of Chicago. "City of Chicago—Environment." 2009. http://ww.cityofchicago.org/Environment (Accessed August 2009).
Johnston, Sadhu. "The Green Development in Chicago." *Economic Development Journal* (Spring 2005).

Julie Cidell
University of Illinois

Cities for Climate Protection

The Cities for Climate Protection program (CCP) is an intermunicipal network that supports municipally based greenhouse gas (GHG) mitigation measures. Municipalities are said to account for between 40 percent and 75 percent of global anthropogenic GHG emissions. In the context of rapid urbanization, these figures are likely to increase. Despite this, climate change (CC) is commonly defined as a global problem, and political responses (such as the negotiations surrounding the United Nations Kyoto Protocol) have played out largely at the national and international levels. The CCP and other partner organizations have worked to bring CC policy down to the local level, while also lobbying for input into the formation of international climate policy.

The CCP was established in 1993 by the International Council for Local Environmental Initiatives (ICLEI). ICLEI was also instrumental in the creation of Local Agenda 21 and has since been central to the formation of a unified municipal presence at international CC negotiations. Under its guidance, the member cities of the CCP have lobbied to have a place at the table to design the structures that will replace the Kyoto Protocol in 2012. Challenging national governments to do more, they have also set ambitious municipal emissions reduction targets that exceed those so far included in international agreements. The CCP has more than 750 members, who collectively account for over 15 percent of global anthropogenic GHG emissions. The network provides members with technical assistance to understand and reduce those emissions.

The core of the CCP is a five-milestone program that guides cities through the steps of mitigating their emissions. The milestones are to:

1. Create a local emissions inventory

2. Set emissions reduction targets

3. Create a local action plan to meet those targets

4. Implement the plan

5. Monitor performance

This program is supported by a specially designed software system and methodology for carrying out local emissions inventories. The system also helps cities to evaluate emissions reduction measures and their effects in terms of both financial savings and reduced GHG emissions. This quantification enables municipalities to benchmark performance and to prove the cost-effectiveness of expenditures on CC-related efficiency programs. The CCP also maintains an extensive library of municipal best practices and coordinates various types of municipal networking opportunities.

Synergies and Cobenefits

Establishing links between CC policy and other municipal objectives is a key part of the CCP approach. The ability to meet multiple linked goals simultaneously is described both in terms of "synergies" and "cobenefits." Key synergies exist between mitigating municipal emissions and improving local air quality, reducing the urban heat island effect, creating jobs and boosting economic development, creating more effective public transportation systems, reducing traffic congestion, increasing overall quality of life, and reducing municipal costs (through energy savings). In 1999, for example, ICLEI estimates that U.S. CCP members saved $70 million in energy and fuel costs. These synergies act as the gateway to membership for many municipalities and help reframe CC action in a locally meaningful way.

Membership

The network is managed by regional and national organizations in Australia, Canada, Europe, Japan, Latin America, Mexico, New Zealand, South Africa, South Asia, Southeast Asia, and the United States. To become a member, cities must pass a formal declaration or resolution committing the city to addressing global CC by reducing local GHG emissions. They must then also commit to moving through the five milestones of the CCP program.

Performance

Results achieved within the CCP have been mixed. Many recognized leaders in municipal sustainability are members of the network. Portland, Oregon, the only U.S. city to have reduced its emissions to 1990 levels (despite rapid population growth) began its CC program with the CCP's support. Other leading municipal members discussed in this volume include London, England; Seattle, Washington; Barcelona, Spain; Malmö, Sweden; and Copenhagen, Denmark.

Beyond the technical support provided by the CCP, for some members the network has been successful in opening up access to financial and political resources. The network also emphasizes the importance of the added legitimacy that membership gives to municipal employees who are championing local responses to CC. The CCP's approach to CC has won international recognition and credibility. Drawing on this credibility in turn makes it easier for local officials to establish that CC action is a legitimate concern within the context of other established municipal priorities. Particularly for those municipalities most engaged with the network, these political and financial benefits have been found to outweigh the value of access to technical information and expertise. Cities in which CC policy has been more successfully established also value the network as a way of promoting themselves as centers of green innovation and leadership.

In terms of real action, member cities often begin by "cleaning" their own shop. This may include energy efficiency retrofits of municipally owned buildings, the installation of light-emitting diode traffic signals and street lighting, the introduction of hybrid vehicles into municipal fleets, or the capture and use of methane from waste disposal sites to generate electricity. Taking advantage of these readily available and relatively simple opportunities allows cities to lead by example. However, moving from these initial actions to address larger sources of municipal emissions is not easy. The ways that cities manage economic development, coordinate transportation and land use planning, or regulate housing standards have proved much more difficult to change.

Overall, progress through the program's five milestones has generally been limited. Studies critique the CCP's focus on technical support for overlooking the political realities that often block action. Of the 155 municipalities in the Canadian program, for example, only a handful have moved into the implementation or monitoring phase. The situation in the United States is similar. In a published interview, ICLEI program officer Timothy Burroughs explained these shortfalls: "Some cities have started out very gung-ho with a strong mayor with political will at top and strong staff, and maybe a year later there's a big staff turnover and a new mayor, and [CC] isn't a priority anymore—it happens."

The CCP approach is based on the assumption that the main barrier to action is lack of information and technical capacity. Increasingly, however, the importance of the local political context is becoming apparent. Rather than focusing solely on building capacity within a small municipal team (or an individual), researchers have argued for the importance of engaging more broadly with members of the municipality, residents, and business, and using those links to build more general support for mitigation initiatives.

Other important limiting factors to the CCP's success include:

- Difficulty of providing the dedicated staff needed to create a solid institutional home for CC programs, often as a result of limited capacity and finances. This is particularly problematic given that CC is a cross-cutting issue that does not fit easily into any of the predefined municipal departments.
- Difficulty finding sustained funding after the initial phase of the program.
- Centralization of participation with the CCP among a small number of municipal employees, and the resulting limits this places on awareness raising and the growth of support within the municipality as a whole.
- Limits imposed by state, provincial, or national jurisdiction in areas such as building codes or regional transportation strategies.

The challenges that the CCP has had in realizing substantial emissions reductions are not unique. The international climate regime has faced similar problems. The reports published by the International Panel on Climate Change, and the Frameworks and Protocols that are based on them, began with a narrow technical focus. In the face of continued political road-blocks, they have moved to include greater discussions of synergies with other national policy objectives and a more direct engagement with the politics of CC policy. We continue, at all scales, to attempt to figure out the complex politics that turn technical capacity into concrete action.

In the 17 years since the 1992 Río Summit on Environment and Development, there has been a rapid growth of intermunicipal groups aimed at increasing urban sustainability. Today there are over 30 organizations around the world, ranging from the United Cities and Local Governments and the European Union–based Energie-Cités to the World Mayor Council on Climate Change and the C40. Of these, the CCP was the first to directly address CC.

Since then, ICLEI's collaborations with some of these other networks has led to the creation of the World Mayors Declaration on Climate Change, as well as a new Climate Roadmap to guide municipal action on CC into the post-Kyoto period. The CCP continues to provide guidance to individual municipal mitigation efforts. Collectively, the CCP and these other, newer structures also intend to influence the course of international climate policy. It remains to be seen how strong their impact will be on the successor to the Kyoto Protocol.

See Also: Agenda 21; City Politics; Mayors Climate Protection Agreement; Mitigation.

Further Readings

Betsill, M. and H. Bulkely. "Transnational Networks and Global Environmental Governance: The Cities for Climate Protection Program." *International Studies Quarterly*, 48:471–93 (2004).

Bulkely, H. and M. Betsill. *Cities and Climate Change: Urban Sustainability and Global Environmental Governance*. London: Routledge, 2003.

Collier, U. and R. E. Löfstedt. "Think Globally, Act Locally? Local Climate Change and Energy Policies in Sweden and the UK." *Global Environmental Change*, 7:25–40 (1997).

ICLEI. "Climate Programs" (2008). http://www.iclei.org/index.php?id=800 (Accessed September 2009).

ICLEI–Local Governments for Sustainability. "U.S. Cities Acting to Protect the Climate: Achievements of ICLEI's Cities for Climate Protection—U.S. 2000." Berkeley, CA: ICLEI, 2000.

Alex Aylett
University of British Columbia

CITIZEN PARTICIPATION

Since the 1950s, citizen participation programs and initiatives have been undertaken and implemented in many different countries at various levels of government, from the national level to the municipal. This thinking about the involvement of citizens in public affairs took many decades to take form and expression. An important share of these actions came from international development agencies looking for the best ways to implement development projects. The rationale for all these actions is that when citizens become actively involved as participants in their democracy, the governance that emerges from this process will be more democratic and more effective.

Citizen participation plays a critical role in the development of green city policies and is explicitly supported by the United Nations' Local Agenda 21, which calls for citizen participation in local sustainability initiatives. Local environmental organizations, empowered through laws and regulations enabling them to participate in decision making, often take the lead in developing new green initiatives.

In general, citizen participation results in better policies because they more closely reflect citizens' preferences and needs. It also embodies the idea that citizens can help themselves and that they can articulate their own needs and find the solutions to address them. So, instead of passive and mere recipients of governance processes, citizens can be active participants in policy making. It works better for them if done "bottom-up," rather than "top-down." The arguments in support of enhancing citizen participation regularly emphasize the benefits of the process itself. Some authors believe that the participation process is a transformative tool for social change. In addition, citizen involvement is intended to produce better decisions, and thus more efficiency benefits to the rest of society.

Citizen participation suggests that citizens can govern themselves by influencing decision-making processes that affect their lives, their livelihoods, their communities, their environments, and their societies. Citizen participation can encompass the ways and means by which citizens influence and take control over the resources and decisions that directly affect them. It requires methods and mechanisms by which ordinary citizens can effectively influence governments to develop responsive policies and to implement responsive programs and services.

Another approach that is widespread in the literature is the one that presents participation as a right. The right to participate in governance is seen as a premise, rather than a favor granted by government. This right allows citizens to claim other rights and entitlements. In this sense, citizen participation becomes more than a technical fix and, in fact, creates a dynamic in which citizens can engage governments for the benefit of the larger population, which is often excluded from formal political affairs. Direct citizen participation in governance promotes a healthy democracy because it enhances active citizenship and government responsiveness in ways far more effective than the traditional forms of representative democracy.

Effective participation means that citizens deepen involvement to the extent that demands are translated into tangible outputs and outcomes, such as new policies or better service delivery. Participation, therefore, cannot be separated from citizens' engagement with government structures and processes. Sherry Arnstein discussed types of participation and "nonparticipation" in "A Ladder of Citizen Participation" in 1969. She defined citizen participation as the redistribution of power that enables citizens, presently excluded from the political and economic processes, to be deliberately included in the future. Multiple other ladders of participation have been presented by several analysts proposing different kinds of engagement that represent different intensities of participation. These can be summarized as follows:

1. *Consultation:* One of the starting points of participation is consultation. It involves getting the state to listen directly to citizens' needs and demands. It may be done through

different means and mechanisms: consultative meetings, surveys, referenda, or home visits. The state may provide mechanisms and resources for these consultations, or in cases where the state is not predisposed to participatory measures, citizens may assert their right to be heard and claim or create space for participation; for example, through protest or mass mobilization. For consultation to be effective, though, its outputs need to be taken up and listened to by those with the power to act on them. It is, therefore, most effective when done in an interactive manner and in an environment of genuine dialogue and information sharing.

2. *Presence and Representation:* Forms of participation that regularize engagements through institutionalized mechanisms are usually more intensive. This means citizens have ongoing access to decision-making processes and are able to engage beyond a mere sporadic presentation of needs and concerns. At this point, citizens are able to negotiate with government for better plans, solutions, and procedures. In many countries, citizens' groups have become increasingly involved in official procedures of planning, budgeting, and monitoring. With presence and representation, government not only listens to but starts to actually work with citizens.

3. *Influence:* Being consulted and having presence does not necessarily lead to influence. Influence occurs when citizens' demands actually find their way into policies, programs, and service delivery. Influence is visible when government begins to act on such demands and begins producing actual outputs. The challenge for citizens, then, is to remain vigilant so that commitments undertaken by governments are fulfilled and carried out in a transparent manner.

It should be noted that the typology given here does not include an analysis of the most significant roadblocks to achieving genuine levels of participation. These roadblocks lie on both sides of the simplistic view. They include racism, paternalism, and resistance to power redistribution. In addition, political and socioeconomic aspects such as access to education and freedom to mobilize are crucial for successful citizen participation.

The widespread engagement of citizen participation in the governance process creates opportunities for redefining and deepening meanings of democracy, for linking civil society and government reforms in new ways, and for extending the rights of inclusive citizenship. At the same time, there are critical challenges to promoting social justice.

See Also: Agenda 21; Sustainable Development; Transit-Oriented Development.

Further Readings

Arnstein, Sherry R. "A Ladder of Citizen Participation." *Journal of the American Planning Association*, 35/4:216–24 (1967).

Gaventa, John. *Towards Participatory Local Governance: Six Propositions for Discussion.* Brighton, UK: Institute of Development Studies, 2001.

LogoLink—Learning Initiative to Strengthen Citizen Participation on Local Governance. *Participatory Planning Topic Pack.* Brighton, UK: Institute of Development Studies, 2002.

Rafael D'Almeida Martins
University of Campinas

CITY POLITICS

As a growing number of cities around the world embrace the goal of becoming green, the role of city politics in the promotion and shape of this phenomenon is becoming increasingly clear. Supportive political leadership, growing support of city greening by businesses, and community activism and involvement are some of the major local political factors that influence the greening of cities. However, although each of these factors may play a crucial role in the green politics of any given city, there is no clear formula or fixed model that maximizes the chance that a city will adopt strong environmental practices. Rather, the initiative for such programs may come from different coalitions of elected officials, nonprofit organizations, business leaders, and so on. Furthermore, political factors beyond the local scale may also have an important influence on green city politics. These include the national political context of which a city is a part, the economic pressures put on cities by globalization, and the growing importance of transmunicipal networks and global civil society.

Local Influences on Green City Politics

Political Leadership

A city is unlikely to embrace green politics unless there is strong support and leadership coming from the local government. In particular, a number of studies have found the existence of one or two key city officials who are willing to take up the cause of city greening to be a crucial factor in the success of green city policies. This key supporter is often the mayor. Many of the cities that have adopted far-reaching city greening policies are led by strong mayors who have made urban greening a priority. Mayors in cities with a strong-mayor system—a political system in which the mayor is elected separately from the city council and in which he or she has significant independence and legal power in determining the direction of city development and policies—are often able to use this power to direct the city in a greener direction. Strong mayoral leadership is seen by many as key to putting greening on the city's agenda, keeping it there, and making sure that there is adequate follow-through to put adopted environmental policies into practice. In cities such as Chicago, for example, the mayor has been by far the loudest voice calling for and ensuring the implementation of strategies of urban greening and, indeed, has made green city politics a cornerstone of his mayoralty. However, many cities with weak-mayor systems—systems in which the mayor holds a largely symbolic position, if there even is one, and the city council possesses both legislative and executive authority—have also moved toward greening. Yet in such cities, many assert, the importance of leadership remains. In cities without a strong mayor there is often an outspoken and consistent leader within the city council who takes up the cause of promoting and implementing green policies.

Business Support

Another important element of the politics of city greening is the role of local businesses. The growing support of business for such programs is a notable change from previous decades. For many years, particularly in the United States, environmental activists saw the national level as the most useful political scale on which to promote their agenda. The local businesses whose environmental practices activists were often trying to change were seen as too powerful and as wielding too much control over local government to be regulated or transformed by policies initiated at the city level. However, this dynamic has changed over recent years in two profound ways. One has been the weakening of some of the most vocal business resistance to local environment regulation. The second and related dynamic is the active and growing support among emerging types of businesses in postindustrial cities for city environmental policies.

Deindustrialization has transformed the local economies of the global North. Though the loss of heavy manufacturing has in many instances created great social and economic hardships on the local level, one of the benefits of the transformation to a postindustrial economy is the elimination of many of the most polluting industries from major cities. With these industries has gone their powerful political opposition to environmental regulation and controls.

Meanwhile, the new businesses that are emerging tend to be either part of the postindustrial service economy or are directly involved in the broader greening of the economy. These emerging business sectors, therefore, often see city environmental programs as in their best interest and are supportive of city efforts to become greener. For high-end service industries such as finance, providers of legal services, and software development, being located in a green city can be beneficial. Plentiful parks and open space, accessible public transportation, clean air, and so on can help attract and retain the skilled labor that such industries need to flourish. As such, these businesses may join political coalitions in support of such policies. Local tourism can also benefit substantially from a cleaner and greener city. Furthermore, one of the growing sectors of the economies in many regions is green technology, such as alternative energy technologies. Local governments' efforts to attract such industries to their cities is an increasingly important part of the politics of city greening. Third, the growing acceptance of green-collar jobs as an important element of economic revitalization, particularly in deindustrializing cities, is broadening support for the argument that economic growth and environmental protection are not only compatible but can be mutually beneficial. Finally, as environmentalism gains increasing support from all sectors of society, many businesses strive to be seen by their customers as socially responsible and concerned about environmental issues, increasing their willingness to actively support these programs, or at least to temper their resistance to them. In some cities, chambers of commerce and other business groups have become outspoken leaders in promoting city greening, and growing numbers of businesses are becoming important members of local environmental coalitions.

Some scholars have gone so far as to claim that the new and growing support for city environmentalism by local politicians and business people can be seen as a shift away from the growth regimes that have long been theorized as dominating local politics and toward a newly emerging phenomenon of sustainability regimes. According to much urban political theory, urban growth regimes are the formal or informal coalitions of developers, local businesses, and city politicians who have historically dominated city politics. For their own financial benefit, these coalitions have promoted city growth—particularly new business and residential developments—above all else. According to this perspective, attempts at city greening, which often include restricting or regulating the kinds of growth and new developments that can happen in a city, will run up against resistance from the dominant urban growth regimes. However, many scholars are now challenging this assumption, arguing that this is changing as local businesses begin to see the financial benefits of controlled growth and city politicians gain increasing political legitimacy from their support of environmental issues. To the extent that sustainability regimes are able to replace urban growth regimes as the dominant force in local politics, the possibility for far-reaching and long-lasting city environmental programs seems great.

Local Activism and Nongovernmental Organizations

Local activist organizations and nongovernmental organizations are another critical part of green city politics. In some cities, such organizations may be explicitly incorporated into policy making and implementation efforts. In others, they may act in more of a lobbying and advocacy position, putting pressure on local leaders to support city greening.

Public participation is seen by many academics and policy makers as instrumental to effective and long-lasting urban greening. The United Nations' Local Agenda 21, for example, explicitly calls for public input and participation in the development of local sustainability initiatives. The rationale is that such a process will increase community acceptance of and commitment to the environmental goals and plans established by the city, thereby increasing the likelihood that the goals will be achieved. In some cities it is local nonprofit organizations that are responsible for implementing many of the elements of city environmental programs, such as providing training for green-collar jobs programs or organizing local Earth Day events.

In addition to helping implement and providing legitimacy for city environmental programs, local environmental activism and political pressure on a city government can play an important role in encouraging city officials to take action to address issues of sustainability and urban greening in the first place. Organizations promoting the expansion and preservation of parks and open space, wildlife conservation groups, urban gardeners, and local chapters of nationwide environmental conservation organizations, among others, are often quite influential and can run effective local advocacy campaigns to put and keep urban greening on the city's political agenda.

Another kind of important and often-influential kind of local-level environmental activism has been that of environmental justice. Environmental justice advocates point to the relationship between race, class, and exposure to pollution and highlight the maldistribution of environmental "bads" and the disproportionate burden of the dirty side of industrial society borne by poor people and people of color. Increasingly, they are also pointing to the need for an equitable distribution of environmental "goods," such as clean water and air, adequate housing, and access to parks and open space. Very much grounded in particular places, such as a specific neighborhood or part of a city, and often responding to a particular environmental threat such as a landfill or polluting power plant, environmental justice activists can bring powerful and specific demands for environmental cleanup and remediation into city politics. Though there has historically been some degree of tension between environmental justice organizations and conservation-oriented environmental activists, coalitions between the two are increasingly being built, leading to ever more effective and inclusive political demands for green cities.

Overall, one of the notable characteristics of the emerging phenomenon of urban greening initiatives is the growing importance of local organizations. Both in their adoption and implementation, many cities' governments are striving to include as wide a variety of civil society and community organizations into the process as possible, and a growing number of environmental activists are prioritizing the promotion of local-level policy changes and pushing their city governments to take action on local and global environmental issues. Many feel that this trend has the potential to increase substantially the effectiveness and longevity of policies of city greening.

National and Global Influences on Green City Politics

Though elements of local politics such as city leadership, business support, and local environmental activism are clearly very important in promoting and implementing policies of city greening, city environmental politics is also influenced by factors beyond the boundaries of the city itself. These include the type of national political system of which the city is a part, the legal powers available to local governments, the influence of transmunicipal environmental networks and global civil society, and the growing necessity for cities to be competitive players in the global economy.

Green Cities and Democracy

Some scholars have examined the relationship between green city politics and the kind of political system that leads to such programs. One common perspective sees democratic governments as the most likely to adopt measures to become more sustainable. There are a number of reasons given for this. First of all, in a nondemocratic political system, elites can isolate themselves from environmental problems and from public pressure to address these problems more easily than can leaders who are directly accountable to a constituency. The lack of accountability and reprimand for inaction that elections provide, this argument asserts, makes governing elites less likely to engage in the often challenging processes of environmental reform. Second, some assert that the complexity of environmental issues is more effectively addressed by the inclusive political process available under democratic systems. This is because the social conflict that is allowed to flourish under democracy facilitates the recognition of problems and the creative identification of possible solutions.

Though a compelling argument can be made for democratic systems creating space in which green city politics can emerge, there are significant counterexamples that need to be recognized. Curitiba, Brazil, for example, offers a significant challenge to this claim. Curitiba is seen around the world as an exemplar of a sustainable city, yet many of its renowned environmental programs were adopted under a mayor appointed by a military dictatorship. China, too, may be offering a counter to the claim of the necessary connection between democratic politics and city greening as the country draws up plans to build green cities from the ground up to meet the needs of its growing population and economy.

National Politics and Local Political Power

The politics of greening in a city may also be strongly influenced by the broader national or regional legal context of which the city is a part. This can serve either to stimulate local environmentalism or limit what policies cities are able to enact. In Great Britain, for example, national mandates require that sustainability be incorporated into local and regional planning decisions. These requirements have presumably encouraged cities that otherwise may not have done so to take action on environmental issues to secure needed funding. However, it is important to note that even in those places where city governments have been legally required to make moves toward greening, local leaders often maintain significant leeway in determining the direction and shape of these efforts, thereby reinserting local level politics as an important factor in their adoption and successful implementation.

In contrast to cities such as those in Britain, the sustainability policies of which have been encouraged by national mandates, in other places cities have acted without such requirements in place. In the United States, for example, particularly for the first decade of the 21st century, city governments enacted environmental programs in the face of overt federal hostility toward environmental regulation. Particularly on the issue of climate change, many local leaders asserted that they were adopting far-reaching greenhouse gas reduction goals explicitly because of federal failure to act on this issue.

In contrast, the broader political contexts of which cities are a part may limit the powers local governments have to enact policies that promote city greening, regardless of the local support and political will to do so. National and state laws may limit the legal powers cities have to determine zoning and growth boundaries, regulate public transportation, stop pollution, and so on. For example, with the exception of London, British cities have very limited control over their bus systems, which were privatized under the Thatcher government. Local leaders, therefore, often cannot easily adjust public transportation services to those that would be most environmentally beneficial if the needed routes are not also the most profitable for the private firms running the bus routes. Studies have cited bus privatization as leading to a notable decrease in bus ridership across the United Kingdom over the past two decades. Another common limitation that cities face is a lack of control over the source of the energy used within the city. Though a city may be able to reduce overall energy consumption by providing residents with incentives for conservation, without controlling its actual energy generation, a city may struggle to reach greenhouse gas reduction goals or other goals that require a shift to cleaner energy sources.

Green Cities and Globalization

In addition to examining the local politics of city greening and recognizing the influence of national level policies on city programs, many claim that the breadth of the phenomenon of urban greening and the often-similar methods used by local governments around the world that are trying to make their cities greener suggests that global factors may also be influencing local environmental politics. Two particularly important global influences on city greening that many scholars highlight are economic globalization and the growing importance of transmunicipal organizations and global civil society.

According to some, globalization has put increasing financial pressure on cities. This pressure has come both from national governments' reducing funding for social welfare and other programs that benefit cities and their residents and from the increasing mobility of financial capital, forcing cities to compete ever more fiercely to attract new businesses and investment. As discussed earlier, a green city can offer the amenities and quality of life that skilled workers and high-end service industries are looking for. As such, many city leaders have found that greening can be one effective strategy for successfully competing with other localities to attract the businesses and skilled labor necessary to support and sustain local economic growth. Cities such as Barcelona and Chicago, for example, have both been quite successful at using their urban greening programs to reinvent their images and their economies to attract tourists, businesses involved in the global knowledge economy, and skilled labor. In these cities, as well as many others around the world, environmental efforts have been explicitly linked to economic goals. As such, some claim that city greening may be as much a marketing and economic tool used to succeed in the globalized economy as it is an indication of any deep commitment to environmentalism on the part of city leaders.

Other scholars who highlight the importance of global forces on local environmental politics, in contrast, argue that it is not the economic incentives of urban greening that are most relevant but, rather, the growing importance and influence of global transmunicipal networks and global civil society. According to these scholars, transmunicipal networks—networks of cities working together, often across national borders—are promoting global, interurban cooperation by providing forums through which municipal governments can share policy tools, technical knowledge, and discursive understandings of environmental problems and appropriate solutions. For example, Cities for Climate Protection, a main project of the organization International Council for Local Environmental Initiatives, or ICLEI, the C40 Cities Climate Leadership Group, and the U.S. Conference of Mayors Climate Protection Agreement provide local leaders with information on what their counterparts around the world are doing to make their cities more green, can help support these local efforts, and serve to spread the idea of city greening around the world. The rise of these networks is given as evidence of a new kind of global cooperation in which subnational governments such as cities are becoming increasingly important to international environmental politics. Similarly, theorists of global civil society point to the growing connections between local environmental organizations located in different cities around the world. The connections between these organizations can, similar to transmunicipal networks, serve as a way to share ideas for greening cities and to share political strategies on how to get there. Global civil society may therefore serve an important role in spreading and strengthening the local politics of city greening.

Conclusion

Examining the influence of both local and nonlocal factors on city environmentalism is necessary for an adequate understanding of green city politics. Though the support of a strong mayor or other powerful local politician may make city greening much more likely to be successfully adopted, even the most powerful mayor is limited by the broader legal framework and economic pressures in which she or he must operate. In contrast, local actors maintain significant agency in determining how they will respond to the challenges and opportunities presented by forces such as globalization. Though no one formula of successful green city politics exists, it is clear that city greening is not merely a technical problem. Rather, it is a highly political process involving many different elements of local society and influenced by factors well beyond the city limits.

See Also: Agenda 21; Cities for Climate Protection; Citizen Participation; Environmental Justice; Green Jobs; Mayors Climate Protection Agreement.

Further Readings

Bulkeley, Harriet and Michele M. Betsill. *Cities and Climate Change: Urban Sustainability and Global Environmental Governance*. London: Routledge, 2003.

Gibbs, David, et al. "Changing Governance Structures and the Environment: Economy-Environment Relations at the Local and Regional Scales." *Journal of Environmental Policy & Planning*, 4 (2002).

Keil, Roger. "World City Formation, Local Politics, and Sustainability." In *Local Places in the Age of the Global City*, R. Keil, et al., eds. Montreal: Black Rose Books, 1996.

Krueger, Robert and David Gibbs. *The Sustainable Development Paradox: Urban Political Economy in the United States and Europe*. New York: Guilford, 2007.

Lake, Robert W. "Contradictions at the Local Scale: Local Implementation of Agenda 21 in the USA." In *Consuming Cities: The Urban Environment in the Global Economy After the Río Declaration*, N. Low, et al., eds. London: Routledge, 2000.

Moore, Steven A. *Alternative Routes to the Sustainable City: Austin, Curitiba, and Frankfurt.* Lanham, MD: Lexington Books, 2007.

Portney, Kent E. *Taking Sustainable Cities Seriously: Economic Development, the Environment, and Quality of Life in American Cities.* Cambridge, MA: MIT Press, 2003.

Raco, Mike. "Sustainable Development, Rolled-Out Neoliberalism and Sustainable Communities." *Antipode*, 37/2 (2005).

Corina McKendry
University of California, Santa Cruz

CIVIC SPACE

Civic spaces, sometimes also referred to as "public places" or "civic centers," are areas of land characterized by the fact that they are managed and (entirely or partially) funded by local municipal, state, and federal governments. Examples include city parks, public libraries, national historic landmarks, and museums. Some broader definitions of civic spaces also encompass government administrative buildings such as post offices, courthouses, and city halls. Typically, however, the designation as a *civic space* implies places that inspire and nurture community interaction. These spaces are intended for cultural programming, protest, relaxation, celebration, and a broad range of community-oriented activities and gatherings. The prevalence and centrality of civic spaces have evolved throughout human history. In the Western world, Greek and Roman city planners first conceived of the notion of a central administrative complex that was the forerunner to today's civic spaces.

In the United States, civic spaces have undergone dramatic shifts over the past century. After World War II, as increasing numbers of urban dwellers moved outside the city center, publicly funded civic spaces were being abandoned in favor of private and semiprivate suburban spaces such as malls and movie theater complexes. Such privately owned public spaces (e.g., malls) offered air conditioning, security, and ample parking and were dominated by commercial and retail activity. Over the past decade, however, as more people sought to return to higher-density living, civic spaces like greenways, parks, and museums have experienced a rebirth. Municipal leaders are also encouraging the return to downtown living and often seek to portray their city's or state's civic space as an emblem for their community. More recently, civic spaces are being renovated and built anew using sustainable green building principles. Civic leaders aim to portray these new sustainable, green civic spaces as economically sound investments and progressive symbols of their communities. In 2009, the American Recovery and Reinvestment Act was signed into law, providing billions of dollars for creating and renovating civic spaces using green building principles. This transformative piece of legislation will provide public funding and investment in creating sustainable civic spaces for the 21st century.

The Evolution of Civic Spaces

Virtually every society in the world has included some form of civic space as part of its community. The history of civic spaces in the United States and much of the Western world is rooted in the urban planning strategies of Greece and Rome. Specifically, the notion of civic spaces as being central to the geography and cultural life of a city is descended from the Greek concept of an acropolis and the Roman use of a forum. The Greeks and Romans extended their civic centers to be all-encompassing spaces for government-related administration and commerce. In modern communities these spaces also often contribute to a city's identity. Clear examples include the Golden Gate Bridge in San Francisco, Grant Park in Chicago, Atlanta's Centennial Olympic Park, and the Boston Public Library. However, rural areas and smaller cities also highlight their civic spaces as the iconic image for their city. Examples in smaller communities include civic spaces such as the World's Fair Park in Knoxville, Tennessee, and the Clinton Presidential Library in Little Rock, Arkansas.

Civic spaces can enhance community livability in a variety of unique ways. They enrich the lives of citizens by providing a place to serve as a community hub where people can interact, share common experiences, exercise and walk, and generally increase the city's social and cultural capital. Many civic spaces exist in the city's traditional center (e.g., the National Mall in Washington, D.C., or Central Park in New York City), though suburbs and exurbs also include civic spaces within their own population centers.

The notion of civic spaces and, by extension, urban centers as the geographical and cultural centers of community life has been eroded in the United States. The collective effect of assembly-line automobile production, expansion of the interstate highway system, and affordable suburban housing made available after World War II resulted in large numbers of middle-class families moving into suburban communities. In 1961, Jane Jacobs published *The Death and Life of Great American Cities* and decried the "endless new developments" reducing urban centers to a "monotonous, unnourishing gruel." Her work had a profound effect on unfreezing prevailing "decentrist" notions of planned communities as "islands" protected from the chaos and terrors of urban life. The book served as a manifesto for proponents of the New Urbanism movement, which argues that the problem was not the idea of the city but, rather, the city planners who had not accounted for all the cultural, social, environmental, and psychological factors that are equally important in helping to renew urban life. Central to Jacobs's ideas about connecting communities with their cities is the notion of trust. Jacobs writes that trust is a necessary element for citizens of any neighborhood to feel safe. She argues that civic spaces (and public spaces) are precisely the places where such trust is formed. The small interactions of everyday life, whether in a community garden, along the sidewalk, in a museum, or on a park bench, combine to create this necessary trust and social capital.

In recent years, many municipalities have placed a special emphasis on reviving decaying urban centers. As suburbanization has drawn communities away from downtown areas, these abandoned urban cavities have often become sites for violent and regressive social problems (gangs, drugs, and alienation). Public officials and leaders are recognizing the importance of civic spaces in sustaining healthy communities and culturally vibrant cities. As a result, a remarkable urban renewal has been taking place across the United States over the past decade. Much of the emphasis has been on civic spaces such as parks, museums, and libraries. In Chicago, for example, entire economies have been generated from renovated civic spaces such as Grant Park, Millennium Park, and the downtown Chicago Public Library. As cities reinvest in their downtown civic spaces, the hope is that more people will choose to live in, work in, play in, and ultimately help renew blighted urban centers.

The phenomenon of urban renewal is a response to the negative externalities of sprawl and excessive suburbanization. As increasing numbers choose to live near their places of employment, shopping districts, and public transit, there is less pressure on cities and states to expand highways. Another positive externality is lower fossil fuel emissions and resultant smog problems. The notion of "smart growth" has clear links with the renewal of down-town civic spaces, which result in a long-term effect on automobile dependence, community health, and expansion of infrastructure.

Sustainable Civic Space

In addition to the sociocultural, psychological, and health benefits of successful civic spaces, these places positively affect the environment of metropolitan areas. Many civic spaces are "green" expanses along riverfronts or parks in concentrated urban areas (i.e., urban forests). These ecosystems provide a rich habitat for a wide variety of species, thereby increasing biodiversity in areas that are often under severe industrial and environmental pressures. For instance, Prospect Park in Brooklyn, New York, is home to hundreds of species of birds, plants, trees, and animals. Numerous city- and state-owned botanical gardens serve their communities in the roles of education, cultivation, and preservation. These green oases also cleanse the air of fossil fuel by-products as well as lower the tem-perature (heat island effect) of areas dominated by asphalt and steel.

Civic leaders are beginning to recognize that investing in green civic spaces possesses both environmental and economic value. Many 21st-century civic center building projects are guided by principles of sustainability such as "new urbanism" and "smart growth." "Greening" civic and public spaces is a key element of the smart growth approach. Large sustainable civic spaces present a progressive image of their city or state to the world. For example, the Downtown Civic Space Park in downtown Phoenix, Arizona, is highlighted as a civic space incorporating many of the sustainable design principles advocated by the U.S. Green Building Council's Leadership in Energy and Environmental Design (LEED) Rating System. This project includes solar panels, rainwater harvesting, extensive shade to reduce the "heat island" effect, and no parking spaces to encourage pedestrian traffic. The LEED Rating System is often cited by government leaders as a useful metric for determin-ing the "greenness" of their building or renovation projects. The smart growth planning philosophy, coupled with the LEED Rating System, drives much of the strategic building of modern civic spaces and urban development.

The Environmental Protection Agency's brownfields project highlights another useful pathway for creating sustainable civic spaces. Rather than developing open land (or "green spaces"), the reclamation and reuse of urban industrial sites such as brownfields can pro-vide city managers and planners with the dual benefit of sustainable site reuse as well as tax-related incentives. The Hank Aaron State Trail in Milwaukee, Wisconsin, is a good example of a vibrant civic space created from a reclaimed brownfield. The brownfield redevelopment project was funded by a partnership between many stakeholders including city, county, private investors, and nonprofit organizations.

In February 2009, President Barack Obama signed into law the American Recovery and Reinvestment Act, providing billions of dollars of economic stimulus funding for public works projects across the nation. Many of the projects being funded by the American Recovery and Reinvestment Act are civic spaces that are being renovated using green building principles. Specifically, the Energy Efficiency and Conservation Block Grant pro-gram awards funding to local, state, and federal projects that meet specified criteria, including energy cost savings, installation of renewable energy capacity, and effect on reduction of greenhouse gases. Many civic spaces are being considered for funding by this federal program, and the act will likely have a transformative effect on reinvesting public funds for a new generation of sustainable civic spaces.

See Also: Community Gardens; Compact Development (New Urbanism); Healthy Cities; Parks, Greenways, and Open Space; Smart Growth.

Further Readings

Benfield, F. Kaid, et al. *Solving Sprawl: Models of Smart Growth in Communities Across America.* Washington, D.C.: Island Press, 2001.
Hester, Randolph T. *Design for Ecological Democracy.* Cambridge, MA: MIT Press, 2006.
Jacobs, Jane. *The Death and Life of Great American Cities.* New York: Modern Library, 1993.
U.S. General Services Administration. "Achieving Great Federal Public Spaces." http://www .pps.org/great_public_spaces (Accessed April 2009).

Ameet D. Doshi
Georgia Perimeter College

Coastal Zone Management

The aim of coastal zone management is to minimize the harmful impacts of human activities on coastal zones. Here, dredge spoils from Baltimore Harbor are contributing to the regeneration of Hart-Miller Island in the Chesapeake Bay.

Source: Mary Hollinger/NOAA

Most civilizations have settled along the edge of rivers, lakes, seas, or oceans, as water is important for the very survival of human civilization, including transportation of goods, and is still of great importance today. Officially the coastal zone is the "interface where the land meets the ocean, encompassing shoreline environments as well as adjacent coastal waters. Its components can include river deltas, coastal plains, wetlands, beaches and dunes, reefs, mangrove forests, lagoons and other coastal features." Approximately 40 percent of the world's population currently lives within 62 miles of a coastline.

Because of the concentration of population living along rivers and oceans, these coastal zones have become highly polluted from human waste, industrial waste, and agricultural waste such as pesticide runoff. The contaminants include even hazardous materials such as persistent organic pollutants. Coastal zones have been also heavily exploited because of their rich resources. This has an effect on coastal marine life and the aquatic ecosystem in general, leading to severely degraded fisheries, drained wetlands, bleached or dynamited coral reefs, and destroyed beaches.

Coastal areas, such as wetlands, offer considerable environmental services, including retention of sediments and nutrients, barriers against hurricanes and surges (reducing flood risk), and protecting coasts from global climate change (especially sea level rise). Since much of urban development is taking place along the coasts, it impacts the coastal ecosystems by polluting the coastal waters (waste water from industries, human waste). Land development also leads to coastal erosion (an ever increasing urban population is leading to more land encroachment and in some cases land reclamation), beach deterioration, a loss of biodiversity, and in some cases losses of habitats of some species, introduction of invasive species into marine systems, and pollution of aquifers. This leads to even more complex problems, such as a decrease in groundwater, depletion of fisheries (overfishing in urban areas), increase of diseases, the loss of biodiversity or some species. This can lead to a decrease in tourism, impacting the economy of the city, and in some cases increasing the risk to flooding. Large urban cities near lakes (e.g., the Great Lakes in North America) face additional threats, including eutrophication, introduction of invasive species, and water pollution.

To control this pollution, *coastal zone management* (CZM) is needed, better known as Integrated Coastal Zone Management. CZM usually refers to all forms of management of human effects on these coastal ocean environments and their flora and fauna. These techniques include fisheries management, conservation management, management of water quality, and management of coastal development. The earliest form of CZM can be seen in the form of coastal engineering in harbors with infrastructure such as docks, breakwaters, and levies to protect the coastal areas (with a main focus on human, rather than ecosystem, protection).

Integrated CZM (ICZM) is a "process of governance and consists of legal and institutional framework necessary to ensure that development and management plans for costal zones are integrated with environmental (including social) goals and are made with the participation of those affected." The main aim of ICZM is to maximize the benefits provided by the coastal areas and to minimize the harmful effects of activities (especially human activities) on the coastal zones and the resources it provides. ICZM starts with setting the objectives for the development and management of the coastal zone involving all the stakeholders. As compared with CZM, ICZM is a more holistic approach, which is analytical, dynamic, and a continuous process of administering the use, development, and protection of the coastal zone and its resources toward democratically agreed objectives. The term *integrated* in ICZM refers to the integration of objectives and to the integration of the many instruments needed to meet these objectives. It means integration of all relevant policy areas, sectors, and levels of administration, as well as integration of the terrestrial and marine components of the target territory in both time and space.

Historically, ICZM started in 1972, when it was first undertaken in the United States with the enactment of the Coastal Zone Management Act by the U.S. Congress. Other countries have now followed suit. It is also required under some international agreements such as Agenda 21, chapter 17, which calls for a prominent role of ICZM in ocean management. The International Panel on Climate Change and the Intergovernmental Negotiating Committee on Global Climate change have also endorsed ICZM to address the effect of climate change on coastal zones.

ICZM focuses on three main operational objectives:

1. strengthening sectoral management, through training or legislation;

2. preserving and protecting the productivity and biological diversity of coastal ecosystems through prevention of habitat destruction, pollution, and overexploitation; and

3. promoting rational development and sustainable use of coastal resources.

As described by the European Union, ICZM is a multi-tiered strategy. However, to implement ICZM, it is important to understand the functioning of the coastal system. This usually requires an understanding and study of certain processes across disciplines—for example, relevant processes in geomorphology, ecology, coastal land use, and the socioeconomical profile of the region—but more important, of the interlinkages between these processes within the coastal zone. At this time, both the understanding of integrated functioning of the coastal system and knowledge of the true integrated approach are lacking. This adds uncertainty to the planning and implementation of ICZM. Another problem in implementing ICZM in most developing countries is the lack of capacity and resources (financial logistic or intellectual).

CZM can be done at local levels, national levels, or regional levels. For the program to be really successful, an effort is needed at a regional level that involves all the coastal countries in the region. Some regions have gone in for joint regional programs to protect coastal zones. For example, the convention for the Protection of the Marine Environment and the Coastal Region of the Mediterranean (also known as Barcelona Convention) is a major milestone for the promotion of environmental protection in general, and coastal zone protection in particular, as well as in integrating the regional policies. The European Commission and European Union Mediterranean Members States are contracting parties to this convention.

There are at least three other articles of legislation for CZM: the Coastal Zone Management Act, MPA Executive Order, and Coral Reef Conservation Act.

See Also: Agenda 21; Ports; Water Conservation.

Further Readings

Otter, Henriette S. and Michele Capobianco. "Uncertainty in Integrated Coastal Zone Management." *Journal of Coastal Conservation*, 6/1:23–32 (2000).
TVLink Europe. http://www.tvlink.org/vnr.cfm?vidlD=53 (Accessed October 2009).

Velma I. Grover
United Nations University–Institute for Water,
Environment and Health (UNU–IWEH)

COMBINED HEAT AND POWER (COGENERATION)

Conventional electric power generation is only about 40 percent efficient—it leaves large amounts of waste heat. *Cogeneration* (also called combined heat and power, or CHP) is the use of this waste heat for heating buildings, providing hot water, and applying to industrial uses. The CHP cycle is about 80 percent efficient, obtaining twice as much energy from the same amount of fuel as conventional electrical generation. In the short to mid-term, cogeneration promises to save significant amounts of fossil fuels. It is a transition technology, making better use of fossil fuel sources that provide very intense heat, but eventually it will become unsustainable as fossil fuel stocks (oil, gas, and coal) decline.

The concept of cogeneration is more familiar than most people think, for every automobile is also a cogeneration plant. An automobile's engine generates motive power for the wheels, but in the process it generates a lot of waste heat. Some of that heat runs the car's heater, warming the occupants of the car in cold weather without using extra energy. There are now schemes planned to generate the electricity needed for the car's accessories from waste engine heat. This would be an example of trigeneration: first is the motive power for the wheels, second is the heat for the passengers, and third is the electricity generated. The car still needs a radiator, however, because the automobile's occupants do not always want extra heat, and the engine usually generates more heat than they can use in any case.

The first commercial electrical generation plant, built by Thomas Edison in New York City, was more efficient than most modern power plants. Edison realized that the waste heat from his steam boilers could be sold profitably to nearby buildings to provide heat and hot water after the high-pressure steam had lost its power but was still very hot. This is the normal application of CHP.

During the course of the 20th century, however, this mode of electrical generation fell out of favor. Rising electrical demand drove utilities to build ever-larger power plants that could not be located in cities because many of them were fueled by coal. Later, regulation of utility rates gave power companies little incentive to operate more efficiently. Usually they were (and are) allowed to charge consumers for the cost of generation and transmission plus a reasonable profit, regardless of what those costs were. As a result, cogeneration stagnated, and utilities found ways to dissipate "waste" heat from electrical generation into rivers, lakes, and oceans. Cooling towers were developed that took in prodigious amounts of water, evaporating about half of it and returning the rest, hot and polluted, to the river or lake from which it came. Urban process heat was generated cheaply and conveniently from fossil fuels in furnaces without the bother of tie-ins to outside heat sources. Labor and maintenance costs were lower per kilowatt-hour produced because a large plant requires only a little more care than a small one.

However, the era of practicality of such inefficient power generation is coming to an end with the approaching decline in the availability of fossil fuels.

In the coming years, however, cogeneration promises to extend scarce fuel supplies because it recovers most of the waste energy from electricity generation for other uses. It takes advantage of the fact that different energy applications have different requirements. Conventional electrical generation powered by fossil fuels or nuclear reactors requires very high temperatures—the higher the better—to run steam turbines or gas turbines. Many other energy needs, however, require less intense heat. Domestic hot water, for instance, needs a heat source only slightly hotter than the water being provided, about 50 degrees Celsius (120 degrees Fahrenheit). Space heating requires a source only slightly hotter than room temperature. In practice, working temperatures range from 80 degrees Celsius to 130 degrees Celsius. Thus, the hot steam coming out of a power-generating turbine, at low pressure but still at a temperature adequate for these uses, represents a "free" source of energy that is otherwise wasted. In fact, in the absence of cogeneration, a power plant is forced to consume energy to cool its water. The same applies to gas turbine electrical generation plants, in which the waste energy appears as the hot combustion gases exiting the turbine.

Some cogeneration opportunities, however, will persist even in a renewable-energy environment. A West Virginia factory, for example, makes high-grade silicon for computer chips in five giant electric arc furnaces, heating quartz to 1500 degrees Celsius. The process consumes more than 120 megawatts of electric power—about one-eighth of a fossil-fuel power plant. The factory operators now plan to use the exhaust from their furnaces, still at 800 degrees Celsius, to generate 40 megawatts of electric power. If the electricity going into the factory were from renewable sources such as wind, hydro, or solar instead of fossil fuels, cogeneration would still be a viable process.

Because the waste heat from electrical power production in a cogeneration plant must be used locally, these plants are usually designed to be smaller than conventional power plants. Their size varies a thousandfold, however, from small units generating as little as 350 kilowatts and servicing a single large building to regional units providing as much as 350 megawatts—about a third as large as a conventional fossil fuel plant. Electricity can be moved over moderate distances in high-voltage wires with little loss, but process heat must flow through large pipes to its end users. Even with good insulation, losses are tolerable only within a range of a few kilometers (1 km = 0.62 miles), and moving steam or hot water for those distances consumes additional power. For these reasons, cogeneration plants are sized to meet the process heat needs of nearby users, usually based on an electrical capacity of a few dozen megawatts or less.

Limitations of CHP

Many of the limitations of cogeneration are not in their engineering but in political and legal arrangements. Because heat and power are generated together, it is usually easier if both kinds of energy go to a single set of users. Here there is a differentiation between arrangements common in Europe and those common in North America. Regulations in many parts of Europe encourage cogeneration, for instance, by requiring owners of buildings near a cogeneration plant to heat their buildings with the plant's waste heat. Furnaces and hot water heaters are not installed in such buildings; only heat exchangers are needed to link regional steam pipes with a building's utility connections. Often the cogeneration facilities are owned by governments or by government partnerships. Cogeneration is more practical in Europe's denser cities, where most people live in apartments rather than single-family dwellings, so that a large number of waste-heat users live within range of a single cogeneration plant. As a result, a significant portion of European electrical power is made with cogeneration.

In North America, and especially in the United States, conditions are generally less conducive to cogeneration, though regulations are changing. Cities are less dense, limiting the number of potential consumers of process heat within a practical distance of a power plant, and private ownership limits cooperation between power generators and nearby potential users. Siting power plants in urban areas is more difficult because of objections from local residents. As a result, cogeneration has a smaller role in the United States than in Europe, providing about 7 percent of electric power, and much cogeneration is in large institutions such as universities and large industrial complexes, where electricity and process heat are consumed by the same user. Such siting tends to minimize "not in my backyard" problems, because those generating the power are also those benefiting from it.

An exception is Manhattan, in which the tradition of cogeneration that began with Edison's first power plant continues, with a number of plants and extensive use of the waste heat for building heat and hot water. Cogeneration is especially practical in dense Manhattan, with the plants disguised among other large buildings. Condensers powered by waste heat also provide air conditioning—an example of trigeneration (electricity, heating, and cooling).

Some U.S. cogeneration plants use unconventional power sources to save on investment. A cogeneration plant at the University of California, Santa Cruz is based on a 2,400-horsepower marine diesel engine the size of a double-decker bus. It runs on natural gas, with a small amount of added diesel oil. The engine is only 38 percent efficient, but waste heat is recovered from both the engine's cooling water and the exhaust gases to provide heat and hot water for many of the campus buildings. Such installations can be used where the need is too small to justify a turbine.

Taking Advantage

In addition to the conventional cogeneration scheme of electricity and building heat, some installations exploit more novel arrangements. A large conventional power plant in Moss Landing, California, generates heat to serve a large array of greenhouses and fish farms in its agriculture-intensive region. Because the power plant was not designed for these uses, however, and is located far from urban areas and generating more than a thousand megawatts of electrical power, only a small fraction of its waste heat can be used.

Another novel cogeneration scheme with promise for the future is in combined waste disposal, electrical generation, materials recovery, and process heat. Two huge installations in Munich, Germany, one to the north of the city center and one to the south, burn all of the household and industrial waste from the city and its immediate surroundings. The plants are not isolated from the city but sit among factories and apartment blocks. Each installation takes in several tons of waste each hour, delivered by garbage trucks and rail cars; the waste stream is first heated and dried with some of the waste heat from the combustion process. Then it is burned at very high temperatures with the addition of a small amount of natural gas. Air pumped into the bottom of the combustion bed guarantees a hot and complete burn. The heat generated runs steam turbines to generate urban electricity, and heat exiting the turbines is piped to the surrounding neighborhoods to provide building heat and hot water. Metals, mostly aluminum and iron, are drained from the hot slag exiting the combustion chambers, separated by their melting points. Glass is also recovered, and all of the recovered products are recycled and sold. A small amount of ash remains. The combustion is so hot that organic garbage, plastics, and chemicals in the waste stream are completely oxidized to carbon dioxide and water. The smokestack effluent consists mostly of carbon dioxide and water vapor; there is no odor in the surrounding neighborhoods. The only materials that cannot be recycled are batteries, and it is illegal to dispose of batteries in household trash in Munich.

Such installations are expensive to install, but they bear little resemblance to the smelly, smoldering incinerators of bygone decades. They pay for themselves by sale of heat, power, and recycled materials, and by disposing of waste without the need for large landfills.

The Future of CHP

Prospects for cogeneration are bright for the short- to mid-term. Fossil fuel costs are certain to rise in the future, increasing incentives for improved efficiency. Further, the reduced greenhouse gas emissions compared with separate power and heating installations will fight global warming and may attract government incentives. Because the electric power is used locally, losses from long-distance transmission and passage through repeated transformers are minimized. Installation of plants is becoming easier, with prefabricated turnkey units available from several manufacturers. A cogeneration plant is just as complex as a large fossil fuel electrical generator, or more complex with the added need to regulate process heat, but automation of many control processes is making it easier to control and administer the facilities. Modern electronics facilitate the tie-in of local cogeneration plants with commercial electricity from large utilities.

See Also: Energy Efficiency; Landfills; NIMBY; Power Grids.

Further Readings

Bluestein, Joel, et al. "Combined Heat and Power White Paper." Western Governors' Association, January 2006. http://www.aceee.org/chp/wga.pdf (Accessed August 2009).

Casten, Thomas and Philip Schewe. "Getting the Most from Energy." *American Scientist*, 97 (January 2009).

Engle, David. "The Gridwise Future." *Distributed Energy* (January–February 2009). http://www.distributedenergy.com/january-february-2009/the-gridwise-future.aspx (Accessed August 2009).

Lindley, David. "The Energy Should Always Work Twice." *Nature*, 458:138–41 (March 2009).

Bruce Bridgeman
University of California, Santa Cruz

COMBINED SEWER OVERFLOW

Combined sewer (CS) systems are integrated underground sewerage systems that carry both domestic and industrial sewage and periodic stormwater flows. Most of these systems were built in the 19th and early 20th centuries, when population densities were lower and extreme weather events less frequent. In American cities, where such infrastructure was often financed through public bond issues, the relative cheapness of building CS systems also worked to favor their proliferation. Most U.S. and European jurisdictions now prohibit the construction of new CS systems, and various strategies are being implemented to make them fit for use in the 21st century. Climate change, the inexorable aging of the underground infrastructure, and continued urban expansion are likely to negatively affect CS systems, increasing the frequency and intensity of CS overflows.

During normal weather conditions such systems function well, with both sewerage and surface runoff being directed through the network to wastewater treatment facilities. Combined sewer overflow (CSO) events occur when the volume of stormwater is greatly in excess of even the combined flow capacity, causing overflows of natural watercourses up through manholes in city streets or, in the worst cases, back up through domestic drains and sumps. CSO events can also occur as a result of network blockages or sudden system inflows.

CSO discharges can cause

- visual pollution resulting from associated litter deposition (e.g., toilet paper, tampons, etc.);
- increased health risks for recreational users, especially if appreciable volumes of raw sewage are involved; and
- damage to the ecology of the river, especially reductions in dissolved oxygen, which can in turn causes fish kills.

Bacterial, viral, and protozoal contamination risks are also associated with CSOs, according to many studies. Persistent pollutants associated with road runoff and industrial effluent such as heavy metals, polycyclic aromatic hydrocarbons, and so on, are also associated with CSOs.

It is important to realize that CSO-related pollution is not usually a function of system failure but, rather, a function of the increasing frequency of high precipitation events causing the CS systems to function exactly as designed and the rising frequency of blockages caused by rising domestic and industrial waste volumes. Flushing of solids such as sanitary napkins and cooking fat also contributes to network blockages to such an extent that many water authorities in the United States and the United Kingdom have mounted public education campaigns on the issue.

Around one-fifth of the U.S. population and more than 750 communities in the northeast, the Ohio and Mississippi River Valleys, and the Pacific northwest are served by CS systems and therefore are at risk of experiencing CSO events. In 2001, a U.S. Environmental Protection Agency review revealed 859 active CSO permits covering 9,463 permitted CSO outfalls in 32 states nationwide. New York City itself has a large CS system (more than 450 outflows) and averages more than 50 CSO events each year, discharging more than 27 billion gallons (710,500,000 m^3) of mixed runoff and sewage into natural watercourses. For such cities, compliance with the 1977 Clean Water Act can be problematic, as the act renders compliance with discharge limits during storm events difficult. Mitigation measures to reduce the frequency and magnitude of CSO events are discussed below.

In Britain, the first major CS system was constructed for London by Metropolitan Board of Works Engineer Joseph Bazalgette between 1858 and 1865 and became a model for similar systems around the world. The technology was, in fact, inspired by the "Great Stink" of 1858. There are around 11,000 CS systems in the United Kingdom, of which approximately 7,000 have been upgraded over the past decade at a cost of around £3.5 billion. Notwithstanding this investment, London still experiences more than 50 CSO events each year, discharging more than 20 million m^3 (5.3 billion gallons) into the Thames River, and there remain over 13,000 permitted CSOs in operation in the United Kingdom.

Overall, in England and Wales one in three beaches—35 percent of 587 beaches—fails to achieve statutory swimming water quality levels, partly as a function of CSO events, and a very high-profile and successful environmental nongovernmental organization, Surfers Against Sewage, was founded in Cornwall in 1990. As a result of better regulation and the efforts of communities, water companies, and nongovernmental organizations, coastal and riparian water quality have been greatly improved in recent years.

In the United Kingdom, CSO overflow events do not necessarily constitute a breach of water quality legislation, as long as average discharges are within agreed (or statutory) tolerances and the CSO events are deemed "exceptional."

Where CSO is deemed a problem requiring mitigation, a number of measures may be undertaken, ranging from the renovation of combined systems into physically separated systems. Unsurprisingly, this is an enormously expensive and disruptive process, and cost alone has caused many water authorities to opt against it. A different approach involves the creation of storage facilities to hold excess stormwater during heavy precipitation events, which is then gradually pumped back into the otherwise unmodified sewerage network once the storm has passed. Another approach involves focusing on the overflow mechanisms themselves, particularly with a view to incorporating some sort of screening structure operated hydrodynamically or using water vortices. The U.K. Water Industry Research Centre (UKWIR) test facility at the Hoscar Sewage Treatment Works in Wigan, Lancashire, has conducted detailed research in the hydrodynamic modeling of CSO flows and the effectiveness of various screening and pumping technologies for reducing the negative effects associated with CSOs.

Other mitigation measures, more orientated toward the sources of CSO events, involve the reduction or diversion of surface runoff through so-called sustainable urban drainage schemes (SUDS), or the reduction of the amount of domestic or industrial sewerage produced in the first place. Thus, for example, systematic replacement of impermeable land coverings in urban areas can reduce inflow to CS systems by promoting infiltration into subsoil and, ultimately, into groundwater. SUDS schemes tend to work best where they are built alongside larger institutions, such as universities, hospitals, and factories. The illustration shows the SUDS scheme in place on the main campus of the University of the West of England in Bristol. On a smaller scale, "green streets" and "street trees" programs can make for a significant reduction in stormwater inflow to sewer networks if they are sufficiently well developed.

See Also: Infrastructure; Stormwater Management; Water Treatment.

Further Readings

Burian, S. J., et al. "Historical Development of Wet-Weather Flow Management." *Journal of Water Resources Planning and Management,* 125/1:3–13 (1999).

Foundation for Water Research (FWR). *UPM Urban Pollution Management Manual,* 2nd Ed. FR/CL0009. Marlow, UK: FWR, 1998.

Graham, S. and S. Marvin. *Splintering Urbanism: Networked Infrastructures, Technological Mobilites and the Urban Condition.* London: Routledge, 2001.

Moffa, P. E. *Control and Treatment of Combined-Sewer Overflows.* New York: Van Nostrand Reinhold, 1997.

Praeger, D. *Poop Culture: How America Is Shaped by Its Grossest National Product.* Los Angeles: Feral House, 2007.

Staddon, C. *Managing Europe's Water Resources: 21st Century Challenges.* Aldershot: Ashgate, 2009.

U.S. Environmental Protection Agency. "Report to Congress: Impacts and Control of CSOs and SSOs." EPA-833-R-04-001. Washington, D.C.: U.S. Environmental Protection Agency, 2004.

U.S. Environmental Protection Agency. "Report to Congress: Implementation and Enforcement of the CSO Control Policy." EPA 833-R-01-003. Washington, D.C.: U.S. Environmental Protection Agency, 2001.

Chad Staddon
University of the West of England, Bristol

Community Gardens

The designation of a community garden includes any urban green space cultivated by a group of people, like this in Melbourne, Australia.

Source: Dahliyani Briedis/Wikipedia

Community gardens are any garden cultivated by a group of people and are the primary space of urban agriculture in the United States. Historically, community gardens emerged in response to political economic crises in U.S. cities. Today, community garden participation is augmented by an interest in organic, local, and "slow" food; an increase in environmental concern; and urban struggles for public space.

Historical and Geographical Context

Community gardens have an extensive history in the United States that is widely understood as seven separate but overlapping "movements" that emerged to try to address specific social problems: Potato Patches (1894–1917), School Gardens (1900–1920), Garden City Plots (1905–1910), Liberty Gardens (1917–1920), Relief Gardens (1930–1939), Victory Gardens (1941–1945), and Community Gardens (1970–present). The first organized community garden program emerged in Detroit in 1893 as a response to urban hunger, when the City of Detroit supported food production through land provisioning. Detroit's Potato Patches promoted self-sufficiency and provided food for hungry urbanites in the midst of economic crisis. In the early 1900s, School Gardens helped children adjust to urban life through a connection to nature. The Garden City Plots of the early 20th century beautified cities, and during the world wars, Liberty Gardens, and later Victory Gardens, provided an outlet for, and promoted the display of, patriotism. Materially, these garden movements emphasized subsistence production to free up resources that were needed for the war effort. Between the two world wars, Relief Gardens helped alleviate urban hunger created by the Great Depression. Contemporary community gardening builds on previous rationales for gardening, including economic necessity and urban disinvestment, but it is also driven by environmental critiques of conventional agriculture.

The Neoliberal Garden

Although the last 30-plus years of community gardening are viewed as one movement, traced back to 1970s environmentalism, it is analytically useful to think of the contemporary community gardening movement as decidedly neoliberal, placed within the current political economic context that emphasizes market supremacy above all else. Community gardens often fulfill various duties—for example, poverty alleviation and food security—formerly met by the state, and provide a sort of buffer against the unpredictability of the market.

New York City provides an example of community garden struggles that are characteristic of the neoliberal era. In 1999, Mayor Rudolf Giuliani exemplified the neoliberal insistence that property ownership be secured via market exchange by placing 114 community gardens on the auction block. The Giuliani administration framed this prioritization of private property over public space as an issue of rights to affordable housing, thereby pitting two common allies—housing advocates and community gardeners—against one another. Gardeners challenged Giuliani through a variety of tactics, including an insistence on their right to cultivate the city. Land trusts eventually purchased the gardens at auction, a move made possible by the effective mobilization of garden advocates to connect their struggle to broader urban politics as well as the fund-raising support provided by singer-actress Bette Midler. Many gardens thus became quasi-public property, ushering in a new landscape of gardening with still-unknown consequences.

Looking Ahead

Today, community gardening in the United States is experiencing a resurgence not seen since the Victory Gardens of World War II. In February 2009, the new secretary of the U.S. Department of Agriculture Tom Vilsack "broke pavement" on a "People's Garden," replacing a parking lot in Washington with a community garden to celebrate Abraham Lincoln's 200th birthday. Shortly thereafter, First Lady Michelle Obama planted an organic kitchen garden on the White House's South Lawn.

Although renewed interest in community gardening accompanies an interest in food politics (e.g., local and organic), community gardening also transforms growing swaths of vacant and public land. These contemporary gardens assume a variety of forms—from guerilla gardening on public land to formal not-for-profit organizations with large budgets and professional staff.

Community gardens often serve as an important tool in addressing urban hunger. Community gardens are employed as a tool for urban beautification, for the creation of green space, and for the provision of recreation and educational opportunities. Gardens also boost local property values and stimulate economic development.

These benefits notwithstanding, some challenges threaten the long-term viability of community gardening. Land tenure remains a primary concern for community gardening, as the recent history of New York City illustrates. By prioritizing private property, neoliberalism effectively undermines community garden efforts frequently located on public property. Furthermore, urban planners often view community gardens as temporary, rather than identifying them as vital parts of the urban landscape. Ecological constraints, such as urban air and water pollution, water shortages, soil contamination, and pests and diseases, also serve as barriers to community gardening. All of these challenges, however, have not hampered the growth of community gardening, and recent trends in the United States provide hope for the future of the movement.

See Also: Civic Space; Food Security; New York City, New York; Parks, Greenways, and Open Space; Urban Agriculture.

Further Readings

Allen, Patricia. *Together at the Table: Sustainability and Sustenance in the American Agrifood System*. University Park: Pennsylvania State University Press, 2004.

Bassett, Thomas. "Reaping on the Margins: A Century of Community Gardening in America." *Landscape*, 25:1–8 (1981).

Lawson, Laura. *City Bountiful: A Century of Community Gardening in America*. Berkeley: University of California Press, 2005.

Smith, Christopher and Hilda Kurtz. "Community Gardens and Politics of Scale in New York City." *Geographical Review*, 93:193–212 (2003).

Staeheli, Lynn, et al. "Conflicting Rights to the City in New York's Community Gardens." *GeoJournal*, 58:197–205 (2002).

Evan Weissman
Syracuse University

COMMUTING

Commuting is the practice of traveling between one's home and one's place of work. In this respect, it is a relatively recent phenomenon. However, it is also one that causes considerable environmental problems, as a large volume of traffic is generated at peak times by this activity. Because commuting increasingly involves the use of cars, it is considered one of our least sustainable practices. In fact, it has been calculated that just one U.S. citizen's average commuting practices represent the equivalent of half the total environmental footprint of the average citizen of India. The solution is often sought in more intensive use of cars through high-occupancy vehicle lanes or carpooling, as well as mass transit or more benign private modes, such as walking and cycling. Such options enable more commuters to be accommodated in the same amount of urban space. Changes in the built environment, particularly denser urban environments, have also been put forward as a solution.

New York City's subway system is so extensive and efficient that many New Yorkers do not have to commute by car. Shown here is one of the hubs of the system, Grand Central Terminal.

Source: iStockphoto

History

In prehistoric societies, when humans roamed over a particular territory for their hunting and gathering activities, there was no fixed place of work, nor often a fixed place for domestic activities. With the agricultural revolution came the rise of towns, and some separation between domestic and work activities began to emerge, at least for some activities. Agricultural workers living in fortified towns or villages might have to travel outside the town gates to the fields, though many villagers in more secure environments lived, and often still live, on the land they farmed. Commuting in its modern form did not really appear until the Industrial Revolution. Industrialization and the mining that supplied its key raw materials—iron and coal—concentrated the means of production in centralized locations and initially, at least, used labor-intensive processes, which employed large numbers of workers. Some early mills, such as the model community of New Lanark in Scotland, provided housing and all other facilities for their workers around the factory. As factories grew in size and number, such arrangements became more difficult to implement, and many mill owners did not regard it as their responsibility to feed, house, or educate their workers. Gradually the need for workers to travel longer distances developed. This was further boosted by the introduction of new public transport modes that made it possible for people to travel farther to their place of work. Horse-drawn omnibuses and horse-drawn trams/streetcars were gradually complemented with steam-powered light rail systems, and steam and internal combustion engines also allowed the development of more powerful buses.

The railway never replaced the horse; instead, it stimulated the growth in the horse economy. However, attempts to stretch the horse system to its limits rapidly led to widespread protest about the treatment of horses among the influential middle classes. In several cities, bus and tram travelers were requested not to make the vehicle stop more often than absolutely necessary to spare the horses the tiring, and therefore damaging, stop–start phase. The car lobby was often in the forefront of these horse protection movements. Horse-drawn vehicles were also increasingly blamed for the growing congestion problem in Europe's larger cities. The car seemed a better alternative—it not only provided quick and efficient transport but also provided the motorized commuter with door-to-door flexibility. And it was more modern. The problems with this solution were not really recognized until nearly everybody started commuting by car.

Suburbanization

Suburbs, which were a deliberate separation of the home from the city, where most places of work were located, were made possible by these new transport developments, and many suburbs initially developed along public transport routes. The development of the bicycle, at around the same time as the internal combustion engine, changed this pattern somewhat, as commuters now had greater freedom in terms of routes, and until the 1960s, shift changes at factories in many industrialized countries generated considerable flows of bicycles, which had by then become one of the most popular commuting modes. However, this dominant role was already being eroded by powered two-wheelers, and particularly by the automobile. Car commuting gradually spread from the United States to other industrialized countries, and today it is spreading across Asia. However, Asia has in many locations also recognized the advantages of the bicycle. Many Chinese cities that introduced antibike policies are busy reinstating the bicycle as a useful commuting mode.

Solutions

Commuting has become a phenomenon of all industrialized and industrializing societies, such that it is now often the principal cause of traffic congestion in many urban areas that are largely uncongested at other times of the day. To limit the effect of the motorized commuter, a number of strategies have been used around the world. Some U.S. cities were among the first to introduce high-occupancy vehicle lanes, whereby one lane of the highway on popular commuter routes was allocated for the exclusive use of cars with two or more people on board. Some Asian cities, such as Jakarta, Indonesia, also adopted the system, even restricting access to the city center at peak times to cars with more than one person. Carpooling is another solution in which a number of people share a vehicle. The system was actively promoted by 3M in the 1970s. Their city center location in Minneapolis, Minnesota, meant they could not expand, and with a growing number of car commuters, they ran out of parking space. In response, the company bought a number of minivans for the use of those employees who were willing to pick up a number of colleagues on their way to work. Instead of having to provide parking for each person's car, only one carpool vehicle had to be accommodated. Other examples exist of employers themselves actively trying to reduce the burden imposed by their employees' commuting.

There have also been many attempts to get commuters out of their car and onto other transport modes. Mass transit is the most popular option for policy makers, whereby dedicated infrastructure-intensive rail systems can often act as a showcase for governments to demonstrate they are tackling the problem. Some, such as the Bay Area Rapid Transit system in the San Francisco Bay Area, have become very effective. New York's subway, too, has created a situation in which not only do many New Yorkers not commute by car, many do not own a car. This pattern is repeated in many large cities in Europe and Asia with efficient underground and light surface rail systems. However, a much lower cost alternative is the bus. To avoid these getting stuck among cars in congested rush hour traffic, many are now used in guided bus systems, or bus rapid transit systems, whereby a narrow route, often with guiderails, is built on which the buses can operate separated from other traffic. Many cities in China are adopting such systems as a low-cost alternative to light rail systems.

An interesting and often-cited example is that of Curitiba in Brazil. Here, in what is to some extent a linear city, a major commuting through-route runs the length of the city. The route is served by a large number of buses that have dedicated access to the central route, with special bus stops allowing quick entry and exit. The system is so efficient that few commuters would consider using a car. In Japan, in contrast, bus services became increasingly inefficient and stuck in congested peak-hour traffic. The often-narrow suburban streets make separate bus lanes difficult. Here commuters adopted the bicycle, which they use to ride to the nearest suburban train station, from where they take one of the many efficient suburban trains into the city center and their place of work. The bicycle has also become increasingly popular with German commuters, and commuting by bicycle has always remained popular in Denmark and the Netherlands. The bicycle is still a private mode and thus provides the commuter with some of the flexibility also provided by the car, in that it can often be used door-to-door. Denmark and France have in recent years introduced public bicycle schemes whereby bikes can be rented from dedicated sites for a nominal fee and returned to a similar site. These have become increasingly popular, with the most recent example being Paris, where the city's Velib scheme has exceeded all expectations in terms of commuter use.

Denser cities are another answer to the problem. Given that commuting is caused by the separation of home and work, bringing the two closer together would reduce the distance over which commuters need to travel, thus making more benign modes such as walking or cycling more viable. Homeworking is another possibility: If a person works from home half the time, the impact of his or her journey to work is also halved. Modern information and communication technology has made this increasingly feasible for many workers, although clearly this solution is not possible for all occupations—many have to be in a particular location to do their job.

To some extent, traffic congestion is a side effect of a dynamic economic and business environment. However, commuting has added an extra burden on the transport system that can also stifle economic development, or at least add to its cost. Rethinking people's need to commute and the necessity to make their commute as efficient as possible is therefore of economic, as well as social and environmental, concern.

See Also: Bicycling; Bus Rapid Transit; Carpooling; Congestion Pricing; Personal Rapid Transit.

Further Readings

Roseland, M. *Toward Sustainable Communities: Resources for Citizens and Their Governments*. Gabriola Island, British Columbia, Canada: New Society, 1998.
Schilperoord, P. *Future Tech: Innovations in Transportation*. London: Black Dog, 2006.
Toth, G. "Reducing the Growth in Vehicle Miles Travelled: Can We Really Pull It Off?" In *Driving Climate Change: Cutting Carbon From Transportation*, D. Sperling and J. Cannon, eds. Burlington, MA: Academic, 2007.

Paul Nieuwenhuis
Cardiff University

COMPACT DEVELOPMENT (NEW URBANISM)

Compact development emphasizes density of residences and businesses. It has become a commonly proposed solution to the urban sprawl and suburbanization that has marked Western development in the postwar era and to the rapid loss of open space and farmland that has accompanied this phenomenon. Compact development is seen to offer environmental and social benefits over the resource intensity, auto dependency, and social isolation of standard suburban development. New Urbanism, a more specific kind of development model that can be understood partly as a subset of compact development, is also credited with offering such benefits. However, many critiques exist as to the actual social and environmental sustainability of both compact development and New Urbanism in practice.

The most basic definition of a compact city is one that has high population density. Beyond this, compact development is generally assumed to include mixed land use (where businesses and residences are located side by side or in close proximity to each other), extensively developed urban infrastructure, multimodal transportation, increased social and economic interactions, and clearly demarcated limits to existing and future urban growth. Promoting higher density through compact urban development has become a policy of the European Union and is one of the principles of the United Nations Agenda 21. Though less widely embraced in the United States, it has made some notable inroads there, as well through the adoption of urban growth management programs and growth boundaries in numerous cities across the country.

Benefits of Compact Development

The proclaimed benefits of compact development are numerous. One is the preservation of open space and farmland from the encroachment of urbanization. Another advantage is that it can reduce automobile dependence by making homes, shops, and places of employment all more easily accessible by foot, bicycle, or public transportation. Third, compact development is seen as reducing per capita resource use, both through reduced automobile dependence and by offering smaller homes and yards than are found in the typical suburban home. In addition to its environmental benefits, compact development is claimed to foster economic benefits by creating revitalized urban economies and by making more efficient provision of infrastructure. Finally, it is seen as improving quality of life by creating more vibrant and safer urban areas; providing urban residents with attractive, convenient places to shop and recreate near where they work and live; improving civic culture by promoting social and economic interactions between people; and increasing social equity.

However, recent research has complicated these widely held assumptions about the benefits of the compact city. For example, although compact cities have been found to increase opportunities for public transportation and to reduce the use of greenfields for new development, these benefits may not be as straightforward as is often asserted. First of all, taken alone, increased urban density has been found to have a much smaller-than-expected effect on single-occupancy vehicle use. Though much of the data are inconclusive, some studies have shown that an individual's income affects vehicle use more significantly than the density of the area in which they live. A second common critique is that compact development often does not live up to its goal of improving social equity and may actually exacerbate inequity, depending on the specifics of the compact form. Both of these critiques raise important challenges for proponents of compact development, many of whom admit that they need to examine the outcomes of existing efforts of compact urban development more carefully and incorporate these findings into future designs.

Two other dilemmas facing advocates of compact urban development are how to successfully blend industrial and residential uses and how to improve people's perception of the desirability of living in a compact city. Much of the initial impetus for suburbanization came from the desire to escape the contamination and pollution of industrial cities. As reurbanization and compact development leads cities to become denser, it becomes harder to separate industrial activities from people's living and work spaces. Even in largely deindustrialized cities, the legacy of pollution left from now-defunct industrial activity can create environmental health concerns regarding the reuse of former industrial land. Finding ways to make industrial and residential land uses compatible is one of the major challenges of compact urban development. Resident perception of the quality and acceptability of compact development has also emerged as an unforeseen challenge to this form of urbanization. Perceived deficiencies include a lack of greenery and open space, a lack of privacy, and a greater exposure to pollution than in suburban areas. It is these deficiencies, critics assert, that maintain popular demand for low-density, suburban areas and that need to be addressed if compact development is to reach its social and environmental potential.

New Urbanism

One of the most visible and arguably influential U.S. groups calling for compact urban development in a way that attempts to address some of the above shortcomings is New Urbanism. In particular, New Urbanists insist that the social and environmental benefits of compact development emerge from careful attention to the particular form that compact development takes and will not be realized merely through greater density.

New Urbanism is built on a critique of the alienating effects of the monotony of urban sprawl and an effort to articulate a new vision for development that can promote community, environmental preservation, social diversity, affordability, public transit use, and walkability. New Urbanist development focuses on redesigning urban neighborhoods and existing towns, as well as designing and building new neighborhoods that facilitate these goals. Principles of New Urbanist development include mixed-use development, building housing of a variety of types and prices in close proximity to promote socioeconomic diversity and equity, plentiful public space, a clear sense of center and edge of a development, access to public transportation, and a pedestrian-oriented area of a quarter to half-mile radius at the core of the development. Theoretically, New Urbanism focuses on the power of design and the built environment to improve the social environment of a community.

New Urbanism is more formalized and institutionalized than the broader movement to promote compact development, of which it can be seen as a part. The organization Congress for the New Urbanism was founded in 1993 and has grown to include several thousand members. Though in many ways an architectural movement, New Urbanism has built alliances with and includes professionals from a variety of fields including planners, environmentalists, developers, politicians, and engineers. Principles of New Urbanism are clarified and articulated through regular congress meetings. New Urbanists claim that they have created a coherent theory of urbanism, as well as the accompanying tools and techniques to help realize the conceptual framework they have established. Yet its proponents also assert that one of the movement's greatest accomplishments has not been the establishment of fixed principles of development and design but, rather, the creation of an arena for the ongoing development of good ideas and their implementation.

Most of the explicitly New Urbanist development projects built to date have been new greenfield developments, rather than urban redevelopments. These New Urbanist developments are generally more compact than conventional suburbs, though sometimes only marginally so. The architectural style frequently adopted by New Urbanists is neotraditional, harking back to premodern town designs, and particularly those of New England.

In addition to designing new developments and neighborhoods, policy change is a major focus of New Urbanist activity. For example, because of the importance of mixed use neighborhoods to New Urbanism, many of its proponents work to change zoning laws, such as those mandating single-use areas, that restrict the development of communities along New Urbanist ideals.

Though explicitly New Urbanist projects remain a small minority of new developments, proponents claim that the principles of New Urbanism have had a far greater effect than the relatively small number of projects completed under its banner suggests. In the last several years, many of the principles of New Urbanism have been adopted by the U.S. Environmental Protection Agency and by the Department of Housing and Urban Development. These principles have been reflected in the major role played by New Urbanist philosophy in the redevelopment of public housing in the United States in the late 1990s and early 2000s. In notable contrast to the public housing of earlier decades, such projects now strive for mixed-income housing that does not allow for a concentration of poverty. Another example of the influence of New Urbanism is the role played by members of Congress for the New Urbanism in cooperation with the U.S. Green Building Council and the Natural Resources Defense Council to develop the new Leadership in Energy and Environmental Design certification category for neighborhood development.

Critiques of New Urbanism

However, a number of critiques of New Urbanism have emerged. One of the major critiques is of its social and cultural exclusivity, despite New Urbanism's stated goal of promoting diversity. Part of this critique is of the neotraditional architectural design that many projects adopt. Critics assert that this type of development reflects a particular and culturally limited version of Anglo-American history and nostalgia for community, therefore restricting its resonance among diverse communities. Even more concretely, critics note that in the context of urban redevelopment projects, a side effect of mixed housing is the displacement of poor people as affordable units are replaced by higher-end complexes. Indeed, it is broadly accepted by critics and proponents of New Urbanism alike that social equity and mixing has been one of the goals of New Urbanism that has been most difficult to achieve.

A second critique emerges from the fact that the primary focus of explicitly New Urbanist projects has been suburban areas, not urban, and that as such, these New Urbanist greenfield developments contribute to the loss of open space that they are supposedly trying to prevent. New Urbanism's defenders counter that the most visible projects have been new developments because it is in new developments that the most thorough imagining can take place. Furthermore, as new developments will continue to be built, it is better that they are built along New Urbanist principles, they argue, than using the standard formula of suburban development. Its proponents do admit, however, that they need to be better at making New Urbanism's relevance for downtown and inner city revitalization clear if they are to truly gain broad-scale relevance.

A third critique is that in the midst of ongoing standard suburban development, New Urbanist projects are too limited to have a significant effect. Designating areas for mixed use, for example, in no way ensures that the mix will work or that the businesses in the area will be viable, particularly if other commercial alternatives are available nearby. Similarly, it is problematic, and generally self-defeating, to just have one New Urbanist subdivision among many other standard developments. In such a situation most people will still rely on cars for the majority of their trips, and walkability does not mean much if it is not integrated with enough people to keep a town center and its businesses vibrant, thereby giving people somewhere to walk to. Though important, rather than a dismissal of New Urbanism in general, this critique can be seen as a call for more development to be done along New Urbanist lines.

Overall, compact development and the New Urbanism have the potential to improve resource efficiency, reduce pollution and single-occupancy vehicle use, protect open space, and stimulate local economic development. It is less clear whether they can address issues of social equity and the promotion of diversity. Proponents of compact development and New Urbanism still have significant work to do for the vision and potential of these ideas to be fully realized.

See Also: Brownfields; Density; Green Communities and Neighborhood Planning; Greenfield Sites; Greening Suburbia; Infill Development; Smart Growth; Transit-Oriented Development.

Further Readings

de Roo, Gert and Donald Miller. *Compact Cities and Sustainable Urban Development: A Critical Assessment of Policies and Plans From an International Perspective.* Burlington, VT: Ashgate, 2000.

Dutton, John A. *New American Urbanism: Reforming the Suburban Metropolis.* Milan: Skira, 2000.

Haas, Tigran, ed. *New Urbanism and Beyond: Designing Cities for the Future.* New York: Rizzoli, 2008.

Kelbaugh, Douglas S. *Common Place: Toward Neighborhood and Regional Design.* Seattle: University of Washington Press, 1997.

Neuman, Michael. "The Compact City Fallacy." *Journal of Planning Education and Research,* 25 (2005).

Williams, Katie, et al., eds. *Achieving Sustainable Urban Form.* London: Spon Press, 2000.

Corina McKendry
University of California, Santa Cruz

COMPOSTING

Composting is a natural process in which organic matter like plant remains, leaves, grass clippings, and fruit and vegetable waste decomposes into an earthy substance that can be used as garden soil or as a natural fertilizer.

Source: iStockphoto

In recent years, our society has developed a hierarchy of waste management. At the top of the hierarchy of integrated waste management, and the most desirable form, is source reduction. This includes the reduction of waste generated through reuse of materials and using reduced levels of packing. The second tier is recycling, including composting. Composting is the aerobic decomposition of organic matter into soil conditioner, and is considered the highest level of recycling. Composting occurs when an organic discarded material can be used to improve soil conditions and plant growth, and reduce the potential of erosion, runoff, and nonsource pollution. A considerable portion of domestic waste is open to composting, such as, for instance, yard waste and biosolids that are organic and easily composted, according to E. Epstein.

Composting has predominantly been used in horticulture and agriculture to improve soil conditions and enhance plant growth, although its utilization depends on product quality and consistency. It has also been used to biologically convert putrescible organisms into a stabilized form to destroy organisms pathogenic to humans. Composting is also capable of destroying plant diseases, weed seeds, and insects and their eggs.

Historical Background

There is scientific evidence that composting has been practiced since ancient times by tribes in Israel, Greece, and in Ancient Rome. The early civilizations of South America, China, Japan, and India also used animal and human waste as fertilizer for intensive agricultural practices. There are also many references to composting in medieval and renaissance literature, which were researched by R. T. Haug. For at least 2,000 years, people have depended on composting to sustain croplands and to feed their civilizations. In fact, it was not until the 19th century that society began to substitute compost with chemical fertilizers.

The history of modern composting began with Sir Albert Howard, a British government agronomist who spent the years of 1905–1934 in India, where he began to recognize was fertile soil contains a high percentage of humus. While working at the Indore Institute of Plant Industry, he developed a composting technique that is considered the first organized plan for composting in modern times.

Defining Composting

Even though societies have been deploying composting techniques for thousands of years, there is no universally agreed definition of composting. Usually it is broadly defined as the biological decomposition and stabilization of organic substrates under controlled, aerobic conditions into a humus-like stable product that is free of viable human and plant pathogens and plant seeds, that does not attract insects or vectors, that can be handled and stored without nuisance, and that is beneficial to the growth of plants.

This biological process occurs when the microbial organisms break down organic matter into carbon dioxide (CO_2), water, minerals, and organic material. The major factors affecting the decomposition of organic matter by microorganisms are oxygen, temperature, and moisture. Different types of aerobic bacteria develop in compost material, depending on the temperature of the compost pile. Psychrophilic bacteria exist at the lowest temperatures (between 55–70 degrees Fahrenheit). Mesophilic bacteria operate when temperatures are between 70–100 degrees Fahrenheit, and produce acids, heat, and CO_2 as by-products. At temperatures higher than 100 degrees Fahrenheit, thermophilic bacteria are activated, and bring the compost temperature up to a range of about 115–160 degrees Fahrenheit. These different types of bacteria will become dominant or dormant as the temperature of the compost pile rises and falls. The CO_2 and water enter the atmosphere, and the minerals and organic material transform into a soil-like substance termed *compost*.

Composting Objectives

Building upon this definition, some major objectives of composting are, according to E. Epstein,

- to decompose organic matter into a stable state and produce a material that may be used for soil improvement or other beneficial uses;
- to decompose waste into a beneficial product that is economically favorable as compared to alternative disposal costs and may be more environmentally acceptable than more conventional solid waste management; and
- to disinfect pathogenically infected organic wastes so that they can be used in a safer manner.

Organic composts have a number of beneficial purposes: When applied to the land, it serves as a source of organic material for maintaining or building supplies of soil humus necessary for proper soil structure and moisture holding capacity. It can also improve the growth and health of crops in intensive commercial agriculture, in home gardens, and in urban garden plots.

Composting has also gained considerable attention as a solution to the municipal solid waste problem in urban areas, and large-scale composting has grown into an important industry since the 1990s. In 2008, for instance, the city of Boston, Massachusetts, proposed building a multimillion-dollar indoor composting center, with the goal of providing energy for 1,500 homes. According to M. Renkow and A. R. Rubin, municipal solid waste composting can be an alternative to the disposal of significant components of the waste stream in sanitary landfills, and cost analyses show that it is a feasible alternative for many geographical areas.

See Also: Adaptive Reuse; Citizen Participation; Density; Green Infrastructure; Green Landscaping; Landfills; Recycling in Cities; Urban Agriculture; Waste Disposal.

Further Readings

Epstein, E. *The Science of Composting*. Boca Raton, FL: CRC Press, 1997.

Haug, R. T. *The Practical Handbook of Compost Engineering*. Boca Raton, FL: CRC Press, 1993.

Lowenfels, Jeff and Wayne Lewis. *Teaming With Microbes: A Gardener's Guide to the Soil Food Web*. Portland, OR: Timber Press, 2006.

Martin, D. L. and G. Gershuny, eds. *The Rodale Book of Composting*. Emmaus, PA: Rodale Press, 1992.

Navarro, Mireya. "Urban Composting: A New Can of Worms." *The New York Times* (February 18, 2009). http://www.nytimes.com/2009/02/19/garden/19worms.html (Accessed December 2009).

Nickisch, Curt. "Boston Wants to Harness Composting Energy." *All Things Considered*, National Public Radio. March 25, 2008. http://www.npr.org/templates/story/story.php?storyId=88163285 (Accessed December 2009).

Renkow, M. and A. R. Rubin. "Does Municipal Solid Waste Composting Make Economic Sense?" *Journal of Environmental Management*, 53:339–47, 1998.

Rafael D'Almeida Martins
University of Campinas

Congestion Pricing

Congestion pricing is a market-based policy instrument that, at its most basic, requires road users to pay for using a stretch or road or for passing through a designated charging zone. Congestion pricing may also be referred to as road pricing or congestion charging. The terminology may differ according to the country using the term and may also be influenced by factors such as public acceptability and the aims of the pricing scheme in question.

Traffic in central London, where congestion pricing has been in effect for a number of years, fell by 21 percent 2002–2006, and an estimated 70,000 fewer vehicles are on the streets each day.

Source: iStockphoto

There are a number of reasons why congestion pricing might be introduced: first, for environmental reasons—to reduce the emissions of greenhouse gases, or local air pollution; second, to ease the traffic flow, especially in cases where congestion is considered to be having an adverse effect on the local economy; third, to pay for improvements to the transport infrastructure; and fourth, to raise revenue for other purposes. Most commonly, congestion pricing is introduced for environmental or traffic flow reasons or a combination of the two. Where the aim of a pricing scheme is to reduce traffic levels, the charge is often higher than if it were simply aimed at raising revenue. This is to provide a disincentive to car users, working on an "unwillingness to pay" principle.

There are several high-profile examples of congestion pricing within cities, most notably Singapore, with the Electronic Road Pricing; Stockholm, with the Stockholm Congestion Tax; and London, with the Congestion Charge. There are a number of ways in which charges can be applied. In all three examples, road users pay as they enter the designated charging zone. In the cases of Singapore and Stockholm, a charge is made every time the road user crosses the zone boundary, whereas in the case of London, a flat daily rate is charged.

Criticisms of Congestion Pricing

Congestion pricing is criticized for its unequal effects on different sections of society, especially where there is a distinction between willingness and ability to pay. The extent of this depends on the setting and level of the charge and on the nature of the pricing zone. Arguably, where an entire area carries a charge, those living or working within it or on the boundaries of it are likely to fare worse compared with those who do not. In contrast, when a charge is made for a particular road, but there are alternative routes, this may be less problematic in this respect, although traffic levels may increase in areas that do not carry a charge, resulting in social and environmental consequences. Peter Jones identifies two significant issues related to congestion pricing—spatial inequality and social equity.

Spatial Inequality

The geographical location of a congestion pricing area may particularly affect living or working in an area. For example, the boundaries of a charge may mean that some individuals and organizations will end up paying charges more often than others, especially if they have to pay every time they cross into a charging area. Some argue that businesses located within charging areas may suffer, as customers avoid entering charging zones. Businesses reliant on regular travel by car, van, or motorbike (e.g., couriers, taxi firms, or those in the building trade) may also suffer if they are required to pay to enter or travel within a congestion zone. Equally, there are arguments that suggest that residential areas on the edges of congestion pricing zones may suffer from increased levels of traffic as motorists attempt to avoid charges. Affects on those living in these areas may include increased levels of pollution, increases in health problems associated with pollution (such as respiratory diseases), increased damage to physical infrastructure, and increases in road traffic, accident-related deaths, and injuries.

Social Equity

Those who have lower incomes but are reliant on access to a vehicle may be most affected by having to pay to use a stretch of road. This is especially the case with charging schemes outside large conurbations that are well served by public transport, as those who rely on using a car to access employment, essential goods, and services may be priced out of the market and may be unable to access these goods and services in any other way.

As a result of such problems, economic instruments such as congestion pricing are often treated with caution by policy makers. However, many of the issues discussed here are often a result of the settings of congestion pricing schemes, rather than the policy instrument itself. Peter Jones argues that social equity can be reconciled with congestion pricing through the following:

- The use of discounts, exemptions, or "free rations" for certain groups; for example, key workers or those living in a particular geographical area who are likely to be most affected by the charge
- The use of exemptions for particular business uses
- The creation of clear linkages to other schemes that discourage the use of a private vehicle and encourage alternative modes of transport
- Incentives to use shops and other businesses once the charge has been paid
- Varying charges for different times of the day
- A clear rationale for the selection of the charging area

Congestion Pricing and Political Acceptability

Proposals for congestion pricing zones are often met with public opposition. In the United Kingdom, when put to a public vote, proposals for pricing zones in Manchester and Edinburgh were both rejected in high-profile, fiercely contested referenda. In contrast, the former mayor of London, Ken Livingston, was elected despite his pledge to introduce such a charge.

Arguably, the manner in which proposals for congestion pricing are presented to the public makes a significant difference in their popularity. Where pricing proposals appear to be for environmental purposes there may be more resistance, especially among the road traffic lobby, as it appears that environmental concerns are being prioritized over economic ones. In contrast, where congestion pricing is promoted as a way of addressing a serious local problem, or as a way of developing a new piece of infrastructure or transport technology that will benefit the local community, it may be more popular. Equally, where concessions, exemptions, and incentives are used to address equity concerns, pricing may be less unpopular.

Conclusions

As a policy instrument, congestion pricing can be a very effective way of reducing traffic in an area—for example, in London, Transport for London, which administers the charge, found that traffic in the area fell by 21 percent between 2002 and 2006, and that 70,000 fewer vehicles were on the streets every day compared with before the introduction of the charge.

However, there are clear disadvantages to congestion pricing that make it politically controversial and potentially socially unjust. Although some of these difficulties can be overcome by the way in which the pricing system is developed, others remain more deep rooted, raising issues about the tension between social, environmental, and economic policy concerns.

See Also: Intermodal Transportation; London, England; Stockholm, Sweden.

Further Readings

Backhaus, J. G. "Increasing the Role of Environmental Taxes and Charges as a Policy Instrument in Developing Countries: Some Conceptual Considerations." *American Journal of Economics and Sociology,* 63:5 (2004).

Foo, T. S. "An Advanced Demand Management Instrument in Urban Transport: Electronic Road Pricing in Singapore." *Cities,* 17/1:35–45 (2000).

Jordan, A., et al. "'New' Instruments of Environmental Governance: Patterns and Pathways of Change." *Environmental Politics,* 12/1:3–24 (2003).

Leitman, J. "Integrating the Environment in Urban Development: Singapore as a Model of Good Practice." Urban Waste Management Working Paper Series 7. Washington, D.C.: Urban Development Division, World Bank, 2001.

Carolyn Snell
University of York

CONSERVATION SUBDIVISION

Conservation subdivision (CS) is one of a number of site-scale design approaches that address the well-documented environmental impacts of traditional subdivision in suburbia and exurbia. As an urban design approach, CS is often associated with the urban development philosophy known as New Urbanism (NU), but it has clear ties to the emerging philosophy of New Ruralism (NR). Perhaps the best-known example of CS in the United States is Prairie Crossing in Grayslake, Illinois, which features both NU and NR components. CS can be found in other parts of the United States, and it may be relatively common in some regions. At the same time, the approach has encountered economic and social-political barriers. In the few published empirical studies on the use of specific CS design features, several concerns about the practice have emerged. These suggest ways in which its use can be improved in the future.

CS, also known as open space subdivision, generally features five characteristics. First, the layout of lots (development pattern) is altered to avoid areas that have sensitive environmental resources (e.g., biodiversity, natural habitats, wetlands, riparian areas), support productive land for the management of natural resources (e.g., agriculture, timber, grazing lands), or offer diverse recreational opportunities (e.g., active and passive forms). Second, the design reduces the lot sizes, often to much smaller sizes than allowed by typical zoning, to maximize open space and thereby reduce infrastructure needs. Third, the project design often clusters these smaller lots together to further maximize open space size and contiguity. This may further reduce infrastructure needs, thereby improving water quality and hydrological dynamics in nearby waterways. Fourth, legal prohibitions against future development are placed on designated open space. This may occur through deed restrictions, conservation easements held by a local government or land trust, dedication or sale of the land to a government entity, or land donation to a land trust. Fifth, the developer creates a homeowners association and specific bylaws that encourage ecologically appropriate resident interactions, both in protected areas and on their own properties.

Although NU principles emphasize compact design, mixed-use development, and walkable neighborhoods that promote high quality of life, NR protects productive farmlands, supports small-to-medium agricultural enterprises, and maintains overlapping wildlife habitats that facilitate sustainable agriculture. Key features of CS predate recent trends toward designing ecologically friendly human settlements. Randall Arendt popularized CS practices in his 1996 book *Conservation Design Subdivisions: A Practical Guide to Creating Open Space*. However, elements of CS, such as ideas about locating development according to biophysical features, can be traced back to the teachings of Ian McHarg, whose 1969 book *Design With Nature* inspired many landscape architects and planners to think differently about nature in the built environment. More recently, emphasis is being placed on using this approach for landscape- and regional-scale efforts to create greenways, including the ecological corridors advocated by conservation biologists.

Prairie Crossing

Exhibiting all of the characteristic CS design features, Prairie Crossing is a 677-acre subdivision with 359 single-family homes, 36 condominiums, and an on-site charter school. Over 60 percent of the land is conserved and supports an organic farm that produces flowers and vegetables, 16 acres of agricultural hedgerows, 20 acres of restored wetlands, and 165 acres of restored native prairie. Individual residents who incorporate native prairie plantings in their yards bolster these restoration efforts. Prairie Crossing includes 10 miles of trails on-site, with a regional trail system providing access to 3,200 acres of legally protected prairie. Nearby, residents can catch the regional commuter rail to Chicago.

In other parts of the United States, CS is part of efforts to protect oak savanna and lakeshore in Minnesota's Twin Cities, conserve a working cattle ranch outside Bend in central Oregon, protect wetland and maritime forest in South Carolina's Lowcountry, preserve natural areas in eastern Michigan, create additional recreation areas and protect habitat in northeastern North Carolina, and preserve farmland in Rhode Island and southeastern Pennsylvania. In southeastern Pennsylvania, the Natural Lands Trust has worked to make CS more common. The trust's website highlights a total of 11 case studies that highlight specific dimensions of the CS planning process and their efforts to develop a "growing greener" model ordinance. This ordinance facilitates CS approval by townships in accordance with Pennsylvania's planning laws. The Atlanta Regional Commission has developed a similar model ordinance.

Potential Barriers

Although CS appears to have clear open space conservation value, potential economic and social-political barriers to its use have been recognized. Among these are the perception of buyers who may be resistant to living in areas with higher density, the reality of economic benefits touted by proponents, and the willingness of local jurisdictions to embrace a different way of doing things. Rayman Mohamed's 1996 statistical analysis of CS use in Rhode Island demonstrated CS lots include a price premium, save on construction costs, and sell more quickly. However, Thomas Daniels cautions that CS's focus on scenic beauty may actually contribute to farmland loss. Likewise, Mark Bjelland and coauthors discovered that some urban jurisdictions in the Minneapolis area voice concerns that CS is promoting inefficient rural densities and contributing to urban expansion.

Concerns about this CS in empirical studies can be broken into two categories: social and ecological management. First, critical urban scholars take issue with the assumption that the spatial prescriptions of NU can remedy social problems associated with suburbanization. For example, Jeffrey Zimmerman critiques Prairie Crossing for its use of a "sanitized" nature that mobilizes nature in defense of the suburban dream, representing a nostalgic defense of the Midwestern frontier and a poor model of sustainability. Research by Mark Bjelland and coauthors in Minnesota demonstrates that CS represents a diversification of the suburban project and not its rejection. More troubling are criticisms by Karen Till that CS represents a "green politics" of exclusion. There is some evidence to support this theoretical criticism. Research in the South Carolina Lowcountry by Patrick Hurley and Angela Halfacre suggests that CS may indeed impinge on the livelihoods of Gullah sweetgrass basketmakers, who rely on access to culturally important local natural resources, given that CS occurs in historic resource commons and gating is common. Arthur Nelson and James Duncan also raise concerns about exclusion, noting that CS residents may impose restrictions on farming inputs or restrict access for farmers altogether.

Second, focusing on ecological management and outcomes, recent research by conservation biologists further tempers some of the boosterism associated with the use of CS. First, research by Liba Pejchar and coauthors challenges the extent to which CS can achieve biodiversity conservation goals, given questions about institutional and management capacity. These correspond well to findings from Michigan that have documented problems with natural area conflicts and management issues, which result from the inability of residents to access appropriate information. Second, using data on mammals, songbirds, ground-nesting birds, and native plant species as measures, B. Lenth and coauthors found no difference between housing developments in Colorado with clustered arrangements and those with conventional layouts for the majority of the variables. Importantly, however, these researchers conclude that with greater attention to ecological guidelines, better regional planning, and active management, CS remains a viable strategy for meeting key biodiversity conservation goals.

See Also: Biodiversity; Compact Development (New Urbanism); Parks, Greenways, and Open Space; Sustainable Development.

Further Readings

Arendt, Randall. *Conservation Design for Subdivisions: A Practical Guide to Creating Open Space Networks.* Washington, D.C.: Island Press, 1996.

Austin, Maureen and Rachel Kaplan. "Resident Involvement in Natural Resource Management: Open Space Conservation Design in Practice." *Local Environment*, 8/2 (2003).

Bjelland, M., et al. "The Quest for Authentic Place-Production of Suburban Alternatives." *Urban Geography*, 27/3 (2006).

Daniels, Thomas. "Where Does Cluster Zoning Fit in Farmland Protection?" *Journal of the American Planning Association*, 63/1 (1997).

Hurley, Patrick and Angela Halfacre. "Dodging Alligators, Rattlesnakes, and Backyard Docks: A Political Ecology of Sweetgrass Basket-Making and Conservation in the South Carolina Lowcountry." *GeoJournal*, Forthcoming.

Kaplan, Rachel and Maureen Austin. "Out in the Country: Sprawl and the Quest for Nearby Nature." *Landscape and Urban Planning*, 69 (2004).

Kraus, Sibille. "A Call for New Ruralism." http://www.sagecenter.org/new-ruralism.pdf (Accessed May 2009).

Lenth, Buffy, et al. "Conservation Value of Clustered Housing Developments." *Conservation Biology*, 20/5 (2006).

McHarg, Ian. *Design With Nature.* Garden City, NY: Natural History, 1969.

Natural Lands Trust. "Growing Greener: Conservation by Design. http://www.natlands.org/categories/subcategory.asp?fldSubCategoryId=26 (Accessed May 2009).

Nelson, Arthur and James Duncan. *Growth Management Principles and Practices.* Chicago: APA Planners, 1995.

Pejchar, Liba, et al. "Evaluating the Potential for Conservation Development: Biophysical, Economic, and Institutional Perspectives." *Conservation Biology*, 21/1 (2007).

Prairie Crossing. "Guiding Principles for Prairie Crossing." http://www.prairiecrossing.com/pc/site/guiding-principles.html (Accessed May 2009).

Randolph, John. *Environmental Land Use and Management.* Washington, D.C.: Island Press, 2004.

Till, Karen. "New Urbanism and Nature: Green Marketing and the Neotraditional Community." *Urban Geography*, 22/3 (2001).

Zimmerman, Jeffrey. "The 'Nature' of Urbanism on the New Urbanist Frontier: Sustainable Development, or Defense of the Suburban Dream." *Urban Geography*, 22/3 (2001).

Patrick Hurley
Ursinus College

CONSTRUCTION AND DEMOLITION WASTE

Construction and demolition (C&D) is one of the many types of waste streams that include municipal waste (household wastes), shipping and boating waste, and commercial/industrial waste. C&D waste is the waste produced during new construction, renovation, and demolition of buildings and structures. This includes building materials such as bricks, plastics and vinyl, carpet, brick and rubble, glass, metal, asphalt roofing, concrete, damaged wood, scraps of insulation, nails, electrical wiring, and rebar, as well as waste originating from site preparation such as dredging materials, and tree stumps. C&D waste may contain lead, asbestos, or other hazardous substances. For example, C&D wastes such as plasterboard are hazardous once broken down in landfill as they release hydrogen sulfide, a toxic gas.

Waste created during new building construction, renovation, or demolition can include bricks, plastics and vinyl, carpet, glass, metal, asphalt roofing, concrete, damaged wood, insulation, nails, electrical wiring, and rebar.

Source: iStockphoto

It has been estimated that C&D waste accounts for from 10 percent to as much as 30 percent of the total municipal waste stream. More specific research has also attempted to estimate the amount of construction and demolition waste in various countries. C&D waste is most often disposed of in landfills, where it takes up much of the space. For example about 65 percent of Hong Kong's landfill space at its peak in 1994–1995 was C&D waste. About one-third of the volume of materials in landfills in the United States is C&D waste. Data available for some European countries also indicate that the amount of C&D waste varies from country to country, depending on how it is defined. In 1996, the amounts of C&D waste in Austria, Denmark, Germany, and the Netherlands were about 300, over 500, about 2,600, and about 900 kg/cap, respectively.

Although C&D waste is a complex waste stream, made up of a wide variety of materials, recent recognition of its increasing volumes and shortage of landfill has led to more attention being paid in most countries to the diversion of waste components from landfills through waste reduction, recovery, reuse, and recycling. Reduction, reuse, and recovery (the three Rs principle) of waste, otherwise known as the waste management hierarchy, has been very widely adopted and has been officially recognized in most countries as one of the principles of sustainable waste management and development. In Denmark, the percentage of recycling is more than 80 percent. Germany, the Netherlands, Finland, Ireland, and Italy each recycle 30–50 percent, whereas the recycling percentage in Luxembourg is 10 percent. Even building rubble (including concrete, brick, tile, and asphalt), can be recycled. In Australia in 2002–2003, approximately 50 percent of all recycled waste was C&D waste, and building rubble was by far the most recycled material.

Recycling, one strategy of waste minimization, offers some benefits: reduction of demand for new resources, reduction of transport and production energy costs, and reduction of space required for landfill. Furthermore, the social effects of solid waste management systems are reflected in the employment, health, and quality of life of residents in the region. Recycling, reusing, and salvaging construction waste can save money and also create employment related to salvaging and recycling of construction waste. In addition, the health and quality of life of residents are improved because less pollution is generated by reducing manufacturing and transportation energy-related emissions when salvaging and recycling are practiced.

C&D recycling is not practiced universally for several reasons: insufficient budgetary allocation for municipal solid waste management, and also inefficient collection of service fees; no active planning to establish common disposal facilities among adjacent communities; no definite regulation and guidelines for C&D waste management hierarchy starting from source separation, collection, transportation, disposal, and monitoring; lack of skilled personnel in operating an efficient waste collection and disposal practice; absence of waste recycling programs in most communities; associated existing legislation does not adequately facilitate construction waste management in an effective direction; lack of public cooperation and participation; and lack of government legal enforcement, among others.

Recycling and Economics

The factors that helped launch the C&D recycling industry—the environmental movement, coupled with a perceived looming landfill shortage—might have reached their maximum point of forward momentum. Although some markets and many recycling techniques have been established because of these initial concerns, economic justification will drive any near-term future growth within the C&D debris recycling industry. Thus, more effort has to be made at stimulating market development for recovered C&D waste. At this time, in many countries, there are no standards for recycled materials. Because of the lack of standard specifications, there is difficulty in overcoming market safety liability concerns by engineers and other designers. Engineers will often not specify recycled material mainly because it is perceived as problematic to specify an inconsistent product. This is an issue that must be addressed if the C&D recycling industry is to continue growing.

Disposal to landfill is cheap, and therefore more competitive than recycling. The other variable that could spur even more attention to C&D recycling is a sharp increase in tipping fees. This could help combat the dumping of C&D waste in landfill because inexpensive landfill tipping fees are the primary barrier to recycling.

Promulgation and implementation of landfill tax and landfill ban have also been shown to be effective when applied as a deterrent to landfill and policy measure for promoting recycling of C&D waste. Many European countries are applying landfill taxes with positive results. Across the European Union, high landfill taxes are associated with higher levels of waste recovery and the lowest reliance on landfills. Some countries are also applying policies such as landfill bans for specific waste streams or regulations on recycling to achieve the high diversion rate from landfill.

Although some of these measures have proven effective from a policy perspective, educating stakeholders (designers, construction companies, etc.) on the importance and benefits (e.g., reduction of transportation and landfill disposal costs) of implementing cleaner production in waste management is very important and can lead to significant reduction in the amounts of C&D waste generated.

Other strategies identified that have been used successfully to reduce C&D waste include reusing existing buildings; extending building lifetime through effective maintenance; designing buildings to accommodate new functions and technologies; incorporating durable, reusable materials into design plans; and deconstructing buildings rather than tearing them down, so that materials can be reused. All of these are strategies that bring down disposal costs, save on landfill space, and reduce the amount of raw materials needed for new projects.

See Also: Green Procurement and Purchasing; Landfills; Recycling in Cities; Sustainable Development; Universal Design.

Further Readings

Brodersen, J., et al. "Review of Selected Waste Streams: Sewage Sludge, Construction and Demolition Waste, Waste Oils, Waste From Coal-Fired Power Plants and Biodegradable Municipal Waste." European Topic Centre on Waste European Environment Agency, 2002. http://reports.eea.europa.eu/technical_report_2001_69/en/tech_rep_69.pdf (Accessed April 2009).

Fishbein, Bette K. "Building for the Future: Strategies to Reduce Construction and Demolition Waste in Municipal Projects." June 1998. http://www.ec.gc.ca/cppic/en/refView.cfm?refId=875 (Accessed April 2009).

Kofoworola, O., et al. "Estimation of Construction Waste Generation and Management in Thailand." *Waste Management,* 29/2:731–38 (February 2009).

Mumma, Tracy. "Reducing the Embodied Energy of Buildings." *Home Energy Magazine Online.* (January–February 1995). http://www.homeenergy.org/archive/hem.dis.anl.gov/eehem/95/950109.html (Accessed April 2009).

Saotome, Tomo. "Development of Construction and Demolition Waste Recycling in Ontario. School of Engineering Practice SEP 704." August 2007. http://msep.mcmaster.ca/publications/Development_of_C&D_recycling_in_Ontario.pdf (Accessed April 2009).

Tam, V. W. Y. and C. M. Tam. "Evaluations of Existing Waste Recycling Methods: A Hong Kong Study." *Building and Environment,* 41:1649–60 (2006).

Tam, V. W. Y. and C. M. Tam. "A Review on the Viable Technology for Construction Waste Recycling." *Resources, Conservation and Recycling,* 47:209–21 (2006).

Oyeshola Femi Kofoworola
University of Toronto

Copenhagen, Denmark

The year 2010 was established as the inaugural year for awarding the title of "European Green Capital," and eligible European Union countries were encouraged to enter their most environmentally conscious cities in a contest to select the awardees for both 2010 and 2011 simultaneously. The award serves to acknowledge work already being done at the municipal level to make environmental stewardship an integral part of urban planning and to provide an incentive for cities to intensify their efforts to improve the quality of life in the urban habitat. Thirty-five European cities vied for the title European Green Capital, and after the first round of elimination, Copenhagen, Denmark, was on the short list of eight finalists.

Copenhagen made the short list of eight finalists for the title of European Green Capital. Wind turbines in Copenhagen Harbor contribute to the sector that produces nearly 25 percent of the country's electricity output.

Source: iStockphoto

Although Hamburg won the award for 2010 and Stockholm for 2011, it is notable that Copenhagen scored higher than both of these cities in the important category of water consumption (conservation of water through water metering and prevention of leakage during transport through water pipes). Ten criteria were used for ranking the cities; in addition to its superior showing in the water consumption category, Copenhagen tied with or had higher scores than at least one of the winning cities in three other categories: local mobility and passenger transportation, sustainable management of the local authority, and sustainable land use. Copenhagen's impressive showing as a finalist is not surprising, given its well-established record of implementing innovative projects that enrich life in the urban milieu for its 1.1 million inhabitants, while fulfilling its commitment to sustainable development. A prime example of the tenaciousness with which Copenhagen addresses the environmental problems that are endemic to urban environs is the rehabilitation of Copenhagen Harbor from a polluted receptacle for the wastewater discharged from sewers and industrial companies. The reclamation project resulted in such a remarkable improvement of water quality in the harbor that starting in 2002, Copenhagen residents have been able to enjoy the wholesome benefits of swimming in seawater. Successfully meeting the challenge of reclaiming the harbor as a public swimming area has spurred Copenhagen to set its sights even higher—the present goal is to make all seawaters in the greater metropolitan area safe for swimming by 2015.

Eco-Metropole of the World

In 2007, Copenhagen announced its vision of becoming the capital city with the best urban environment in the world—"the Eco-Metropole of the world"—by 2015. To meet this target, 14 new parks are in development. Already, 8 of 10 residents have a park within 300 meters of their homes. The plan is that by 2015, 9 of 10 residents will be within a 15-minute walk of a recreational area. Also in progress is the planting of 3,000 additional street trees. The streets of Copenhagen are already pedestrian friendly, with many streets closed to car traffic. As an example, the Strøget, a world-renowned shopping street in downtown Copenhagen, is a 1.1-kilometer stretch of pedestrian-only shopping streets and squares. Known as the "City of Cyclists," Copenhagen has made great strides in ending the glut of cars on the road. As of 2009, 55 percent of its population uses a bicycle for all trips, with commuters and students making 37 percent of their trips by bike. Moreover, the number of cyclists is bound to increase as the city continues to expand its vast network of bicycle routes and enlarges the fleet of public bicycles that are made available for use in downtown Copenhagen for a small returnable deposit.

In addition to being a Baltic seaport, Copenhagen has an abundance of canals and lakes, which have fueled its ambition to become a "Green and Blue Capital City," in which waterscapes are as abundant as park greenery and equally accessible. In this vein, the redemption of Amager Beach merits mentioning. Since its creation in the 1930s, Amager Beach had required almost annual additions of sand to keep it from disappearing into the shallow waters of the Baltic channel in which it lies, called Øresund. To resolve the problem of the shallow waters washing away each year's sand addition, Amager Beach was moved deeper into Øresund by constructing an artificial island from 1.5 million cubic meters of raw material. In its new locale, Amager Beach is self-preserving because the waves are large enough to pull the sand both onto and away from the beach. The new and improved Amager Beach Park ("Amager Strandpark") is both an island and an activity-filled lagoon, and it is only seven minutes from the center of Copenhagen by subway and is also accessible by a bike trail from the center of the city.

A Green Tourist Destination

Although not as well known as the bronze sculpture of the Little Mermaid in Copenhagen Harbor, the Tivoli Gardens amusement park in the heart of Copenhagen (directly across from the central railroad station) attracts approximately 3 million visitors a year. Established in 1843, Tivoli is situated on approximately 15 acres and contains thrill rides, game arcades, restaurants, concert locales, a fairy-tale garden with 400,000 flowers, and 110,000 lamps. However, the energy demands of this attraction will be greatly reduced in 2010 when Tivoli becomes the first amusement park in the world to run on renewable energy. In the spring of 2009, Tivoli and DONG Energy (a leading energy company in northern Europe) announced the formation of a "climate partnership" with the goals of reducing Tivoli's energy consumption and ensuring that the energy consumed is provided by carbon dioxide–neutral production facilities. Tivoli will have its own wind turbine housed at a nearby high-efficiency multifuel power plant.

Another site attracting visitors to Copenhagen, the Bella Center—Scandinavia's largest convention center—recently had a 75-meter-high wind turbine constructed on its property that is capable of producing 1.6 million kilowatt hours of energy a year. The Bella Center hosted a number of climate conferences in 2009 and was the site for the United Nations Climate Change Summit (COP 15). The holding of COP 15 in an environmentally sustainable milieu will be the subject of a white book published in 2010 to assure stakeholders that this historic summit in Copenhagen was planned and implemented in full compliance with BS8901, the new international standard for sustainable event management. A consortium composed of the City of Copenhagen, the Danish Ministry for Foreign Affairs,

Wonderful Copenhagen, Visit Denmark, and MCI Copenhagen entered into a contract with MCI Sustainability Services to write the white book, titled *Copenhagen Sustainable Meetings Protocol*. The consortium anticipates that the book will enhance Copenhagen's reputation as a green city and propel it toward its goal of becoming the ecological metropolis par excellence.

See Also: Hamburg, Germany; Malmö, Sweden; Stockholm, Sweden; Sustainable Development.

Further Readings

European Green Capital. "Environment." http://ec.europa.eu/environment/europeangreen capital/about_sumenus/background.html (Accessed September 2009).

"A Green and Blue Capital City." *ECO-Metropole.* http://www.kk.dk/PolitikOgIndflydelse/ Byudviling/Miljoe/Miljoemetropolen/Eco-metropole/AG (Accessed September 2009).

Meetings International. "FN:s Klimatkonferens Ska Bana Väg För Fler Gröna Möten Och Event i Danmark" ("UN Climate Conference Shall Pave the Way for More Green Meetings and Events in Denmark"). September 18, 2009. http://www.meetings international.se/news.php?id=159 (Accessed September 2009).

Gwendolyn Yvonne Alexis
Monmouth University

Curitiba, Brazil

Often referred to as the ecological capital of Brazil, Curitiba has legislation that provides tax incentives for green spaces in building projects. This has allowed the expansion of parks and green areas, like the Public Botanical Garden.

Source: iStockphoto

The city of Curitiba is the capital of Paraná, a mainly agricultural state in the south of Brazil. The city is located 900 meters above sea level and 87 kilometers west of the Atlantic Ocean, near the coastal mountain ridge. Its land area covers approximately 435 square kilometers, and its population is around 1.8 million. It is known internationally for its progressive transportation system, social and environmental programs, preservation of the city's cultural heritage, and expansion of parks and green areas, and the city also has achieved international recognition for having charted an effective route to becoming a sustainable city.

Despite its achievements, Curitiba faces the same problems that metropolises around the world do, including social inequality, overcrowding, pollution, and limited public funding. What is different about Curitiba is that its planners have deployed creative and inexpensive ways to respond to these problems. This includes the investment in an extensive transportation system using buses, recycling programs that clean up the environment and also address poverty, and attracting new industry while expanding green spaces and using preserved historical areas to revitalize neighborhoods and to grow tourism.

The city's straightforward approach to city planning, unique not only within Brazil but also within the developing world, began many decades ago. The French planner and architect Alfred Agache developed the first plan to direct urban growth in Curitiba in 1943. At that time, the region's blooming agriculture industry attracted new settlers and immigrants, mostly from Europe. The government did not implement the plan at the time, but its main legacy was to introduce the concept of urban planning to Curitiba's citizens and government.

Urban Planning

In 1964, the municipal government put forward a call for proposals to prepare Curitiba for rapid and ongoing growth. A team of young, idealistic architects and planners from the Federal University of Paraná answered that call. In 1965, the government created the Curitiba Research and Urban Planning Institute, which is responsible for developing the city's master plan, studies, and projects, and for coordinating the urban planning activities. Their proposal addressed measures to minimize urban sprawl, reduce downtown traffic, preserve Curitiba's historic district, and provide easily accessible and affordable public transport. This proposal was adopted and came to be known as the Curitiba Master Plan.

During the 1970s, the development of the city was influenced by three main elements: the design and implementation of an integrated transport system, the development of the road network system, and land use legislation to allow environmental preservation, cultural services, and quality of life. These elements were later complemented by the development of the industrial district.

Curitiba's Transportation System

The creation of the road network in coordination with the master plan and land use legislation began construction and implementation in 1974. It encompassed the design of an urban plan that emphasized the linear growth along five structural axes. Each axis was designed as a triple road system. The central road has two exclusive express bus lanes. Parallel to the express bus lanes are two local roads running in opposite directions— one for traffic flowing into the city and the other for traffic flowing out of the city. The land use legislation encouraged large, high-density buildings along these corridors, mainly for services and commerce. Moving out of these central corridors are the residential neighborhoods.

The transportation system is made up of three complementary levels of service that include feeder lines, express lines, and interdistrict routes. The feeder lines pass through outlying neighborhoods and make the system easily accessible to lower-density areas. Sharing the roads with other vehicles, these feeder lines connect with the express system along the structural corridors. The express system then uses these dedicated bus lanes and transports large numbers of passengers to various locations along these structural corridors, thus operating much like a surface subway system. The interdistrict routes allow passengers to connect to the axis of the express lines without entering the central city area. Passengers can pay one fare and travel throughout the system.

Curitiba's transport system is used by more than 1.3 million passengers every day and attracts nearly two-thirds of the city's population. The city is among the less-air-polluted state capitals in Brazil.

Waste Management and Recycling

Curitiba was the first city in the country to practice selective garbage collection by separating organic from inorganic waste. The "Garbage that is not garbage" program recycles approximately 70 percent of the city's waste. Trucks collect paper, cardboard, metal, plastic, and glass. In addition to the environmental benefits, the program generates financial resources to support social programs for unemployed people and create job opportunities in the recycling plants.

Another waste management program was implemented in Curitiba. The "Purchase for garbage" program was created to benefit the 10 percent of the population living in squatter settlements (*favelas*). These settlements are located in the outskirts of the city and lack infrastructure and access to roads. These residents used to dump their garbage in the river or next to some unoccupied area, bringing health and environmental problems to these communities. To overcome these problems, residents can exchange their waste bags for bus tickets and agriculture and dairy products. Far from being a long-term solution, the program has proven to be cost-effective in the short term, however, and has improved the quality of life of the urban poor living in vulnerable conditions.

Cultural Heritage and Green Areas

Curitiba is referred to as the ecological capital of Brazil. The land use planning has allowed the expansion of parks and green areas and preserved the cultural heritage of the city. Legislation provides tax incentives for builders to include green spaces in their projects. There are also tax incentives for the preservation of buildings with historic value.

Through the green areas, rainfall is diverted into lakes in parks, avoiding flooding while also protecting floodplains and riverbanks, acting as a barrier to illegal occupation. These green areas are spread among 30 forests and parks that have become highly valued meeting places for the population. Several of these areas are allusive to important local history, honoring the arrival of immigrants and the native habitants, the Tingui Indians.

A Few Lessons from Curitiba

The integration of different elements into the urban planning provided a framework for the city's development. The close relationship between the public transport system, the land use legislation, and the hierarchy of the urban road network highlights that solutions for cities are not specific and isolated but interconnected.

Creative solutions can be cost-effective and fulfill the city's needs. The urban transportation system is one of Curitiba's best-known planning successes—a model for cities around the world that want to implement eco-efficient transportation networks that are well integrated with urban form and produce environmental benefits. The city pioneered the idea of an all-bus transit network with exclusive express bus lanes created along well-defined structural axes that were also used to channel the city's growth. The transit system is rapid and cheap, and its efficiency encourages people to leave their cars at home.

See Also: Bus Rapid Transit; Historic Preservation; Parks, Greenways, and Open Space; Recycling in Cities; Sustainable Development; Transit-Oriented Development; Urban Forests.

Further Readings

Newman, Peter and Isabella Jennings. *Cities as Sustainable Ecosystems: Principles and Practices.* Washington, D.C.: Island Press, 2008.

Rabinovitch, J. "Curitiba: Towards Sustainable Urban Development." *Environment and Urbanization,* 4/2:62–73 (1992).

Rabinovitch, J. "Innovative Land Use and Public Transport Policy: The Case of Curitiba, Brazil." *Land Use Policy,* 13/1:51–67 (1996).

Rabinovitch, J. and J. Hoehn. *A Sustainable Urban Transportation System: The "Surface Metro" in Curitiba, Brazil.* The Environmental and Natural Resources Policy and Training Project, Michigan State University, 1995.

Register, Richard. *EcoCities: Rebuilding Cities in Balance With Nature.* Gabriola Island, British Columbia, Canada: New Society, 1996.

Schwartz, Hugh. *Urban Renewal, Municipal Revitalization: The Case of Curitiba, Brazil.* Reston, VA: Higher Education, 2004.

Rafael D'Almeida Martins
University of Campinas

D

DAYLIGHTING

The electricity used to light interior spaces comprises a significant portion of the energy cost for buildings. Daylighting uses natural light to illuminate interiors, which reduces energy costs.

Source: U.S. Department of Energy

Daylighting refers to the theory and applied methods for allowing natural sunlight to enter into interior building spaces. Various daylighting techniques have been used in architectural design for much of human history. During the 20th century, however, natural daylighting was replaced by the widespread use of electric fluorescent lighting. Today, as increasing numbers recognize the environmental, economic, and psychological benefits of daylighting, more interior spaces are being lit by the sun. Green building advocates promote daylighting as an example of an integrative approach to sustainable design. Successful daylighting carefully integrates passive and active design elements for a holistic effect on the whole building. Such a "whole building" approach takes variables like glare, heat gain, seasonal change, cost, paint color, and other elements into account.

Three main benefits ascribed to daylighting of interior building spaces include:

1. Environmental

2. Economic

3. Psychological

Daylighting and Sustainability

The recent resurgence of interest in daylighting is driven by the increasing cost of fossil fuels and the recognition of global warming as a tangible threat to environmental ecology and human quality of life. The electricity used to light interior spaces makes up a significant part of the overall energy cost for buildings. In the United States, electricity used to light interior spaces accounts for 20–25 percent of the total electrical energy used nationally. In colder climates and during winter months, the practice of naturally heating spaces via solar heat gain can greatly reduce heating costs. Naturally heated buildings require less energy and help minimize the burning of fossil fuels. The U.S. Green Building Council estimates that buildings that successfully use daylighting can conserve 50–80 percent of their normal energy use.

A comprehensive approach to daylighting includes the use of automated or electronic management systems to balance lighting between natural sunlight and electrical lighting. A 2005 study conducted by the Center for the Built Environment at the University of California, Berkley, concluded that installing control systems to properly manage daylit environments resulted in a 65 percent decrease in electrical lighting use.

Cost

Prospective owners should view such green building technologies using a life cycle analysis approach. This type of approach emphasizes long-term energy cost savings, rather than a singular focus on up-front costs of daylighting techniques. Daylighting improvements, similar to all properly calibrated green building measures, should be viewed as investments, rather than sunk costs. In addition, there are numerous positive externalities to daylit environments that can positively affect the bottom line. Some of these externalities include lower rates of employee absenteeism, a healthier workforce, increased productivity, and higher morale. The effect on worker morale and well-being is especially valuable, as it is typically more costly to continuously hire and train new employees than to make investments in creating a pleasant working environment that would help retain the current workforce.

Psychological Connections

In addition to reduced energy use, daylighting affects human psychology in profound ways. The U.S. Green Building Council's Leadership in Energy and Environmental Design reference manual notes that a visual connection with the outside, natural world can result in greater appreciation and respect for nature by building occupants.

In a widely cited 1999 study conducted by the Heschong Mahone Group, a group of design consultants and daylighting experts aimed to quantify the effect of daylighting on student performance and retail sales. Their findings indicated a statistically significant correlation between increased daylighting and improved standardized test scores for reading and math. Heschong Mahone Group researchers also found a positive correlation between retail stores with skylights and sales performance. The research even suggests that the amount of daylight in a retail store can be an accurate predictor for total sales performance.

Daylighting Methods

Daylighting strategies are often tested using computer models that simulate the arc of the sun from sunrise to sunset over the course of the year. Advanced models can also simulate the actual illumination that occurs in interior spaces with added windows or other design elements. In general, daylighting is divided into two approaches: active and passive. Passive approaches consist of static, nontracking fixtures and design to allow natural light to enter into interior spaces with little or no adjustment. *Active daylighting* refers to devices that move in relation to the sun (i.e., tracking solar movement) or shift openings based on the amount of solar light available. For example, heliostats use electronic sensors to adjust large mirrors to direct sunlight into interior spaces that would not receive daylight via passive means. There are many solutions available to architects and designers, but most successful daylighting projects apply a combination of design elements including clerestories, skylights, light tubes, louvers, heliostats, overhangs, glazing, and absorptive tinted glass.

For projects in the preconstruction or design stage, site selection can play a critical role in the effectiveness of a daylighting strategy. Leadership in Energy and Environmental Design guidelines emphasize construction that includes south-facing windows, allowing for maximum solar heat gain during the winter months when the sun's trajectory is lower. The result is less reliance on fossil fuels for heating, resulting in lower carbon emission into the atmosphere. South-facing building orientation is perhaps the most critical and effective passive element for any daylighting project. Such a strategy creates a more controlled and balanced solar heat gain for the entire building, particularly during the summer months. Ideally, any daylighting methodology will maximize the amount of soft, reflected, diffuse natural sunlight and minimize direct sunlight. Direct sunlight can create too much heat gain, which, in turn, requires greater cooling during warmer weather.

Glare control is the most common problem with many daylighting projects. If not properly controlled, glare can offset many of the benefits gained by daylighting elements. Best management practices for reducing or controlling glare include exterior shading devices, light shelves, blinds, or glazing.

Best Management Practices

The National Renewable Energy Laboratory created a list of six best management practices for daylighting based on case studies culled from high-performance buildings around the United States. These best practices for daylighting (as outlined by the laboratory) are:

- design that provides daylight to all interior spaces,
- accounting for glare mitigation early in the design stage,
- using automatic dimming in daylit areas,
- selecting interior furnishings that maximize natural light,
- integrating electrical lights with daylighting system, and
- accurately determining postoccupancy energy savings.

Daylighting is an effective and cost-efficient way to minimize energy use and create holistic residential and commercial spaces. When best management practices are properly applied, successful daylighting projects can help create greener buildings and happier, more productive occupants.

See Also: Energy Efficiency; Green Design, Construction, and Operations; LEED (Leadership in Energy and Environmental Design); Sustainable Development.

Further Readings

Ander, Gregg D. *Daylighting Performance and Design*. Hoboken, NJ: John Wiley & Sons, 2003.

Phillips, D. *Daylighting: Natural Light in Architecture*. Burlington, MA: Architectural, 2004.

Torcellini, P. A., et al. *Lessons Learned From Case Studies of Six High-Performance Buildings*. June 2006. http://www.nrel.gov/docs/fy04osti/36290.pdf (Accessed April 2009).

U.S. Green Building Council. *LEED Reference Manual for Operations and Maintenance*. Washington, D.C.: U.S. Green Building Council, 2009.

Ameet D. Doshi
Georgia Perimeter College

DENITRIFICATION

Denitrifiction refers to a microbial process in which nitrate is converted to nitrogen gas. The process is part of the nitrogen cycle, a series of biogeochemical processes that transforms nitrogen and nitrogen-containing compounds in the environment. Denitrification is an important pathway in the nitrogen cycle because it removes nitrogen from an ecosystem and emits it into the atmosphere. Excess nitrogen (particularly nitrate) in water and aquatic systems has become a serious environmental problem that can affect human health and severely degrade ecosystems by stimulating plant and algae growth in receiving waters, reducing oxygen levels and causing algal blooms that are toxic to aquatic life. Cities increasingly use denitrification to remove nitrogen from sewage and wastewater before it is discharged into waterways. Because wetlands support conditions for denitrification, they are increasingly being used on the landscape to remove excessive nitrogen from water.

Most soils support microbes that function in aerobic environments. They use oxygen as their primary electron acceptor to complete cellular respiration. However, in anaerobic environments (where oxygen becomes unavailable), some microbes are capable of using alternative electron acceptors. Anaerobic conditions naturally occur in wetlands and other aquatic environments when soils become flooded. As water fills soil pore spaces, the diffusion of oxygen from the atmosphere into the soil environment is drastically reduced. As organisms deplete the remaining oxygen, soils can quickly become anaerobic.

Denitrification is an anaerobic process in which microbes use nitrate as an alternative electron acceptor in the absence of oxygen. The conversion to nitrate is performed by a variety of facultative bacteria that can use either oxygen or nitrogen oxides as an electron acceptor, making them well adapted to fluctuating aerobic/anaerobic conditions such as wetlands. Denitrification normally occurs in the absence of oxygen; however, the availability of nitrates in the soil is ultimately dependent on nitrification, a strictly aerobic process in the nitrogen cycle. Nitrification is the conversion of ammonium to nitrate. Ammonium is the form nitrogen normally takes when it becomes mineralized (broken down) from organic matter. Even when wetlands are flooded, anaerobic nitrification usually occurs because of a thin oxygenated layer at the soil surface. Ammonium will slowly diffuse to the soil surface, where nitrifying bacteria oxidize it in a two-step process:

ammonium to nitrite, and nitrite to nitrate. The nitrate generated at the soil surface then diffuses back into the anaerobic soil layer, where it can undergo denitrification.

The denitrification process is also dependent on an energy source (an electron donor), normally in the form of organic matter such as glucose ($C_6H_{12}O_6$). During the process, nitrates are converted to nitrogen gas (either molecular nitrogen or, less often, nitrous oxide), which is then released to the atmosphere. The denitrification process can be shown as

$$C_6H_{12}O_6 + 4NO_3 \rightarrow 6CO_2 + 6H_2O + 2N_2$$

or

organic matter + nitrate $\rightarrow$ carbon dioxide + water + molecular nitrogen gas.

Molecular nitrogen is considered the predominant product of denitrification; however, it can also result in considerable release of nitrous oxide, which is concerning because of its status as a greenhouse gas. Recent studies have shown that the proportion of nitrous oxide emitted by denitrification in wetlands may be controlled by soil temperature, hydrology, and the presence of vegetation. These factors seem to combine to influence nitrate availability and microbial activity, which combine to favor nitrous oxide production. The production of nitrous oxide does contribute to the pool of other greenhouse gases in the atmosphere but is a minor contributor compared with others (carbon dioxide, methane).

Nitrogen: A Limiting Nutrient

Nitrogen is one of the most common elements in the biosphere. Molecular nitrogen constitutes 78 percent of the atmospheric gases present; however, despite its prevalence in the atmosphere, it is often a limiting nutrient, particularly in aquatic environments. Nitrogen is an important nutrient for living organisms because it is a key component to several important organic molecules. For instance, nitrogen is present in all amino acids, proteins, and nucleic acids. It is also a critical component in the chlorophyll molecules that allow plants to photosynthesize.

Because nitrogen is naturally a limiting nutrient, when its levels are artificially increased it can lead to eutrophic conditions. In aquatic environments, excess nitrogen often results in rapid growth of algae and phytoplankton, resulting in tremendous organic matter production. After the algae dies, the decomposing organic matter sustains tremendous microbial activity that consumes dissolved oxygen in the water. Oxygen can quickly become depleted, leading to anaerobic or anoxic conditions in the water column, which are inhabitable for most organisms. Fish kills are a common result of waters that rapidly become anoxic because of eutrophic conditions.

For the purposes of maintaining and improving water quality, denitrification is an important process. Denitrification in wetlands is part of the nitrogen cycle that returns previously assimilated organic nitrogen back into the atmosphere. It is not the only process that removes nitrogen from surface waters; however, most of the other major pathways (e.g., sedimentation, immobilization, plant absorption) only temporarily remove the nitrogen from water—the nitrogen is still present in the ecosystem.

Denitrification can be an important component in municipal wastewater treatment. To achieve effective nitrogen removal from wastewater via denitrification, four conditions must be met: facultative bacterial mass must be present, nitrate must be present in an anaerobic environment, suitable conditions for bacterial growth are needed, and an electron donor (organic matter) must be available. Wastewater treatment plants seeking to reduce nitrogen levels often use methanol as an electron donor. Methanol is a simple alcohol molecule (CH_3OH) that is readily degraded by bacteria, has a neutral pH, and does not contain other nutrients (e.g., nitrogen and phosphorus). Where denitrification is incorporated into the treatment process, water is passed through a series of aerobic and anaerobic reactors. Aerobic treatments are used to remove organic matter and enhance nitrification. Water is then treated with methanol or some other organic substance before being treated in an anaerobic reactor to elicit denitrification and the emission of nitrogen gas.

Although nitrogen is normally a limiting nutrient in many ecosystems, humans have nearly doubled the conversion of atmospheric nitrogen to biologically available nitrogen. This conversion has occurred through a combination of legume cultivation (plants capable of fixing atmospheric nitrogen), the processing of chemical fertilizers, and nitrogen-based pollution emitted from vehicles and industry. Excess nitrogen (particularly nitrate) can accumulate in aquatic systems. Nitrate is an extremely mobile compound because it is a negative ion. As a consequence, it does not readily bond to soil particles, and if it does not get absorbed by plants or microbes, it readily moves with water. Because of its mobility, the use of excess nitrogen-based fertilizers is increasingly being scrutinized. Nitrogen fertilizers are commonly applied to row crops beyond the capability of plants to absorb them. What does not get absorbed by plants often moves with runoff into receiving streams and rivers.

An Environmental Problem

Excess nitrogen—and in particular, nitrate—has also become a persistent environmental problem. In its latest nationwide water quality assessment, the Environmental Protection Agency indicated that excess nutrients such as nitrates were the top cause of impairment in lakes, ponds, and reservoirs in the United States. Agriculture is considered the primary source for excess nitrogen in the environment. Nitrate fertilizers have been used extensively in agriculture and have been shown to cause considerable damage to downstream aquatic systems, particularly in coastal waters. In the United States, an annual hypoxic zone has been monitored at the mouth of the Mississippi River. This "dead zone" occurs every spring and coincides with spring snow melt and precipitation in the upper Midwest. During this time, excess nitrogen is moved down the Mississippi River and its tributaries, where it eventually enters the northern Gulf of Mexico. Nitrogen is the primary limiting nutrient in coastal estuaries, and consequently, when large amounts enter the Gulf, it results in a tremendous algae growth. As the algae die off and begin to decay, decomposing microbes consume the oxygen in the water, resulting in hypoxic conditions. Because of the lack of oxygen, these areas become void of most life forms—organisms either move out of the area or die. In the Gulf of Mexico, the seasonal occurrence of the hypoxia zone varies in its timing and size, based on precipitation patterns; however, zones as large as 20,720 km^2 (or 8,800 mi^2) have been documented.

The worldwide increase in biologically available nitrogen has also led to related human health problems. Infants that ingest waters high in nitrates may become susceptible to a condition called methemoglobinemia, or "blue baby syndrome." Bacteria present in infant digestive tracts converts the nitrate to nitrite, which in turn reacts with hemoglobin to form methemoglobin. Methemoglobin cannot carry oxygen through the blood stream like hemoglobin, and therefore the infant can become oxygen deprived (and take on a bluish skin color—hence the term *blue baby syndrome*). The condition can be very serious, but if treated, most infants make a full recovery. As infants mature, their stomachs produce an increasing level of acid that kills much of the bacteria responsible for blue baby syndrome, but until then, infants remain susceptible to excess nitrate.

To counter problems associated with excess nitrogen in water, there is increasing interest in providing an environment for denitrification to occur. Wetlands and other anaerobic environments represent a cost-effective method to remove excess nitrogen from aquatic systems back into the atmosphere. Wetlands are now being prescribed as a means to treat wastewater before it is released back to natural waterways.

Numerous strategies have also been offered to reduce nitrate loading from agricultural fields. To tackle the problem of excess nitrogen in the Gulf of Mexico, multiple strategies will need to be employed, including appropriate fertilizer applications and better management of cropland drainage. Restoring wetlands on the landscape has also been offered as a method to reduce nitrogen runoff into streams. Many streamside wetlands have been filled or bypassed in the management of agricultural areas. If wetlands were restored between agricultural fields and streams, they could potentially intercept runoff and support environments conducive to denitrification, resulting in the removal of nitrogen before it even enters the stream.

In agricultural and natural settings, the use of wetlands for denitrification potential has focused on intercepting groundwater. Urban areas can also represent significant sources of nitrogen. Common urban sources of excess nitrogen include septic systems, automobile combustion, and fertilizers. However, riparian wetlands and other urban ecosystems in which denitrification may occur are often restricted to surface flow because of the prevalence of impervious surfaces. Urban riparian wetland soils have been shown to have a high denitrification potential, so if urban runoff can be channeled into these areas, there is the potential for significant nitrogen removal.

See Also: Ecosystem Restoration; Water Pollution; Water Treatment; Wetlands.

Further Readings

Groffman, Peter M. and Marshall Kamau Crawford. "Denitrification Potential in Urban Riparian Zones." *Journal of Environmental Quality,* 32 (2003).

Mitsch, William J., et al. "Reducing Nitrogen Loading to the Gulf of Mexico From the Mississippi River Basin: Strategies to Counter a Persistent Ecological Problem." *Bioscience,* 51/5 (2001).

U.S. Environmental Protection Agency. "National Water Quality Inventory: Report to Congress, 2002 Reporting Cycle." EPA 841-R-07-001. Washington, D.C.: U.S. Environmental Protection Agency, 2007.

Christopher Anderson
Auburn University

Density

Building green cities means creating human environments that sustain natural habitats, regenerate socioeconomic viability, and maintain healthy standards of living. Part of this formula is examining how the population can be distributed in the available space to best achieve this goal. Residential design, forms of transportation, and communication infrastructure greatly affect social and economic mobilization, physiological and psychological health, and the ecosystem. With rates of urbanization intensifying across the globe and an increasing awareness of climate change, managing human populations in sustainable urban environments is an increasing challenge. Population density is a measure of the number of individuals occupying a given area of space. In the study of cities, a correlation is drawn between densely populated urban regions and energy consumption: The greater the population density, the lower the amount of energy consumed. As such, the concept of density is relevant in building green cities.

There are various measures of urban density. Transport density can be used to measure the number of miles of train tracks per square mile, for example. Residential density measures the number of housing units per square mile. Floor Area Ratio measures the amount of floor space built on a particular unit of land. This article, however, focuses on urban population density as a key measure of how city design can regulate life.

Urban Density: A Historical Perspective

The question of whether human populations should live in close quarters or spread out across green pastures is an old debate in North American and European urban discourse, and both of these spatial designs are tightly knit with sociopolitical and economic patterns, as well as transportation technology. Until the late 19th century, most cities were situated along seafaring passages. Old maps of continental Europe show that the largest cities sat along rivers or on the coast. Large cities were built around harbors, which formed nodes along trade routes. Later, with the introduction of trains, cities emerged along railways. Densely populated urban centers were places of commerce, trade, and political administration. Country properties, in contrast, were sparsely populated, and lands were ruled by wealthy landowners.

John Sewell has documented the history of urban form in his book *The Shape of the City*. He notes that as early as the 11th century, the elite of Florence, Italy, fled to the countryside to escape the city, which was viewed as the site of disease and crime. For centuries onward, cities were often seen as evil necessities. To live in the country was to live the good life. It was not until the late 19th century that cities in Europe and North America began to expand beyond traditional city limits, as intellectuals began conceptualizing new urban forms. Ebenezer Howard, one of the originals of the British Garden City movement, began designs for a revolutionary urban form for what he saw as the modern man. Instead of being crowded within urban districts, emancipation from the shackles of Victorian-age city dwelling was to be found by opening up the countryside as a place of residence. In England, examples that were built along these ideals were the Garden Cities of Letchworth and Welwyn. In Barcelona, Antoni Gaudi designed the famous, yet unfinished, Park Güell, and across the ocean, in Toronto, Rosedale was developed. These neighborhoods marked the beginning of a new era in urban design and signal some of the first modern suburban flights from the city.

Later, the values enshrined in the British Garden City movement would be echoed in designs of suburban developments of the mid-20th century. This transformation was, however, not just a product of design but also the consequence of changing attitudes toward women and labor. Women were to take care of the household in the countryside, and men commuted into the city for work. Work and home became spatially separated; hence, their other name, "bedroom communities." This socioeconomic split enabled family living in green cleaner environments, and the city remained the node of economic activity and the place of labor. The 20th-century arrival of mass-produced automobiles gave rise to the post–World War II American suburban sprawl, as is commonly known today. Living in single-family homes with a private garden and a private means of transportation became available to middle and lower classes, as well as the elite.

In the 20th century, many different forms of suburbs were established. Population densities were lower and ranged from 2 to 30 people per square hectare. They may have emancipated some from the ills of urban living, but it is equally debatable whether or not this is a sustainable model of urban design. Consistent throughout North American suburban spatial form, for example, is the lack of public services. Public transit systems and social programs are seldom financed by taxes collected from areas sparsely populated, and thus are often absent from such districts. In addition, residences in suburban districts are usually limited to those who have the means to purchase a private home and car. Suburban developments of single-family homes depend on the concept of individual capital and the ideology of car-oriented planning, which in turn depends on affordable oil prices, as well as access to and maintenance of widespread water, sewage, and communications infrastructures.

Modern high-rises emerged in the postwar years as a form of housing. This urban form of housing was also a reaction to previous urban patterns. Inspired by Swiss architect Le Corbussier, 30-story buildings (or higher), surrounded by green fields, were designed to hold hundreds of families. Unlike suburban sprawl, however, the goal of this design was not expansion of the city but, rather, the replacement of the inner city. Architects had shown that thousands could be housed on small properties while the surroundings remained green. In Toronto, St. Jamestown was built following these ideals. According to Sewell, older homes were demolished and apartment blocks built in their place. With approximately 17,000 residents, St. Jamestown remains one of the most densely populated communities in North America. In the postwar years, many high-rise developments were built. They were originally designed for the middle class, and they were praised in capitalist as well as socialist cities. In East Berlin, these apartments were hailed for their modern heating and plumbing facilities. High-rise apartments also became a popular design of social housing in both North American and European cities.

High-rise apartment buildings have the capacity to house thousands of people on a small piece of land, and as such, they are potentially a very efficient design in terms of the delivery of infrastructure and resources. They are criticized, however, for several reasons. First, they are seen as unpractical for families and dangerous for small children. Second, if the buildings are not maintained, they may be hazardous and/or extremely cumbersome to live in. Third, it is debatable whether or not the common grounds around high-rise apartment complexes are effectively used, or even safe. As Jane Jacobs, for example, once noted, crime tends to occur in parks, not in busy streets. Despite these possible setbacks, however, there has been a return to high-rise living in many downtown cores, evidenced by the condominium booms in many cities across the globe.

Urbanization, Density, and the Environment

At the beginning of the 21st century, and for the first time in history, more than half of the world's population of 3.3 billion lives in cities. According to the United Nations, the most heavily populated city is Tokyo, with 33 million residents, followed by Mexico City and New York, with 19 million inhabitants each, followed by São Paulo, with a population of 18 million. In 2005, at the time of their report, there were over 20 megacities on the planet with populations greater than 10 million. Cities in the new millennium are also interconnected in ways never before seen in history. Saskia Sassen has written extensively about the transnational flows of capital and labor between global cities, and Michael Peter Smith and Geraldine Pratt have also each discussed the transnational character of local communities in cites. Further, S. Harris Ali and Roger Keil have also observed that cities are networked in new ways, as is evidenced by the rapid spread of infectious diseases—a virus in rural China, such as the SARS vector, is only a plane ride away from downtown Toronto. Ali and Keil argued that the Earth, therefore, is not simply a collection of articulated urban centers and isolated ecosystems: It is a global village. In this context, it is of interest to planners and city dwellers to conceptualize the city and create ways of sustainability. Thus, designing and building livable, green, and sustainable cities is on many planning agendas worldwide. Density, as a specific mechanism of controlled urban growth, is on the agenda of Vancouver, Canada, for example. Although other cities around the globe may not target density, specifically, the concept of population density is implicit in all planning decisions.

In 2008, the City Council of Vancouver accepted the EcoDensity charter as their platform for sustainable urban development. Vancouver has undergone unprecedented population growth in recent years and, as a result, is looking for new ways of planning the expanding city. The guidelines outlined in the charter called for urban growth that protected the natural environment while providing affordable and livable environments for its current and future residents. A denser city, they argued, requires less energy to operate, promotes public health, and is more affordable. The charter aimed to reduce the ecological and carbon footprints of its residents. That is, it aimed to build a city that consumed fewer resources and emitted fewer greenhouse gases. The city cited transportation as their main source (87 percent) of greenhouse gas emissions. As a consequence, the charter encouraged more biking and public transit infrastructures, more pedestrian shopping, the smart zoning of employment and amenity areas, access to local food producers, and more recycling programs. If ideally instituted, these programs would reduce the need for automobiles within the city and would also promote community.

The population density of Vancouver's downtown core reaches 28,000 persons per kilometer square, whereas the outlying areas of the Vancouver Metropolitan Area are as low as four persons per square kilometer. The average density is 736 persons per square kilometer. EcoDensity is a work in progress.

Whether or not Vancouver will achieve their goals remains to be seen. In the meantime, it is a process of reconciling two facts: On the one hand, increased density can lead to more efficient transportation systems and improved delivery of services and infrastructure; and on the other hand, heavy densification can overextend the delivery capacities of urban infrastructure, leading to reduction in water and air quality and the overall quality of services.

The International Awards for Livable Communities, organized by the United Nations, might be one measure of effective urban practices. In 2004, the City of Münster, Germany, was the winner. Among the several factors that were accounted for were the green promenade that encircles the Münster's old city; the plan that reuses the foundation of the old city as the cultural and economic center point, resulting in an absence of high-rise developments; the extensive pedestrian zone; and the omnipresence of bicycle transport. Sixty-five percent of Münster residents use an environmentally friendly mode of transportation: walking, bicycle, bus, or train. Thirty-five percent use bikes. The city boasts 25 bicycle lanes, nine bike-only roads, a regional bicycle sign and traffic light system, and underground bicycle parking facilities for up to 3500 bikes. With a total population of approximately 270,000, the density reaches 15,000 persons per square kilometer in the city center and averages 900 persons per square kilometer in total. Unlike in Vancouver, density was not a specific planning mechanism, yet the net result of Münster's urban design is a livable city that is an example to the world, according to the United Nations.

Shanghai is the most densely populated area of China, with over 2,000 people per square kilometer, measured across a region of 8,000 square kilometers. Urbanization in China has occurred at a spellbinding rate. John Pucher and colleagues have studied this growth, as well as that of India, which has also undergone massive urban expansion. Between 1978 and 2003, urban populations in China rose from 178 million to 524 million. China has now 174 cities with populations of over 1 million. In 1981, 160 million lived in Indian cities. Twenty years later, this number rose to 285 million, and today India has 35 cities with populations over 1 million. Between 1980 and 2005, China's real per capita income quadrupled, and the number of automobiles increased tenfold. In India, per capita income doubled, and the number of cars has tripled.

In both countries, modernization and an expanding middle class have favored the emergence of the private automobile. These changes have pressured the environment and the availability of resources in urban cores, as well as at the peripheries. Pucher and colleagues observed that their main problems concerned heavy traffic and its effect on air and noise pollution, traffic safety, and energy use. The sheer magnitude of urbanization, they argued, is reason enough to take such concerns seriously. If car-oriented planning continues, the carbon footprint of India and China would exceed and overshadow any reduction efforts made by Western nations, and thus a geopolitical element is woven into the picture. As it is, China's urbanization is primarily in the east and southern arable lowlands of China. As a result, they have acted to ensure that 80 percent of the farmland is preserved. Thus, suburban sprawl has been curtailed, and Pucher and colleagues have observed that Chinese suburban districts do not sprawl, as is characteristic of the North American model. Indian cities, in contrast, as a result of limits on population density within central urban districts, are expanding at the fringes. Unlike China, however, Pucher and colleagues observed that these developments were unplanned and are not well serviced. Within denser city cores, numerous new highways have been constructed, and the presence of cars increased manifold. Still, in China's megacities, 80 percent of all trips made throughout the city are by foot, bike, or public transit. Pucher and colleagues would like to see continued support for environmentally sound traffic. They note, however, the absolute absence of bicycle infrastructure in India, with the exception of Chandigarh, and the plans of Chinese governments to scale back the number of cyclists on the road, as they are increasingly seen as a nuisance.

With intensifying urbanization and expanding cities, it remains as important as ever to create urban designs that best accommodate the population. Overpopulated urban regions continue to pose health threats to the population, just as they did a millennium ago. There are lessons to be learned from the history of North American sprawl that can inform the discourse on densification. As new regions in Asia modernize, the discourse becomes more and more relevant there as well. Sparsely populated areas depend on an abundance of resources. Addressing the question of density in urban areas is essentially a question about land and resource distribution and use. Is there enough space for everyone to have his or her own private country house and live happy and healthy lives? Or must we all crowd into downtown high-rises and coordinate the distribution of services and resources? What standards of living are to be met? How will quality of life be measured, and who will decide? How much ecology will be sacrificed for the support of humankind? These are the wider questions that stand at the core of densification. Green cities find a winning balance.

See Also: Barcelona, Spain; Beijing, China; Bicycling; Bus Rapid Transit; Carbon Footprint; Cities for Climate Protection; Commuting; Environmental Planning; Smart Growth; Vancouver, Canada; Walkability (Pedestrian-Friendly Streets).

Further Readings

Ali, S. Harris and Roger Keil. *Networked Disease: Emerging Infections in the Global City.* Oxford: Blackwell, 2008.

Jacobs, Jane. *The Life and Death of Great American Cities.* New York: Random House, 1961.

Pratt, Geraldine. *Working Feminism.* Edinburgh, UK: Edinburgh University Press, 2004.

Pucher, John, et al. "Urban Transport Trends and Policies in China and India: Impacts of Rapid Economic Growth." *Transport Reviews,* 27/4:379–410 (2007).

Sassen, Saskia. *Cities in a World Economy.* Thousand Oaks, CA: Pine Forge, 1994.

Sewell, John. *The Shape of the City: Toronto Struggles With Modern Planning.* Toronto: University of Toronto Press, 1993.

Smith, Michael Peter. *Transnational Urbanism: Locating Globalization.* Oxford: Blackwell, 2001.

"Vancouver EcoDensity Charter." Vancouver, British Columbia, Canada: City of Vancouver, 2008.

Constance Carr
Humboldt University, Berlin

Distributed Generation

Distributed generation is a term used to describe any process for the production of electricity that is localized, with minimal transmission and distribution systems. Distributed generation stands in contrast to centralized methods of energy generation that provide electricity across a large region by means of a network (often referred to as "the grid") of high- and low-voltage transmission lines with substations, and associated voltage loss and high infrastructure costs. Distributed generation generally falls within the range of 3–10,000 kW of generation and involves less initial investment, greater reliability (fewer outages) and consistency (fewer power dips or surges), and in almost all instances, less environmental impact than centralized generation. A combination of circumstances including the development of alternative energy sources, environmental regulations, utility restructuring, and a rapidly developing marketplace for electrical energy have made distributed generation an increasingly important strategy for the development of new energy capacity. The primary defining element of distributed generation is that the power source is located close to the load being served. Distributed generation can supplement the grid by reducing peak electricity demand (peak shaving) or providing energy independent of the grid.

Centralized energy generation typically relies on nuclear fusion, coal, or other fuel sources dependent on economies of scale for their efficient use. Distributed energy resource systems can employ smaller scale technologies such as:

- Solar panels
- Wind turbines
- Biodiesel
- Reciprocating engines
- Microturbines
- Combustion gas turbines (including miniturbines)
- Fuel cells
- Photovoltaics

The Electric Power Research Institute estimates that for the near term, 20 percent of all new power generation will use distributed, as opposed to centralized, technologies. Where new energy markets require rapid infrastructure, an array of small gas-fired power systems have become available to meet this demand. These systems provide fuel flexibility and clean generation, using natural gas, propane, or fuel gas from coal or biomass. Additional fuel sources include waste from refineries, municipalities, and the forestry and agricultural industries. An estimated 2 billion people around the world currently live without electricity. Distributed generation constitutes their best hope for a transition to an energy-based economy.

Distributed generation can also provide needed capacity in areas where demand is approaching available supply from the grid. *Peak shaving* is a term used to describe any practice that reduces generation at primary plants during peak demand periods. By providing additional capacity, distributed generation sources can contribute to peak shaving and thereby help avoid the economic and social costs of new centralized plant construction. This approach is efficient in terms of time as well as resources because the costly pursuit of permitting and approvals for transmission-line rights-of-way is avoided. An environmental as well as economic benefit is provided by the avoidance of transmission line costs and associated electrical losses. Energy losses in power generation constitute a significant source of carbon emissions that can be reduced by means of distributed generation. The savings realized by distributed generation represent improved efficiency and productivity and promote economic growth.

Distributed generation is also able to address power quality concerns by removing the variables associated with extended transmission and distribution networks. This is of particular importance to computer microchip manufacturers and other high-technology industries, for which power surges or dips pose quality control problems. Multiple local generation sources overcome transmission bottlenecks resulting from grid deficiencies and are more responsive to tighter emission standards. An additional benefit cited by the U.S. Department of Energy (DOE) is the advantage of dispersed distributed generation facilities in addressing energy security concerns.

Many forms of distributed generation can employ cogeneration, also known as combined heat and power. This is a process whereby waste heat recovered from energy generation processes is employed for site-specific purposes, such as space or water heating. Through the use of an absorption chiller, the cogeneration process can also provide space cooling. The DOE estimates that by 2020, 5 percent of all energy consumed in the United States will be recycled thermal energy. The efficiency of an energy-generation system is a factor of the percentage of input fuel converted to useful energy in the form of electricity and heat. Cogeneration can increase many distributed generation technologies, including fuel cells, from 40 percent to 80 percent efficiency. By using what would otherwise be waste heat for useful purposes, cogeneration provides an environmental as well as economic benefit.

The DOE is actively supporting research in distributed generation technologies. Among these are the following:

- *Industrial gas turbines:* Improved efficiency is sought through advanced materials research in areas such as composite ceramics and thermal barrier coatings. Low-emission technologies research and development seeks to improve the combustion system by reducing the nitric oxide and carbon monoxide production.
- *Microturbines:* Planned activities focus on specific performance targets for the next generation of microturbine product designs, including fuel-to-electricity conversion efficiency of at least 40 percent, nitric oxide less than 7 ppm (natural gas), 11,000 hours of reliable operation between major overhauls, and a service life of at least 45,000 hours, with system costs less than $500/kW and costs of electricity that are competitive with alternatives (including the grid) for market applications and options for using multiple fuels including diesel, ethanol, landfill gas, and biofuels.
- *Thermally activated technologies (TATs):* TATs optimize energy delivery systems by replacing electric with nonelectric devices when the electric distribution system is at peak demand. TATs such as absorption cooling/refrigeration and desiccant dehumidification can reduce the demand peak driven by air conditioning and refrigeration in the United States.
- *Gas-fired reciprocating engines:* The target for fuel-to-electricity efficiency for a new generation of gas-fired reciprocating engines is 50 percent by 2010, along with engine improvements in efficiency, combustion, strategy, and emissions intended to substantially reduce overall emissions to the environment. Natural gas–fired engines are to be adaptable to future firing with dual-fuel capabilities. Dual-fuel options may be considered in the design. (The target for busbar energy costs, including operation and maintenance costs, is 10 percent less than current state-of-the-art engine systems, while meeting new projected environmental requirements.)

The DOE has also invested in fuel cell production, with specific targets for commercial introduction of new technologies, including molten carbonate and solid oxide fuel cells, solid-state composition and advanced fabrication, and fuel cell/gas turbine hybrids. An additional goal is the introduction of an advanced gas-fired reciprocating engine, with 50 percent efficiency and nitric oxide emissions of 5 ppm or less, by 2010.

The economic and environmental advantages of distributed generation are associated with significant social benefits. Where distributed generation is connected to the grid, decentralization of supply provides greater resilience and redundancy to the power network, and greater autonomy and self-determination for the user. A study by the American Center for an Energy Efficient Economy concluded that savings realized from distributed generation are retained in the local economy, resulting in greater economic benefit than energy fees. Any reduction of reliance on foreign oil has multiple beneficial consequences for balance of trade, energy security, and international relations.

Yet it may be the environmental benefits of distributed generation that prove to be the most profound. Renewable energy sources represent the ultimate low-impact form of distributed generation, but even forms that employ fossil fuels or nonrenewable resources wield a smaller environmental impact than centralized energy generation. The environmental consequences of coal and nuclear power and the depletion and political costs of oil are strong arguments for policies that encourage distributed generation.

Such policies have been enacted in the Energy Policy Acts of 1992 and 2005, as well as Executive Order 13423 and the Energy Independence and Security Act of 2007, and will likely receive further reinforcement in future legislation. Decentralization of generation marks a profound paradigm shift in the economics of energy and may transform the nature of world economies and the structure of societies. Through an emphasis on local generation and the specific needs of individual communities, distributed generation has the potential to address the triple bottom-line concerns of "people, planet, and profit," as opposed to the single bottom line of the balance sheet. Although at this time coal supplies 50 percent of all electricity generated in the United States, the combined environmental concerns of carbon emissions and toxic sludge from coal ash will result in growing pressure on the coal industry to alter generation processes. Similarly, the nuclear industry is burdened with enormous start-up costs and waste storage concerns. A concerted effort to address both the environmental concerns of a carbon-based energy system and the projected increase in demand over the next several decades must employ all potential sources of distributed generation if disaster is to be avoided.

The recent influx of capital into the realm of alternative energy is a further indication of the paradigm shift represented by distributed generation. As technology advances reduce costs and improve efficiencies, the potential for reductions in energy costs, combined with environmental and social benefits, may smooth the way to a post-oil era of political stability, with increased social and economic equity.

See Also: Carbon Footprint; Combined Heat and Power (Cogeneration); Power Grids.

Further Readings

Borbely, Anne-Marie and Ian F. Kreider, eds. *Distributed Energy: The Power Paradigm for the New Millennium*. Boca Raton, FL: CRC Press, 2001.

Chambers, Ann, et al. *Distributed Generation: A Nontechnical Guide*. Tulsa, OK: PennWell, 2001.

"Prospects for Distributed Electricity Generation." Congressional Budget Office, 2003. http://www.cbo.gov/doc.cfm?index=45528&type=O (Accessed October 2009).

A. Vernon Woodworth
Boston Architectural College

DISTRICT ENERGY

The dominant model for home heating and cooling consists of highly centralized and expensive generating or distribution facilities powered by natural gas, hydroelectric, coal, or nuclear energy sources that send energy, gas, or fuel to many thousands of homes. This model is becoming increasingly strained because of its inherent inefficiencies, social inequities, and overall environmental unsustainability.

Now many are advocating a return to an older model: decentralized, or district, energy production that distributes space heating, domestic hot water, and cooling within a complex of buildings or over a range of city blocks. District energy (DE) has the potential to address issues related to network vulnerability in terms of disrepair, accidents, catastrophic weather, and even acts of terrorism. However, the particular value of this approach in terms of sustainability is that it allows communities to transition from less sustainable (or more expensive) fuel sources to "greener" alternatives.

Strictly speaking, DE does not refer to the production of the energy itself but, rather, to a thermal network and the way it is managed on a community level. Instead of piping gas into a housing unit from a central location to be combusted in a furnace, or electricity to be consumed by an air conditioner, a DE system uses a medium (usually water or steam, but also synthetic fluids) to carry and transport the heat (or coolant) from the point of origin to the user through a network of underground pipes. Heating and cooling needs are met in individual buildings, and the medium is then returned to the DE plant to be once again heated or cooled as needed.

Far from being a new technology, DE was first made commercially available in Lockport, New York, in 1877, and by the late 19th century it was in wide use in North America and Europe. Much of Manhattan is still heated by Con Edison's steam plants, with new structures added to the system all the time. However, following World War II, Europe continued to expand its DE services, whereas North American cities turned instead to coal, oil, and natural gas.

Common DE markets include municipalities, or clusters of buildings within a municipality. Campus settings such as hospitals, universities, and industrial parks are ideal potential customers. DE systems work well within an industrial ecology model in which waste products from one industrial process may be used by another. In the case of DE, this is referred to as *cogeneration*, or combined heat and power, wherein the waste heat from one industrial process is used to heat the medium in a DE system to generate electricity, rather than being discharged into the environment. This flexible model extends to taking advantage of other natural and built assets. DE cooling systems can pump already-chilled water from deep lakes to cool buildings or draw warm water from abandoned and flooded mines to heat nearby towns. Alternately, they can be teamed with other green energy solutions, such as thermal solar energy collectors.

Best Uses of Direct Energy

DE systems are best suited for conditions where there is a large institutional energy consumer, so an essential consideration in DE systems is the nature of the urban form. They are not generally well suited for low-density suburban neighborhoods unless there is a significant institutional user nearby because they require a certain level of building density. Large hospitals, businesses, universities, or industries are desirable primary customers that can house the power plant, and then surrounding residential users may be incorporated into the system.

For the institutional customer, taking advantage of a DE system offers many benefits, as they can forego the expenses and maintenance responsibilities required for privately owned on-site heating and cooling equipment. The only ongoing expense is for the heat or cooling actually delivered, instead of the capital and ongoing expenses involved in purchasing, maintaining, and replacing furnaces, boilers, and centralized air conditioning. This then frees up space that can be used for other—potentially revenue-generating—uses. This consideration carries over into the wider urban environment: With massive and polluting centralized power plants no longer needed, the quality of the cityscape may be dramatically improved.

Adopting DE approaches can, under certain conditions, also help meet social sustainability goals. The growing social problem of fuel poverty (or low-income energy burden) is a result of the fact that every homeowner is responsible for paying for his or her own heating systems and fuel sources, and renters must pay for their energy use plus contribute to their landlord's ongoing capital costs. Household energy demand is relatively inelastic; that is, a certain level of energy consumption is necessary to meet minimum comfort and utility, and so regardless how energy units are priced in a particular jurisdiction, they will be consumed according to need. Lower-income households also generally share certain characteristics that exacerbate their energy burden: They are rented, use more expensive electric space and water heating rather than natural gas, and are equipped with older heating units. In an energy regime with escalating prices, energy consumers have little choice: They must adjust their household expenses so as to afford a minimal level of heat.

In contrast, a DE system distributes the costs for energy infrastructure across an entire community, so households are freed from the possibility of catastrophic expenses. Furthermore, the underlying flexibility of DE systems allows them to take advantage of fuel-switching strategies: When certain sources become too expensive, the utility can heat the DE medium with another, less expensive source. Heating and cooling costs are thus stabilized for the entire community, including institutional users.

Environmental Sustainability

For purposes of environmental sustainability, the attraction of DE systems is this ability to "fuel switch" that allows for the use of a wide range of fuel sources, which can include municipal waste or otherwise valueless industrial by-products, such as wood chips. The real benefits accrue when a DE system is used as a platform for transitioning from fossil fuels to renewable sources such as biomass.

Because of the diversity of energy consumers in a given community, each with different energy needs, energy conservation measures are highly dependent on social marketing approaches geared toward obtaining buy-in and cooperation from each individual energy user. In contrast, DE systems connect all these potential users together and control the consumption of energy at a central source. This offers utilities a greatly enhanced ability to manage energy consumption and promote wider sustainability goals in the community. Unlike standard systems, in which it is up to the resident or owner of each building to regulate fuel flows, these systems are centrally managed and can even take the form of thermal storage during off-peak hours. The levels of efficiency possible with DE are such that emissions can be significantly reduced when compared with standard systems, even if the fuel source (e.g., natural gas) is the same.

In its ability to make use of available waste heat, to regulate output according to community demands, to manage energy consumption, and to switch to more sustainable fuel sources, DE is becoming an increasingly popular choice for community energy planning initiatives and for emerging "green" urban developments.

See Also: Combined Heat and Power (Cogeneration); Distributed Generation; Power Grids; Renewable Energy.

Further Readings

Arkay, Kathereine E. and Caroline Blais. *The District Energy Option in Canada.* Ottawa: Natural Resources Canada, 1999.
Church, Ken. "District Energy: Is It Right for Your Community?" *Municipal World* (November 2007).
Gochenour, Carolyn. *District Energy Trends, Issues, and Opportunities: The Role of the World Bank.* Washington, D.C.: World Bank Publications, 2001.

Michael Quinn Dudley
University of Winnipeg

Dongtan, China

The master plan for what has been called the world's first sustainable city—Dongtan, Chongming Island, in Shanghai, China—is perhaps the best known of the ecocity designs that swept professional and popular imaginations in the first decade of the 21st century. In late 2005, the investment arm of the Shanghai municipal government, the Shanghai Industrial Investment (Holdings) Co., Ltd. (SIIC), contracted with the U.K. design and engineering firm Arup to draw the strategic master plan for 86 square kilometers of land adjacent to a nationally protected migratory bird park. The signing of the agreement drew international accolades and led to the launch of a collaboration between industry and academia in the United Kingdom and China that was focused on sustainable cities. Controversy arose around two points: First, the potential of protecting a wetland of international importance for migratory birds adjacent to a newly built city of 500,000 people, and second, how to ensure an equitable future of those persons physically or financially displaced by its construction. Less obvious to outside observers were the conflicts between various government offices within China over the development of Dongtan, and the economic and national interests in the United Kingdom for championing the project.

Despite construction on the city having never begun, both its planners and reporting media spoke of Dongtan as if it had already overcome the juggernaut of urbanization and energy usage that is conceived as imperiling the 21st century. Heralded as the world's first ecocity, Dongtan's first phase of development was to be completed by 2010, so that visitors from around the globe could come to Shanghai's idyllic Dongtan marina to sip coffee, watch yachts glide by, and contemplate their bets on the afternoon's horse races. Dongtan was to have a planned population of 50,000 by 2010, have 90 percent of its waste recycled, and obtain 95 percent of its energy from alternative sources.

Dongtan: Conservation Area and Ramsar Wetlands

Sitting at the far end of Chongming Island, Dongtan ("Eastern Sandbank") extends farther each year into the Pacific Ocean, growing with the Yangtze River's deposits of silt gathered along its run through central China. Deforestation and other causes of increased silting of the Yangtze resulting from rapid economic development led Chongming Island to double in size between 1950 and 2005 to 1,229 square kilometers, making it China's third-largest island after Hainan and Taiwan. It is on the detritus of economic growth elsewhere that the Dongtan ecocity is planned to be built.

Dongtan is divided into two sections: first, an 86 square kilometer conservation area of farmland and aquaculture enclosed between the 1968 and 1998 dykes; and second, the exterior wetlands on the sea side of the 1998 dyke. Although critics of the SIIC/Arup project argue against the depopulation of the site, the area of development does not currently include permanent residents, as recognized by Chinese law, and lands now being farmed have been legally managed by SIIC since 1998.

The wetland area exterior to the 1998 dyke was listed as a nationally protected area in 1992, and in 2001 it became an Important Bird Area listed under the Ramsar Convention on Wetlands. Although best known as a passage area for the critically endangered black-faced spoonbill, the Dongtan wetlands also provide passage for the spotted greenshank and winter grounds for the hooded crane.

World Ambitions

In late 2002, the government of Shanghai hosted a conference on how to transform Shanghai into a world city. According to a book jointly published by SIIC and Arup, Dongtan is a place with a wealth of natural capital of extraordinary value that is not being developed. The spoondrifts and spoonbills that populate Dongtan are tacitly negatively contrasted with the 500 buildings that have risen in Pudong's Liujiazui neighborhood in the past 10 years. Although Chongming Island occupies 20 percent of Shanghai's total area, the book laments that Chongming Island contributes just 1 percent of the city's gross domestic product.

In the 1980s, Pudong was selected over Chongming as the site for central government–supported economic development. The plan for Dongtan and the obstacles facing it demonstrate that the exact meaning and implementation of ecological modernism remains unclear, however. It is clear that the impetus for developing Dongtan was to build a city that would not only put Shanghai on a global map but also enable it to redraw that map with Shanghai and the world's first ecocity at its center.

Dongtan is reflective not only of China's ambitions in the 21st century but also of the United Kingdom's ambitions. National ambitions are often congruent with industry's ambitions, and such was the case with the development of Dongtan. The United Kingdom has been seeking to reemerge as a center of technological innovation and, with the lack of leadership in the United States over climate change, has made ecological innovation—and export of expertise, technology, and financing—part of its development strategy. The congruence of national and industrial interest is evidenced by the signatories to a 2008 memorandum of understanding between the governments of the United Kingdom and China on sustainable cities. The memorandum reaffirmed support for the implementation of Arup's master plan for Dongtan and sought cooperation on matching U.K. technology and expertise with Chinese public works projects, as well as fostering a Sino–United

Kingdom Institute for Sustainability; the signatory representing the government of the United Kingdom is Arup's director for global research. That Dongtan is to be commercially profitable, and that strategic relationships between national governments are made through financial and trade relationships between private firms, is customary, yet such monetary incentives are often downplayed or ignored in favor of portraying sustainable development as being solely of ecological benefit.

The Master Plan

Arup's master plan is the latest of the more than 60 development strategies proposed between 2000 and 2005. SIIC's first solicitation was for town and land use planning of an expanded area of 174 square kilometers that included Chenjia town and Dongtan. Plans were submitted by the Atkins Group (United Kingdom), Philip Johnson (United States), and the Architecture Studio (France) in 2001. The Shanghai Urban Planning Research Institute was hired to execute a comprehensive plan, which was approved in 2002. SIIC again sought international design collaboration in 2002 for an international leisure park from EDAW (United States), Kuiper Compagnons (the Netherlands), and Groep Planning (Belgium), and for a wetlands park from Sasaki (United States), Groep Planning (Belgium), Alterra (the Netherlands), Japan Comprehensive Planning, and Fleming (United States). Arup became involved in 2004, after McKinsey & Co. was hired to consult on the development of Dongtan and sought out Arup to analyze the effect of development on the Ramsar-protected Dongtan wetlands. Arup, an engineering and design firm, made its response clear in its proposal: increase the planned population of the city from 50,000 to 500,000. A development of 50,000 was argued to be an environmental disaster. According to Arup, a denser, more-populous city was not only more environmentally sustainable but also offered greater commercial opportunities.

After the completion of the strategic plan for 86 square kilometers of the Dongtan conservation area, Arup was contracted to create a master plan for urban development. In contrast to the earlier plans proposed in 2001 that were integrated with Chenjia town, in response to SIIC's request Arup proposed a new, purpose-built ecocity at the southern end of the site.

Arup's master plan for Dongtan is a modular plan intended to be built over the span of 45 years, to be completed in 2050. The plan provides 29 square meters of green space per person—more than four times the amount in Los Angeles—and ensures that no place in the city is farther than 540 meters from a bus stop. Dongtan's ecological footprint is modeled as being less than half that of a typical Chinese city. Ninety percent of all waste is to be recycled, and 95 percent of all energy is to come from renewable sources. Biogasification of rice husks will supplement wind and solar power. Public transportation will be made viable by ensuring a density of 160 people per hectare. Only cars with zero tailpipe emissions would be allowed inside the city; all others would have to be left in a parking lot. Through the conversion of agricultural lands into parks, green roofs, and waterways, Arup projects that biodiversity will increase in the conservation area.

The city of Dongtan would be formed through the integration of three towns, referred to as villages in the SIIC/Arup literature. Marina Village is designed as a place of leisure and luxury. Sailing yachts, riding and betting on horses, and playing golf serve as the area's tourist draws. Lake Village is planned as a center for ecological- and tourism-minded agriculture; home-made cheese will be available for purchase here. Plans build on SIIC's existing farming ventures in Dongtan and propose several international joint-venture farms, including a Sino–German ecofarm, a Sino–Japanese ecofarm, a Sino–Italian ecofarm, and a Taiwan Folk Village. What makes these farms specifically ecological is not addressed. Conference centers, restaurants, and hotels will also populate this village. Pond Village is planned as a center for information technology industries, science research institutes, and smaller-scale residential neighborhoods. Canals would be a prominent feature. Each town will have its own central downtown that will serve as a secondary medium-density area to a high-density town center at the periphery. Each of these three towns' peripheries will eventually overlap to form the downtown of Dongtan City.

In addition to the town and city plan, SIIC's comprehensive master plan includes three leisure parks, each focused on a different theme: International Leisure Center for equine and water sports (Arup), New Energy Park for water sports and science education (Arup), and National Wetland Park for water sports and vacation villas (Fleming). Marina Village and each of the three leisure parks were to be finished by 2010, and Marina was to provide residence for 50,000.

SIIC/Arup's presentation of the future of Dongtan seeks to emulate and combine the development models of Florida, United States, and Bangalore, India: capitalizing on beaches and wetlands to turn them into pleasure resorts and focusing on information technology for development to create a benign ecological environment. Such comparisons indicate that although the master plan for Dongtan proclaims it as an ecological city, its focus remains solidly on becoming a leading economic city. According to the Natural Resources Defense Council, the Everglades' wading bird population has decreased 90 percent in 20 years as a result of the rapid development of Florida's East Coast. Bangalore is beset by poor air quality and waste management. Each, however, demonstrates half of SIIC's vision for Dongtan: an international leisure destination and an information technology center.

Arup's master plan for Dongtan may never be built, yet lessons have been learned. The knowledge gained through the SIIC/Arup contract to design a pleasant, commercially viable city that uses near 100 percent renewable energy has been leveraged into new business for Arup in locations such as Destiny, Florida. Arup also shares some of the technical knowledge gained with other professional audiences in the United States, United Kingdom, and China.

With the first phase of ecocity construction expected to cost more than 1 billion GBP, financing and return on investment remain an obstacle.

Corruption and Conflict

Although construction of the Dongtan ecocity was to begin in 2006, it was widely reported that the arrest and sentencing to 18 years of Shanghai Communist Party secretary Chen Liangyu for extensive bribery and abuse of real estate transactions caused the SIIC/Arup project to come to a standstill. It is less often noted that SIIC was intimately involved in this $4.8 billion corruption scandal. Because real estate played an extensive role in the corruption scandal, after Chen's arrest in September 2006 all large construction projects were put on hold pending further review.

In a move to reinvigorate the project and demonstrate its financial viability, in January 2008 Arup announced that its plans were moving forward and that HSBC and Sustainable Development Capital Limited would be working on financing the real estate development with a new business model. Seeking to emulate the Cambridge, Massachusetts, development base of Harvard University and Massachusetts Institute of Technology, a memorandum of understanding announced that a new research facility would ground the development of Dongtan ecocity: the Dongtan Institute for Sustainability. As of fall 2009, no construction has begun on the ecocity, and there has been no physical development of an institute.

Future Possibilities

As of late 2009, the path of future development of Dongtan remains unclear. SIIC still holds the rights to develop the land, Arup is still advocating its ecocity master plan, and Chongming County is still putting development priority on Chengqiao New City and Chenjia Town. The tunnel and bridge connecting central Shanghai to Chongming Island will open by the end of 2009, likely bringing dramatically increased traffic to Dongtan and the rest of the island. Speculation about the construction of various large theme parks on the island was rampant in the summer of 2009. Such construction would raise land values and make the political decisions necessary to protect the Dongtan conservation area and Ramsar wetlands even more fraught with economic and ecological conflict. Whether development of Dongtan will prove to be a lesson in ecological management success or failure remains to be seen.

See Also: Adaptation, Climate Change; Carbon Footprint; Ecovillages; Green Design, Construction, and Operations; Natural Capital.

Further Readings

Brown, Gordon. "Prime Minister Gordon Brown's Speech at UK China Business Summit: January 18, 2008." http://www.cbbc.org/GordonBrownspeech.pdf (Accessed August 2009).
Castle, Helen. "Dongtan, China's Flagship Eco-City." *Architectural Design*, 78/5:64–9 (2008).
Chongming County. "Chongming Ecological County Construction Plan 2007–2020." http://cmx.sh.gov.cn/gb/node2/node4089/node4091/userobject1ai10510.html (Accessed August 2009).
Chongming County. "The Master Plan for Development of Chongming Three Islands." 2006. http://cmx.sh.gov.cn/cmwebnew/node2/node21/node23/node826/userobject1ai9753.html (Accessed August 2009).
Natural Resources Defense Council. "Florida Everglades." http://www.nrdc.org/water/conservation/qever.asp (Accessed August 2009).
Shanghai City. "Chongming Plans Ambitious Development." April 8, 2004. http://en.shac.gov.cn/nyxw/t20040408_98884.htm (Accessed August 2009).
Shanghai Industrial Investment (Holdings) Co., Ltd. and Arup. *Shanghai Dongtan: An Eco-City*. Shanghai: Shanghai Sanlian Shudian, 2006.
Wright, Chris. "Dongtan: China Firms Up Eco-City Proposal." *Euromoney*, March 4, 2008. http://www.sdcapital.co.uk/src/Euromoney.pdf (Accessed August 2009).

Shannon May
University of California, Berkeley

DZERZHINSK, RUSSIA

Dzerzhinsk, Russia, is one of several cities in the former Soviet Union named after the first head of the Soviet secret police, Felix Dzerzhinsky. The city is located approximately 250 miles east of Moscow, in Nizhny Novgorod Oblast. Dzerzhinsk was formed in 1929 because of its proximity to the railroad, proximity to the river Oka, and proximity to Nizhny Novgorod, the Oblast capital. Dzerzhinsk has been recognized as the second most polluted city (after Chernobyl) in the world. Recently, its prominence has been replaced by a few cities in China, although it continues to be listed in the top 10 most polluted cities in the world. As a result of improper disposal of chemical weapons that were produced from the mid to late 1940s until the late 1960s, the city of Dzerzhinsk has numerous health problems. Throughout the city, over 1,150 sources of toxic pollution have been identified.

The city was initially organized to produce enamel, phosphorous, and poison gas. However, the city primarily manufactured chemicals for use in wartime, including the ingredients of mustard gas and lewisite (from arsenic and chlorine), which irritates the lungs and causes blisters. The city area contains a karst topography that provided sinkholes into which the industries of Dzerzhinsk dumped chemicals. In fact, when one company discharged ammonium sulfate into a sinkhole, it appeared quickly in the entire region's groundwater.

Over 300,000 tons of chemicals produced by Dzerzhinsk's factories were dumped in local bodies of water and the sinkholes. This occurred in earnest as the Soviet Union broke apart in the late 1980s and early 1990s, when funding dried up for proper disposal. This dumping has produced a harsh discolorization of the water (nicknamed the White Sea) and has, in turn, affected the drinking water of the surrounding communities. It is rumored that birds that touch the water die. According to a 2003 BBC report, in one village outside of the city, a cow's growth was halted after drinking from the town's well. Reports of increased cancer have surfaced, and many gravesites of those younger than 40 years have been noted. In 2004, the Blacksmith Institute funded the installation of water treatment systems of nearby Pyra and Gavirolvka, whose drinking water is highly polluted and which have only one water source. Because many chemical factories now lay idle, the water table has risen and threatens the Oka river basin, which supplies fresh water to the nearby city of Nizhniy Novgorod, a city of 1.3 million.

Environmental groups have taken soil samples and have found all kinds of toxins high above the legal limits. The major dangerous chemicals in the contaminated water and soil around the town are phenol and dioxin. The levels of these two chemicals are 17 million times above the safe limit. Phenols can be found in common household disinfectants, but the concentration levels around Dzerzhinsk are very high, which can cause harm to the central nervous system, the heart, and kidneys. Phenols can also have harmful effects on blood pressure. Dioxins, a result of chlorine production, cause a cancer of the connective tissue in the body (named sarcoma). Moreover, dioxins can cause reproductive problems in both men and women. A recent report from the BBC told of the extent of reproductive problems in men from Dzerzhinsk that have caused local women to look elsewhere for husbands. In the same report, Dr. Grachya Muradian, the chief of the Dzerzhinsk's maternity ward, noted the negative changes in hormonal and immune systems of Dzerzhinsk's newborns resulting from the contamination of the city's groundwater and land.

The smell of the city is quite putrid. Many citizens could tell which stop they were at in the city on the local tram simply by the smell of the factory nearby. This gave rise to a local ditty:

> The express train doesn't come to my native town
>
> We give all the kids gas masks
>
> I know you my beloved Dzerzhinsk
>
> The capital of chemicals and affliction.

Dzerzhinsk's residents (which number approximately 260,000) were reported to have a life expectancy of the mid-40s, whereas the overall life expectancy in Russia is in the mid-60s. The death rate of people in the city far outstrips the birth rate (2,600 births to 4,400 deaths, according to 2008 statistics).

Dzerzhinsk is currently seeking foreign investment to improve its factories and put its people to work. Dzerzhinsk has one Wella shampoo factory that employs about 200 people, but 25 percent of the town still works at factories producing toxic chemicals. In addition, the Joint Stock Company, Research and Design Institute of Urea and Organic Synthesis Products provides technical guidance on running chemical factories and provides turnkey projects. However, economic progress remains weak, although the city has a presence on the web at http://www.dzr.nnov.ru.

See Also: Norilsk, Russia; Water Pollution; Watershed Protection.

Further Readings

Anderson, Dennis. "Years of Industrial Pollution Take a Toll on Russians: Albertan Leads International Team Helping Communities Overcome the 'Silent Killer.'" *Calgary Herald* (April 4, 1998).

Bolotova, Alla. "Ignorance of Environmental Risks in the 'Capital of Soviet Chemistry.'" http://www.allacademic.com//meta/p_mla_apa_research_citation/0/2/1/2/1/pages21216/p21216-1.php (Accessed September 2009).

Rudnitsky, Jake. "Toxic Felis: A Visit to Dzerzhinsk, Russia's Most Polluted City." (June 2004) http://www.exile.ru/articles/detail.php?ARTICLE_ID=7377&IBLOCK_ID=35 (Accessed September 2009).

Samuals, Timothy. "Russia's Deadly Factories." *BBC News* (March 7, 2003).

Charles R. Fenner, Jr.
State University of New York, Canton

ECOINDUSTRIAL PARKS

Ecoindustrial parks (EIPs) are defined as communities of businesses that cooperate with each other and their local community to efficiently share resources (information, by-products, wastes, materials, water, energy, infrastructure, and natural habitat), leading to economic and environmental gains for the participating businesses and the surrounding community. The concept emerged in the early 1990s as a concrete manifestation of the argument that the traditional model of industrial activity, in which individual manufacturing processes take in raw materials and generate products to be sold plus waste for disposal, needs to be transformed into a "closed-loop" industrial ecosystem, in which used materials (wastes), excess energy (e.g., steam), and by-products (unwanted products resulting from the production of other products; e.g., asphalt from petroleum refining) substitute for virgin materials during production processes. The resultant new ecoindustrial system would significantly reduce the overall environmental impact of production and consumption compared with current practices. This article discusses the evolution of EIPs, their purported benefits, the challenges in establishing a successful EIP, and how the concept might evolve in the future.

Evolution of the Concept of EIPs

Much of the early impetus behind the development of EIPs was derived from the now classic, and thoroughly documented, example of the town of Kalundborg, Denmark. Here, a complex web of 18 by-product material and energy exchanges among a relatively tight-knit cluster of companies and the local community emerged over the last 30 years. For example, the local power station pipes steam to a local biotechnology company (energy), which provides surplus yeast to local farmers as pig food (waste) and excess calcium sulfate (by-product) from the sulfur dioxide scrubbers to a local wallboard company. Such linkages evolved gradually and bilaterally (i.e., between individual firms), and without an overall design plan, as individual firms sought to make economic use of their by-products and to minimize the cost of compliance with new, ever-stricter environmental regulations. The network continues to evolve, with some linkages disappearing and new ones forming.

The experience at Kalundborg and the identification of other waste and by-product networks around the world stimulated calls for the development of EIPs that would establish by-product exchanges and energy cascades to form the nucleus of a future closed-loop industrial ecosystem. Further motivation toward EIPs in the United States was driven by the President's Council on Sustainable Development, which designated (but did not fund) four EIP demonstration sites in 1994—Baltimore, Maryland; Cape Charles, Virginia; Brownsville, Texas; and Chattanooga, Tennessee. Of the four, only Cape Charles became operational, but it abandoned its EIP focus in the early 2000s. The strategy of constructing EIPs was also adopted in Europe, Australia, and most recently, as an essential element of Chinese attempts to "green" their economy.

Although the term *EIP* uses the word *park*, the community of businesses does not necessarily need to reside within the physical confines of an industrial park. Rather, the key elements are proximity and collaboration among the firms, which leads to the mutual benefit (economic and environmental) of the firms involved, rather than a common physical location. Many developers use the term EIP relatively loosely, and more as a marketing tool, for facilities ranging from recycling centers, to groups of environmental technology companies, to companies making green products, to industrial parks with environmental infrastructure, and to mixed-use developments. However, the original conception of EIPs was that the firms would use each other's wastes and by-products as inputs for their products. For example, one model conceives of EIPs as involving at least one major anchor firm exporting raw or processed materials, which is connected with one or more firms capable of using significant portions of the anchor firm's waste stream and by-products, which in turn, would be linked to several "satellite" enterprises converting those wastes into usable products.

Promise and Reality of EIP Developments

The development of EIPs was purported to generate a variety of positive economic, environmental, and social advantages. Lower environmental impacts would accrue from the reduction of virgin material and energy inputs and the reduction of waste and emissions from outputs. By using wastes and by-products rather than virgin input, companies would have lower input costs, and firms that generate wastes would have lower disposal costs, which would lead to a better competitive position and greater profits. EIPs were also heralded as a new local economic development tool in the sense that new markets, and therefore new firms and jobs, would emerge from the new opportunities to exchange wastes and by-products. In addition, the clustering of ecoindustrial firms in a locality and the subsequent interactions among those firms should result in greater economic competitiveness because the shared access to information, network suppliers, distributors, markets, resources, and support systems would result in faster problem solving, ultimately leading to new product and production innovations.

Despite a wide variety of different ecoindustrial park experiments, however, most of the planned EIPs failed. For example, reviews of various EIPs in the United States, Western Europe, and parts of Asia found that only a few parks had operational by-product exchanges, although a number had plans to do so in the future. Other collaborative practices such as shared greywater recycling, waste disposal, and other utility sharing were more common. However, although these practices have led to a reduction in the environmental impact of industrial parks, the original core concept of by-product or waste exchange has remained elusive.

To some extent, the lack of success for many planned EIPs is not surprising. Firms sometimes get large shares of their inputs locally and/or find markets for their outputs within the confines of the local region; more often, firms are linked to the outside world, with the majority of their suppliers located in other places. In addition, EIPs face a series of barriers to the establishment of waste, by-product, and energy exchanges, including technical, where an exchange is technically not feasible; economic, where an exchange is economically unsound or risky; informational, where the appropriate people do not have the relevant information; organizational, where exchanges may not fit with corporate plans; and regulatory or legal, where environmental laws and regulations prohibit exchanges from occurring. Although these types of barriers exist for many types of interfirm transactions, EIPs face an additional hurdle, in that the fundamental transactions within an EIP are defined by specific types of physical exchanges (material, energy waste, and by-products) among the firms. In most cases, these are likely to be of lower value and less interest to a firm than its core products. Finding someone to buy or just take a company's waste and substitute it for a virgin input may be more time- and knowledge-intensive than simply hiring a waste management company to dispose of it (particularly if negative external costs are not included in the prices of virgin materials or waste disposal). It is this need for specific types of transactions that make EIPs so difficult to achieve.

This is not to say that planned EIPs do not work. Rather, the point is that achieving a fully functioning EIP may take a long time and require a steady commitment and openness to environmental improvement from the firms involved. Successful EIPs work best when waste and by-product exchanges are conceived of as a complementary strategy to a solid and financially sound economic development strategy for a park or a locality. In addition, ethnographic analysis of successful EIP experiments such as the Devens Ecoindustrial Park (Devens, Massachusetts) point to the central importance of a major advocate or champion that can coordinate and guide the process and help build the necessary social connections between the firms to solve the technical, economic, and environmental issues involved in developing effective by-product and waste exchanges.

Future of EIPs

These difficulties have led some to ask whether such exchanges need to be the defining feature of EIPs. Some have argued that a more incremental approach is needed, whereby by-product and waste exchanges might come at the end of a long process of adoption of pollution prevention measures, energy efficiency measures, "green architecture," and resource recovery. From this perspective, materials and energy exchanges may be the ultimate goal of EIPs, but given the difficulty in achieving these, they do not need to be a key part of the initial development strategy.

Others suggest that self-organized EIPs work better than planned EIPs. For example, many of the more successful examples of by-product and waste exchanges are found both among firms that were already located close to each other and where firms developed the exchanges among themselves in response to greater potential profit or as a method to reduce waste disposal or environmental compliance costs; for example, piping pretreated wastewater from an oil refinery directly to a power station allows the oil refinery to avoid the cost of disposing of the pretreated wastewater, and the power station gets cheaper water for electricity generation. The strategy then is to develop (grow) these exchanges into more complex networks, rather than to plan new ecoindustrial parks. In addition, because these self-organizing networks emerge from decisions by private actors operating in the marketplace, each exchange faces an immediate market test and, if successful, is likely to be more robust and lead to further exchanges in the future.

The EIP concept has come a long way in the 15 years since the initial workshops and President's Council on Sustainable Development experiments. Although many of the initial and subsequent experiments failed, the essential notion of energy, waste, and by-product substitution and interfirm collaboration on environmental impacts has spread around the world. Furthermore, the early failures have informed more recent attempts to find the right way to anchor ecoindustrial development. For example, integrated waste management systems between firms, combined water recycling systems, ecocenters for environmental services, and so on, although not pure examples of exchanges in the strict sense of the concept, can all be traced back to the initial EIP concept. In other cases the EIP exchange strategy is evolving into elements of national policy. For example, EIPs are a central feature of the Chinese concept of a circular economy. Although it is impossible to predict the ultimate form of future EIPs, it is highly likely that the essential ideas that grew from the initial experiments will continue to be part of the larger effort to move cities toward a greener future.

See Also: Combined Heat and Power (Cogeneration); Environmental Planning; Green Jobs; Natural Capital; Recycling in Cities.

Further Readings

Chertow, Marian. "Uncovering Industrial Symbiosis." *Journal of Industrial Ecology,* 11 (2007).
Côté, Raymond and Edward Cohen-Rosenthal. "Designing Eco-Industrial Parks: A Synthesis of Some Experiences." *Journal of Cleaner Production,* 6 (1998).
Gibbs, David and Pauline Deutz. "Reflections on Implementing Industrial Ecology through Eco-Industrial Park Development." *Journal of Cleaner Production,* 15 (2007).
Lowitt, Peter C. "Devens Redevelopment: The Emergence of a Successful Eco-Industrial Park in the United States." *Journal of Industrial Ecology,* 12 (2008).

Donald Lyons
University of North Texas

ECOLOGICAL FOOTPRINT

The ecological footprint is a measure of the physical area required to provide the resources consumed by an individual, city, or nation. It is a measure of sustainability based on the recognition that when resources are consumed faster than they are produced, the resource is eventually depleted. Both populations and the economies supporting them depend on materials and services provided by nature, and accounting for resource use is one of the key elements in developing sustainable economies. Without an effective metric, these assets and their contributions to society cannot be systematically assessed or included in strategic planning.

The methodology to calculate an ecological footprint focuses specifically on the human use of renewable natural resources in relation to the rate at which such resources are regenerated by the world's ecosystems—what nature provides and how much we use. Because these resources are drawn from the productive areas of the planet's surface, the ecological footprint calculates the supply and demand in terms of the land required to replenish our annual use. The footprint of a city or nation is therefore defined by the total land and sea area required to supply it with all the food and other renewable resources used by its population, as well as the area required to absorb their wastes. The ability to express these demands on nature in a single standard unit of measurement allows equivalencies to be tracked and compared across a wide range of activities.

It should be noted that this is only an accounting measure of the use (or overuse) of nature's ability to produce materials used by humans or to sustainably absorb human waste materials; it does not address the health of ecosystems or requirements for the survival of other species. This approach also does not directly address the use of nonrenewable resources such as minerals or fossil fuels, nor does it address pollution in the form of accumulation of persistent toxic substances that are incompatible with health, as each of these categories is inherently unsustainable.

Background

Concern over whether many large-scale human activities are environmentally sustainable has received increasing attention as depletion of natural resources has become broadly apparent. In 1983, the United Nations established the World Commission on Environment and Development to help define shared perceptions of long-term environmental issues in relation to the developmental goals of the world community. The commission's 1987 report, *Our Common Future*, established the broadly accepted definition of sustainable development as "development that meets the needs of the present without compromising the ability of future generations to meet their own needs"; in other words, not degrading the resources we depend on beyond the level needed to continue to meet anticipated use.

Determining whether or not any given activity meets this criterion requires some means to measure "sustainability." Although many approaches to this problem already existed for specific classes of resources, it is problematic to compare very different activities. Governments must choose between policy alternatives, and a major challenge arises in determining the relative effectiveness of actions in support of sustainability; for example, how to directly compare the consumption of a pound of beef to a pound of fish or firewood.

Seeking a common unit of measure to allow direct comparison between the use of different environmental resources, Mathis Wackernagel and William Rees developed the idea of an ecological footprint as "appropriated carrying capacity" in the early 1990s at the University of British Columbia. Carrying capacity, as it relates to biological species, is the population size that the environment can maintain over time, given the limits in food, water, habitat, and other necessities available in the environment; *appropriated carrying capacity* refers to the portion of these available resources used by a particular population or species such that it is not available for others. Wackernagel and Rees's concept recognized that the ecosystem exists on the surface of the planet, and any use of resources from the ecosystem therefore requires the use of a measurable area of land or sea. This allows a direct and consistent comparison between entirely different resource use options by comparing the amount of area required for each.

The concept was first published in its full form in their 1996 book *Our Ecological Footprint: Reducing Human Impact on the Earth*. The approach has been refined and updated continually since and applied in a wide range of settings internationally. Inconsistent application of multiple versions of the footprint method by a variety of agencies led to the establishment of the Global Footprint Network in 2003 and the release of formal methodological standards in 2006, focusing on its application in relation to populations. A further update of these standards, including specific application of the methodology to organizations, products, processes, and services, was released in December 2009.

Application to Cities

With the majority of the world's population residing in urban and suburban municipal districts, this is where resource demand occurs, and action at the city level is central to achieving overall sustainability. The effect of a city's resource consumption and emissions extends far beyond its own boundaries. As an example, the land required to meet the current resource use in the city of London is greater than the total land area of Great Britain, meaning that a quantity of resources equivalent to the needs of the rest of the country must be continually imported. As cities make long-term decisions on infrastructure, growth patterns, and economic development, choices affecting resource dependence are locked in, influencing quality of life and economic competitiveness for generations.

Ecological footprint accounting provides information about a city's use of and dependence on a wide variety of natural resources. Reliance on resources that are being used at unsustainable rates creates vulnerabilities. Tracking resource stocks and their use allows risk prediction and supports policy responses toward both reducing dependence on declining resources and avoiding continued overuse. Data from the established National Footprint Accounts provide accurate per capita estimates based on specific average values for each country in all resource categories. Where local data are available, a customized accounting can be created representing unique regional qualities, as has been done for regionalized heating and cooling energy use for the United States and some other countries.

As part of a national effort, ecological footprints have been calculated for every city in England, Scotland, and Wales. This study combined national data and spending data for each of the local authorities to assess per capita resource use. The results show demand in each of eight sectors: housing, transport, food, consumer items, private services, public services, capital investment, and miscellaneous other. Such results, measured in global hectares and using the standardized methodology, allow direct comparison among cities and tracking of changes within a city in response to applied policies.

Footprint accounts incorporate and add value to existing data sets on production, trade, and environmental performance by providing a comprehensive means of interpreting them with respect to sustainability, trade-offs between different policy options, and possible unintended consequences. Such accounting promotes awareness of resource dependence, depletion, and relative efficiencies. In so doing, risk management may be improved by guiding local economic development to better use resources and reduce reliance on unsustainable resource streams. Illuminating these connections aids decision makers to more accurately and equitably shape policy in support of societal goals.

Methodology

National Accounts databases have been developed on the basis of international trade and production data to establish human use of the world's ecosystems for the years since 1961.

Using the Earth's total productive surface area as a baseline, the annual capacity for biological production is calculated, establishing an upper limit for available regenerative supply. Productive surface area is defined as all land and water areas currently generating the types of timber, pasture, food, fuel, and fiber resources used in human economies. These areas are classified as one of five use types: cropland, grazing land, forest (including a subcategory of carbon uptake land), fishing grounds, and built-up land covered by human structures and facilities. The average productivity is determined by dividing the total capacity by the total productive area, and this average establishes a standard unit of measure: a global hectare (1 hectare = 2.47 acres).

Because different climates and types of land or sea have varying productivity, the measure of each specific region's actual area is adjusted by a yield equivalence factor, allowing them to be compared on an equal basis. The yield data used to establish equivalence factors are based on the Global Agro-Ecological Zones model, combined with measures of the actual areas of cropland, forestland, and grazing land area from United Nations Food and Agriculture Organization statistics. Because biological reserves, agricultural practices, and productive capacity change over time, the determination of a standard global hectare is specific to a given year and adjusted accordingly.

Consumption or demand figures for national populations are drawn from Food and Agriculture Organization statistics and additional data, as available, and converted to global hectares of productive area required, thereby establishing National Footprint Accounts for each country. For example, consumption of 195 categories of crops is tracked, and the cropland required for each is calculated and converted to standardized global hectares. Imports and exports of crops, meats, fish, wood, and fiber products are also included in the National Accounts, allowing a calculation of the ecological footprint for a population's domestic consumption regardless of where the resources originated, or alternatively, for the footprint created locally in production of products consumed elsewhere. The footprint can be expressed as a single number or disaggregated to reflect use in each of the categories individually, allowing a more precise comparison against available resources.

Limitations of Ecological Footprint Analysis

Footprint accounts do not provide a complete assessment of sustainability, which should also include social, economic, and other nonecological components, and the approach does not address practices that are inherently unsustainable. In addition, the use of a single number to express sustainability will mask important details. Although footprint analysis will provide a single overall figure, attention to each subcomponent is essential for evaluation of actual performance.

Water, minerals, toxic waste, fossil fuels, and energy use in general do not fall within the categories measured by the ecological footprint because they are neither produced nor decomposed by renewable biological systems. However, fossil fuels are derived from ancient biological sources, and their combustion releases stored carbon dioxide, which is directly absorbed in the carbon cycle of the biosphere. The ecological footprint does account for the natural rate of absorption of these emissions, and the production of carbon dioxide from direct and indirect fuel use makes up a large portion of most footprint calculations. As with other factors, the carbon footprint component is measured in forest area required, after deducting for ocean sequestration effects. This is the only component of the footprint based on absorption instead of production rates. As with each component in the footprint, new research on the sustainable rates of ocean and biogenic soil sequestration of carbon is incorporated in periodic updates of the methodology.

Assessment of water supply and use is an important factor in determining sustainable development practices and should be considered in addition to the biological use measured by the ecological footprint. As water is not created by biological processes, and the land area that provides watershed and aquifer recharge is already counted based on biological capacity, water use is not a direct component of the methodology. Energy use related to water use, and the land used for artificial production, treatment, transportation, and storage, are each components of the footprint because these place measurable surface area demands on the biosphere.

The footprint does not differentiate between the quality of practices occurring at any given time, although the effect of detrimental practices that lead to increased soil salinity, loss of topsoil, acidification, and so on, is captured in reduced productivity rates once the damage has occurred.

The ecological footprint also does not directly measure biodiversity; however, the size of a population's footprint represents the pressure placed on the ecosystems from which it draws resources. As such, the Convention on Biodiversity has adopted the ecological footprint as an indicator of pressure on biodiversity.

See Also: Carbon Footprint; Cities for Climate Protection; Ecosystem Restoration; Environmental Impact Assessment; Environmental Risk; Natural Capital; Sustainability Indicators; Sustainable Development.

Further Readings

Best Foot Forward. *City Limits: A Resource Flow and Ecological Footprint Analysis of Greater London.* The Chartered Institution of Wastes Management, 2002. http://www .citylimitslondon.com (Accessed March 2009).

Ewing, B., et al. *Calculation Methodology for the National Footprint Accounts, 2008 Edition.* Global Footprint Network. http://www.footprintnetwork.org/en/index.php/GFN/ page/publications (Accessed March 2009).

Global Footprint Network. *The Ecological Footprint Atlas 2008.* http://www.footprint network.org/en/index.php/GFN/page/publications (Accessed March 2009).

Hails, Chris, ed. *Living Planet Report 2008.* WWF International. http://www.panda.org/ what_we_do/footprint/one_planet_living/documents (Accessed March 2009).

Wackernagel, Mathis, et al. "The Ecological Footprint of Cities and Regions: Comparing Resource Availability With Resource Demand." *Environment and Urbanization,* 18/1 (2006).

Wackernagel, Mathis, et al. "Tracking the Ecological Overshoot of the Human Economy." *Proceedings of the National Academy of Science,* 99/14 (2002).

Wackernagel, Mathis and William Rees. *Our Ecological Footprint: Reducing Human Impact on the Earth.* Gabriola Island, British Columbia, Canada: New Society, 1996.

Kenneth S. White
San Jose State University

ECOSYSTEM RESTORATION

Ecosystem restoration is generally understood as the practice of interfering in an ecosystem and taking specific actions to return that ecosystem to a prior state, thus restoring what was previously in existence. Efforts at restoration often occur because an environment is degraded or altered in an undesirable manner. Ecosystem restoration is a strategy used by cities to improve biodiversity and ecosystem services in areas that have been polluted or otherwise damaged.

Human activity can be to blame for ecosystem degradation or alteration. This could be in the form of deliberate alteration of an environment for a specific aim; for example, the removal of predators from an area to ensure that ranchers' stock is not compromised. A famous instance of this type of ecosystem alteration is the removal of wolves from North American landscapes. The recent reintroduction of wolves is one effort at ecosystem restoration. Human activity can also inadvertently radically alter ecosystems, as with the often-unintentional introduction of "exotic" and "invasive" species. A recent and problematic example of this is the case of zebra mussels entering waterways—often by attaching themselves to boats and stowing away in ballast tanks—where previously there were none. The mussels introduce radical change and new pressures to local ecosystems. There are numerous cases such as these around the world.

Ecosystem restoration efforts are not conducted solely in rural environments, however, as many cities pursue restoration efforts with goals as diverse as providing cleaner and healthier water to its citizens, to providing aesthetically pleasing recreation areas, to protecting and providing spaces for local wildlife to flourish. For example, the city of Dallas, Texas, is currently implementing the Trinity River Corridor Project. This multifaceted project demonstrates a confluence of common motivations that cities have for embarking on restoration efforts. The major anticipated outcomes of restoring this urban river include providing flood protection to the city's inhabitants, installing new design features that provide enhanced recreation activities (e.g., Olympic-class boating; newly installed courses for kayakers and canoeists; parks and trails, etc.). In addition, the efforts aim to provide a variety of educational opportunities and facilities for citizens young and old, as a means for inhabitants to better understand and connect with the river. The efforts are not only aimed at human benefit, however; there is also a focused effort to improve the habitat for local birds and other wildlife, in addition to improving the overall health of the river ecosystem itself. These types of efforts pose a number of interesting and challenging conceptual and ethical issues.

Conceptual Issues

Ecosystem restoration discussions can differ greatly depending on one's understanding of the nature of nature. For example, if the goal is primarily to restore the ecosystem to a properly functioning state, then the introduction of exotics, presuming they do not compromise the ability of the ecosystem to function at a healthy level, theoretically does not pose a problem. If the ecosystem is degraded, then the question is to what state the ecosystem should be restored. If one believes there is a balance or harmony in nature, then the goal would be to restore the balance to that ecosystem, either by attempting to revitalize the previous organisms to restore the necessary ecosystem functions and services, or alternatively, substituting exotic species that otherwise perform more or less the same functions. If one believes that nature is in a state of constant flux, the present state may be seen as one further change in the ecosystem that is no better or worse than previous states, and thus not in need of restoration. Or it becomes open for debate which prior state the ecosystem should be restored to, as it is not necessarily obvious that one prior state is "better" than another.

The issue could be said to be a matter of choosing which ecosystem type members of a community desire—be it for ethical reasons, conservation reasons, or aesthetic preference. We might lament the loss or degradation of an ecosystem because it is a rare or threatened type and wish to restore it for its biological diversity and conservation value, or we might lament the loss of a recreational spot and wish to restore it for the pleasure it provides to members of a community. There are many different and polarizing objectives in ecosystem restoration.

Ethical Issues

Beyond the conceptual issues at play in ecosystem restoration, there are difficult ethical issues at stake. Take the example of the North American beaver in sub-Antarctic Chile. The beaver is an invasive exotic species in this ecosystem, and there are no natural predators to keep the population in check. This ecosystem is one of the last "pristine" ecosystems left on the planet, home to a great array of biological diversity. The beaver introduce dramatic changes to this unique ecosystem, threatening its continued existence. One reaction might be the conservationist response to eliminate the beaver for this rare and unique ecosystem to persist. However, an alternative response may question the right humans have to eliminate a species in this manner. If a species establishes itself in a new home and thrives there, it can be difficult to morally justify its total eradication.

Entire fish populations are occasionally poisoned out of lakes or ponds for the species that once inhabited that body of water to be given a chance to reinhabit the area. Proponents of restoring the ecosystem in this way argue that the ecosystem is being restored to its prior state, and in some sense a victory is achieved for the prior species that were victimized. Recreationists may happily embrace the action on the grounds that the prior species provided better fishing opportunities than the current invasive one does. Opponents see such actions, however, as cruel and unnecessary violence inflicted on one species to favor another.

Ecological restoration discussions bring to light how influential human beings are within any given ecosystem. Through our actions, both deliberate and inadvertent, we introduce ecosystemic change. Depending upon how we view ourselves as a species and our role within ecosystems, we can choose either to act and restore ecosystems to a prior state (recognizing perhaps that this is an arbitrary choice), to restore ecosystemic functions to a healthy but new ecosystem, or we can choose inaction, allowing the ecosystem to change and adapt on its own (which is its own form of action). Not all people involved in ecosystem restoration feel burdened by these ethical matters—for some, ecosystem restoration is a straightforward activity of eliminating threats to a given ecosystem, whether those are pollution or invasive species, and restoring the previously given ecosystem.

A final example ties the ethical and conceptual issues together with issues of commerce. Ecosystem restoration is frequently framed within discussions of business plans to extract resources but later "restore" the ecosystem to its normal state. The policy makers argue that the final result will be an ecosystem that is indistinguishable from that which was originally present. Opponents often contest these plans, arguing on practical grounds that the company may fail to restore the ecosystem as a result of unforeseen financial or practical and technological problems. On a deeper level, the question of whether humans possess the requisite knowledge and ability to reconstruct an ecosystem is raised. Finally, even if we are able to fully reconstruct ecosystems, opponents argue that something still is lost in restoration efforts, and thus procedures or extractions that would necessitate restoration should be avoided. The reason given is that once we know an ecosystem is constructed by human hands, it loses some of its value and is inferior to what was naturally present. A restored ecosystem, some contest, is mere artifice, and cannot compare in value to a "wild" ecosystem that has evolved independent from human intervention. There is an authenticity factor that is compromised. For others, the fact that an ecosystem was constructed by human hands may not diminish its authenticity at all but, on the contrary, enhance one's appreciation for the ecosystem. Given all these difficulties, the personal and communal understanding of the nature of nature, and the role of humans within nature, goes a long way toward shaping how ecosystem restoration discussions and actions proceed.

See Also: Biodiversity; Habitat Conservation and Restoration.

Further Readings

Anderson, C. B., et al. "Exotic Vertebrate Fauna in the Remote and Pristine Sub-Antarctic Cape Horn Archipelago Region of Chile." *Biodiversity and Conservation,* 10:3295–313 (2006).
Elliot, Robert. *Faking Nature.* London: Routledge, 1997.
Hull, R. Bruce. *Infinite Nature.* Chicago: University of Chicago Press, 2006.
Nature in the City. http://www.natureinthecity.org (Accessed January 2010).
Trinity River Corridor Project. http://www.trinityrivercorridor.org (Accessed January 2010).

Jonathan Parker
University of North Texas

Ecovillages

Efforts toward intentional communal living have a long history in the United States and the world. An ecovillage is a specific type of intentional community that is guided by ecologically relevant organizing principles, among other things. The ecovillage movement is largely motivated by a desire to be proactive, rather than reactive. Recognizing the existing and potential problems stemming from current unsustainable practices, ecovillages are intended to help develop and model sensible alternative ways of organizing residential life. Moreover, ecovillage advocates see these communities as offering a higher quality of life than conventional settings in terms of providing a meaningful sense of community and opportunities for genuine involvement. Contrary to the somewhat narrow connotation of the term *village*, ecovillages exist in rural, urban, and suburban settings. The term *ecovillage* came into common usage in the early 1990s. The most common definition of an ecovillage includes the following characteristics:

- *Human-scale:* usually between 50 and 500 members, though there are exceptions
- *Full-featured:* providing major needs for food provision, manufacture, leisure, social life, and commerce in balanced proportions (though an ecovillage is not meant to be totally

self-sufficient; the goal, rather, is to be linked to other communities in networks of economic, social, and political ties)

- *Minimize ecological impact:* human activities are integrated into the natural world in ways that demonstrate a recognition of their interdependence with other forms of life
- *Supportive of healthy human development:* including the physical, emotional, mental, and spiritual development of individuals and the community as a whole
- *Sustainable:* successfully able to continue into the indefinite future

It is more accurate to say that these characteristics describe the "ideal" ecovillage. Ecovillage founders are careful to point out that there is no perfect ecovillage; rather, these communities are works in progress. What is distinctive about the ecovillage is that carefully thought out and explicit intentions guide their development processes. As a consequence, ecovillages regularly engage in the assessment, reassessment, and refinement of their practices in relation to their goals.

Although the goals, missions, and visions of ecovillages vary in form, they are similar in spirit, and similarly exhibit a commitment to fostering ways of thinking and acting that reflect an awareness of human–ecosystem interconnectedness. These statements, and efforts to realize them, are a fundamental component of ecovillage success.

The intentions of an ecovillage are manifest in a number of ways that encourage ecologically conscious behavior in everyday life. Important among them are rules and policies that limit or encourage particular practices, physical design, and infrastructure that reflect and contribute to achieving community goals, social interaction that brings people together in formal and informal ways to reinforce intentions and practices, and interaction with people outside the ecovillage—by opening the community to visitors and/or reaching out to others—aimed at sharing information, ideas, and services.

The ecovillage concept gained an organizational home with the formation of the Global Ecovillage Network (GEN) in 1995. GEN has since split into three regional divisions: GEN Oceania and Asia (including Asia, Australia, and the Pacific Islands), Ecovillage Network of the Americas (including North, Central, and South America), and GEN Europe (including Europe, Africa, and the Middle East). GEN has recently created a branch for youth, NextGEN, and is also affiliated with a number of international organizations, including the United Nations (the Economic and Social Council; GEN has also received a UN-HABITAT Best Practices designation and is listed in its database), EU Pologne, Hongrie Assistance à la Reconstruction Economique, and European Youth for Action.

It is difficult to estimate the number of ecovillages in the United States or the world because many are still in formation or were started as local initiatives and are not yet connected to formal networks. At the time of this writing, there are 435 ecovillages formally registered with GEN, 105 of which are in the United States. Information varies, however; the Intentional Communities Directory shows 162 in the United States. Both of these numbers are necessarily conservative estimates. Communities must self-register with these and similar organizations; neither GEN nor the Intentional Communities Directory attempt to count existing ecovillages. Those involved in the ecovillage movement report far higher numbers of ecovillages worldwide. One author puts the estimate between 4,000 and 5,000, and another's approximation is closer to 15,000, using a more inclusive definition.

Ecovillages face a number of challenges in both formation and maintenance. Initially, it can be difficult for founding members to find the land, money, and people with which to begin building a community. Other obstacles include legal barriers to sustainable development (prohibitions on natural building, water catchment, and composting toilets, for example), conventional zoning practices, a lack of alternative financing options, and difficulties in building diverse communities, in terms of members' ethnic, economic, religious, and other characteristics.

The ecovillage movement is premised on the fact that a sustainable society depends not only on what we do but also on how we think. Ecovillages are motivated by the attempt to encourage more holistic ways of thinking about our place in the world. The organization of ecovillage practices, social relations, and physical settings strives to consciously articulate and embody this way of understanding the world, bringing ways of living into alignment with ecological thinking.

See Also: Dongtan, China; Green Housing; Malmö, Sweden; Universal Design.

Further Readings

Bang, Jan Martin. *Ecovillages: A Practical Guide to Sustainable Communities.* Gabriola Island, British Columbia, Canada: New Society, 2005.
Christian, Diana Leafe. *Creating a Life Together: Practical Tools to Grow Ecovillages and Intentional Communities.* Gabriola Island, British Columbia, Canada: New Society, 2003.
Global Ecovillage Network. http://gen.ecovillage.org (Accessed April 2009).
Intentional Communities. http://www.ic.org (Accessed April 2009).
Jackson, Hildur and Ross Jackson, eds. *Ecovillage Living: Restoring the Earth and Her People.* White River Junction, VT: Chelsea Green, 2002.
Jackson, Ross. "The Ecovillage Movement." *Permaculture Magazine,* 40/4 (2004).
Kasper, Debbie Van Schyndel. "Redefining Community in the Ecovillage." *Human Ecology Review,* 5/1 (2008).

Debbie V. S. Kasper
Sweet Briar College

EMBODIED ENERGY

Also called "emergy," *embodied energy* is defined as the available energy that is used in the work of making a product. It has also been said to be an accounting methodology that aims to find the sum total of the energy necessary for an entire product life cycle. This life cycle encompasses raw material extraction, transport, manufacture, assembly, installation, disassembly, deconstruction, and/or decomposition.

The associated environmental implications of embodied energy such as resource depletion, the production of greenhouse gases, maintenance of biodiversity, and environmental degradation are embedded in the measurement of embodied energy. The energy is expressed in units of megajoules (MJ) or gigajoules (GJ) per unit of weight (kilogram or tonne) or area (square meter).

Embodied energy plays a large role in the choice of any product and is an important factor to consider when assessing the life cycle of a product because it relates directly to the concept of sustainability of that product. Closely related to the embodied energy of materials or products is the concept of life cycle assessment (LCA). LCA is the assessment of the environmental impact of products, buildings, or other services throughout their lifetimes. The assessment includes the entire life cycle of a product, process, or system, encompassing the extraction and processing of raw materials; manufacturing, transportation, and distribution; and use, reuse, maintenance, recycling, and final disposal. LCA is a much-explored concept and has been used as an environmental management tool worldwide since the late 1960s.

A derivative of LCA, life-cycle energy analysis (LCEA) ties in closely with embodied energy of materials because it considers energy as the only measure of environmental impact of any product. Emerging in the late 1970s, the purpose of LCEA is to present a more detailed analysis of energy attributable to products, systems, or buildings to enable decision-making strategies concerning energy efficiency and environmental protection. LCEA was not developed to replace LCA but, rather, to compare and evaluate the initial and recurrent embodied energy in materials, energy used during the operational phase, and during recycling and disposal. It is often used to estimate the energy use and savings over the product's or building's life, and more important, to find out the energy/carbon dioxide payback period (the time spent for the initial embodied energy cost to be paid back by energy savings during operational and disposal/recycling stages).

The concept of embodied energy has become very central in the design of buildings. When sustainability is the driving force in the creation of a building, quantity surveyors and architects need additional knowledge to assist them in making decisions about the choice of building materials and the way in which they are used. Some standards exist for making this comparison. The two most common are the U.K. Code for Sustainable Homes and the U.S. Leadership in Energy and Environmental Design standards. These standards rate the embodied energy of a product or material, along with other factors, to assess a building's environmental impact.

Types of Embodied Energy

Embodied energy is divided into two main areas; namely, "initial" embodied energy and "recurring" embodied energy. Initial embodied energy is nonrenewable energy consumed in the process of construction, from raw materials acquisition to the construction of the building. It is influenced by the source and type of building materials and the nature of the building. Recurring embodied energy, in contrast, is nonrenewable energy consumed to maintain, repair, restore, refurbish, or replace materials, components, or systems during the building's life span. It is influenced by the durability and maintenance of building materials, systems, and components installed in the building and the life span of the building.

Research application of the embodied energy concept has shown that to minimize the operation energy of buildings, the initial embodied energy must be quite substantive and represent a large percentage of the operating energy. This has led to increasing efforts that will probably be directed toward measuring and reducing the amount of embodied energy in buildings. Some options for achieving this include reuse and recycling.

The embodied energy in recycled materials is generally less than that contained in new materials. Recycling provides easily obtainable manufacturing feedstock. There is very low extraction energy associated with recycled materials. Although manufacturing with recycled feedstock can involve transporting, cleaning, and sorting the recycled materials, this often requires far less energy than manufacturing from a virgin resource that must be extracted and refined before use. Reusing materials, reducing the energy required at any stage of production, choosing durable, long-lived building materials, and using indigenous—or local—materials (which usually requires lower transportation energy costs) are also measures that can help to reduce the embodied energy of any material or product.

The idea of embodied energy has come to stay and is continuing to have effects in other fields in which it had not been traditionally considered. For example, the concept of embodied energy has also found application in other research areas such as water. In the 2000s, drought conditions in Australia generated interest in the application of embodied energy analysis methods to water. This has led to use of the concept of embodied water, which measures how water is embedded in the production and trade of food and consumer products.

A Complex Process

The process of calculating embodied energy is very complex, and a variety of data sources are used. Factors such as geographical location and technology employed in the manufacturing process, as well as methods of manufacture, play a very important role in the final value of the embodied energy of any material or product. These factors have contributed to the lack of international agreement on what should be appropriate data scales and methodologies.

Because of the lack of appropriate international data scales and methodologies, embodied energy calculations may omit important data on, for example, infrastructure and services needed to transport and market a product. Such omissions have resulted in a wide range of embodied energy values for any given material. Nevertheless, most scientists agree that products can be compared with each other to see which has more and which has less embodied energy. Typical embodied energy units used for making comparisons are MJ/kg (megajoules of energy needed to make a kilogram of product) and tCO_2 (tons of carbon dioxide created by the energy needed to make a kilogram of product).

It has been established that there exist strong correlations between the embodied energy and environmental impacts of any product/material. Embodied energy can be a very useful measure when environmental improvements are sought in any process or product. However, it should not be viewed in absolute terms. Its limitations must be borne in mind whenever and wherever it is applied. Therefore, it should not be used as the sole basis of material, component, or system selection.

See Also: Environmental Impact Assessment; Smart Growth; Sustainability Indicators; Sustainable Development.

Further Readings

Chenoweth, Jonathan. "Looming Water Crisis Simply a Management Problem." *New Scientist,* 28:28–32 (August 2008).

Holtzhausen, H. J. *Embodied Energy and Its Impact on Architectural Decisions.* Hampshire, UK: First International Conference on Harmonization Between Architecture and Nature, 2006.

McCormack, M. S., et al. "Modeling Direct and Indirect Water Consumption Associated With Construction." *Building Research and Information,* 35/2:156–62 (2007).

Menzies, Gillian, et al. "Lifecycle Assessment and Embodied Energy: A Review."
    Construction Material, 160/4:135–43 (November 2007).
Mumma, Tracy. "Reducing the Embodied Energy of Buildings." *Home Energy Magazine
    Online* (January–February 1995). http://www.homeenergy.org/archive/hem.dis.anl.gov/
    eehem/95/950109.html (Accessed April 2009).

Oyeshola Femi Kofoworola
University of Toronto

Energy Efficiency

Within the context of sustainable design, energy efficiency is regarded as an important key to reducing fossil fuel use by improving the energy performance of equipment and appliances, buildings, and municipal infrastructure. Many energy experts today regard continued dependence on coal, oil, and natural gas to be unsustainable because supplies are not limitless and combustion of fossil fuels results in a variety of negative environmental impacts including air and water pollution and significant carbon dioxide emissions.

Houses designed for a specific site, like this solar-heated, cement house built into a hillside, can employ topography and the path of the sun to increase efficiency.

Source: iStockphoto

Energy efficiency describes the rate at which energy is used to accomplish a particular task (e.g., the miles that a car can travel per gallon of gasoline). Improving energy efficiency means getting more work from the same or less energy input. A hybrid sedan is more energy efficient than a large pick-up truck because it will travel farther on the same amount of gas.

Improvement of energy efficiency has traditionally been considered a production engineering strategy that increased profits by reducing fuel costs. Improvement of the energy efficiency of equipment and machinery played a key role in the phenomenal growth of industrial productivity in the early 20th century. Since the oil embargo of the 1970s, federal, state, and utility programs have laid a solid foundation for improving the efficiency of energy consumption across all sectors of the economy. However, the new sense of urgency surrounding carbon emissions and other environmental impacts has led cities to identify and support the many opportunities that remain to improve energy efficiency in homes, businesses, schools, government buildings, and industrial facilities.

The Legacy of the 1970s

The consumer side of energy efficiency has its roots in the oil embargo of the 1970s. When political upheaval in the Mideast in 1973 temporarily cut the oil supply flowing to the United States, Americans were made aware for the first time that the fossil energy supply was neither free nor endless. This brief supply disruption illustrated America's energy vulnerability and caused a subtle but permanent change in the way the United States would regard all its fossil energy resources from then on. Energy efficiency as a national policy concern began with President Jimmy Carter in the late 1970s, and it has continued to remain on the agenda to some degree with every presidential administration since.

State Energy Offices

By the end of the 1970s, every state had established an energy office to administer federally mandated programs to promote energy efficiency through financial incentives and public education. These programs were funded by the Petroleum Violation Escrow funds that were settlements from lawsuits brought by the federal government against several U.S. oil companies that took advantage of the oil crisis and overcharged consumers at the gas pump. Also referred to as the Oil Overcharge Program, this was the first time energy efficiency was ever promoted to the public by the federal government. Even though most of the original funding settlements have been spent, many states have retained their energy offices. Some have expanded into helping communities with sustainable energy planning.

Appliance Efficiency Standards

In 1987, the federal government began mandating energy efficiency standards for home and commercial appliances and equipment through the National Appliance Energy Conservation Act. The appliances in the initial law included residential room air conditioners, water heaters, refrigerators, dryers, ovens, and ranges. The law was drafted to allow improved standards as technology made more efficient appliances economical to manufacture.

Since then, many additional appliances have come under federal standards. In September 2009, Secretary of Energy Steven Chu continued this tradition by announcing new residential appliance standards that include dishwashers, general service incandescent lamps, and microwave ovens, as well as commercial standards for boilers, air conditioning equipment, and vending machines. Appliances manufactured under all federal standards since 1987 continue to save energy. According to the American Council for an Energy Efficient Economy, U.S. residential and business energy consumers will save approximately $186 billion in energy costs from 1990 to 2030 because of the federally mandated appliance and equipment efficiency standards.

Energy Star

Another successful federal program promoting energy efficiency is Energy Star, established by the U.S. Environmental Protection Agency. The program was first established for the purpose of reducing air pollution from energy-related causes. So far, over 9,000 manufacturers have signed on with this widely recognized and trusted program, voluntarily exceeding the federal appliance standards by program-mandated amounts. For example, to qualify for Energy Star, a refrigerator must perform 20 percent more efficiently than a conventional model.

Starting with computers and office equipment, the Energy Star program now includes home appliances, commercial equipment, and standards for both new and remodeled homes. The program partnered with the U.S. Department of Energy in 1996 and continues to expand to include more and more appliances and equipment—often those that have not yet been covered by the federal standards.

Utility Programs

In the late 1970s, electric utilities also began to promote energy efficiency through "demand-side management" programs that helped electricity customers reduce or shift their electricity load. Rising fuel costs and increasing environmental regulation were making electricity more expensive to generate, and customer demand was rising rapidly.

To slow the utilities' requests for rate increases to build new power plants, state regulators began to require them to consider demand-side efficiency to be an energy resource that could delay or eliminate the need to build a new power plant. As a result, electric utilities began incentive and education programs to help customers cut energy use in residential, commercial, and industrial settings.

It is interesting to note that the natural gas utilities that provide gas to many of the same customers that buy electricity have not, as a rule, been involved in promoting energy efficiency. On the contrary, the price of natural gas had been relatively inexpensive until recently, and "fuel switching," or replacing an electric water heater or furnace with a gas-fired model, has often been encouraged to save household electricity bills.

As the energy picture has changed over the years, the creation of state energy offices and the advent of federal appliance standards, utility demand-side management programs, and Energy Star–qualified buildings and equipment have formed a solid foundation for moving toward a sustainable energy future. Although renewable energy resources such as solar and wind power are important for moving away from the world's unsustainable dependence on fossil fuels, using energy more efficiently in all applications is vital if these renewable sources are to become practical and economical substitutes.

Energy efficiency is now seen as part of an integrated "sustainability" package for homes, businesses, and cities. Building sustainability standards similar to the U.S. Green Building Council's Leadership in Energy and Environmental Design standard incorporate energy efficiency as a vital element of the sustainability of the completed building. State and local building officials are recognizing the need to adopt and enforce building energy codes such as ASHRAE 90.1 (developed by the American Society of Heating, Refrigeration and Air Conditioning Engineers).

Energy Efficiency Technologies and Strategies in Buildings

There are two basic opportunities to introduce energy efficiency into a structure, whether it is a home, a commercial or institutional building, or an industrial facility. The first is when the structure is initially constructed (or remodeled), and the second is when the occupants move in and make choices about the equipment and furnishings they bring with them. A few strategies for the building envelope (walls, roof, and foundation) apply to all structures, taking local climate conditions into account.

Building Envelope Measures for Energy Efficiency: All Buildings

- *Insulation:* Walls, roof, and foundation should be well insulated. In colder climates, insulation methods can be combined to reduce heat loss through framing and structural materials. Types of insulation include bats, insulation board, foam, and loose material that is blown in. Environmentally friendly insulation materials such as cotton bats or loose-fill cellulose are available. Green roofs (with live plantings) or white reflective roofs are often used on larger commercial or institutional buildings to reduce heat gain from the sun.
- *Windows:* Double-paned, low-emittance windows are becoming standard for improving building efficiency performance. Low-emittance coatings on the glass help reduce heat transfer between the panes. These coatings can also control the level of solar gain, offering options for which windows work best in different climate zones.
- *Air Sealing:* Construction methods for energy-efficient buildings include careful sealing of cracks, seams, pipe penetrations, and other openings in which air might infiltrate a structure. A well-sealed building can be heated or cooled with less energy. However, ventilation is required for all well-sealed buildings, whether residential or nonresidential, to provide a healthy indoor environment for occupants. A heat exchanger installed in the ventilation system can reduce the need to use additional energy to heat (or cool) incoming fresh air.
- *Climate Control System:* For a home or a small commercial building, climate control can be as simple as a setback thermostat that can be programmed to run the heating or cooling system only when it is needed. Larger buildings usually require more complex systems that control temperatures in different zones or areas, depending on the uses of the rooms in the building. Computerized systems can control building lighting, as well as the heating, ventilating, and cooling systems, offering an additional efficiency opportunity.
- *Building Orientation:* When structures are designed for a specific site, they can be oriented to take advantage of the path of the sun to increase their potential energy efficiency. Use of south-facing windows and properly sized overhangs can make use of the sun's heat during cooler months to reduce the heating load. Using fewer or smaller windows on the north side of the building reduces heat loss. Cross-ventilation from east to west can reduce cooling loads in the warmer months.
- *Daylighting:* Daylighting design, used primarily in large commercial and institutional buildings, takes advantage of light from the sun to reduce the need for electric light during daytime hours.

Additional Energy Efficiency Strategies and Tools: Commercial Buildings

Achieving optimal energy efficiency in larger buildings such as schools, office buildings, or government buildings requires engineering the complex heating and cooling systems they require from the initial stages of the design process. Once installed, these systems must be tested and calibrated to ensure they are operating as intended. After a building is occupied, it is necessary to operate the equipment properly so that maximum energy efficiency is maintained. The following tools are typically used:

- *Energy Modeling:* Software programs that allow engineers to enter a description of the building design, detailing materials and dimensions, and the climate of its location to determine projected energy use. Modelers can use the software to change building configurations and specifications to find optimal combinations before final design decisions are made.
- *Building Commissioning:* When a building is completed, an independent commissioning engineer tests the heating, cooling, and ventilating systems to see that all are performing to the manufacturer's specifications.

- *Operation and Maintenance Protocols:* To maintain the maximum efficiency of the mechanical systems, building managers follow the operation and maintenance procedures recommended by equipment manufacturers.

Plug Load: Energy Used by What Gets Plugged In

The energy efficiency of the building envelope is only part of the story. The energy used by building occupants offers further opportunities to cut energy use and costs. On average, lighting is the greatest energy user in commercial-scale buildings. Using efficient lighting in both new and existing buildings is usually the most cost-effective energy-efficiency strategy, and it can save on the cooling load. Heating, ventilating, and air conditioning systems that are properly sized, calibrated, and maintained are the most energy-cost effective. Efficient office equipment such as copiers and computers with an Energy Star rating can be purchased, and efficiency practices can be instituted among staff.

In homes, if natural gas is used for heating, cooking, and clothes drying, lighting is the greatest electricity load, and compact fluorescent lights are recommended to cut costs. Appliances using the most energy in the home are furnaces, air conditioning systems, and refrigerators, which are all available with Energy Star ratings. Utilities report rising per capita use of electricity primarily because of computers, televisions, and other entertainment electronics. Flat-screen televisions are beginning to compete with refrigerators as the greatest home energy users. The Environmental Protection Agency has recently announced that for flat-screen televisions to get an Energy Star rating, they must use at least 40 percent less energy than conventional models.

Energy Efficiency in Transportation

Surprisingly conservative in its vision, the Energy Independence and Security Act of 2007 produced Corporate Average Fuel Economy standards that require an average of 35 miles per gallon by 2020. However, hybrids such as the Toyota Prius and the Honda Insight are already averaging 50 miles per gallon and above, and plug-in electric vehicles such as the Chevrolet Volt are nearly ready for the mainstream market. Recent volatility in oil prices has meant new efforts to redesign other vehicles as well, from airliners to semi trucks and cruise ships.

However, the efficiency of the transportation sector goes beyond the fuel efficiency of individual vehicles. Transportation systems require land for roads and parking, and these take their environmental toll. Commuters are spending an increasing amount of time behind the wheel in the midst of greater and greater traffic congestion. To contain this unsustainable expansion of urban sprawl, municipalities are looking at altering traditional zoning practices to encourage more compact development, transit-oriented development that encourages density around transit stops, and the integration of different forms of public transportation.

Municipal Energy Efficiency

Cities are inherently more energy efficient than low-density suburbs or rural communities because urban residents live in multifamily homes and are able to walk, bicycle, or use transit instead of driving. However, municipalities are energy consumers and benefit from energy efficiency efforts. Often, a city's greatest energy expenditure is for pumping water and sewage. High-efficiency motors can be vital in reducing this expense. Government buildings follow the pattern of other commercial-scale buildings, with lighting and mechanical systems being the largest energy users. Other types of municipal energy expenses include outdoor lighting and traffic controls, transportation fuels for fleets, and heating systems in maintenance buildings and garages. Traffic engineers can synchronize traffic lights to keep traffic moving and reduce idling time when drivers are achieving zero miles per gallon. New technologies like light-emitting diode outdoor lighting will reduce the cost of operating these fixtures.

Difference Between Energy Efficiency and Energy Conservation

Energy efficiency is a useful and necessary tool in sustainability planning, but it cannot control the ultimate use of energy. People can install compact fluorescent lights in their lamps, but will they turn off the lamps when they leave the room? Energy-efficient equipment must be coupled with energy conservation to take full advantage of its attributes. Energy conservation, or using energy in just the amount needed, is an even more difficult challenge than energy efficiency.

See Also: Compact Development (New Urbanism); Daylighting; Green Infrastructure; Green Roofs; LEED (Leadership in Energy and Environmental Design).

Further Readings

American Council for an Energy Efficient Economy. http://www.aceee.org (Accessed July 2009).
Kelley, Ingrid. *Energy in America: A Tour of Our Fossil Fuel Culture and Beyond.* Burlington: University of Vermont Press, 2008.
U.S. Department of Energy. "Energy Efficiency and Renewable Energy." http://www.eere .energy.gov (Accessed July 2009).
U.S. Environmental Protection Agency. "Energy Star." http://www.energystar.gov (Accessed July 2009).

Ingrid Kelley
Energy Center of Wisconsin

ENVIRONMENTAL IMPACT ASSESSMENT

Environmental impact assessment (EIA) is a process that identifies and evaluates environmental consequences of development projects and suggests measures for preventing and mitigating the potential adverse effects before projects commence. EIA is perceived as a means to promote sustainable development by facilitating environmentally sound decision making, securing better environmental planning, improving the design of project proposals, ensuring compliance with environmental standards, helping avoid excessive costs of unexpected impacts, and increasing decision-making transparency and project acceptance by the public. Given these benefits, awareness about the nature, key actors, and stages of EIA, as well as barriers to it, is essential for creating a green society.

EIA originated from the U.S. National Environmental Protection Act (1969) in response to the concerns about environmental degradation resulting from major developments and the associated lack of accountability. The widely proliferating sustainability concept has pinpointed EIA as a tool to achieve the integration of environmental protection with socioeconomic development. EIA has gained international acknowledgment and become institutionalized in more than 100 countries, as well as in subnational units of government, including states and cities.

Initially focused on ecological impacts, EIA has expanded to encompass the broader "environment"; that is, social, economic, cultural, and health impacts. It has also evolved from applying purely technocratic approaches (e.g., expert-driven assessments) to combining those with participatory approaches such as stakeholder panel reviews. Although EIA is viewed as a predominantly technical-rational process, its political nature is also recognized, in that EIA findings are used for political decisions on whether to proceed with or reject a project.

The main parties to EIA are the proponent, who owns a proposed project; the consultant, who conducts EIA on behalf of the proponent; the authorities, who ensure the quality of EIA and grant approvals; and the affected public/nongovernmental organizations. Public involvement is essential to the EIA process and should occur at several stages of EIA.

The EIA process varies from country to country depending inter alia on legal requirements and planning traditions; however, its key elements are similar in many jurisdictions. EIA starts with screening, which aims to determine whether a proposed project should undergo an EIA or not and, if so, the level of assessment needed. Screening can be guided by a formal prescriptive method (e.g., a list of projects that legally require EIA) and/or by a case-by-case approach at the discretion of the authorities and based on certain criteria, such as significant environmental effects.

Scoping is the next and crucial stage in EIA, aimed at identifying the key issues and significant impacts to be thoroughly examined during EIA. Scoping also delineates the appropriate time and space boundaries of the EIA study and specifies the information necessary for both EIA and decision making. To identify the key concerns of the affected parties, scoping should be an open and interactive process involving the relevant stakeholders. The findings and conclusions of the scoping stage are usually documented in a concise outline of the EIA study describing both the EIA process issues, such as the planned steps of EIA, the key issues to be considered, provisions for public involvement, and EIA-related management issues, for example, roles and tasks of multidisciplinary EIA teams, time schedules, and budgets.

Once scoping is completed, the impact prediction and evaluation stage starts with identifying potential environmental and socioeconomic impacts of the proposed project. Those include not only direct impacts resulting from the project (e.g., air pollution) but also indirect impacts resulting from various activities arising from the project (e.g., noise-related lower property prices) and cumulative impacts caused by a combined pressure of a particular and other existing, past, or planned projects in the area. EIA typically also assesses the impacts of not proceeding with the proposed projects and one or more alternatives that might result in smaller adverse impacts than the proposed project.

To not omit any significant impacts, EIA takes a systematic approach to impact identification, deploying various methods such as matrices (tables of potential interaction between project actions and receptors), networks (visualized cause–effect chains of impacts on the environment), and expert judgments. Next, the parameters of the identified impacts are predicted, preferably in a quantitative manner; for example, the magnitude of impacts (severe, moderate, low) and their likelihood of occurrence (probable, uncertain, etc.). Any uncertainties in knowledge and prediction should be acknowledged. Afterward, the impacts are evaluated to determine their significance. For this, a distinction is drawn between the "predicted" and "residual" impacts. First, the significance of predicted impacts is evaluated to determine the mitigation measures, and then the significance of residual impacts—after mitigation—is considered. The significance test may be run against the predefined criteria or thresholds (e.g., air quality standards) to determine the environmental acceptability of the project.

The next EIA stage is proposing mitigation measures to prevent, minimize, or remedy potential adverse impacts and to keep residual impacts within acceptable limits. Developing mitigation measures entails a search for better design alternatives and ways to implement the project to enhance its environmental and social benefits. Mitigation measures are usually included in the approval conditions of projects (issued by the authorities, see following) to be implemented during the project delivery; for example, simultaneously with constructing a factory, a noise fence should be constructed to protect the local community from noise during the construction and operation of the factory.

Then, the results of the EIA study—a report, referred to as an Environmental Impact Statement (EIS)—are submitted to the authorities and other relevant parties. The purpose of the EIS is to provide a coherent statement of the potential impacts of a proposal and the measures needed to mitigate them. The format and contents of the EIS are usually defined by laws or regulations.

Further, a review of the EIS is conducted to secure and improve its quality for informed decision making. The authorities examine the adequacy, comprehensiveness, and quality of the EIS and produce recommendations and approval conditions that are to be addressed before or during project implementation. At this stage, public hearings and consultations are also held, and the results of those are incorporated into the EIS.

Drawing on the EIS review, as well as other political or economic considerations, the authorities make decisions as to whether to reject or approve the proposed project. Rejections means the project might need to be redesigned, reassessed, and resubmitted; otherwise, once the approval is granted, the implementation of the project may commence in compliance with the approval conditions.

Finally, EIA follow-up is instigated alongside the project implementation and lasts over its whole life cycle. It includes monitoring/auditing to identify the actual effects and observe the delivery of mitigation measures, evaluation of monitoring/auditing data against the predicted values and formal standards, evaluation-based management (e.g., to cope with unpredicted effects), and communication to provide feedback to the project and to future EIAs and to communicate the EIA follow-up results to the stakeholders.

The typical barriers to a successful EIA include weak legislative and enforcement frameworks, no EIS quality assurance mechanisms, poor administration of the EIA process, lack of baseline data, gaps in knowledge and uncertainties, deficient EIA follow-up, and lack of understanding of EIA benefits. "Learning-by-doing" is crucial to improve EIA practice, as well as for all actors to understand the added values of EIA to the sustainable society.

See Also: Environmental Planning; Environmental Risk; Green Design, Construction, and Operations; Sustainable Development.

Further Readings

Canter, L. *Environmental Impact Assessment,* 2nd Ed. New York: McGraw-Hill, 1996.

Glasson, J., et al. *Introduction to Environmental Impact Assessment,* 3rd Ed. London: Routledge, 2005.

Sadler, B. and M. McCabe, eds. *Environmental Impact Assessment Training Resource Manual,* 2nd Ed. Geneva: United Nations Environment Programme, 2002.

Maia Gachechiladze
Central European University

ENVIRONMENTAL JUSTICE

The term *environmental justice* refers to the fair and equitable treatment and participation of all peoples, regardless of race, ethnicity, national origin, color, or socioeconomic class, with respect to the development and implementation of environmental laws, regulations, and policies. It aims to ensure that all people enjoy the same degree of protection from environmental risks and hazards, equal access to natural resources and environmental benefits, as well as equal access to information and decision making, thus ensuring a healthy environment in which to live, learn, and work.

It is possible to identify two critical dimensions of environmental justice: (1) protecting human health and the environment in places where people live, learn, and work; and (2) developing grassroots leadership and community-based planning and policy approaches that meet community needs. Self-determination and participation in decision making are considered to be key factors in achieving grassroots leadership in environmental justice. The following can be identified as the main principles behind the notion of environmental justice: respecting environmental limits; sustainability; combating social exclusion, inequality, and poverty; information sharing and participation; and a transparent enforcement of environmental justice.

There are however two aspects of environmental justice which should be considered as well as distinguished. They are (1) procedural injustices (i.e., laws or regulations that cause communities of color to bear disproportionate environmental risk); and (2) de facto injustices (i.e., the disproportionate risk borne by communities of color regardless of institutional—or legal—procedures that cause risky facilities to be located in communities of color). This means that environmental justice needs to be done both at regulatory and de facto levels. Environmental injustice exists when certain members of society (e.g., indigenous persons, ethnic minorities, and other disadvantaged groups) suffer disproportionately from any one or more of the following: environmental risks and hazards; violations of fundamental human rights as a result of environmental factors; denial of access to environmental resources and benefits; denial of access to information; denial of the right to participate in decision making; and denial of access to justice in environment-related matters. Environmental injustice also exists when there are inequitable laws that cause ethnic minorities to face disproportionate environmental hazards. Thus both overt racism as well as disproportionate effect may lead to environmental injustice.

The concept of environmental justice emerged through the work of grassroots environmental activists in the United States. The environmental justice movement began in the early 1980s when multiracial coalitions of activists fought against pollution and toxic waste dumpings in areas adjacent to African American communities in Warren County, North Carolina, and Dickson County, Tennessee. In 1987, the United Church of Christ Commission for Racial Justice produced an innovative study titled "Toxic Wastes and Race in the United States" that showed race, more than class or wealth, was the most significant variable when examining areas where toxic waste sites were located. The study suggested that the higher the percentage of African American or Hispanic residents in a neighborhood, the more likely the neighborhood is to be located near a hazardous waste facility.

The activities of environmental justice groups in the United States have generally focused upon protesting the location and operations of polluting factories and waste sites in predominantly African American and Hispanic neighborhoods and on the lands of indigenous peoples. Accusations of environmental racism and discriminatory practices in facility location, regulation, enforcement, and decision making have been key components of the claims and discourses of environmental justice.

On February 11, 1994, President Bill Clinton signed Executive Order No. 12898, the Federal Actions to Address Environmental Justice in Minority Populations and Low-Income Populations, an order considered a major milestone for the environmental justice movement. It requires federal agencies to achieve environmental justice by identifying and addressing disproportionately high and adverse human health and environmental effects, including the interrelated social and economic effects of their programs, policies, and activities on minority populations and low-income communities in the United States.

Proponents of environmental justice seek to redress inequitable distribution of environmental burdens and to equitably distribute access to environmental goods such as nutritious food, clean air and water, recreation facilities, healthcare, education, transportation, and safe jobs. They identify several causes of environmental injustice, which include institutionalized racism; the commodification of land, water, air, and energy; unaccountable government policies; and lack of resources and power in affected communities. However, critics of environmental justice contend that any such "unjust" effects are unintentional, attributable to a variety of factors, and that environmental justice is a vague concept that may stand in the way of economic benefits brought by industrial projects.

On an international level, there is evidence that much environmental damage is the result of the actions of more affluent nations or more affluent groups within nations, with impacts on poorer nations and poorer groups within nations. There is a growing concern that poorer peoples internationally experience multiple environmental harms including— but not limited to—depletion of natural resources, pollution, and health hazards. Public awareness of the issues surrounding environmental justice is having some effect in terms of policy shifts along with increased attention to issues of equality and human rights.

Internationally, there are several examples of regions and cities that are implementing environmental justice programs. A case in point is the expansion of East–West Expressway near the central business district of Durham, North Carolina. An unbuilt segment of the expressway was slated to be within a small African American neighborhood known as Crest Street. The road was built after due consultation with the local community and is an example of the successful implementation of the mitigation and enhancement plan that preserved a cohesive community, as well as demonstrated a collaborative process with community residents and committed professionals both inside and outside the transportation community.

In Europe, the public is gaining more access to information on environmental harms through various policy mechanisms, such as the United Nations Economic Commission for

Europe's Convention on Access to Information, Public Participation in Decision-Making and Access to Justice in Environmental Matters (the Aarhus Convention, enforced October 30, 2001). For example, in 2005 the Scottish Government announced its Sustainable Development Strategy titled "Choosing Our Future," identifying a number of intended actions and desired outcomes vis-a-vis the promotion of environmental justice:

- *Decision Making:* supporting greater public participation, modernizing the planning system, and providing more accessible, useful information on the environment
- *Flooding:* protecting communities from flooding, which may have a disproportionate impact on low-income families
- *Greenspace:* providing parks and greenspaces, making neighborhoods healthier and more attractive places in which to live and work
- *Health:* directing a new focus on the connections between environmental conditions and health outcomes
- *Housing Improvements:* tackling fuel poverty, and building safety and security into new housing at the design stage
- *Learning:* developing a new program to support learning through doing, with a focus on deprived neighborhoods
- *Quality-of-Life Improvements:* tackling problems on the ground leading to improvements in the local environment, personal well-being, and safer communities
- *Regeneration:* building environmental objectives in regeneration and simplifying access to funding in this area
- *Transport:* pursuing improvements in transport to tackle growth in road traffic, reduce congestion, and improve air quality
- *Waste:* moving away from over-dependence on landfill, which can have a significant negative impact upon some communities

In summary, environmental justice means ensuring that everyone has a right to a healthy environment irrespective of who they are, what they do, and where they live. This includes the right of individuals and communities to have secure access to the law in resolving environmental concerns, and the ability of relevant agencies to implement a range of effective enforcement methods, including powers to ensure that those who harm the environment or cause environmental injustice bear the cost of putting things right.

See Also: Environmental Impact Assessment; Green Communities and Neighborhood Planning; Sustainable Development.

Further Readings

Matsouka, Martha. "Building Healthy Communities From the Ground Up: Environmental Justice in California." Asia Pacific Environmental Network. Berkeley, CA: Inkworks Press, 2003).

Mock, Brentin. 2008. Will Environmental Justice Finally Get Its Due? *The American Prospect,* December 22, 2008. http://www.prospect.org/cs/articles?article=will_environmental_ justice_finally_get_its_due (Accessed January 2009).

The Scottish Government. "Choosing Our Future." *The Government's Sustainable Development Strategy.* December 2005. http://www.scotland.gov.uk/Topics/Sustainable Development/environmentaljustice (Accessed February 2009).

Stephens, C. and P. Stair. "Charting a New Course for Urban Public Health." In *State of the World 2007: Our Urban Future*, L. Stark, ed. New York: W. W. Norton, 2007.

United Nations Economic Commission for Europe (UNECE). *Aarhus Clearinghouse for Environmental Democracy.* 2007. http://aarhusclearinghouse.unece.org (Accessed October 2008).

United Nations Environment Programme. *Global Environmental Outlook 2007.* Nairobi, Kenya: UNEP, 2007.

Jawad Syed
University of Kent

ENVIRONMENTAL PLANNING

Environmental planning consists of a set of theoretical tools and management practices that help humans make informed decisions about how to protect and conserve shared environmental resources. Like all types of planning, environmental planning attempts to translate social values and goals into policies aimed toward protecting public health, welfare and safety. Environmental planning adds to these goals an express concern for protecting and conserving the natural environments found in urban, rural, and wild landscapes. These goals, though, are often contested at all levels of government and society. As such, environmental planning is intricately woven into social and political processes.

Environmental planning addresses a broad array of issues that confront contemporary society. For example, the following topics are central to environmental planning practice: protecting air and water quality; conserving resources, such as farmland, forests, and open space; reducing the impact of environmental hazards on humans; managing the disposal of solid and hazardous waste; and protecting and developing the environmental amenities that make communities attractive places to live. Additionally, environmental planners work to remediate and redevelop brownfields; find innovative strategies to preserve energy resources; and protect natural resources, such as scenic wilderness, wildlife habitat, coastal areas and wetlands. Effective environmental planning practice attempts to foster sustainability in each of these endeavors. By approaching these problems with the intent to create environmental health, economic development, and social equity, environmental planning practice is a key input for a more sustainable society.

Environmental planners often receive professional training in a variety of disciplines, such as planning, geography, biology, landscape architecture, geology, hydrology, economics, law, or a number of other related fields. They work for local, state, or federal government agencies, nonprofit environmental associations and resource user groups, as well as private corporations.

In many ways, any human that has ever made a decision about the use of a piece of land has engaged, in one way or another, in a form of environmental planning. However, as urbanization and natural resource use increased in the 19th and 20th centuries, more people became concerned about environmental planning. The historical development of environmental planning theory and practice in the United States reflects the country's changing environmental and social conditions.

During the progressive era in the late-19th and early-20th centuries, environmental planners reflected the national concern regarding poor urban conditions and the country's growing appetite for natural resources. Planners like Frederick Law Olmsted, who designed New York's Central Park, sought to create beautiful urban greenspaces that would combat the increasingly ugly and unsanitary conditions of cities in the turn of the century. Outside of cities, naturalist John Muir fought for wilderness protection, while forester Gifford Pinchot contended that efficiency and a sustained yield of timber resources were paramount concerns for wilderness areas. The federal government became involved in environmental planning during this period and created several new federal organizations, such as the National Parks Service and the Soil Conservation Service. Pinchot established the newly formed National Forest Service's utilitarian mission as creating "the greatest good for the greatest number"—a conservation focus that continues to inform the planning practices of many government land use agencies.

The era of modern environmental planning in the 1960s and 1970s helped the federal government solidify its role as an environmental planner. By this time, the United States was beginning to feel the effects of decades of poor environmental decision making. Events such as toxic landfill disasters, increasingly poor urban air quality, and the rapid expansion of urban sprawl gripped media headlines and prompted all levels of government to action. New federal legislation ensured the federal government would take the lead in environmental planning. For example, the National Environmental Policy Act (1970) mandates that all federal development projects prepare an environmental impact statement (EIS) for review prior to development; the Coastal Zone Management Act (1972) mandates that state coastal management programs meet minimum federal guidelines; the Endangered Species Act (1973) protects endangered or threatened flora and fauna from development; and the Resource Conservation and Recovery Act (1976) ensures proper hazardous substances disposal by tracking hazardous materials from "cradle to grave." Moreover, the Environmental Protection Agency (EPA) acted as the federal agency responsible for environmental planning throughout the United States. U.S. states also began to be more involved in environmental planning. Hawaii, Vermont, Florida, and Oregon all began state level environmental planning efforts to combat suburban sprawl and other environmental ills.

A new era of sustainability currently defines environmental planning and practice. Environmental planners are beginning to approach sustainability, or long-term economic, social, and environmental viability, primarily at the local urban level. Environmental planners aim to ensure that the cities where residents live, work, and recreate are healthy, efficient, and attractive places. To produce these effects, planners strive for high levels of air quality, walkability, social equity, energy efficiency, water conservation, and green spaces. These and other approaches all help a city become more sustainable. Effective urban environmental planning practices also help to reduce a city's ecological footprint, thereby contributing to a solution to the impacts of global climate change.

Climate change, perhaps more than any other environmental problem the world has faced, has forced cities and countries around the world to recognize a key principle for environmental planning: natural processes do not easily conform to government boundaries, programs, or initiatives. Global environmental challenges such as the loss of tropical forests, species extinction, freshwater shortages, and climate change all require a global environmental planning approach. Similarly, more local and regional issues, such as watershed protection, transportation planning, or species migration also require a more cooperative approach in order to plan for a solution.

See Also: Ecological Footprint; Green Communities and Neighborhood Planning; Sustainable Development.

Further Readings

Daniels, Thomas. "A Trail Across Time: American Environmental Planning From City Beautiful to Sustainability." *Journal of the American Planning Association*, 75 (2009).

Daniels, Thomas and Katharine Daniels. *The Environmental Planning Handbook for Sustainable Communities and Regions*. Chicago: APA Planners Press, 2003.

Madu, Christian N. "Introduction to Environmental Planning and Management." In *Environmental Planning and Management*. http://www.icpress.co.uk/environsci/p460.html (Accessed May 2011).

Selman, Paul. *Environmental Planning: The Conservation and Development of Biophysical Resources*. Thousand Oaks, CA: Sage, 2000.

Jeremy Bryson
Syracuse University

ENVIRONMENTAL RISK

Environmental risk refers to the likelihood of adverse effects resulting from various environmental hazards that occur in urban areas. In the narrow sense, this term refers primarily to the negative health effects and environmental degradation that occur as a result of pollution generated by anthropogenic activities. More broadly, environmental risk encompasses the likelihood that physical hazards—such as rising sea level, severe weather, and natural disasters—will affect a community, and to what degree they are expected to occur. Urban planners and policy makers can employ a range of mitigation and adaptation measures to manage these risks, including innovative infrastructure design, urban planning, and implementation of new technologies.

Pollution and Environmental Risk in Cities

Industrial, domestic, and transportation activities in cities generate waste and by-products that lower air and water quality, with resulting public health impacts. Air pollution in many cities around the world poses near- and long-term hazards to public health. In the United States, for example, high concentration of fine particulate matter—emitted as a result of industrial processes, transportation, and home heating—has been linked to asthma episodes, increases in emergency room visits for respiratory problems, pulmonary disease and asthma. There are over 1,000 chemical compounds and elements that have been identified by the Environmental Protection Agency as potential hazardous air pollutants. These pollutants are released into the air by a variety of sources and can have lasting long-term health effects, potentially including some forms of cancer.

Indoor environments also expose occupants to potentially hazardous chemicals as a result of building materials, by-products from cleaning equipment, emissions from cooling or heating, or presence of dust and particulate matter. This issue has only recently received attention, and it will pose an additional challenge to city planners striving to design safe buildings for places of both business and residence.

Water quality and availability can be affected by urban activities in part related to the generation of heat and organic waste. In fact, cities play a significant role in nutrient cycling locally, and potentially globally. Waste nitrogen is a particularly harmful nutrient that can cause eutrophication and can support toxic algal blooms. In cities, this can reduce surface water quality, taint potable supplies, limit recreational opportunities, and reduce biodiversity. Developed cities often have denitrification technology to reduce leachate of nitrogen from industrial runoff and waste disposal, but continued population growth and development of industry could limit the capacity of existing infrastructure to handle increasing volumes of waste in the future.

Physical Risk Factors and Climate Change

Physical climate change impacts have multifaceted direct and indirect implications for urban settlements. These impacts often affect multiple sectors of human activities (e.g., public health, the economy) and can accentuate vulnerabilities associated with, for example, overcrowding or development in hazardous areas. Floods, severe storms, landslides, and earthquakes claim lives and can cause lasting damage to the livelihoods of urban residents by destroying assets and causing physical impairments that require time away from work. In this way, environmental risk factors can make it difficult to address issues of poverty in cities, because the poor are often most economically impaired by these factors.

Damage to physical infrastructure, particularly during flooding, can cost billions of dollars in repair. Economic damages can be long-lasting when transportation is disrupted, with losses occurring not only in the city itself but also in other economically interrelated locations around the world. For example, damage to seaports during Hurricane Katrina caused trade disruption in global markets.

Slow-onset climate changes including sea level rise, subsidence, and increasing air temperatures can also act as risk factors on cities. Over time, they can cause weathering of roadways, bridges, and railways that forces more frequent repairs and traffic disruptions. Sea level rise can cause saltwater intrusion into drinking supplies, which creates both an economic and a public health concern. Increasing air temperatures and extreme heat events—including higher nighttime temperatures and more frequent heat waves—have been linked with increases in heat-related mortality and illness, particularly for the young and the elderly.

The risk of heat-related effects can be exacerbated in cities, which tend to have higher air and surface temperatures compared with rural areas as a result of the combined effects of structural interference with thermal radiation, low albedo of impervious surfaces, and reduced evapotransporation. This "urban heat island" phenomenon can also impose costly energy demands on urban systems. By disrupting energy flows, these effects can cause formation of photochemical smog and reduce air circulation that could diffuse the concentration of air pollutants.

For many cities around the world, environmental risk as a result of physical climate changes is increasing or is likely to increase in the future. This is in part because many large urban settlements are located in coastal zones, where storm surges and hurricanes have been observed to occur with greater intensity compared to historical averages. In addition, urbanization is on the rise, and population growth that outpaces urban planning tends to force populations to settle in hazardous areas. This is particularly true in developing cities, where urban slums are increasing in size and prevalence. These settlements face very high levels of environmental risk—despite lacking polluting industrial facilities—because they lack the infrastructure and institutional capacity to prepare for climate-related risk factors.

Distributional Impacts of Environmental Risk

Environmental risk factors do not have equivalent impacts on different cities around the world. Cities facing the same physical risk factors can have different vulnerabilities based on their socioeconomic situation, as well as the strength of local and national political institutions. Economic capital and urban planning allows cities to institute protective physical infrastructure, to limit development in hazardous areas, and to adequately prepare for and recover from disasters. Those urban areas with limited capacity to implement such strategies often experience the greatest losses as a result of environmental risk factors. In developing nations in particular, urban areas are highly vulnerable to risk.

Within cities, too, the effects of environmental risk are distributional. Low-income classes tend to experience greater losses compared with wealthier groups within the same urban area. The poor have limited mobility and fewer resources to protect themselves from environmental hazards and to recover from disasters. They are also excluded from planning processes and are typically located in the most hazardous areas of a city; for example, near polluting facilities and in high-risk areas for flooding and landslides.

Other groups that face distributional impacts include women, racial and ethnic minorities, and indigenous peoples. These groups often face the disadvantages associated with low income and may also face exclusion from insurance, limited education, low access to information regarding environmental hazards, and tacit or outright discrimination during emergency planning or disaster relief procedures.

Managing Environmental Risk

There are varied integrative approaches to managing current and anticipated environmental risk factors. Environmental risk assessments can be used at the levels of individual firms, cities, or regions to examine, prioritize, and manage local hazards. Approaches for pollution management can involve cleaner technologies and stricter standards for industrial and domestic practices. Adaptation measures to combat risk include development practices that limit construction in high-risk areas and include consideration of future population increases.

Planners, managers, and policy makers around the world are now capitalizing on opportunities to upgrade infrastructure to plan for the potential for increasing risk factors as a result of climate change. Greening or climate-proofing mechanisms for transportation include improved drainage systems on roads and protective covers and overhangs over portions of rails and roads. Given the extent to which most aspects of urban life depend on water, floods will also challenge the ability of planners and managers, who must ensure that reliable services such as wastewater treatment and clean water provision are available in the face of a highly variable climate.

See Also: Adaptation, Climate Change; Green Infrastructure; Infrastructure; Sea Level Rise.

Further Readings

De Sherbinin, Alex, et al. "The Vulnerability of Global Cities to Climate Hazards." *Environment and Urbanization,* 19/1 (2007).

Pope, Arden C. and Douglas Dockery. "Health Effects of Fine Particulate Air Pollution: Lines That Connect." *Journal of the Air & Waste Management Association,* 56 (2006).

Ruth, Matthias and Rebecca Gasper. "Water in the Urban Environment: Meeting the Challenges of a Changing Climate." Milan, Italy: Presentation at OECD International Conference: Competitive Cities in Climate Change, 2008.

Rebecca R. Gasper
University of Maryland

F

FOOD DESERTS

Residents living in a so-called urban food desert, where shops selling healthy food are more than a 10-minute walk away, would not have ready access to an open-air farmers market, like this one in Detroit, Michigan.

Source: iStockphoto

The term *food desert* was coined in the late 1990s by policy makers to describe low-income neighborhoods with poor access to fresh and affordable food. One criterion for defining a food desert is when people have to walk more than 500 meters to a shop selling healthy food—or, expressed another way, the distance it takes a fit person to walk in 10–15 minutes. In reality, the concept extends beyond how retail geography and planning affect access to healthy food and into wider debates about social exclusion and its links to poor diet and health. A complex combination of factors has emerged that present obstacles to eating a healthy diet for those surviving on a low income.

In the 1960s, major changes in food retail systems led to the rapid growth of supermarkets in out-of-town locations, which in turn led to the decline of local shops and traditional markets that had previously supplied food locally. Deprived neighborhoods were left with a limited choice of expensive fresh fruit and vegetables in the remaining shops as they struggled to compete with supermarket pricing and a growing number of fast-food outlets.

Studies in the United Kingdom and the United States have examined the assumed links between food retail provision and social exclusion, poor diet, and health and have sought to clarify whether food deserts actually exist and, if they do, how they can be tackled. U.K. policy has concentrated on

185

retail-led interventions as a solution to the problems of food access in both urban and rural areas. However, research evidence is mixed about the benefits of supermarkets opening in deprived areas. For example, car ownership remains a key determinant in choice of food store, and carrying shopping and food storage remain barriers to accessing supermarkets for those living on low incomes. Others have looked at socioeconomic factors and food retail and have found that even when access to healthy food is good for those living on low incomes, dietary knowledge and a healthy lifestyle play a critical part in eating a healthy diet. Some initiatives are linking local health education programs and local retail regeneration to balance the benefits to health and community cohesion. For example, farmers markets are being funded in food deserts in some parts of the United States, using partnerships between community health organizations and grassroots community groups.

As governments have begun to focus on urban regeneration, there has been a renewed focus on the importance of the city center and the mix of retail provision and vibrant local neighborhoods. U.K. planning guidance has resulted in the major supermarket chains' opening new, smaller-format stores in city and town centers, but higher pricing and limited choice of fresh produce still pose a problem for those living on low incomes. There has also been a renewed recognition of the importance of public space for community cohesion and the valuable role that traditional food markets can play in these spaces by offering opportunities for social interaction within local communities, by fostering social inclusion, and by supporting disadvantaged members of the community through the provision of low-priced food.

It is estimated that 53 percent of children living in inner-city London live below the poverty line, and inner London boroughs contain sizable areas that some consider to be food deserts. The Capital Growth initiative championed by London's mayor, Boris Johnson, aims to boost the amount of locally grown food in London to help improve access to nutritious food in the capital by turning 2,012 pieces of land into thriving green spaces to grow food by 2012.

In the United States, underutilized space is also being used for infill developments to improve access to fresh, healthy food in urban neighborhoods. In New York, where it is estimated 750,000 people inhabit food deserts, initiatives include more licenses for carts selling fruit and vegetables, food stamps for those on low incomes to use at farmers markets, and zoning and tax incentives to encourage new grocery stores in poor neighborhoods. First Lady Michelle Obama also has recently begun a new initiative to encourage healthy eating in urban areas by turning part of the White House garden into an organic kitchen garden, and the U.S. Department of Agriculture has funded grassroots initiatives to increase the availability of fresh fruits and vegetables in low-income neighborhoods.

More recent work has started to consider the expansion of supermarkets into cities in developing countries and the effect on local food retail provision, which is currently dominated by small shops and local markets. Planners and policy makers are focused on avoiding the experiences of industrialized countries to prevent the exacerbation of urban food access problems in cities in developing countries.

Opinion still seems divided over whether food deserts are a myth or a reality. However, global crises have put food security firmly back on the agenda, and governments must address their food policy goals as the development of a sustainable food system becomes a more pressing issue. One thing is clear—people with low incomes need places to buy fresh and affordable food if they are to be successfully encouraged to eat a healthier diet.

See Also: Civic Space; Food Security; Walkability (Pedestrian-Friendly Streets).

Further Readings

Cummins, S., et al. "Reducing Inequalities in Health and Diet: Findings From a Study on the Impact of Food Retail Development." *Environment and Planning*, 40:402–22 (2008).

Kyle, R. and A. Blair. "Planning for Health: Generation, Regeneration and Food in Sandwell." *International Journal of Retail and Distribution Management*, 35/6:457–73 (2007).

"Welcome to the Food Desert Website." http://www.fooddeserts.org (Accessed April 2009).

Wrigley, N., et al. "Assessing the Impact of Improved Retail Access on Diet in a 'Food Desert': A Preliminary Report." *Urban Studies*, 39/11:2061–82 (2002).

Julie Smith
Countryside and Community Research Institute

FOOD SECURITY

Food security, as defined by the World Food Summit of 1996, is "when people at all times have access to sufficient, safe, nutritious food to maintain a healthy and active life." Usually, the concept of food security includes both physical and economic access to food that meets people's dietary needs and preferences. Cities employ various strategies to ensure that their populations are food secure, including supporting urban agriculture, food banks and other charities that feed people, and the elimination of food deserts by supporting food retailers in communities that lack sources of healthy food.

The definition of food security has evolved over a period of time. In 1974, food security was defined by the World Food Summit as the "availability at all times of adequate world food supplies of basic foodstuffs to sustain a steady expansion of food consumption and to offset fluctuations in production and prices." This was expanded by the Food and Agriculture Organization of the United Nations in 1983 to include securing access by vulnerable people to available supplies. This was to balance the demand and supply side of the food security equation.

In 1986, the World Bank Report on Poverty and Hunger brought attention to the temporal dynamics of food insecurity and introduced distinction between chronic food insecurity and transitory food insecurity. Chronic food insecurity is usually long term, occurring when people are unable to meet their minimum food requirements over a sustained period of time, and is related to (or to the result of) issues of low incomes and continuing poverty, whereas transitory food security is short term and temporary and refers to the food shortage resulting from a sudden drop in the ability to produce or access enough food because of natural disasters, conflict, or maybe some sort of economic collapse.

By the mid-1990s, food security had become a major concern at a global level, and access to food now included sufficient food, indicating concern for protein-energy malnutrition. Later, the definition was expanded to incorporate food safety, nutritional balance, and food preferences. In 1994, the UN Development Programme Human Development Report included food security in the concept of human security (which is related to the human rights perspective on development). In 2001, the definition of food security described above was modified in *The State of Food Security 2001*, now reading, "a situation that exists when all people, at all times, have physical, social and economic access to sufficient, safe and nutritious food that meets their dietary needs and food preferences for an active and healthy life."

In addition to chronic and transitory food insecurity, there is a third type of food insecurity: seasonal food insecurity. Just like chronic food insecurity, it is relatively easier to predict because it follows a sequence of known events. Seasonal food insecurity is usually of limited duration and can be seen as recurrent transitory food insecurity. This is usually associated with seasonal fluctuations in weather, climate, cropping patterns, and diseases at certain times.

The Pillars of Food Security

The concept of food security has three pillars: food availability, food access, and food use. Food availability means that there is enough quantity of food available at all times; food access refers to the available resources to obtain the amount of food needed for a nutritious diet; and food use means the appropriate use of a variety of food items based on the knowledge of basic nutrition and care. Food use also includes the knowledge or application of adequate water and sanitation.

There have also been studies to define food security at a household or individual level. The concept of household food insecurity originated with research among low-income women in upstate New York by Radimer (and team) in the early 1990s, when four dimensions of household food insecurity were identified:

- *Quantitative:* Referring to not enough quantity of food available
- *Qualitative:* When the quality of food is sufficient (i.e., reliance on inexpensive food that is nonnutritious)
- *Psychological:* This is related more to the stress related to meeting daily food needs when there is not a surety that food (all meals) will be available every day (i.e., there is an anxiety about food supply)
- *Social:* This dimension includes getting food from charitable assistance or incidents in which food was bought on credit or stolen

In a way, these four dimensions of food insecurity are linked to the five conceptual components of food security. The existence of food-insecure households are examples of absence of *universal* access to food by all people, inability of households to have stable *access* to food because of resource limitations, leading to a *psychological* effect on adults if there is a constant anxiety for feeding themselves and their children. Household food insecurity can also lead to issues with access to food with human *dignity* if the household has to rely on charities or *socially* unacceptable ways to meet their dietary needs.

There has been a slight shift in the focus by many agencies, including public health agencies, from household to community level—the main difference being in the sense that the focus of household food security is mainly physical and economic access to food, whereas community food security adds the importance of economic, environmental, and social aspects of the food system, in addition to the two goals of household-level food security.

So far the definitions of food security have focused on universal access to availability of food. There are other definitions of food security in which security is taken to a global level and refers to the self-sufficiency and security of a country. The Organisation for Economic Co-operation and Development, for example, defines food security as a "concept which discourages opening the domestic market to foreign agricultural products on the principle that a country must be as self-sufficient as possible for its basic dietary needs."

For any intervention in food insecurity, it is not enough to know just about the length of insecurity, it is also important to measure the intensity or severity of the effect of the identified problem on the overall food security and nutrition status. The Food and Agriculture Organization measures hunger in terms of "undernourishment," which refers to the proportion of the population whose dietary energy consumption is less than a predetermined threshold (which is country specific). The severity of undernourishment indicates the extent to which dietary energy consumption is below the predetermined threshold level of that region/country. Another way to look at food security is to look at people who are vulnerable to experiencing food insecurity in the future. "Vulnerability" is defined in terms of three critical dimensions: vulnerability to an outcome, vulnerability from a variety of risk factors, and vulnerability resulting from inability to manage risks. The two interventions suggested include reducing the degree of exposure to the hazard and increasing the ability of the vulnerable population to cope with the risks and hazards.

The U.S. Agency for International Development measures the food security of a country based on the gap between projected domestic food consumption (which is a difference between food produced domestically and food imported from nonfood use) and a consumption requirement. All the aid commodities are converted into grain equivalent based on calorie content to allow aggregation. Food gaps are projected using two consumption criteria: status quo target (it takes three-year average for per capita consumption target to eliminate short-term fluctuations) and nutrition-based target (in which the objective is to maintain daily caloric intake standards, as recommended by the Food and Agriculture Organization).

According to the Food and Agriculture Organization (FAO), "food security depends more on socioeconomic conditions than on agroclimatic ones, and on access to food rather than the production or physical availability of food." It also states that, to evaluate the potential impacts of climate change on food security, "it is not enough to assess the impacts on domestic production in food-insecure countries. One also needs to (i) assess climate change impacts on foreign exchange earnings; (ii) determine the ability of food surplus countries to increase their commercial exports or food aid; and (iii) analyze how the incomes of the poor will be affected by climate change."

As pointed out by FAO, "in the 'century of cities,' a major challenge will be providing food for urban inhabitants, especially the poor." A major challenge in urban areas is that open land is replaced by buildings and pavement, meaning that land is not available for agricultural production, leaving residents dependent on food grown elsewhere. Food insecurity in urban areas also relates to income inequality, problems of inequitable distribution, an increasing urban, land degradation, and institutional and governance shortcomes. A few factors directly influencing urban food (in)security are listed below:

- *Income Insecurity:* either the household or individuals do not have enough or a stable income to buy sufficient food, or a disproportionate amount of their income is allocated for other living expenses, such as rent
- *Spatial Factors:* these may leave certain areas or neighborhoods without access to affordable grocery stores or markets
- *Isolation:* lack of a social network

Food Security and Climate Change

Climate change has an effect on agriculture, on food production, and as a result, on food security. There is a direct effect on food production resulting from changes in temperatures, precipitations, land suitability, crop yields, and agroecological conditions, and the indirect impact comes from the effect on growth and distribution of incomes (and its link to demand for agricultural products). Change in global and regional weather conditions (an increase in the frequency and severity of extreme events such as droughts, floods, hailstorms, and cyclones) has an effect on crop yields and local food supplied, affecting the stability of food supplies, and thus food security. Climate change will also have an effect on the ability of individuals to use food effectively. This will alter the conditions for food safety and change the disease pressure from vector-, water-, and food-borne diseases. "The main concern about climate change and food security is that changing climatic conditions can initiate a vicious circle where infectious diseases causes or compounds hunger, which, in turn, makes the affected populations more susceptible to infectious diseases." This can result in decline in labor productivity (if labor is unable to work because of illnesses or lack of nutrition), an increase in poverty, and even an increase in mortality rates.

Some of the different strategies to deal with these issues include strengthening resilience (which involves protection of existing livelihoods of vulnerable people, diversifying their sources of income, and changing their livelihoods) by changing consumption patterns and food preparation practices, raising productivity through improved agricultural water management (this is really crucial to ensuring global food supply and global food security and can be done by either conservation agriculture or promoting agro-biodiversity among other things), and adapting sustainable livestock management practices.

To address the issue of food security and hunger, the first Millennium Development Goal seeks by 2015 to reduce by half the proportion of the world's population experiencing hunger. According to the World Bank's estimates in 2005, 1.4 billion people in developing countries were living in extreme poverty. According to recent studies, even though the proportion of people worldwide suffering from malnutrition and hunger has dropped since the early 1990s, the absolute number of people lacking access to food has risen. With recent economic crises and increases in food prices, it is estimated that about 1 billion people will go hungry, and another two billion will be undernourished. Studies have shown that eastern Asia, and China in particular, have been successful in halving the proportion of underweight children between 1990 and 2006 compared with southern Asia, where 50 percent of children are still underweight. This region accounts for the more than half the world's undernourished children, whereas sub-Saharan Africa represents the region in which there has been the least progress made in meeting the target of reducing child malnutrition.

Food Security and Biofuels

Food security issues are also compromised by the production of cash crops to maximize profit, or by growing agricultural products geared toward biofuels. Because of increased demand for coffee, corn, and other foodstuffs in North America, there has been a trend in some developing countries to grow more cash crops for exporting to these countries than growing food to meet local demand. This has been an ongoing issue in some developing countries for a while, threatening food security in the region.

Biofuel development was seen as a positive development in terms of mitigation of climate change (as compared with use of traditional fuels) to alleviate global energy concerns, and also to foster rural development. However, the rapid growth in biofuels production has raised concerns about the threat it poses to food security. The UN Secretary General, in his opening remarks to the High-Level Segment of the 16th session of the UN Commission on Sustainable Development, noted that "[w]e need to ensure that policies promoting biofuels are consistent with maintaining food security and achieving sustainable development goals." In addition, integration of the agricultural and energy sectors (caused by growth and demand of biofuels) is something new and has an effect on food policy and the concept of sustainable development. As a result of this integration, the prices of agricultural commodities went up for the first time in decades. The potential effect of this large-scale expansion of the biofuel market and reduction of food production for human consumption and consumers, especially in low-income countries, is a challenge for food policy planners and also raises a question of sustainable development and whether the Millennium Development Goal on sustainable development can be achieved. However, most of the studies have concluded that although the first generation of biofuels did threaten the availability of adequate food supplies by diverting resources away from food and feed crops into biofuel production, the second generation of biofuels has taken sustainability issues (e.g., social, economic, and environmental) and the effect on land use, and so on, into consideration. This would imply that the second generation of biofuels is actually greener than the first generation of biofuels.

Food security is a complex sustainable development issue that is linked to health through malnutrition and is also linked to sustainable economic development, environment, and trade. Food security happens usually when food is produced in a way that it is safe for people and the environment; farmers can earn a fair income; local, regional, and community food production is encouraged and supported; social justice is a priority; and all people work together to create positive change in the food system and communities.

See Also: Carrying Capacity; Millennium Development Goals; Sustainability Indicators; Sustainable Development.

Further Readings

Food and Agriculture Organization of the United Nations. "Climate Change and Food Security: A Framework Document." http://www.reliefweb.int/rw/lib.nsf/db900sid/PANA-7KADCQ/$file/fao_may200888.pdf?openelement (Accessed October 2009).

Food and Agriculture Organization of the United Nations. "Food Security Concepts and Measurements." http://www.fao.org/docrep/005/y4671e/y4671e06.htm (Accessed October 2009).

Food Security Network of Newfoundland and Labrador. "What Is Food Security?" http://www.foodsecuritynews.com/What-is-food-security.htm (Accessed October 2009).

"Global Food Security Under Climate Change." *Proceedings of the National Academy of Sciences.* http://www.pnas.org/content/104/50/19703.full.pdf (Accessed October 2009).

OPEC Fund for International Development. "Biofuels and Food Security." http://www.ofid.org/publications/PDF/pamphlet/ofid_pam38_Biofuels.pdf (Accessed October 2009).

Toronto Public Health. "Definitions of Food Security." http://www.toronto.ca/health/children/pdf/fsbp_ch_1.pdf (Accessed October 2009).

United Nations. "Report of the World Food Conference, Rome, November 5–16, 1974." http://www.fao.org/docrep/005/y4671e/y4671e06.htm#fn25 (Accessed August 2009).

U.S. Department of Agriculture, Economic Research Service. "Appendix 1—Food Security Model: Definition and Methodology." http://www.ers.usda.gov/publications/GFA16/GFA16f.pdf (Accessed October 2009).

Velma I. Grover
United Nations University–Institute for Water,
Environment and Health (UNU–IWEH)

Garbage

Solid waste is generally known as trash or garbage in layman's parlance. In general it refers to domestic waste—waste generated from households that includes, among other things, food scraps, paper, newspaper, clothes, packaging, cans, bottles, grass clippings, furniture, paints, and batteries. In most developing countries, it is mixed with hospital waste, industrial waste, and other hazardous waste contaminating it. Different categories of waste include municipal solid waste, industrial waste, hospital waste, hazardous waste, and toxic waste.

Although the above definition defines waste as trash, in the world of entrepreneurship, one person's waste can be another person's treasure. Another apt definition of wastes is "matter in the wrong place," implying that a material becomes waste only when a specific owner ceases to have a use for it. For example, yesterday's newspaper is waste to the man who bought it, but it could become raw material for a paper mill.

P. V. Tebo (1997) states, "Waste is really a mindset. All we really have are ingredients that haven't yet found a home." The World Health Organization defines waste as "something that the owner no longer wants at a given place and time, and which has no current perceived value." This implies that

- waste is a burden to the one who generates it;
- waste occupies useful space that would otherwise be used as storage for value giving materials; and
- waste has no value (i.e., it cannot, in its existing form, bring the much-needed monetary gain to the owner).

The definition of waste by the European Commission, as adopted by the U.K. government is "Waste consists of any materials, substances or products which the producer or holder discards or intends to discard or is required to discard."

Waste is defined by United Nations Educational, Scientific and Cultural Organization as "useless, unwanted, or discarded materials, normally in the solid state that arise from human activities." According to White and colleagues (1995), waste often contains physically the same materials found in useful products, and it only differs from useful production by lack of its value.

Solid wastes is the term used internationally to describe nonliquid waste materials arising from domestic, trade, commercial, industrial, agricultural, and mining activities and from public services. Solid wastes comprise countless materials: dust, food wastes, paper, metal, plastic, glass, discarded clothing and furnishings, garden waste, agriculture waste, industrial waste, and hazardous and radioactive waste, to name a few.

In most cases, however, *solid waste* is the term used for municipal solid waste and is also used for the heterogeneous collection of wastes generated from urban areas. Its composition and quantity vary from region to region, depending on the living standard and lifestyle of inhabitants and the abundance and availability of natural resources. Literally, municipal solid waste includes wastes that result from municipal functions and services such as street waste, dead animals, and abandoned vehicles. However, in waste management practices, the term is applied in a wider sense to incorporate domestic wastes, institutional wastes, and commercial wastes that arise in an urban area. The quantity of waste generated depends on socioeconomic conditions, cultural habits of the people, urban structure, density of population, extent of commercial activity, and degree of salvaging at source.

Where Garbage Comes From

Some of the factors that contribute to an increase in solid waste generation are a growth in gross domestic product, rising disposable incomes, and a structural change in the pattern of production. In fact, the kind of waste generated, and hence, the way it should be handled, changes with modernization, urbanization, and globalization—mainly because people in rural areas have a different kind of waste generation. There is always an increase in paper and plastic waste and a decrease in ash and earth content as a society moves from rural to urban. Another difference experienced between more Western societies and those in still-developing countries is the difference in the amount of organic waste generated. Western countries relying on packaged food and canned food have shifted organic waste production from domestic to industrial waste.

With the change in their gross domestic product, some developing countries have also experienced a change in the type of waste produced. For example, where previously goods were transported in cloth bags or baskets from stores and markets, the trend is now moving toward using plastic bags (influencing how waste can be handled). Another of the major shifts concerns the type of food waste generated—there has been a global shift from cooking at home to either buying food from restaurants (shifting waste from the household to the commercial level) or buying canned or packaged food (shifting waste from the household level to the industrial level).

Mankind is producing more garbage and waste than ever before. This can be attributed to two legacies of the 20th century—an unprecedented population explosion and the industrial revolution. However, an increase in population is not the only cause of the increase in waste generation—it can also be attributed to a higher rate of consumption. For example, car manufacturers introduce a new model every year, tempting the consumer with a more attractive exterior and advances in automotive technology. The result is that many older but still usable goods are discarded, filling up landfills and dumps. Another shift in waste produced is that much of the trash produced by earlier generations was organic in nature, and thus decomposed easily, whereas most materials used in the modern age are inorganic and deteriorate or degrade slowly, or sometimes not at all.

The use of plastic bags also has changed tremendously over time. Consumers used paper bags for carrying goods like groceries and vegetables in the 1970s, until plastic bags largely replaced them. Recently (especially in Ontario, Canada), reusable bags are replacing nonbiodegradable plastic bags. A surcharge of 5 cents to 35 cents per plastic bag is being imposed in some cities to encourage consumers to use reusable bags.

Waste can be disposed of in many ways, including the following:

- A shift from dumping refuse in rural areas and developing countries to more sophisticated sanitary landfill sites: Dumping on land was a common method of waste disposal in urban communities because it was easy to haul solid wastes to the edges of towns and deposit them there. This problem has been controlled in developed countries via engineered or sanitary landfills.
- Reuse, recycle, and/or composting: Recycling is a "process by which materials that would otherwise become solid waste are collected, separated or processed and returned to the economic mainstream to be reused in the form of raw materials or finished goods." Composting is a controlled aerobic process carried out by successive mesophilic and thermophilic activities, leading to the production of carbon dioxide, water, minerals, and stabilized organic matter.
- Incineration and recovering energy from waste: In a way, incineration of raw waste has been practiced throughout the world for decades in the crudest form of incineration— indiscriminate burning of waste. However, the technology has now come a long way, from open burning to sophisticated incinerators, and the purpose has changed from simply getting rid of the waste, to reducing the volume of the waste, to recovering energy from waste. The process takes place inside an incinerator called a pyrolysis, which achieves a thermal decomposition of waste at high temperatures with an absence or near absence of oxygen.

The increase in consumerism and the limited land available to dispose of waste has brought the realization that there is a limit to the availability of natural resources, encouraging the present generation to preserve resources for future generations. A trend has been put in motion to reduce consumption, and an era stressing reusing and recycling goods has begun.

Getting Rid of It

Because waste has become a part of society, it needs to be handled and disposed of efficiently. Integrated waste management is one of the recommended ways to handle waste effectively. It is a complex, multistage process that includes generation, collection, storage, transportation, and disposal of wastes. In other words, effective waste management includes planning for and managing waste from cradle (where it is generated) to grave (how it is disposed of). Any change in policy or dealing with waste requires a change in the behavior of people; hence, municipalities are involving communities while exploring options for dealing with the growing challenge of increase in waste generation.

The new waste hierarchy is based on the principle of three Rs—reduce, reuse, and recycle—but involves five stages in solid waste management, starting from prevention of excess waste generation, waste minimization (when waste generation cannot be avoided, the attempt should be to minimize the amount of waste generated), waste reuse, waste recycling, energy recovery, and the final option, disposal.

Designing an effective solid waste management system includes striving for both environmental sustainability (reducing environmental impact) and economical sustainability (driving costs down). To achieve these two goals, the system should be integrated to manage how waste is generated, should identify sources of waste, and should determine which type of disposal or treatment methods are best for that type of waste. A waste management system should also be market-oriented (material and energy have end users) and flexible (for constant improvement and adaptation). Solid waste management systems should also define the objectives clearly, design a total system to meet these objectives, and operate on a large-enough scale.

This holistic approach has several advantages. For example, it provides an overall view of the waste management process. This is important for strategic planning, as handling each stream differently is not a very efficient system. Second, environmentally, all waste management systems are part of the same system—the global ecosystem. Looking at the overall environmental burden is the only rational approach, as otherwise, reductions in the environmental impacts of one part of the process may result in greater environmental impacts elsewhere. Third, economically, each individual unit in the waste management chain should run at a profit or at least a break-even cost. It is even more important to see whether the entire system can run at a breakeven or a profit and not just the individual components. Only then can all the parts be viable, provided that the income is divided approximately in relation to the costs.

A solid waste management system should not only ensure human health and safety but also be both environmentally and economically sustainable. To be environmentally sustainable, the system must reduce as much as possible the environmental impacts of waste, including energy consumption; pollution of land, air, and water; and loss of amenity. To be economically sustainable, waste management options should be such that the cost is acceptable to the community, including private citizens, businesses, and government. Because it is usually the case that both environmental and economic objectives cannot be fully achieved at the same time, there needs to be a tradeoff. A balance needs to be struck to minimize the overall environmental impacts of waste management as far as possible within an acceptable level of cost.

The latest research has also linked solid waste management policies to affect on climate change. For example, when waste is disposed of in landfill sites, methane, a greenhouse gas, is produced. A better solution would be to either flare off the gas or use this gas to generate electricity. Another option that can be considered is composting. Even if the compost cannot be used in agricultural practices, it reduces the amount of waste and the quantity of methane in the landfill site. Similarly, incineration with energy recovery is a better option than just incinerating waste. The advantages and disadvantages of recycling or recovering metals are also being studied.

More Attention Needed

Solid waste management is not always given the attention it deserves, but with an increase in the quantity of waste generated and not enough land, most municipalities are facing the urgent need to focus their attention on waste management. Poor solid waste management, especially uncontrolled dumping, can cause health problems and environmental problems, such as pollution of surface and groundwater from leachate production. These unhygienic conditions created if waste is not managed properly put waste workers at risk of acquiring infections of the skin and gastrointestinal and respiratory tracts. Poor waste management can also trigger epidemics of some vector-borne or foodborne infections. Accumulated garbage leads to problems such as infections caused by pathogens, vector-borne diseases, and groundwater pollution, among others. Health hazards also occur in the presence of human excreta, hospital and clinical waste (including medicines, syringes, and infected body parts), and hazardous wastes from small-scale industries. Moreover, uncovered and mismanaged disposal of waste attracts flies, mosquitoes, and rodents, leading to a spread of diseases and infections. Appropriate solutions—such as minimizing waste at the source and appropriate disposal through recycling, reuse, composting, incineration, or disposal at landfill—should be sought.

See Also: Composting; Landfills; Recycling in Cities; Sustainable Development; Waste Disposal.

Further Readings

Diaz, F. L., et al. *Solid Waste Management for Economically Developing Countries*. Hercules, CA: ISWA, General Secretariat, Denmark, and CalRecovery, Inc., 1996.

Grover, V. I., et al.. *Solid Waste Management*. Delhi: Oxford and IBH Ltd., 2000.

Holmes, J. R., ed. *Managing Solid Wastes in Developing Countries*. New York: John Wiley & Sons, 1984.

Suess, M. and J. W. Huismans. "Management of Hazardous Waste." WHO Regional Publication, European Series No. 14, 1983.

Tchobalogious, G., et al. *Integrated Solid Waste Management*. New York: McGraw-Hill, 1993.

Tebo, P. V. "Going Green." *Chemical Engineering News* (August 4, 1997).

UN Educational, Scientific and Cultural Organization. "Module on Solid Waste Management—A Module." Unpublished Paper.

White, P. R., et al. *Integrated Solid Waste Management*. Blackie Academic & Professional. Glasgow: Chapman and Hall, 1995.

Velma I. Grover
United Nations University–Institute for Water,
Environment and Health (UNU–IWEH)

GREEN BELT

Greenery functions here as a conservation buffer in this green belt in suburban Des Moines, Iowa. Green belts are a popular tool for planners and citizens to foster a higher quality of sustainable city life.

Source: Lynn Betts/Natural Resources Conservation Service

Many cities use green belts, or legally protected open space around a city, to create recreational opportunities for residents, preserve agricultural viability, and attempt to contain urban sprawl. Some green belts are held as undeveloped land, such as wetlands, hillsides, or forested lands, whereas others are protected as rural agricultural land. Green belts, although not an unproblematic solution, are an increasingly popular tool for planners, public officials, and citizens working toward fostering a higher quality of life in a more sustainable city.

Maintaining open space around the city has long been an attractive solution for urban decision makers trying to combat the perceived ills of an increasingly urban society. For instance, as early as 1580, Queen Elizabeth I effectively created a green belt around London that prohibited any new building within three miles of the city wall. Turn-of-the-century British planner Ebenezer Howard, however, is considered the father of the modern green belt movement. With his Garden Cities concept, Howard theorized that a network of densely settled 1,000-acre cities surrounded by 5,000 acres of rural green belt could help to reform urban society. The rural agricultural belt would serve to keep each city distinct from neighboring settlements while also providing visual and spatial relief from potential urban congestion. Although Howard and his associates built only two garden cities in early-20th-century England, the planning ideas, particularly the green belt concept, had an important influence on new town planning elsewhere. In the United States' New Deal era, for example, the Resettlement Administration created three small "green belt towns"—including the aptly-named Greenbelt, Maryland—that were modeled after Howard's concepts and featured prominent green belts.

Green belts emerged as a fashionable urban planning tool in the second half of the 20th century. By this time, evidence was mounting that, if unchecked, cities would expand rapidly into the surrounding countryside. This urban expansion into far-flung greenfields drove a quickly receding recreational, agricultural, and species-habitat frontier and contributed to a culture of increased automobile dependence, along with an economy based on even greater residential and infrastructure advancement. By the 1970s, forward-looking cities began to turn to green belts to combat the environmental, social, and economic challenges of urban sprawl. Green belts were designed to protect agricultural land, maintain species habitat, protect urban watersheds, provide scenic views and recreational spaces, and promote compact urban development.

Unless legally protected, however, the green belt of undeveloped land that naturally surrounds a city is likely to be constantly under threat as a result of urban expansion. Adopting an urban growth boundary is among the most common strategies for creating a secure green belt. An urban growth boundary is an officially adopted and mapped line that marks the spatial extent of urban or suburban development. The urban growth boundary typically is set in place for a considerable amount of time to discourage speculation at the rural–urban interface. Open space such as rural agricultural land, watersheds, parks, and golf courses are often allowed within the green belts formed outside the urban growth boundary. Green belts that have been protected through urban growth boundaries exist in London, Copenhagen, Vancouver, and other cities throughout the world. In the United States, the Pacific Northwest states of Oregon and Washington require cities to adopt urban growth boundaries. Other cities such as Boulder, Colorado, and several cities in the San Francisco Bay Area have voluntarily adopted urban growth boundaries to protect their green belts.

In addition to urban growth control, green belts provide environmental and economic benefits for urban areas. Green belts help to provide clean water by protecting urban watersheds from development pollution, provide protection for threatened agricultural land, and create useful species habitat and transportation corridors for wildlife on the urban–wild interface. Increasingly, however, the quality of life and economic benefits of green belts are presented as being equally important as the environmental benefits. The Greenbelt Alliance in the San Francisco Bay Area is an example of a local advocacy group that has been very effective at promoting the economic advantages of green belts. A selection from a recent report describes some of the reasons why green belts are so widely used: "As the greenbelt of hillsides, farmland, and forests around the region's cities is developed, the region loses the very things that make it special. When sprawl development replaces the region's spectacular landscapes and inviting cities with subdivisions, strip malls, and freeways, the Bay Area loses the high quality of life that makes it a center of innovation." Green belts not only challenge unsustainable urban sprawl but also act as a critical element for promoting a creative and innovate urban setting.

Green belts, however, are not perfect tools for creating greener cities. Green belts can encourage leapfrogging—or noncontiguous development beyond the green belt. Moreover, if jobs stay in the central city while residences leapfrog to outlying satellite communities, then the green belt may simply make the urban system less sustainable by encouraging driving through the green belt. Another criticism of green belts is that they artificially limit housing supply, which then increases real estate costs and contributes to a scarcity of affordable low-income housing. More theoretically, green belts are often criticized for codifying the unproductive separation of rural land and urban space. Because the city's fringe can be protected as green open space, the interior of the city can be more fully urbanized—more grey and less green. Only slowly is this rigid urban–rural dichotomy shifting to a more active relationship between city and country. Increasingly, planners use concepts such as green wedges, green corridors, and greenways in conjunction with green belts to simultaneously guide urban development and create more ecologically sound networks of urban open space.

See Also: Environmental Planning; Parks, Greenways, and Open Space; San Francisco, California; Watershed Protection.

Further Readings

Amati, Marco. *Urban Green Belts in the 21st Century*. Burlington, VT: Ashgate, 2008.
Freestone, R. "From Garden City to Green City: The Legacy of Ebenezer Howard." In *Greenbelts in City and Regional Planning*, K. C. Parsons and D. Schuyler, eds. Baltimore, MD: Johns Hopkins University Press, 2002.
San Francisco Greenbelt Alliance. "At Risk: The Bay Area Greenbelt." http://www.greenbelt .org/resources/reports/atrisk_2006/index.html (Accessed April 2009).

Jeremy Bryson
Syracuse University

GREEN COMMUNITIES AND NEIGHBORHOOD PLANNING

Global concerns over climate change and the associated search for greener urban lifestyles have created new challenges for urban planning theory and practice, most notably in the field of neighborhood planning. The neighborhood is the territorial scale at which people develop specific behaviors and attitudes in domains of crucial environmental relevance such as housing, transportation, and consumption. Thus, if urban planning is to contribute to urban environmental sustainability, it is expected to do so by acting on the physical environment at the neighborhood level. This endeavor lies at the heart of contemporary green communities, with several recent groundbreaking planning models seeking to show the way forward. To avoid the limitations of physical determinism, however, these models of neighborhood planning must acknowledge that achieving behavioral sustainability cannot exclusively rest on top-down alterations of the built environment but requires involving local residents in the "greening" of their communities.

The Neighborhood as a Social Fact

Recognition of the neighborhood as an appropriate geographical unit for urban planning may be traced back to the first half of the 20th century. In the 1920s, urban planners slowly began to acknowledge that the neighborhood scale would provide a compelling complement to citywide, comprehensive planning by offering a window into the social organization of urban society. First, the neighborhood is the spatial container in which citizens spend most of their time, and with which they are most concerned. Second, planning at the neighborhood level allows taking into account the physical, social, and economic distinctiveness of communities whose specificity goes unnoticed in comprehensive planning.

This interest for the neighborhood as a "social fact"—as prominent American planner Lewis Mumford once put it—very much conditioned the various models and practices of neighborhood planning from the 1920s onward. It was hoped then, and to a large extent it still is today, that social problems may be tackled through adequate neighborhood design. Good physical design, according to this perspective, should not be rated for its architectural, infrastructural/technological, and aesthetic standards alone but, most crucially, for its ability to attain social objectives by influencing people's choices and behaviors in socially efficient ways. By applying appropriate standards of physical planning, the argument goes, one can act on society through the physical environment to produce desired social, economic, and environmental outcomes. As a consequence, every model of neighborhood development devised throughout the history of neighborhood planning represented a reaction to the social context and urban conditions of the time. A famous example is the "neighborhood unit formula" presented by Clarence Perry in 1923. Perry's formula, which has remained one of the most influential endorsements of neighborhood planning, intended to fight the problems of social anomy and urban criminality that were crippling U.S. cities. Through the design of self-sufficient neighborhoods promoting the use of common facilities, neighborhood units were expected to foster a spirit of community and neighborliness. Another influential urban planner, Patrick Abercrombie, played a somewhat similar role in the 1940s in the United Kingdom by establishing the neighborhood as the basic planning unit. For the first time, planners on both sides of the Atlantic were urged to take into consideration the design of residential communities in a comprehensive, integrated fashion.

Green Communities as a Response to Global Environmental Concerns

Today's most discussed and experimented models of neighborhood planning may be subsumed under the heading "green communities." They constitute the latest attempt to address societal concerns through altering and shaping the physical environment. Global climate change and its likely consequences for human and natural life represent new challenges for local planners, policy makers, and their constituencies. The global nature and repercussions of climate change and environmental degradation are driving urban planning into a much wider realm than before. By designing communities in ways that encourage environmentally sustainable behaviors, it is hoped that cities will be able to curb the global production of greenhouse gas emissions and to adapt to the consequences of global warming.

Contemporary green communities thus constitute a new, particular type of neighborhood embodying distinctive historical, scientific, and societal assumptions. These assumptions are different from those underlying previous conceptualizations of green communities. In the early 20th century, for instance, the widespread interest of planners and policy makers in Ebenezer Howard's "Garden City" was underpinned by romanticized notions of the need for human settlements to function in harmony with nature. The Garden City represents an old type of green community, which Howard designed in reaction to conditions of overcrowding and pollution associated with the cities of the Industrial Revolution. Garden cities, in contrast, were imagined as planned, self-sustaining, and well-balanced communities surrounded by large areas of undeveloped wild or agricultural land. Howard's model had a lasting influence, as various social preoccupations such as housing shortages in the aftermath of World War II or, later on, the need to contain suburban sprawl did regularly revive interest in certain elements of the Garden City throughout the 20th century.

Although the Garden City and subsequent models of green communities all shared concerns for environmental preservation, the contemporary approach to green communities is the first to place environmental protection at the forefront of its planning preoccupations. This constitutes a reaction to the idea that urban systems leave an ecological footprint far larger than their actual size—in other words, they are consuming a bigger share of the world's resources than they are entitled to. Green communities of the 21st century, therefore, are urged to focus their efforts on reducing their ecological footprint through planning, technological, and behavioral strategies. For instance, planning strategies include promoting compact developments and higher building densities so as to limit the use of land and reduce car travel, technological strategies refer to the application of state-of-the-art technologies in the field of renewable energies and waste recycling, and behavioral strategies may include taxation of car travel and incentives to buy and use locally produced consumer goods.

Planning Green Communities: State-of-the-Art Models and Initiatives

With these objectives at heart, several innovative projects have been formulated and implemented in recent years. Whereas the highly industrialized world initially took the lead on the development of green communities, many initiatives have started to flourish in less-developed countries. The green communities brought together under the label "One Planet Living" (OPL) constitute an interesting example from a global perspective. OPL is the result of a partnership between the World Wildlife Fund and the British environmental consultancy BioRegional. The latter is famous for having conceived Beddington Zero Energy Development, a pioneering green community in South London aimed at achieving carbon-free construction and housing standards. OPL communities are asked to devise strategic planning solutions tailored to meet the following 10 requirements: zero carbon; zero waste; sustainable transport; sustainable materials; local and sustainable food; sustainable water; natural habitats and wildlife; culture and heritage; equity, fair trade, and local economy; and health and happiness. There is an OPL community in the making on every continent. The first OPL development in the United States, Sonoma Mountain Village, is a planned green community of 1,900 new homes currently under construction in California.

In addition to complying with OPL green requirements, Sonoma Mountain Village is featured among the pilot projects of another important example of the growing connection between neighborhood planning and environmental protection. This program, which is called Leadership in Energy and Environmental Design for Neighborhood Development (LEED-ND), is the product of a partnership between the U.S. Green Building Council, the Natural Resources Defense Council, and the Congress for the New Urbanism. LEED-ND is a certification system whereby "certified" neighborhoods are required to comply with a series of standards designed to achieve sustainable communities. LEED-ND provides a response to criticism against its smaller-scale counterpart, the Leadership in Energy and Environmental Design scheme for green buildings (LEED). Certification programs for green buildings have become popular in recent years, as many countries in the highly industrialized world have adopted their own. Those programs, however, are often faulted for consisting of mere "checklists" of technical norms and green specifications, ignoring both the physical and human dimensions of environmentally responsible development. LEED-ND, in contrast, seeks to integrate the use of technical solutions with principles of sustainable urban and transportation planning through the use of two preferred planning models: New Urbanism and "smart growth."

New Urbanism is a neotraditionalist version of 19th-century small European towns. Also known as "traditional neighborhood development," this model of neighborhood planning emerged in the 1980s in reaction to the growing disengagement of planning practitioners from neighborhood design and efforts at shaping urban communities. Although compact, mixed-use, and pedestrian-oriented development is the hallmark of traditional neighborhood development schemes, advocates of green communities have blamed their exaggerated emphasis on the architectural aspects of buildings and public space designs. From a green community standpoint, these concerns are considered to have often overshadowed more important issues such as suburban sprawl and car travel. The association of traditional neighborhood development with smart growth, a planning theory oriented toward the promotion of public transportation, pedestrian travel, and the restoration of vitality and community spirit in derelict suburbs and old city centers, may thus prove to be a useful strategy. Through this combination of different planning instruments, LEED-ND may be able to overcome the traditional limitations associated with each of these procedures when taken in isolation. The scheme is, nonetheless, in its infancy, and whether it will bring meaningful benefits to the planning of green communities remains to be seen.

Planned Communities Versus Community Planning

Although LEED-ND and similar initiatives in other countries may provide stimulating contributions to the development of green(er) communities, an essential distinction has to be overcome if those schemes are to ever produce truly sustainable neighborhoods: the distinction between "planned communities" and "community planning."

Community planning is a participatory process that does not have a fixed, predetermined objective whose attainment is conditioned to the use of a programmatic set of standards. Rather, it treats planning as a process in constant evolution that should rest on a significant involvement of local residents and stakeholders. Planned communities, in contrast, are founded on the hypothesis that the application of certain planning standards will bring wished-for outcomes. This approach, which falls in line with the philosophy underlying Perry's neighborhood unit formula, bypasses the question of the desires, attitudes, and behaviors of the users and residents of the concerned neighborhoods. Although it is well recognized that the environmental performance of communities depends as much on behavioral sustainability as on physical and technological sustainability, the belief that sustainable behaviors may be enabled through the built environment is an assumption underlying many contemporary experiments in green community planning.

A paradigmatic example is that of Masdar City, a large community under construction in Abu Dhabi that is expected to deliver carbon-free, car-free, and waste-free lifestyles. Although the highest standards of planning, design, and technologies are being used to minimize its future environmental impact, this planned community is located in a country with the highest per capita levels of carbon dioxide emissions. The initiative is expected to serve as a replicable model and pave the way for greener lifestyles in the United Arab Emirates, but it does not address the fact that an essential part of the task of "greening" communities is dependent on citizens' daily choices and behaviors, both as residents and as consumers. Even in the areas of our lives in which environmental protection may be embedded in physical design and green technology solutions, most of these solutions require environmentally minded attitudes, efforts, and maintenance, but without involving local citizens in the planning process, it is difficult to cultivate among them the necessary values and behaviors.

These questions have been regularly debated in the scholarly literature and have translated into various attempts at opening neighborhood planning systems to local populations, though not necessarily in relation to green communities. The first notable initiative of popular involvement in neighborhood planning in the United States was the Community Action Program launched in 1964 by the Kennedy administration. The program, which was managed by the federal government, was aimed at engaging local citizens in neighborhood regeneration initiatives so as to design the initiatives according to the specific social and economic needs of each community. The experience, however, fell short of securing significant participation and was even blamed for engendering greater stigmatization of the targeted neighborhoods. More comprehensive—and arguably successful—experiments of participatory community planning followed in the late 1960s and throughout the 1970s. In the United Kingdom, a famous government-sponsored document known as the Skeffington Report (1969) laid out the bases for a participatory approach, allowing the public to take an active part throughout the plan-making process. In the United States, growing distrust of the federal government surrounding antiwar and civil rights struggles led municipalities to look at new avenues for participation in planning at the municipal level. However, on both sides of the Atlantic, the reality of participatory community planning did not live up to its political rhetoric. Broad-based participation was continually difficult to secure, thereby undermining the representativeness of community planning schemes, and the ability of municipalities to dedicate economic and human resources to participatory planning processes was generally not sufficient to ensure their viability.

Since the early 1990s, the adoption of the idea of sustainable development as the single most pervasive objective of urban development has brought discourses about the importance of citizen involvement back into the limelight. However, the implications of such a discourse still have to translate into concrete implications for planning practice. An interesting example is the Green Communities program launched in the United States by Enterprise Community Partners, an organization dedicated to providing affordable homes to low-income people. The Green Communities initiative is taking Enterprise's traditional affordable housing activities into the environmental realm by adding green criteria (which are aligned on LEED indicators) to its housing programs. Yet the program's requirements do not include public participation. In the brochure establishing the criteria for green classification and funding by Enterprise, no reference is made to the question of public participation.

Academic commentators have become increasingly critical of the top-down character of the planning processes presiding over the creation of many green communities. They warn that this lack of popular involvement may undermine the social and environmental performance of these communities over time. The sustainable paradigm development has created a natural framework for expanding and improving the relationship between neighborhood planning and green communities. The success of this relationship will nevertheless depend on its ability to fully address the nonphysical dimensions of urban sustainability.

See Also: Compact Development (New Urbanism); LEED (Leadership in Energy and Environmental Design); Smart Growth.

Further Readings

Barton, Hugh, ed. *Sustainable Communities: The Potential for Eco-Neighborhoods.* London: Earthscan, 2000.
Berke, Philip R. "The Evolution of Green Community Planning, Scholarship, and Practice: An Introduction to the Special Issue." *Journal of the American Planning Association*, 74/4 (2008).
Goltsman, Susan and Daniel Iacofano, eds. *The Inclusive City: Design Solutions for Buildings, Neighborhoods and Urban Spaces.* Berkeley, CA: MIG Communications, 2007.
Rohe, William M. "From Local to Global: One Hundred Years of Neighborhood Planning." *Journal of the American Planning Association*, 75/2 (2009).

Laurence Crot
University of Neuchâtel

GREEN DESIGN, CONSTRUCTION, AND OPERATIONS

Buildings are responsible for more than half the world's energy use, thus thinking sustainably requires attention to how buildings can be designed, constructed, and operated to use less energy and fewer material resources. The processes of design, construction, and building operations are inextricably linked: Design decisions determine in large part the methods, materials, and sequencing of construction; and the type of building systems selected in design and installed during construction determines the costs (dollar costs as well as energy costs) of operating systems both in the short-term (operational) and in the long-term (replacement). Traditionally, these three processes have been sequential and separate: performed by different teams of people at different times and places. More recently, the concept of performance engineering has been used in order to integrate the expertise of maintenance workers and contractors from the earliest points of building design in order to produce buildings that will be appropriately designed for prevailing construction practices in particular locales, and effectively and properly maintained by owners or maintenance staff.

Two central objectives in green design are to reduce energy use, especially the use of fossil fuels, and to minimize the wasteful use of resources. The aim of designing green buildings is to produce zero net energy; this is achieved through conservation measures and by using the building or site to produce energy with renewable sources, such as geothermal, solar, or wind power. Any surplus may be sold back to the local energy grid. The latter goal, minimizing use of resources, may be achieved by using smaller amounts of materials, using recycled or salvaged materials, using materials that are produced in a sustainable way or without harmful chemicals, or by some combination of these.

Reducing Energy Use

Before modern (usually fossil-fuel generated) methods of heating, cooling, and lighting became standard in the Western world, builders designed their structures to respect and work with climate and site conditions. In northern climates, for example, buildings were designed to minimize exposure on the northerly side through use of berms and limitations on the size and amount of windows. Exposure was maximized on the south-facing sides, allowing buildings to take advantage of passive solar gain. Southern overhangs were calculated using sun angles to allow low angled winter sun to penetrate windows, yet block solar radiation during the warmer months. Deciduous trees were planted to shade buildings in the summer and allow sun to reach roofs and walls during the winter. Evergreen hedges provided windbreaks on the northern sides. Windows and doors were situated to maximize natural ventilation and natural lighting where most needed during the daylight hours. In colonial America, a large masonry chimney absorbed heat from fireplaces and radiated it into adjacent spaces. In climates with moderate to heavy precipitation, roofs were sloped to prevent water or snow from collecting and leaking into interior spaces. Natural, renewable materials, such as wood, sod, and straw were the building materials of necessity.

Many of these techniques and materials began to receive renewed attention in the 1970s, when oil shortages and spiking prices refocused design attention on energy conservation. In particular, this period focused attention on efficient building envelope designs and materials that would conserve energy, such as new forms of insulation with greatly increased insulation capacity (measured in higher R-values, the industry metric).

More recently, concerns about global climate change through increases in atmospheric concentrations of greenhouse gases, coupled with geopolitical and technical limitations on mining petroleum reserves, have renewed interest in green building. Green-oriented websites and magazines provide a wealth of information for the do-it-yourself renovator or small contractor, while architectural and engineering firms working on large industrial and office buildings have also accepted the challenge of designing buildings with zero or low net energy loads. A 2003 exhibit titled "Big and Green: Toward Sustainable Architecture in the 21st Century," at the National Building Museum in Washington, D.C., showed some of the ways that large buildings can be made sustainable. Designers have adapted techniques used in smaller buildings, like shading and natural ventilation; they have also employed newer technologies like solar photovoltaic paneling, green roofs, and greywater systems to reuse roof and site runoff. Moreover, tall buildings potentially use heating and cooling more efficiently than small buildings. With many occupants in a single building, the volume-to-building-envelope ratio is larger and thus the building is more energy-efficient.

In commercial buildings, particular attention has been paid to making maximum use of natural light. This strategy is particularly important in office and retail buildings, where lighting is a large proportion of the energy load. Large expanses of windows, integral window shades, louvers, skylights and light wells, and building layouts that minimize windowless spaces are all design techniques to use natural light more efficiently. Given the social prestige of workspaces with windows, such techniques may also result in higher property values and lease-up ability for building owners and managers.

Research on new technologies and metrics for increasing energy efficiency is a priority in the design and construction industries as well as in government. The U.S. Congress, in the Energy Independence and Security Act of 2007, authorized the U.S. Department of Energy (DOE) to launch the Net-Zero Energy Commercial Buildings Initiative, which has set voluntary targets for energy use in commercial buildings (50 percent zero energy by 2040 and 100 percent zero energy by 2050). The DOE also provides technical assistance through its research laboratories. Net zero energy buildings (NZEBs) are highly energy-efficient buildings that produce their energy needs through renewable sources (e.g., solar, wind, geothermal, and biofuel), thus they do not add to the fossil-fuel-generated energy load, and may even contribute positively to the grid by selling back surplus energy.

Building owners and their architects and engineers may certify the sustainability of their designs through the U.S. Green Building Council's Leadership in Energy and Environmental Design (LEED) privately-run certification process. Such certification is voluntary, but adds cachet and can command higher prices for real estate in markets that are particularly sensitive to the social benefits of being perceived as green. However, some have criticized the LEED certification process as favoring pricey high-tech add-ons at the expense of the types of low-tech, low-cost common sense solutions for energy conservation previously described. Moreover, LEED certification, because it deals primarily with buildings independently of their site context, is not equipped to assess the negative externalities of design for greenfield sites (such as the energy consumption of single-occupancy vehicles that are required to access suburban or exurban sites), although it should also be noted that the Green Building Council has begun to address such issues in its new regulations for neighborhood development (LEED-ND).

Regardless of how conventional or innovative a building design is, the construction process offers additional ways to reduce energy use. The simplest and most effective way to reduce energy use is through a well-sealed, tight building envelope, with sufficient insulation at all surfaces that meet the exterior (walls, roof, and slab), energy-efficient windows and storm windows, weather-stripped doors and storm doors, and insulation around pipes that feed through unheated spaces. Constructing a tight building envelope requires constant vigilance to make sure there are no gaps or thin spots in insulation or caulking. Utility companies in most locations will conduct energy audits that can identify the steps necessary to retrofit dwellings and calculate the payback period for the expenditure. Such retrofits may not be especially glamorous, but experts recommend that these basic steps be taken first before building owners or contractors proceed with more radical options, such as installing alternative energy sources.

The most energy-efficient appliances and building systems should also be selected. Even if the initial cost is slightly higher, energy-efficient devices will pay off in reduced energy costs over their lifespan. The Environmental Protection Agency (EPA) instituted the Energy Star program in 1992 in order to help contractors and other buyers compare the energy use of appliances, boilers, air-conditioning units, and other energy-consuming devices.

Further energy-reducing strategies are available once the building has been designed and constructed. Making sure that owners or operators of building systems understand the optimal settings and capabilities of their equipment and use it to reduce energy use is essential when new buildings or building systems are put online. For example, thermostats should be programmed for the lowest comfortable temperature (or highest during the air-conditioning season), and the timing of automatic systems should be coordinated with the hours of the day and days of the week so that heating and cooling are at full strength only when the building is occupied.

Automatic lighting controls that turn off lights or dim them when spaces are unoccupied need to be properly calibrated to the correct sensitivity, and awareness campaigns can be used in commercial buildings to encourage users to turn off the lights when exiting a room. Proper maintenance schedules should be followed so that components, such as filters, that need to be replaced when worn out or used up are checked and replaced at appropriate intervals.

Additionally, building owners or managers can often choose the source of their electricity from their local utility provider (opting in on wind power or solar-generated power), even if it costs more, in order to build demand for alternative energy and to increase efficiencies of scale for its production and distribution.

Resource Use

Rethinking the use of materials is the second key strategy in green design, construction, and operations. One objective is simply to use fewer material resources. For example, engineered structural products (like engineered joists and beams) use less wood to produce the same design strength. Thinking critically about the amount of space that humans actually need is another way to reduce materials use. The "not so big house" movement pioneered by architect Susan Saranka and others challenges homeowners to think creatively about how small spaces can be made attractive and multifunctional; such ideas may be applied to commercial spaces as well. American houses have steadily grown in square footage since the 1960s, with foyers, atria, master suites, restaurant-style kitchens, and special purpose spaces like fitness rooms and home entertainment areas. Thinking "not so big" not only reduces building costs and use of materials at the construction phase, but also reduces operating costs for heating, cooling, lighting, and general maintenance over the life cycle of the dwelling.

A second strategy in materials use is to substitute recycled or salvaged materials where possible. Using salvaged materials keeps them from contributing to an already overburdened solid waste system, helps to sustain and increase markets and distribution systems for such products, and reduces the energy and materials needed to manufacture new construction products and transport them long distances to building sites. Salvaged lumber is one such product: It is generally of higher quality and available in larger dimensions than newly produced lumber. Recycled glass can be used in tiles and countertops for kitchens and baths; these products are made from postconsumer glass waste as well as manufacturing waste. There is also an emerging manufacturing sector that uses other parts of the recycling stream to produce building materials, such as "plastic wood," which is dimensional lumber that is resistant to rot and splintering.

A third strategy for rethinking materials is to use products that are made from renewable resources. Timber may be certified to be from renewable sources. Bamboo, a fast-growing (even invasive) plant, is manufactured into flooring, cabinets, and other building materials. Carpets made from wool or other natural fibers are installed in place of synthetics made from petroleum derivatives. With product information widely available, it is also possible to select materials with less energy-intensive manufacturing methods, or materials manufactured near the building location, so that less energy will be used in transportation.

Construction waste is one of the largest material impacts of new construction. Wrappings, packaging, and scraps of material fill dumpsters, and ultimately, landfills or incinerators. Contractors can work to minimize waste on the jobsite, and consumers can insist that manufacturers of building products work to reduce wrapping and packaging.

Last, building owners, managers, and contractors can consider other environmental impacts of the products they select and install. For example, consumers concerned about the toxic chemicals and fumes generated by conventional paints and solvents have spurred the development of many types of low-VOC (volatile organic compound) products that are much less hazardous to workers and to the atmosphere. Carpets and other textiles that are required to have flame-retardant coatings are being manufactured with alternative chemicals to reduce the health effects of off-gassing after installation.

Thinking about materials use extends to the building landscaping as well. Rainwater and greywater from buildings may be directed from impervious surfaces (like roofs and paved areas) into the landscape in order to save potable water, reduce runoff, and recharge the water table. Paved areas themselves may be constructed with pervious pavers to allow rainwater to infiltrate. The amount of grass, which is highly maintenance-intensive both in terms of chemical inputs (fertilizers, herbicides, pesticides) and irrigation, can be reduced in favor of native species that adapt well to local conditions. Although rooftop landscapes have additional costs for structural requirements and are more maintenance intensive than on-the-ground landscaping, they are helpful at filtering or slowing the rate of runoff, as well as providing additional roof insulation.

Broader Issues

More information than ever is available on building green. However, it is important in green design, construction, and operations to select materials and methods that respond to the locale of the building, generally accepted local construction practices, and the realities of what can reasonably be expected in operations and maintenance. Popular media coverage of green industries tends to universalize strategies without regard to what works best where.

Finally, principles of green design and construction also extend to how neighborhoods and communities are designed. Small houses that are close together, with small yards and narrow streets, reduce expenditures for construction and operation of infrastructure, such as water and sewer lines, and encourage walking because land uses are closer together. The mix of land uses within walking distance is critical: When shops, workplaces, and entertainment or cultural venues are within walking distance, vehicle trips can be made shorter, combined, or eliminated altogether. After buildings, transportation is our largest energy user. Integration of transportation planning with building design, construction, and operations at the neighborhood scale as well as the individual scale will result in the greatest energy savings and material reductions.

See Also: Construction and Demolition Waste; Green Communities and Neighborhood Planning; Green Housing; Green Roofs; Indoor Air Quality; LEED (Leadership in Energy and Environmental Design); Sustainable Development.

Further Readings

Gissen, David, ed. *Big and Green: Toward Sustainable Architecture in the 21st Century* (Museum Catalog). New York: Princeton Architectural Press, 2002.

Kibert, Charles J. *Sustainable Construction: Green Building Design and Delivery*. Hoboken, NJ: John Wiley & Sons, 2008.

McHatton, Barb. "Building for the Future: Green Building Solutions." http://www.green buildingsolutions.org/s_greenbuilding (Accessed March 2009).

Saranka, Susan and Kira Obolensky. *The Not So Big House: A Blueprint for the Way We Really Live*. Newtown, CT: Taunton, 2008.

U.S. Environmental Protection Agency. "Components of Green Building." August 25, 2008. http://www.epa.gov/greenbuilding (Accessed March 2009).

Judith Otto
Framingham State College

GREEN ENERGY

Green energy is power that is derived from natural energy flows, such as solar radiation, whereby there are no harmful, toxic, or other emissions that are deleterious to the environment. Green energy sources are ultimately more sustainable than fossil fuel–based power generation, as they are harnessing "ambient" energy and, as such, cannot "run out." It must be noted that although green energy technologies represent an improvement on the technologies they supersede, no source of power is free of environmental impact.

Although green energy is usually used in the context of electricity generation, "green energy" can in some contexts of application be used to provide mechanical drive directly and, in some cases, thermal energy.

Green power refers to renewable resources, including solar technologies. The Sandia National Laboratory presented this new design for solar panels in 2009.

Source: Randy Montoya/Sandia National Laboratories

The term *green energy* may also be used to refer to technologies such as cogeneration and combined heat and power, which, although still producing some emissions and undesirable environmental effects, represent a substantial improvement in the efficient use of energy over the orthodoxy of centralized electricity generation and local heat production.

Green energy may also be applied to technologies that are used as energy vectors; for example, hydrogen and fuel cells that, although not producing any energy themselves, can be used to transport, store, and transform power in cleaner ways than the technologies they aim to replace.

Before the Industrial Revolution, other than heat from combustion, use of green energy was widespread and accepted as a part of everyday life. For example, the power of the wind was harnessed to mill grain and pump water, and flowing water was used to turn waterwheels to provide motive power.

Many argue that a shift to greater proportions of green energy will require a different type of energy network, or "grid," for distribution. This arises out of the fact that green energy is more "diffuse" and needs to be harnessed over a wide area; this scenario is commonly referred to as *decentralized generation*. This contrasts with traditional generation, in which large plants are often centrally located. It has also been argued that a shift to green energy will require a "smart grid"; that is, a power network that has a degree of automation and "intelligence" built into it to sensibly manage the fluctuating power output from a large number of individual generators.

Defining Green Energy

The term *green energy* is open to different interpretations; however, it is generally used to refer to renewable resources and power derived from human effort (e.g., hand-crank generators).

Renewable resources include the following:

- *Solar technologies:* power is harnessed from the sun to provide heat, electricity, or a combination thereof.
- *Wind technologies:* sails or turbine blades are used to capture the movement of the wind and transform it into useful motive power or electricity.
- *Wave technologies:* the movement of the waves is captured to produce electricity, using wave generators.
- *Tidal power:* the power of the tides is captured using tidal stream turbines, barrages, or tidal lagoons.
- *Hydro technologies:* the potential energy stored in the weight of water acting against gravity is harnessed by allowing the water to fall, turning a turbine or wheel at the end of the process and producing mechanical work and/or electricity at the other end. (Typically, a distinction is drawn between large hydropower, in which in many cases a dam is constructed and large land-mass flooded, and "small hydro," which has a much lower environmental impact as small rivers and streams are diverted to produce power.)
- *Biomass technologies:* living matter that grows as a result of inputs of energy from the sun, carbon dioxide from the atmosphere, and nutrients from the ground (and/or agricultural chemicals) are used to produce heat and, in turn, motion. These can be classified into three categories: solid biomass, liquid biofuel, and gaseous biogas.
- *Geothermal power:* the heat that is trapped deep underground; in areas where the Earth's crust is thin the heat can be profitably extracted and used to produce electricity, using a thermal generation plant.
- *Osmotic power:* the energy that can be electrochemically extracted from the difference in concentration of salt in water that is saline and water that is fresh.

Green Nuclear Power?

Some advocates of nuclear power would like to classify nuclear power as "green." In 1983, Bernard Cohen proposed that uranium reserves would be inexhaustible if used in concert with fast breeder reactor technology. A number of those involved in political circles, including former president George W. Bush, Charles Crist, and in the United Kingdom, Baron Sainsbury of Turville, have tried to classify nuclear power with "renewables." However, this neglects a number of points:

- "Scientific" definitions of "renewable energy" exclude nuclear power.
- Globally, no legislative body includes "nuclear power" in its definition of renewables.
- Although no carbon emissions are produced at the point of use of nuclear power stations, vast quantities of carbon dioxide are emitted in the mining, transportation, processing, and reprocessing of uranium. Furthermore, these processes and nuclear power plants often require the use of industrial gases, whose global warming potential is very much greater than that of carbon dioxide, and thus, although emitted in smaller amounts, cause a much greater environmental impact.
- Fast breeder reactor technology is still in an early stage of development, and although there have been prototype models built in the United States, France, the United Kingdom, and former Soviet Union, they have not proven themselves to be a viable technology.
- It has been suggested that uranium reserves will follow a "peak uranium" depletion cycle similar to the "peak oil" scenario advocated by Marion King Hubbert. Uranium reserves are finite, and using present, proven thermal reactor technology, the world's uranium reserves would be consumed much more rapidly if there were a major international move to use nuclear power. Thorium has been suggested as a possible alternative to uranium; however, it too only exists in finite quantities.
- Nuclear waste poses a serious environmental hazard that has not fully been addressed. Waste must in some cases be guarded for many thousands of years, and at present, all strategies for the handling and disposal of nuclear waste are at best short- to mid-term solutions.

It must be borne in mind that even renewable energy has some degree of "impact." Some solar cells require rare and/or toxic chemicals in the manufacturing process, and large-scale hydrogeneration causes large-scale environmental impact (and so some would exclude large-scale hydro from the definition of *green energy*). In addition, renewable energy devices require energy and materials to manufacture that must be offset against the energy generated over the lifetime of the device.

Energy Return on Energy Invested

One way of appraising the "green-ness" of an energy source is to calculate its EROEI value; that is, energy returned on energy invested. To do this, an estimate is prepared of how much energy the device is likely to generate over its useful service lifetime. Then the "embodied energy" of the device is calculated. This accounts for all of the energy used in production and manufacture of the device. It can then be appraised how much energy must be "invested" for a given return of green energy, enabling different green energy technologies to be compared. If the EROEI figure is less than one, the device is an "energy sink"—it consumes more than it produces and therefore cannot be considered a green power source.

Other Innovative Green Energy Terms

Hydrogen is not an energy source in itself, but an energy carrier, sometimes referred to as an *energy vector*, which can be used for a range of applications, both mobile and stationary. Hydrogen must first be produced. Renewable energy can be used to create hydrogen through a process called electrolysis, in which water is split into hydrogen and oxygen. This is referred to as *green hydrogen*. Fossil fuels can also be reformed into hydrogen; however, a by-product of this process is carbon dioxide emissions—the same as if that fossil fuel were burned. For this reason, it is a contentious issue as to whether hydrogen from fossil fuels can be classed as green.

One of the arguments surrounding green energy sources for next-generation vehicle technologies is known as "the long tailpipe argument." It argues that with the development of new vehicle technologies, such as electric vehicles and hydrogen vehicles, they can only be considered "zero carbon" where they are powered by nonpolluting energy sources. Where fossil fuels are used in the production of electricity for electric vehicles or hydrogen for hydrogen-powered vehicles, although there may be some improvement in "carbon emissions per vehicle mile traveled," the effect is the same as merely "extending the vehicles tailpipe to a centralized location."

Although those promoting some variants of fossil fuel (coal, oil, gas) and nuclear power would try and market them as green, ultimately they all produce by-products and emissions.

See Also: Distributed Generation; District Energy; Renewable Energy.

Further Readings

Boyle, Godfrey. *Renewable Energy*. Oxford: Oxford University Press, 2004.
Freris, L. and D. Infield. *Renewable Energy in Power Systems*. New York: Wiley, 2008.
Sørensen, Bent. *Renewable Energy*, 3rd Ed. London: Academic, 2004.

Gavin D. J. Harper
Cardiff University

GREENFIELD SITES

Greenfield sites are created when land presently used for agricultural, recreational, or amenity purposes is developed for urban uses. These sites are usually adjacent to existing urban areas, and as such, the development of greenfield sites leads to sprawl. For many centuries the development of greenfield sites was regarded as a normal development pattern that was a sign of civic progress. However, the realization of the financial and environmental costs of this style of development in the late 20th and early 21st centuries led to an effort to reduce the reliance on greenfield sites through the greater use of former industrial sites (brownfields) and urban infill sites for development.

Greenfield development and associated urban sprawl became a political issue in Britain and Europe in the 1990s because of its close association with the "crisis" in housing provision. This led to a series of policy initiatives such as the British Urban Renaissance Programme that tried to redirect development from greenfields to existing developed urban areas. The conversion of greenfield land into urban development is largely irreversible and has a number of environmental and social impacts. If the land concerned, such as a wetland, has environmental values or provides landscape values, then the environmental costs of development can be substantial. Equally, if the land is being used productively

Greenfield sites are created when land presently used for agriculture, like these cornfields on the edge of Des Moines, Iowa, are developed for urban uses.

Source: Lynn Betts/Natural Resources Conservation Service

for agriculture, then the development of greenfield sites will result in the loss of that productive capacity. This serves to push agricultural production of everyday food supplies such as market garden crops and milk supplies farther from the users of those products. This has consequences for retail prices and the energy and environmental costs associated with feeding city dwellers. It also reduces the amount of land available for agricultural production and makes the remaining agricultural land difficult to use.

The planning response to urban growth was often to establish a greenbelt around cites in an attempt to delineate the extent of urban growth. More recent planning approaches such as smart growth policies retained the concept of a city or metropolitan limit, which defined the boundaries for growth of that city but included at least a modest area for greenfield site development. The idea is that some greenfield development should be provided to allow for some choice within the housing market. Depending largely on the speed of growth, these greenfield areas may remain undeveloped for some years and face a number of challenges. The continuation of their present use for purposes such as farming may be impossible because of a number of factors. Urban neighbors may be unwilling to accept the noise, smells, and environmental issues associated with farming, and the farmers are faced with meeting the costs of city services that they may or may not have access to. Alternatively, the land is used for recreation or amenity purposes, which is usually more acceptable but that can produce community outrage when the land identified as having community values is revealed as land awaiting development. The land may also become derelict in a rural version of urban blight.

Greenfield developments by their very nature also expand the area of the city, partly because most greenfield sites are used for low-density housing developments. This is understandable because the suburbs were always regarded as a place to build a home suited to a family; that is, a house set on a site with some open space around it to be used as a garden or for children's play. Despite efforts to promote higher-density living, this type of low-density living is what many urban dwellers still aspire to—even New Urbanist developments may provide for the traditional single-family home sites. The desire for larger greenfield sites can also be traced to the emergence of what Jack Nasar and colleagues call "McMansions or tract mansions"; that is, oversized and expensive houses that are out of keeping with existing urban development. One of the controls to address this issue is to restrict greenfield site developments or to place strict controls on the percentage of such sites that can be covered with buildings.

The development of greenfield sites can also lead to increased infrastructural costs, as power lines, sewer, and water connections have to be established and linked to existing systems. This can result in higher costs for the municipal authority that has responsibility for the supply of infrastructure but only limited means to recover those costs. The location of greenfield sites can also increase the transport and energy costs as home owners commute to workplaces that are located in the already-developed city. Although some developers, particularly if they are a public or civic authority, attempt to undertake integrated greenfield site developments, in reality this is difficult to achieve for a number of reasons. Often the new greenfield site owner is in an established job that cannot be replicated in the new area, or employment opportunities develop more slowly than housing. Thus, it is not unusual for greenfield sites to be overwhelmingly dominated by housing, with only limited retail and employment provision, and still strongly economically linked to the existing urban area. Planning authorities may attempt to overcome this problem by requiring the development of a structure or master plan for any greenfield development to produce more balanced development.

From a development perspective, greenfield sites are often preferred because they offer the greatest development flexibility at what is perceived as the lowest cost. This is based on the costs of the alternative—usually infill or brownfield sites.

Brownfield sites are the planning opposite of a greenfield site, in that they have been used for another urban purpose—often a relatively heavy industrial activity—that has now ceased, allowing the site to be developed for a new purpose. This results in an expectation of higher costs, as a brownfield site may require extensive and expensive remediation to remove toxic industrial residues or to remove buildings and infrastructure. Such cost comparisons may not be accurate, as they are based on some specific assumptions and consider only direct costs. For instance, if the greenfield sites are established on land formerly used for farming, market gardening, or orcharding, there is a strong possibility that the soil may contain significant residues of commonly used agrochemicals such as fruit tree sprays or soil fumigants. Remediating these types of contamination can be both difficult and expensive. Economic calculations are usually based only on direct costs and make no calculations of, for instance, the ecological services the area provides or the energy costs associated with use of the new sites, which may be substantial.

Greenfield developments will always form part of urban development, but probably a lesser proportion as the full costs of developing them are better assessed.

See Also: Brownfields; Compact Development (New Urbanism); Green Belt; Infill Development.

Further Readings

Ganser, Robin and Katie Williams. "Brownfield Development: Are We Using the Right Targets? Evidence From Germany and England." *European Planning Studies*, 15/5 (2007).

Greenfield Sites: Sustainable Build (UK). http://sustainablebuild.co.uk/GreenfieldSites.html (Accessed July 2009).

Nasar, Jack, et al. "McMansions: The Extent and Regulation of Super-Sized Houses." *Journal of Urban Design*, 12/3 (2007).

Caroline Lomax Miller
Massey University

GREEN FLEETS (VEHICLES)

Making individual car users opt for cleaner vehicles has always met with a degree of resistance. For this reason, regulators—those interested in reducing vehicle pollution—have often favored targeting larger fleet operators to adopt greener vehicle technologies. Unlike individual consumers, fleet vehicle operators can make decisions that affect a large number of vehicles. Fleets also tend to refuel at central locations, making it possible to use alternative fuels that are not available in commercial refueling stations. Fleets are often used for specific purposes, and those used for short trips can rely on vehicles with limited ranges, such as electric vehicles. Furthermore, many fleets are owned or regulated by public agencies with responsibilities to reduce pollution, making it easier for them make commitments to greener vehicles.

Cleaner, alternative fuel vehicles, like this hydraulic hybrid UPS delivery truck, are often introduced through fleets.

Source: U.S. Environmental Protection Agency

One of the problems many alternative fuels present is a lack of infrastructure. However, for fleet operators with vehicles returning to a central depot, a centralized fueling or electric vehicle charging facility is often much more feasible than providing individual private vehicle users with access to fuel. For this reason, cleaner alternative fuels and powertrain are often introduced through fleets. In the United Kingdom, milk was delivered for many years by means of small, battery-electric "milk floats," which were charged up at the depot, did their milk-round of limited miles, and thereafter returned to the depot for a recharge. The fact that many fleets operate in urban environments, where pollution and air quality problems are particularly serious, has also been an argument to target fleets. Thus, taxis in many Asian countries are run on liquefied petroleum gas or propane gas. Though well-established in Japan and South Korea, in Hong Kong, most of the taxis converted to this fuel in the early 2000s. In a similar move, India mandated the conversion of all taxis and buses used in major urban areas to compressed natural gas.

Other environmental issues can also be tackled by fleets. In the United States, for example, the courier company UPS uses dedicated vans for much of its distribution service. These vans are designed for a long service life of around 25 years. The vans produced by Grumman for the U.S. Postal Service are designed along similar design parameters. Planned obsolescence is far removed from this approach to product design, and such fleet users show how a nonconsumer, business focus can in some circumstances lead to more environmentally optimized solutions.

Fleets can also be used on a more experimental basis to test new fuels or powertrain solutions. Thus, small electric buses are used in places like downtown Santa Barbara, California, both as a showcase for visitors and to ensure zero emissions from local transit. Similarly, fuel-cell buses have been used experimentally in a number of cities including Vancouver, Chicago, Perth (in Western Australia), London, Amsterdam, and Iceland's capital Reykjavik. Not only do such experiments allow the testing of new technologies in real operating environments and real operating conditions but they also allow these novel technologies to be introduced to the general public, allowing them to familiarize themselves with such alternative approaches. A relatively new type of fleet user is the car-sharing scheme or car club. These are now increasingly receptive to new cleaner technologies, as they tend to benefit from a membership that already includes many environmentalists.

Zipcars in San Francisco

United States–based Zipcars is the world's largest car-sharing operation, and it has long included hybrids in its fleet of vehicles, introducing the first hybrid in its Seattle fleet in 2003. In 2009, it announced a pilot program for plug-in hybrids in San Francisco, in conjunction with the local authorities. Moves to introduce cleaner vehicles are supported by 80 percent of the Zipcars membership, according to a survey. San Francisco's mayor, Gavin Newsom, was responsible for adding plug-in hybrids to the city's vehicle fleet and saw the partnership with Zipcars as a means of allowing members of the public direct hands-on experience with the new technology.

This highlights one of the least-discussed aspects of such programs—the ability to reduce the risk of new technology introduction, particularly when public authorities are able to cooperate with appropriate private-sector providers. When confronted with a choice between tried-and-tested technology or any novel technology, typical car buyers will usually opt for the lower-risk option of tried-and-tested technology. Fears about reliability, and particularly residual values on resale, are the motivating factors. One of the reasons GM's radical EV-1 electric sports car of the 1990s was only available on a lease basis is that with so many new technologies, the risk to customers—and also to GM itself—of selling the cars would have been too high. By not selling the cars, that risk is taken away—fears of reliability and residual values are then the responsibility of the vehicle owner (in this case, the manufacturer), not its user. Car-sharing schemes are therefore a perfect vehicle for the introduction of such radical new technologies, as the risk is collectivized to the car club, rather than burdening individual users. Zipcars' move to plug-in hybrids is a perfect example, and perhaps a more deliberate use of car clubs in this way should be considered by government, vehicle manufacturers, and suppliers of alternative-technology vehicles.

One of the first car-sharing schemes was run in the Dutch capital of Amsterdam in the 1970s. This scheme, dubbed *Witkar*, used unique and rather novel battery electric vehicles. Although the motivation was their zero-emissions nature, they also introduced members to electric vehicle technology. More recent is the "Move About" concept in Norway. This is a car-sharing scheme, or car club, linked with Norwegian battery electric vehicle–producer Th!nk and is designed to use Th!nk vehicles to deliver an urban mobility package based around this electric vehicle. The principles of the scheme are that it has to be clean in terms of energy supply. In conjunction with public- and private-sector partners, Move About provides charging points at key locations and aims to be "affordable and available" to the largest number of users with a minimum of hassle. The smart card used for the system can also be used to access public bicycles that are part of the bike share system.

Electric Cars in London

London's mayor Boris Johnson is introducing a public–private electric car rental program for London. Johnson has been partly inspired by the French Velib scheme, based on bicycles, and intends to make London a leader in electric vehicle technology. He is building on the foundations provided by the high number of electric vehicles already in use in the city to avoid paying the congestion charge, from which zero-emission vehicles are exempt.

Although bus, truck, and taxi fleets can be used to test new greener technologies, car-sharing schemes enable such technologies to be placed directly in the hands of the public. This allows ordinary car users to become familiar with such novel technologies, allowing a broader introduction of such technologies with less consumer resistance. Fleets can thus play a real role in making transportation more sustainable.

See Also: Commuting; Green Procurement and Purchasing; Personal Rapid Transit; Reykjavik, Iceland; San Francisco, California; Vancouver, Canada.

Further Readings

Keiser, Bethscheider. *Green Designed: Future Cars*. Ludwigsburg, Germany: Avedition GmbH, 2008.

Meijkamp, R. *Changing Consumer Behaviour Through Eco-Efficient Services: An Empirical Study on Car Sharing in the Netherlands*. Delft, Netherlands: Delft University of Technology Press, 2000.

Schilperoord, P. *Future Tech: Innovations in Transportation*. London: Black Dog, 2006.

Sperling, Daniel and Deborah Gordon. *Two Billion Cars: Driving Toward Sustainability*. New York: Oxford University Press, 2009.

Paul Nieuwenhuis
Cardiff University

GREEN HOUSING

Green housing relates to the practices used to build the house and the products used in the building process, as well as the long-term savings and effect on the health and well-being of the users and the environment. The objectives of green homes are to create and maintain home environments that are sustainable for users, the community, and the environment. Green homes are part of the sustainability movement that aims to meet needs without compromising future generations' ability to meet their needs.

Achieving the Objectives of Green Housing

Government Mandates, Programs, Building, and Material Standards

Advancements in green building have prompted governmental programs and standards. For example, the U.S. Environmental Protection Agency and Department of Energy implemented the Energy Star program, which has revolutionized consumers' ability to access information about energy-efficient products and services from a trusted source. In 2008, Americans bought 550 million Energy Star–certified products, and households are experiencing energy use and savings (use cut by one-third, saving $750 annually). In addition, there are 940,000 Energy Star–certified new construction homes in the United States; 100,000 of those homes were built in 2008.

Other standards such as the International Energy Conservation Code help establish commercial and residential building code efficiency standards. ASTM International (formerly known as the American Society of Testing and Materials) is the main standards organization in the United States. ASTM E2114 sets standards for sustainable building as performance requirements that minimize disturbances to and improve the ecosystem both during and after construction and during their life span. The International Organization for Standardization is a nonprofit standards organization that creates standards for international sustainable development.

Nongovernmental Third-Party Building and Material Standards

Nongovernment third-party certification systems have benefits and disadvantages. Third-party certification systems can be either for profit or nonprofit. A well-known nonprofit, nongovernmental, third-party organization is the U.S. Green Building Council and its building rating system, Leadership in Energy and Environmental Design (LEED). LEED certification has gained industry approval as the leading benchmark for sustainable buildings.

There are established nonprofit third-party certification systems for processes and products that are related to the building process. The Forest Stewardship Council (FSC) is a nongovernmental, nonprofit organization that is responsible for management of the world's forests. The LEED rating system recognizes FSC as the only sustainable wood certification label. The FSC has been perceived as a successful partnership between nongovernment and private industry. Nevertheless, there are critics of the FSC and its ability to establish global forestry standards.

The Cradle to Cradle program from McDonough Braungart Design Chemistry was established in 2005 and is an example of a propriety certification system. Cradle to Cradle has industry recognition because it challenged the cradle-to-grave approach that has dominated for years. Cradle to Cradle has altered the way we approach manufacturing by evaluating the life cycle of a product from production to product end life, once it is no longer in use. Although the Cradle to Cradle sustainable paradigm is respected in the industry, few manufacturers have sought certification from McDonough Braungart Design Chemistry. This could be because of the proprietary nature of the certification system, even though Cradle to Cradle is a certification label that LEED accepts as an innovation point, and the Environmental Protection Agency recognizes it as a requirement for government purchasing of Environmentally Preferable Products.

Tougher Industry Standards

In 2001, the U.S. Green Building Council founded the Green Building Certification Institute to develop and administer the accreditation program for LEED-certified professionals. Eight years later, the council reports over 77,000 LEED Professional Accreditations, or LEED APs, awarded. In response to interest, in 2009 the Green Building Certification Institute made substantial changes to the credentialing of LEED APs. The changes reflect the increased standards for green building in the marketplace and the knowledge growth as practices become established.

Competition will also grow in the industry for establishing standards that will challenge all existing building standards. The Green Building Initiative, founded in 2005, is a 501(c) (3) nonprofit educational organization that is challenging LEED with the National Association of Home Builders (NAHB) rating system for residential and Green Globes rating system for commercial properties. The NAHB has an accreditation program for the designation of Certified Green Professional. Similar to LEED building ratings, the NAHB founded the National Green Building Certification, which meets the requirements of the NAHB Model Green Home Building Guidelines or the ICC 700-2008 National Green Building Standard.

Selecting Nongreenwashed Materials and Resources

The most recognized and trusted product certification for nonindustry professionals is Energy Star. Americans have broadly accepted Energy Star as the standard for energy-efficient products. In 2007–2008, the U.S. Green Building Council partnered with the American Society of Interior Designers to launch the REGREEN program. Unlike LEED, which scores and certifies buildings, REGREEN is focused on residential remodeling guidelines for green renovation projects. The objective of the REGREEN guidelines is to provide homeowners and builders with practices to improve energy savings and indoor air quality and decrease homes' impact on the environment. Unique to REGREEN is the guideline approach, instead of certification through a rating system. Even though this standard is not stringent, it has benefits. Because REGREEN is a guideline and not a rating system, households may implement REGREEN suggestions with more ease than following stringent guidelines, such as LEED for Homes (LEED-H).

Applications of Green Housing

Establishing a complete understanding of green housing objectives requires acceptance that green principles and practices vary based on the project. Nevertheless, the basic applications of green building are similar to the requirements of the LEED-H building rating system, which provides a logical checklist for assessing the sustainability of a project from inception to completion. Understanding the LEED-H credit or point breakdown provides an applied example of the principles and practices of green housing. The credit system for achieving LEED-H certification varies depending on the desired certification level of platinum, gold, silver, or basic. The home is evaluated for credits in the following areas: Sustainable Sites, Water Efficiency, Energy and Atmosphere, Materials and Resources, and Indoor Environmental Quality. The management of the project is evaluated for credits in the following areas: Innovation and Design Process, and Awareness and Education. The following text discusses material selection, waste management, water conservation, and energy in relation to green housing.

Material Selection

Selecting or specifying green materials can improve indoor air quality, respond to consumer desires, and offer compliance to applicable regulatory requirements. Even though the benefits of selecting green materials are established, the actual act of selecting the right materials for a green home project can be overwhelming and challenging for both industry professionals and homeowners. There is not one right approach to selecting material, which makes green product selection a challenge. The simplest starting point is to learn about the different green building and material standards. The next step is to ask the right questions, because an educated green consumer does not need to rely solely on government or third-party labeling to identify the legitimacy of products. Being a wise green consumer requires the ability to identify greenwashing. The consumer must also learn the types of things that establish sustainability legitimacy, such as the product's place of origin, its composition, and strategies for disposal, among other things.

Waste Management

Crucial to the success of any green building project is waste management and reduction, which help improve overall cost-effectiveness. The successful implementation of the three Rs (reduce, reuse, and recycle) requires the transmission of accurate waste management information and easy access to alternative waste management facilities. Continued education regarding consumption behaviors and enforcement of reduced consumption as a tenet of green housing are critical tools to improve waste management and reduction.

Water Conservation

Residences consume nearly half of the water in municipalities. Water conservation can be achieved by using less water through use of water-efficient equipment and by water reclamation and reuse. In 1992, the Energy Policy Act established standards for equipment that required water use including toilets, faucets, and showerheads. Standards for appliances have been beneficial for conservation; however, water conservation is largely influenced by user behaviors. A faucet may be low-flow, but the user can still run the faucet for long periods of time. A water-conserving trend for the exterior of the home is the landscape. Xeriscaping is natural landscaping for the local habitat that minimizes or eliminates the use of water. Other watering options for landscape and gardens include greywater systems that reuse wash water (nontoilet and food water).

Energy

Energy has been at the forefront of the sustainability building movement. Greenhouse gases are created by the burning of fossil fuels (petroleum, coal, and natural gas). The more energy saved, the fewer fossil fuels are burned, which reduces air pollution. Second, fossil fuels are nonrenewable resources, and the objective is to lower dependency on nonrenewable resources. Energy conservation through insulation, high-quality construction, and Energy Star windows and doors is critical. Viable renewable energy options for green homes and energy consumption include use of Energy Star appliances and solar and wind energy resources.

Solar options are diverse, feasible, and most important, renewable. Renewable energy resources include direct solar sources such as passive solar heating solar thermal generation, photovoltaic solar cells, and solar-generated hydrogen. Solar water heating systems are glazed, flat-plate collectors mounted on a roof that absorb sun heat and that power approximately two-thirds of the hot water produced by a water heater. Solar transmittance options include shades and screen solutions that decrease the need to counteract undesired temperatures. In addition, daylighting techniques capture daylight for optimum natural light use by lowering the dependency on simulated lighting.

Other indirect solar energy options include hydropower, tidal energy, geothermal, and wind energy. Wind-powered energy is created by harvesting wind through turbines. At this time, no state in the United States offers 100 percent wind-powered home energy; however, in some states residents can pay an increased energy fee to receive supplemental wind-powered energy at their home. Wind turbines could benefit households in rural communities, where a single turbine on the property could supplement dependency on fossil fuel energy.

Future Efforts to Raise Awareness of Green Housing

Increased Education

In 2008, the "Change the World, Start With Energy Star" campaign was a six-city touring energy-efficient home exhibit. The campaign asked Americans to make a pledge to opt for energy-efficient choices at home to reduce greenhouse gas emissions. The campaign reached 28 million people and had nearly 500,000 pledges. In 2008, the U.S. Department of Energy and the Ad Council, a private nonprofit organization that promotes public service campaigns, released the "Lose Your Excuse" educational campaign, which focuses on the ease of altering home energy consumption while simultaneously providing an interactive learning experience for educators, parents, and children.

Tax Incentives

Tax incentives tend to foster public interest in implementing new behaviors. Since the 1970s, the U.S. federal and state governments have offered minor incentives to foster energy-efficient building. The majority of the programs have been focused on energy conservation. In 2005, consumers of Energy Star products or home certification were eligible for federal tax incentives. For existing homes only, 2009 and 2010 tax credits were available for 30 percent of the cost, up to $1,500 for windows and doors, insulation, roofs, heating–air conditioning, water heaters (nonsolar), and biomass stoves. For existing homes and new homes, tax credits of 30 percent of the cost with no upper limit before 2016 include geothermal heat pumps, solar panels, solar water heaters, small wind energy systems, and fuel cells. In addition, tax credits of $2,000 are available for homebuilders for new energy-efficient homes that meet the 50 percent energy saving for heating and cooling as defined by the International Energy Conservation Code. A $1,000 tax credit is available to homebuilders for new manufactured homes that meet 30 percent energy savings on heating and cooling.

Opportunities and Challenges for the Future of Green Housing

The post-fossil-fuel world will reflect innovative building principles and practices; improved material and resources; easily accessible green housing information; the development of government mandates, programs, and educational resources; and changing lifestyles. Innovation in building practices will challenge how we understand building homes. Innovative homebuilders are perfecting methods of energy-efficient mass-produced green homes in factories, out of weather, and with stringent quality controls. Materials and resources will become more readily available and more trusted. Improved ease of use and access to building guidelines and rating systems will foster community interest in pursuing green housing options. Even though there is information available to the public regarding the cost analysis benefits of home energy savings, personal household experience with cost benefits will continue to be the best catalyst for implementation.

Meanwhile, challenges within the industry do still persist that stall the development of green housing. Apartment owners and developers seem to lag considerably behind the industrial, commercial, and retail property sectors in terms of maximizing their green potential. These multifamily owners and operators opt instead for a minimum threshold of sustainability necessary to obtain LEED certification, ostensibly for marketing purposes. In addition, the siting decisions and capital expenditures associated with green apartment living suggest that this particular type of niche housing is more suited to upper-income households. This particular inequity sparks outcries of social injustice, especially as the populations that could most benefit from the savings inherent within green housing are low-income. Finally, the appraisal industry has been accused of not recognizing the benefits inherent within a green property, thus creating wide disparities in the pricing and capitalization of green housing features. Ironically, such inefficiencies in pricing can only be remedied through continued recognition of the benefits of these green features by consumers.

See Also: Daylighting; Green Design, Construction, and Operations; Green Landscaping; Green Roofs; Greywater; Indoor Air Quality; LEED (Leadership in Energy and Environmental Design); Sustainable Development; Water Conservation; Xeriscaping.

Further Readings

Carswell, A. T. and S. W. Smith. "The Greening of the Multifamily Residential Sector." *Journal of Engineering, Design & Technology,* 7/1:65–80 (2009).

Dingwerth, K. "North-South Parity in Global Governance: The Affirmative Procedures of the Forest Stewardship Council. *Global Governance,* 14/1:53–71 (2008).

Dunn, C. "REGREEN: Remodeling Guidelines to Re-Green Your Interior Design," March 20, 2008. http://www.treehugger.com/files/2008/03/regreen-remodeling-guidelines-interior -design.php (Accessed April 2009).

Fortmeyer, R. "(Mis)understanding Green Products." November 2007. http://www .construction.com/CE/articles/0711edit-1.asp (Accessed April 2009).

Foster, K., et al. *Sustainable Residential Interiors.* Hoboken, NJ: Wiley & Sons, 2007.

Gutterman, S. "Leveling the Playing Field." *Green Builder,* 10 (December 2008).

Heikkinen, R. "Sustainability in the Buildings Industry: How ASTM Standards Are Addressing the Trend." *ASTM Standardization News* (August 2001 http://www.astm.org/ SNEWS/AUGUST_2001/heikkinen_aug01.html (Accessed April 2009).

Kopec, D. *Health, Sustainability and the Built Environment.* New York: Fairchild Books, 2009.

Maddock, M. G. and R. L. Vitn. "Why Not Mass Produce Green Homes?" *Business Week* (October 22, 2008).

McDonough, W. and M. Braungart. *Remaking the Way We Make Things.* New York: North Point, 2002.

Parker, D. S. "Very Low Energy Homes in the United States: Perspectives on Performance From Measured Data." *Energy & Buildings,* 41/5:512–20 (2009).

Spiegel, R. and D. Meadows. *Green Building Materials: A Guide to Product Selection and Specification,* 2nd Ed. Hoboken, NJ: Wiley & Sons, 2006.

U.S. Department of Energy and Ad Council. "Lose Your Excuse." (2008). http://www .loseyourexcuse.gov/index.html#/index (Accessed April 2009).

Winchip, S. M. *Sustainable Design for Interior Environments.* New York: Fairchild, 2007.

Megan Lee
University of Georgia

Stephen Smith
Gainesville State College

Andrew T. Carswell
University of Georgia

GREEN INFRASTRUCTURE

One organic metaphor envisions urban parks as the lungs of a city, as suggested by this bucolic scene in Brooklyn, New York's Prospect Park.

Source: iStockphoto

Green infrastructure is the interconnected systems of soils, water, air, vegetation, and animal life that constitute a healthy ecosystem. It provides services to humans that would otherwise need to be provided by constructed infrastructure. Examples of green infrastructure include wetlands that provide stormwater filtration and trees that cool buildings by providing shade. The concept of green infrastructure serves as a corrective to common understandings that urban landscapes exist at the expense of natural ones. Emphasizing the services that green infrastructure provides reconnects urban systems to the natural landscape, no matter how transformed, and advocates for preserving, restoring, and replicating natural systems in considerations of urban policy and planning. The concept of green infrastructure arises in a number of interrelated discussions, as urban planners, green builders, city managers, and the public wrestle with what it means to create a Green City.

Components of Green Infrastructure

In the simplest sense, green infrastructure has been used to describe networks of parks thought to moderate the social and physical ills of modern urban life. Organic metaphors for the city envisioned parks as the lungs of the city, allowing spaces for residents to recreate away from traffic and factories. A more dynamic understanding argues that elements of green infrastructure perform ecological functions that benefit people, plants, animals, and natural processes and systems. Natural and constructed forms provide services to the entire ecosystem, moderating microclimates; mitigating pollution in water, air, and soils; controlling invasive species; maintaining biodiversity; detaining water runoff; and serving as habitat for native species.

Critical components of green infrastructure provide multiple functions and operate at multiple scales. For example, trees provide shading for adjacent buildings, lowering cooling costs and reducing urban heat island effects. Trees take up large amounts of rainwater via their extensive root systems, capture airborne pollutants, and improve air quality, while providing habitat and food sources for terrestrial, avian, and aquatic species. A number of studies have documented the economic and psychological benefits of the urban forest as a whole and beyond the aesthetic values attached to individual trees as landscape elements. Establishing buffer zones along streams and creeks keeps new development at a distance from water bodies, providing water-quality protections, preserving habitat, and creating green space that can help avoid the localized areas of flooding and erosion associated with increases in impervious surfaces.

At larger scales, forms of green infrastructure include preserved or restored floodplains and wetlands, forests, and farmland. In addition to providing habitat for a number of species assemblages based on the size, shape, and connectivity of the network, these spaces also function as recharge areas for local aquifers, with rainwater percolating through the surface soils and into the groundwater system. During high-water events, low-lying areas adjacent to water bodies disperse and store water, lessening downstream surges. Once classified as undesirable land, wetlands play a critical role in maintaining regional water quality and species diversity. Clean air, carbon sequestration, and flood control are dependent on the healthy functioning of these once-marginalized spaces. Restored wetlands have been used as a final step in sewage treatment, with biodegradation, filtering, and dispersed outflow performed by carefully designed and managed wetland ecosystems.

Because many cities have sewer and stormwater management systems that are combined, overflows of raw sewage occur when high-volume rain events exceed the capacity of the system. Efforts to minimize increases in stormwater generated by new development require preserving as much of the existing capacity of the landscape to capture and infiltrate rainwater, in combination with designing and constructing facilities to mimic these natural processes. Considered early in the design process, a site and proposed uses can be holistically analyzed to retain as much of the existing functions on the site, minimize the loss of vegetation and soils, and then replace or improve on functions through thoughtful design, construction, and maintenance of green facilities. As a part of new construction, green infrastructure often takes the form of site-specific stormwater management including green roofs; pervious paving for sidewalks, driveways, and parking lots; stormwater swales; and areas set aside for natural resource protections. Diverting increased runoff to swales detains stormwater before it can surge into the existing piped system, provides an opportunity for evaporation, and slows the water so that suspended solids and pollutants drop out. Appropriate plantings take up water and pollutants, providing additional water-quality benefits. These elements are no less constructed than the elaborate pipe and outfall system of conventional stormwater management; however, the facilities prioritize on-site mitigation, using natural processes. The green facilities reduce the immediate need to dramatically increase the size of the traditional piped stormwater system.

At the downstream end of stormwater management, restoring the outfalls from culverts halts the flow of warm, dirty water that otherwise moves directly from streets into rivers. Many urbanized rivers have been placed into pipes that run underground, with stormwater runoff augmenting the flow before a culvert ejects the water into a larger river. Removing the end pipe and restoring the riverbank redirects the flow through a series of engineered meanders to slow the water, allowing sediment to drop out, lowering the water temperature, and giving native vegetation an opportunity to absorb and filter some of the water. This ecological restoration of riverbanks returns urban waterfronts to a highly designed and constructed version of a natural confluence but is critical in restoring habitat for aquatic, avian, and terrestrial species that use the river's edge. These restoration activities recognize and increase the capacity of riparian areas to provide water quality benefits and habitat. This strategy also helps municipalities comply with requirements to reduce pollutant levels and volume of stormwater discharged into regional waterways.

Green Infrastructure and Public Policy

This holistic approach to green infrastructure is an emerging component of urban land use planning and practice. Regulatory requirements related to clean air and water and endangered species legislation often provide an impetus for local government to consider green infrastructure. Preserving existing green spaces maintains previously unacknowledged ecological service capacity. At a time when many municipalities are experiencing budget cuts, green infrastructure is frequently a low-cost alternative to expanding traditional infrastructure systems. An increased interest in creating sustainable, livable communities has encouraged the recognition and enhancement of green infrastructure as a response to the problems of rapid urban growth and sprawl. No longer simply an aesthetic or recreational site, components of green infrastructure can be understood as underpinning the health of the entire urban ecosystem and also reforming the ills of the modern industrial city.

Contrast With "Grey" Infrastructure

Urban infrastructure has traditionally been thought of as the networks in an urban landscape that manage the flows of energy, food, materials, waste, information, and people on which cities depend. Economic activity requires the movement of people and products on roads and rails and in ports. As cities grew, the provision of clean water and the orderly removal of wastes became a significant health concern. Large networks of pipes conveyed clean water into the city and sewage out of it, defining the orderly, efficient, and well-governed modern city. The process of urbanization was thought to replace natural landscapes with built environments, and the resulting problems of the modern city required technological solutions that could handle the increasingly large flows in and out of the urban landscape. In contrast with this traditional view, green infrastructure encompasses existing and constructed natural spaces that replicate not only the services that modern, piped, "grey infrastructure" was designed to provide but also the maintenance and resiliency of the broader ecosystems in which cities are located.

This comparative definition of green infrastructure has provided new opportunities for articulating urban environmental policy. Instead of contesting the removal of natural areas in only environmental or aesthetic terms, identifying them as green infrastructure elevates the services natural spaces provide that have previously been unmeasured. Accounting for these benefits provides greater credibility for preservation in policy analyses when green infrastructure is placed on par with other components of urban infrastructure. In addition, there is an opportunity to articulate the value of creating or augmenting natural landscapes as a method of extending the usefulness of traditional infrastructure when water, sewer, stormwater, and energy networks are over capacity and expensive to expand. The incorporation of environmental concerns into public facilities planning has been accompanied by an effort to create indicators and benchmarks or place a monetary value on the services provided by undeveloped land to inform the cost-benefit analysis of environmental preservation and development. For example, the costs borne by residents and governments as a result of the flooding along a canyon road can be more directly tied to the loss of vegetation and pervious areas in the adjacent hills, raising questions about whether the loss of ecosystem services and resulting flood mitigation costs are outweighed by the benefits of new construction.

However, the distinction between green and grey infrastructure is not precise. The green features built to augment ecosystem services are clearly constructed and can be thought of as part of the built environment. Few, if any, natural areas remain clear of human-induced changes, whether through historic logging, hydrologic changes from adjacent development, or proliferation of invasive species. Parks, as well, are constructed to replicate specific ideals of urban green space. Grey infrastructure also plays a role in natural systems; rivers are diverted into pipes through industrial areas but remain a part of the watershed. Salmon populations may decline as a result, but other species, such as rodents, proliferate in the new habitat. The contrast with traditional grey infrastructure that gives green infrastructure its wide-based appeal depends on the conceptual dualism between city and nature, even as it tries to erase it.

The Role of Green Infrastructure as Metaphor

In the role of metaphor, green infrastructure functions as a component of the understanding of the form and function of the city, providing a way of framing a chaotic, unruly, and unhealthy city and what potential solutions may be. Analogues drawn from biology and ecology have frequently served as models for organizing society and the city's physical landscape. The city has been previously conceptualized in terms of areas of ecological succession, as neighborhoods transitioned from one social group to another; as a site of metabolic transformation of nature, as raw materials were brought into the city and manufactured into consumer goods; and now, through the concept of green infrastructure, as urban systems are understood as enmeshed within larger ecological systems.

The green infrastructure approach views the modern city's process of destroying existing ecosystem capacity and then using a technological solution to capture waste within piped systems as flawed. Green infrastructure instead privileges the services ecosystems already provide in contrast to the dismissal of nature that typified the modern city. The definition of a city restored to a virtuous relationship with natural systems depends on the contrast with a modern city, destructive of nature and ignorant of holistic ecological understandings. The return to nature as the model for organizing urban life provides the panacea for urban ills.

In this antimodern view, green infrastructure projects have increasingly taken on important roles in the symbolic urban landscape previously held by monumental grey infrastructure projects. The dams and fountains of previous eras have been replaced by stormwater swale–lined streets and restored wetlands as key public works projects in the contemporary green city. Fountains in central social spaces celebrated a city's ability to bring clean water vast distances to urban residents. Current green infrastructure projects, such as a park built around a large rainwater filtration swale built on the site of a formerly polluted industrial site, mark a city as being on the forefront in public service provision and progressive green urbanism. In postindustrial spaces, the reformation of nature through green infrastructure projects performs not only ecosystem functions but also a symbolic role in articulating a future of green urbanism. In an era of increased competitiveness between cities, high-profile green infrastructure projects contribute to a city's identity and claims to livability. These multiple functional and symbolic roles can give rise to new conflicts that suggest that the appeal to natural systems as a corrective to urban problems is not as straightforward as it might appear.

Difficulties in Implementing the Concept of Green Infrastructure

The holistic ecosystems approach has been difficult to interject into the administration of public services. It has been difficult to incorporate green infrastructure elements into existing urban financial plans. Infrastructure is typically paid for by capital funds, shown on a municipality's balance sheet, and depreciated according to well-established schedules to recognize deterioration over time. Problematically, green elements such as trees do not deteriorate like a sewer pipe but, instead, increase in water retention capacity, shade potential, aesthetic value, and habitat provision as they age. To further compound the problems of accounting for this infrastructure, most of the urban forest and many other components that make up the green infrastructure of a city are located on private property. It is difficult for a public entity to include these components as a capital asset on its balance sheet and direct funds to their care and maintenance when they do not have direct ownership or control. Without access to capital funds, the traditional method for constructing piped systems and roads, it is difficult to secure the funds needed to invest in widespread tree planting, large-scale restoration, or even proper maintenance of public trees. Regulating these elements in the urban landscape on private and public lands runs afoul of the common problems of land use planning: Policies may exist for which no rules are put in place; rules may be in the city code, but enforcement is not an urban priority; or inspection and enforcement fail to educate and enroll homeowners as participants in managing their individual components of green infrastructure.

Although preserving or replicating natural processes to provide ecosystem functions, green infrastructure still requires monitoring and maintenance. This reality often runs counter to expectations that a more natural systems approach would require less oversight. The degree of management becomes another avenue for debate—should floodplains be reengineered after large floods to optimally disperse waters in the future, or left alone for fluvial reshaping? Should invasive plant removal, tree thinning, and burning be a regular component of natural area maintenance? When it is not possible to optimize all ecosystem functions, which should prevail? Water quality, carbon sequestration, habitat, or other considerations?

By introducing ecosystem functions into the accounting and analysis methodology of contemporary urban governance, a green feature is often reduced to indicators or measures of its service functionality: gallons of water infiltrated or volume of waste processed per acre of wetland. This reductionism is an outcome of the functionalist approach of cost-benefit analysis but is often a departure from the holistic approach that green infrastructure champions. Defining a built form as a piece of green infrastructure that provides stormwater management functions may be at the expense of other elements, such as maintaining urban forest canopy for shade and habitat. Measured narrowly in terms of stormwater retention, a vegetated hill slope can be replaced with a series of constructed swales to manage rainwater produced by new development. One measure can gain primacy in analyses. For example, tree canopy coverage over the urban area as the primary indicator of urban forest health can come at the expense of retaining and planting large native trees that provide vital habitat and forage sites for embattled native species.

Broadened Definitions of Green Infrastructure

To further complicate implementation of the concept of *green infrastructure*, use of the term has broadened in some cases to include alternative transportation facilities such as bike lanes, pedestrian pathways, light rail, dedicated bus corridors, aerial trams, and streetcars. Although these public projects do not provide direct ecological services per se, they do provide alternatives to the dominant mode for moving people and goods that requires low-efficiency, oil-consuming vehicles. In the wake of government stimulus funds intended to grow a green economy, there has also been increased policy and investor and public interest in initiatives to promote green building and create local alternative energy sources. Developing green building products as a part of expanding a new green economy can also be construed as developing a region's green infrastructure, but they may come at the expense of a holistic ecosystems analysis. These efforts gain credibility by being captured under the definition of green infrastructure but hamper efforts to recognize, preserve, and improve ecosystem services.

Criticism of Green Infrastructure Concepts

The mobility of meanings associated with green infrastructure lends itself to applying it to any number of endeavors and projects that have little to do with establishing ecosystem health. Instead, economic growth stimulated by the joined efforts of government and finance privileges economic growth, suggesting it will aid in making cities more sustainable. In not addressing this underlying mechanism, green infrastructure, as discussed, has been criticized for not addressing social change or the role of capitalist urbanization in creating environmental degradation alongside social inequality. Conceptually, although green infrastructure appears to contrast with the modern city's dismissive attitude toward the natural environment, it depends on many views of nature that are of the modern era: a belief in the redemptive power of nature, pastoral idealism, and organic metaphors that view the city as a self-organizing organism. These each use nature idealistically as something outside the city, society, or human activity that can be a source of inspiration to correct problems. However, this dualism falls to acknowledge how understandings of nature are themselves mediated by society and the experience of cities. This criticism suggests that the concept of green infrastructure may play a more important symbolic or rhetorical role but will only selectively amend existing urban practices without halting or reversing the destructive patterns of urbanization.

See Also: Combined Sewer Overflow; Environmental Planning; Stormwater Management; Urban Forests.

Further Readings

Ahern, J. "Green Infrastructure for Cities: The Spatial Dimension." In *Cities of the Future: Towards Integrated Sustainable Water and Landscape Management*, Vladimir Novotny and Paul R. Brown, eds. London: IWA, 2007.

Benedict, Mark A. and Edward T. McMahon. *Green Infrastructure: Linking Landscapes and Communities*. Washington, D.C.: Island Press, 2006.

Gandy, Matthew. "Urban Nature and the Ecological Imaginary." In *In the Nature of Cities: Urban Political Ecology and the Politics of Urban Metabolism*, Nik Heynen, et al., eds. Questioning Cities Series. London: Routledge, 2006.

Hagerman, C. "Shaping Neighborhoods and Nature: Urban Political Ecologies of Urban Waterfront Transformations in Portland, Oregon." *Cities: The International Journal of Urban Policy and Planning*, 24/4:285–97 (2007).

Kaika, Maria. *City of Flows: Modernity, Nature and the City*. London: Routledge, 2004.

Portland Bureau of Environmental Services (Oregon). "Stormwater Management Manual." http://www.portlandonline.com/BES/index.cfm?c=47952 (Accessed 2009).

Chris Hagerman
Portland State University

Greening Suburbia

The greening of suburbia has garnered significant interest over the last decade from municipal planners, scholars, and lay practitioners, mostly in direct response to a set of socioeconomic, political, and ecological critiques around suburban form. This article provides a brief historical overview of suburbia (including its characteristic features and diffusion across North America) and summarizes the charges made by its most ardent critics. It then examines efforts to green suburbia, including the greening of new suburban developments, as well as the transformation of existing suburbs toward greater sustainability.

Brief Historical Overview

Suburbs—politically independent jurisdictions—lie within commuting distance of an inner urban area and have come to describe a variety of different residential communities and landscapes that, together with the city core, make up the larger metropolitan area. The suburbs first developed in the early 19th century as people fled densely populated, industrialized city centers to seek a less-congested and cleaner environment, sense of space, improved family life, and upward social mobility (subsequently spurring the growth of the middle class). Urban decentralization was augmented post–World War II, when suburban development provided a convenient housing solution for returning troops and their families. Since the postwar period, changes in transportation infrastructure, personal automobile ownership, mortgage financing, and the resocialization of women toward domesticity fueled the rapid diffusion of the Anglo-American suburb. The desire to own a single, detached home in the suburbs became uncritically accepted as the North American dream.

Characteristic Design Features

The suburbs purport to offer the best of town and country, typically vis-à-vis single, detached homes laid out along crescents, gently winding streets, and cul-de-sacs—each plot individually landscaped with trees, hedges, bushes, and lawns. Low population densities make the provision of public transportation economically challenging, which, combined with the spatial separation of residential, commercial, and industrial land uses, results in an almost complete reliance on the automobile. Housing expectations have changed as well—the average suburban home in North America has increased in size from 800 square feet in 1950, to 1,500 square feet in 1970, to 2,266 square feet in 2000. Today, homes with up to 4,000 square feet of living space are not uncommon. Suburban neighborhoods are typically serviced by shopping malls and commercial strips (complete with big box stores) offering "landscapes of consumption," and resulting in increasing per capita resource use.

Low-density suburban development—colloquially known as "sprawl"—has been achieved at significant ecological cost. New suburban developments typically envelop prime agricultural land, woodlands, and open space, resulting in the loss and fragmentation of wildlife habitat, disruption of ecosystems, and decline in arable land. With the increase in house size and commuting distance, per capita energy consumption and carbon emissions similarly increase, exacerbating climate change and affecting air quality negatively. For many, lawns—the now ubiquitous archetypal design feature of the suburbs—require high inputs of fertilizers and pesticides to maintain nonnative grasses. Municipal and, increasingly, statewide and provincial bans on cosmetic pesticide use represent changing social mores in this regard. Lawns require an inch of water each week, so water use represents a significant ecological cost as well, particularly in arid regions.

Although cultural homogeneity has given way in part to increased social diversity and racialization of the suburbs, minorities still live disproportionately in multifamily dwellings, older housing close to urban cores, or segregated sections of existing suburbs. Critics argue that many developers meet new density requirements by situating multifamily dwellings on the periphery simply to act as a buffer between less-dense development and a commercial strip or highway—effectively creating a "social apartheid." Critics also charge that more affluent suburbanites move to upscale urban condos, gated communities, or larger properties at the urban fringe.

Greening New Suburban Developments

The suburbs are far from monolithic, and planners, architects, city officials, scholars, and environmental activists have for some time attempted to reenvision suburban form through land use planning that seeks to protect critical habitat and arable land, increase population density, and create walkable, human-scaled communities. Efforts aimed at decreasing reliance on the automobile typically seek to facilitate access to public transit and to promote pedestrian connectivity. Pedestrian-friendly design features include buffer foliage between sidewalks and roadways and the use of traffic-calming devices such as traffic circles, speed bumps, diverters, crosswalks, and yield strips. Both on-street parking and tree-lined streets further decrease traffic speed, increasing pedestrian safety. Higher-density, mixed-used activity nodes in which, ideally, shops, restaurants, and offices form a streetscape (with parking underground or at the rear of buildings) form a local transit hub.

Postwar planned communities such as Columbia, Maryland, and Reston, Virginia, emerged in the 1960s as part of the New Town movement, characterized by mixed-use, cluster housing; preservation of open space; self-contained villages; and a town center. Similarly, Miami-based architectural firm Duany Plater-Zyberk & Company and others, such as Peter Calthorpe (The Pedestrian Pocket), promoted the concept of neotraditional towns, drawing on influences of urban theorist Jane Jacobs. These architects sought to achieve higher densities through a return to the design of the American small town (including more efficient land use, diversity of class, and preservation of open space, with development linked to mass transit). These influences developed in the 1980s into the New Urbanism design movement, which advocated socially diverse, pedestrian-friendly neighborhoods with good access to public transit, public spaces, and community institutions. New Urbanism is one variant of a broader urban planning and transportation movement known as Smart Growth, which seeks to achieve many of the same socioeconomic and ecological objectives.

The LEED for Neighborhood Development Rating System, developed collaboratively by the U.S. Green Building Council, the Congress for New Urbanism, and the Natural Resources Defense Council, combines principles of Smart Growth, New Urbanism, and green building to inform a more holistic and sustainable neighborhood design. The program—still in the pilot phase—seeks to promote healthy living through "compact, walkable, vibrant, mixed-use neighborhoods," to reduce urban sprawl through strategic siting close to existing developments and with easy access to public transit, and to protect threatened species through minimal disturbance to wildlife habitat.

One Planet Communities—a global network aimed at reducing humanity's footprint to a sustainable (i.e., one planet) level—purports to endorse the Earth's "greenest neighborhoods." The program delineates the following 10 principles for achieving sustainable community development: zero carbon, zero waste, sustainable transport, local and sustainable materials, local and sustainable food, sustainable water, natural habitats and wildlife, culture and heritage, equity and fair trade, and health and happiness. To date, it has officially recognized four such communities, including One Brighton and One Gallions in the United Kingdom, Mata de Sesimbra in Portugal, and Sonoma Mountain Village in California. In the case of the latter, Codding Enterprises and BioRegional North America aim to redevelop a 200-acre site (40 miles north of San Francisco) into a neighborhood that will be zero carbon and zero waste by 2020.

Greening Existing Suburban Communities

Notwithstanding extensive critiques, the suburbs remain the predominant land use pattern for new developments and, as such, a critical environmental challenge. Some argue suburban land use is incapable of natural succession—that its grid is not readily scalable—but others argue that density and diversity, self-sufficiency and sustainability, exist along a continuum, and that transformation is always possible. Several examples of progress are noteworthy in this regard.

Dan Chiras and Dave Wann detailed several potential options for retrofitting suburban development in their 2003 work *Superbia! 31 Ways to Create Sustainable Neighborhoods*. Changes range from traditional examples of creating more sustainable neighborhoods by sponsoring potlucks, carpooling, and organizing community events to bolder suggestions such as reclaiming shopping malls to create mixed-use spaces, tearing down fences and planting edible landscapes, establishing community orchards, creating community energy systems, and retrofitting garages into spare apartments. Other suburban visionaries suggest replacing single, detached homes with a cluster of townhomes; building workplaces into front setbacks to effectively urbanize the suburbs; and reclaiming the private space of defunct shopping malls to create new town centers featuring mixed residential, recreational, and commercial uses.

See Also: Adaptive Reuse; Density; LEED (Leadership in Energy and Environmental Design); Smart Growth; Walkability (Pedestrian-Friendly Streets).

Further Readings

Chiras, Dan and Dave Wann. *Superbia! 31 Ways to Create Sustainable Neighborhoods.* Gabriola Island, British Columbia, Canada: New Society, 2003.

Duany, Andrés, et al. *The Rise of Suburban Sprawl and the Decline of Nation.* New York: North Point, 2000.

Greene, Gregory and Barrie Silverthorn. *The End of Suburbia: Oil Depletion and the Collapse of the American Dream.* Toronto: Electric Wallpaper, 2004.

Kunstler, James Howard. *The Long Emergency: Surviving the End of Oil, Climate Change, and Other Converging Catastrophes of the Twentieth Century.* New York: Grove/Atlantic, 2006.

Leccese, Michael and Kathleen McCormick, eds. *Charter of the New Urbanism.* New York: Congress for the New Urbanism: McGraw-Hill Professional, 1999.

One Planet Community. http://www.bioregional.com/oneplanet (Accessed June 2009).

Peters, Adele. "Sonoma Mountain Village: Is Green Suburbia Possible?" http://www .worldchanging.com/archives//009448.html (Accessed June 2009).

Pratt, Geraldine. "Suburb." In *The Dictionary of Human Geography,* R. J. Johnston, et al., eds. Oxford: Blackwell, 2000.

U.S. Green Building Council. *LEED for Neighborhood Development.* 2008. http://www .usgbc.org/DisplayPage.aspx?CMSPageID=148 (Accessed June 2009).

Patricia Ballamingie
Carleton University

Sherrill Johnson
Independent Scholar

Green Jobs

A bio-energy scientist conducts crop research on switchgrass at this U.S. government laboratory.

Source: Roy Kaltschmidt/Berkeley Lab, U.S. Department of Energy

In recent decades, global economic growth has been fueled largely by manufacturing industries. Technology advancement in manufacturing has improved production capacity in terms of output and connected the world through efficient supply chain processes. The benefits of global economic growth through an industrial economy are indeed far reaching, providing opportunities for increased quality of life for millions, particularly in developing nations.

The boom of an industrial economy does come at a price. Most manufacturing requires extensive use of fossil fuels to operate efficiently. Research strongly suggests that the use of fossil fuels produces excessive carbon emissions, resulting in significant environmental degradation including climate change. Beyond the effect of growing carbon emissions on the environment, the industrial economy produces excess waste that threatens the Earth's ecosystems. The increased energy usage and excessive waste discharge from the industrial economy has created an urgent need for change across the globe to protect the Earth's natural resources for future generations.

In recent years there has been a movement to create so-called green jobs to address the issues attendant to an industrial economy. The broadest view of green jobs includes those that facilitate any process important to the future of our environment, such as those that conserve energy, reduce natural resource usage, and manage or avoid the generation of waste and pollution. Governmental sources present a similar view, categorizing green jobs as those that provide products and services that use renewable energy resources, reduce pollution, conserve energy and natural resources, and reconstitute waste. Many believe that any job supporting a process focused on sustainability of our natural resources should be considered green.

Despite the various interpretations, there is general agreement that green jobs are those that improve environmental quality and build sustainable enterprises that do not harm the Earth's ecosystems. To date, reference to green jobs includes multiple industries, from manufacturing to agriculture, and includes any focus on improving the environment. Many in government and in the private sector believe these green jobs will launch the next revolution in the global economy.

It is important to note, however, that an inclusive notion of green employment makes measuring the specific impact on the environment a challenge. For example, most would agree that a firm dedicated to installing solar panels is considered to have green jobs. However, some might question the jobs at the solar panel manufacturing plant if the plant itself releases carbon emissions from the use of fossil fuels.

As concern for the environment has grown, so has the interest in measuring the specific impact certain jobs have on the Earth's natural resources. Early environmentalists thought of "end of the pipe" jobs as green, which included only those that focused on cleaning pollution after it had been produced. More recent interpretations view production, distribution, sale, and disposal as a holistic process, attempting to track the full effect on the environment from a systems perspective.

Defining what constitutes a green job, then measuring the impact on the environment of such jobs, continues to be a challenge. Most agree that new industries exclusively dedicated to improving the environment are relatively easy to identify and measure. However, the impact from existing industries that integrate new technologies to reduce their energy consumption or recycle waste output is difficult to quantify. In those instances, "green" is relative to a previously employed process versus a well-defined standard. One approach to defining green jobs is to measure the impact of the industry in terms of its overall impact on the environment.

See Also: Carbon Footprint; City Politics; Ecological Footprint; Environmental Planning.

Further Readings

American Solar Energy Society and Management Information Services, Inc. "Defining, Estimating and Forecasting the Renewable Energy and Energy Efficiency Industries in the U.S. and in Colorado." (December 2008). http://www.greenbiz.com/research/report/ 2009/01/15/defining-estimating-and-forecasting-renewable-energy-and-energy-efficienc (Accessed December 2009).

United Nations Environment Programme. "Green Jobs: Towards Decent Work in a Sustainable, Low-Carbon World." (September 2008) http://www.ilo.org/global/What_we_ do/Publications/Newreleases/lang--en/docName--WCMS_098503/index.htm (Accessed December 2009).

U.S. Environmental Protection Agency. "State Climate and Energy Technical Forum on Clean Energy Workforce Development: Growing Green Jobs to Achieve Climate and Energy Goals." (February 2009). http://www.epa.gov/RDEE/energy-programs/state-and-local/state-forum_date.html (Accessed December 2009).

Timothy P. Keane
Saint Louis University

Green Landscaping

Green landscaping is an emerging landscape design approach that, in contrast with conventional landscaping, focuses on environmentally-friendly principles. Conventional landscaping represents any alteration of visible features of land, in various scales, and with different management concerns; for example, daily gardening work for a homeowners' backyard, a poolside garden in a holiday resort, or an urban park in a downtown area with ornamental and recreational facilities. Landscaping has traditionally been designed to enhance the visual beauty of the surroundings from a human's point of view. However, some conventional landscaping, gardening routines, or practices posing heavy resource demand impair ecosystems or have other negative effects.

Principles of green landscaping can be concluded from perspectives of design and maintenance. They include the following:

- Mimicking the natural processes in gardens and urban green spaces for both maximum environmental (promoting ecological diversity and the habitat's connectivity in urban areas) and economic efficiencies (saving time, resources, and money) in a sustainable manner
- Selecting appropriate plants (preferably native plants), and using water efficiency (e.g., xeriscaping) and conservation measures to consciously reduce maintenance and water requirements, as most native plants can thrive with minimal care
- Avoiding massive clear-cutting of indigenous plants when planning a landscaping project, and incorporating and preserving original landscape characters and their seasonal interest and "sense of places" as much as possible (e.g., preserving mature trees, terrains, bodies of water, etc.)
- Using recycled materials thoroughly, such as mulch, other soil amendments, and landscape construction materials (e.g., using recycled plastic bender board, pieces of used concrete, paving stones from abandoned or old gardens, or recycled brick)

Traditional landscaping is conducive to environmental degradation, including greenhouse gas and volatile organic compound emissions by lawn and garden equipment, water pollution by fertilizers and pesticides, increased chance for flooding because of erosion, habitat loss originated from invasion of exotic plants, and consumption of natural resources. The traditional approach emphasizes a greening effect by adopting short-lived flowering plants, transplantation of mature trees, and massive turf—this kind of design often creates a monotonous landscape that requires intensive maintenance and thus depletes resources.

Green landscaping has several synonyms, including sustainable landscaping, ecofriendly landscaping, and natural landscaping. As a rule of thumb, green landscaping encourages careful sustainable design with natural landscaping. It can bring the feel of wilderness to urban environments and attract birds, butterflies, and other animals. Such landscapes can sustain their own microecosystems once established. Native plants do not need extensive fertilizers, herbicides, pesticides, or watering.

William Thompson and his colleagues wrote a comprehensive guidebook on how to construct outdoor, built environments based on principles of sustainability, and they put forward 10 principles for practitioners to follow and shed insights to landscape architectural practices:

1. Keep healthy sites healthy: Incorporate a baseline survey as an essential and integral part of planning, avoid utility damage by easement and trenches, minimize the adverse degradation of the site during constr]uction and preserve healthy topsoil, save existing trees, and use appropriate construction machinery

2. Heal injured sites: Restore landscapes structurally, restore damaged soils on-site, and restore regionally appropriate vegetation

3. Favor live and flexible material: Hold sloped land in place with biotechnical erosion control, make vertical structures habitable with green walls, turn barren roof spaces into green roofs, and construct for and with plants

4. Respect natural water sources: Restore natural wetlands, rivers, and streams; collect and conserve water; irrigate intelligently and sparingly; reuse greywater; purify water at every opportunity; and let constructed wetlands treat water

5. Pave less: Plan and design to reduce paving by taking advantage of context-sensitive road design, use techniques that reduce runoff from paving, use porous paving materials, and cool asphalt with planting and albedo

6. Consider the origin and fate of materials: Reuse and use local salvaged or recycled materials, evaluate environmental costs when choosing suppliers, avoid toxic and nonrenewable materials, become familiar with toxicity issues by material type, and avoid polyvinyl chloride, preserved wood (wood preservatives), and fertilizers (heavy metals, toxic chemicals, and radioactive waste)

7. Know the costs of energy over time: Understand how landscape energy use is different with the use of different machines, tools, and labor; use life-cycle costing to justify sustainable design; and apply guidelines for landscape energy conservation

8. Respect the need for darkness and use lighting efficiently: Use lighting only if necessary, try low-voltage lighting for flexibility, and evaluate lamp performance and adopt light-emitting diode lighting

9. Quietly defend silence: Push for quieter landscape tools and modify pavement to reduce road noise

10. Maintain to sustain: Know the resource costs of traditional landscape maintenance, switch to bio-based maintenance products, apply integrated pest management, use fertilizer sustainably, use on-site resources, consider alternatives to mowing, evaluate life-cycle costs of maintenance options, and coordinate design, construction, and maintenance

Green landscaping is not only applicable to backyards, public parks, or commercial resorts; it also provides substantial improvement to environmental quality along roads and highways, golf courses, and bridges. Golf courses and roadside environments are conditioned with composts to control erosion and facilitate revegetation. Some parks use recycled tires and plastic lumber in pathways, playground surfaces, and flooring tile.

The following four examples illustrate some applications of green landscaping practices in the United States:

- Los Angeles County, California, has recently discovered the beauty and utility of old tires. Plagued by 10 million waste tires a year, the county has successfully saved landfill space by recycling tire rubber into pathways, playground surfaces, and flooring tile throughout its district. In two state parks in Indiana and Minnesota, structures made from recycled milk jugs are proving to be an environmentally preferable—and economic—alternative to wood products.
- In two Environmental Protection Agency facilities in North Carolina and Colorado, innovative landscaping techniques conserve water while helping the facilities meet federal environmental goals. The need for irrigation has been reduced by using water-efficient designs and native plants suited to local conditions. A more self-sustaining landscape was created by preserving existing trees, by converting an acre of irrigated turf into a wildflower meadow or drought-tolerant perennial bed, and by reducing pesticide use. These measures could reduce water usage by around 50 percent and save more than 650,000 gallons of water and $1,600 to $1,900 per year over the life of the project. Instead of nonnative ornamental grass and plants that need extra water to cope with summertime conditions, the design uses native plants, which are accustomed to hot summers. Native plants require little irrigation or maintenance—in fact, there are no water sprinklers on the entire campus.
- Texas struggled with roadside erosion. Erosion of roads is generally caused by rainwater flowing near paved surfaces. This problem can be prevented by surrounding the pavement with vegetation, but if vegetation does not take root immediately after road construction, the soil will wash away, and plant growth will become extremely difficult. Repeated attempts at revegetation are timely and costly and often do not work because of poor soil conditions. The Texas Department of Transportation found that compost helps prevent the erosion of roads by helping vegetation grow quickly. Because compost also has the ability to absorb large amounts of water, it has experienced a reduction in surface runoff and soil loss and increased rates of percolation. Using compost saves the state money through lowered construction expenditures and avoids future erosion maintenance.
- The Queens Botanical Garden in New York City merges new technology with natural elements to reduce the environmental footprint of its newest buildings. Its building design incorporates components like a green roof, composting toilets, and a rainwater harvest system to lessen the effect of its construction and operations. Landscaping is also an important component of the green design at the garden. Horticultural staff, trained in sustainable landscaping, will plant woodland, wetland, and prairie plant communities that contribute to the stormwater management system. The staff will incorporate native plants that are adapted to the local environment, which will reduce the need for watering. After the completion of the construction, the garden will move to total organic management of its gardens, transitioning from little to no synthetic pesticide use. The garden also uses an in-house system called "smoothing" to make sure different departments are informed of the correct procedures to follow in certain situations during the construction process. This system includes providing information ranging from how to dispose of materials in environmentally responsible ways to how to purchase recycled-content supplies. Through the smoothing system, employees at all levels appreciate the value of the environmental features of this construction project and are helping to recycle more than 90 percent of the waste generated during construction. They also follow a recycling program for everyday operations.

Green landscaping provides a momentum for concerned industries to innovate new solutions and cost-effective ways for sustainable landscape designs. This involves a complete paradigm shift in designing and managing the landscaping projects or homeowners' gardens. Efforts and initiatives from government, business sectors, nongovernmental organizations, interest groups, and community organizations are essential to inform all interested parties about the latest available technology and options that are under development or commercially available, and also to showcase successful designs and management practices. Genuine benefits and positive changes from green landscaping can only be captured with solid and accurate information flow, an exchange and sharing of ideas, and convenient platforms for promoting new ideas.

Green landscaping is a trend flourishing worldwide with ample market opportunity. It remains important to distinguish and to avoid greenwashing and to report inconsistencies in practices by public entities and companies.

See Also: Green Design, Construction, and Operations; Greywater; Habitat Conservation and Restoration; Xeriscaping.

Further Readings

Greenacres, U.S. Environmental Protection Agency. "Innovative Landscaping Techniques at Two EPA Facilities." http://www.epa.gov/epawaste/conserve/rrr/greenscapes/projects/epa-fac.htm (Accessed March 2009).

Greenacres, U.S. Environmental Protection Agency. "Landscaping With Native Plants." http://www.epa.gov/greenacres (Accessed February 2009).

Greenacres, U.S. Environmental Protection Agency. "Queens Botanical Garden Uses Innovative Methods to Promote Sustainable Design." http://www.epa.gov/epawaste/conserve/rrr/greenscapes/projects/qbg.htm (Accessed March 2009).

Greenacres, U.S. Environmental Protection Agency. "Texas Roadside Composting." Available online at http://www.epa.gov/epawaste/conserve/rrr/greenscapes/projects/tx_road.htm (Accessed March 2009).

Greenacres, U.S. Environmental Protection Agency. "These Tires Were Made for Walkin' Los Angeles Recycled Tire Parks." http://www.epa.gov/epawaste/conserve/rrr/greenscapes/projects/la_tires.htm (Accessed March 2009).

Green Living Ideas. "Latest Trends in Eco Landscaping." http://greenlivingideas.com/landscaping/latest-trends-in-eco-landscaping.html (Accessed February 2009).

Thompson, J. William, Sorvig Kim and Craig D. Farnsworth. *Sustainable Landscape Construction: A Guide to Green Building Outdoors,* 2nd Ed. Washington, D.C.: Island Press, 2008.

Caroline Man Yee Law
University of Hong Kong

GREEN PROCUREMENT AND PURCHASING

Green procurement and purchasing (GPP) could be broadly defined as procurement and purchasing that are consistent with the principles of sustainable development, such as ensuring a strong, healthy, and just society, living within the environmental limits, and promoting governance. According to C. McCrudden, by doing so, governments participate in the market as purchaser and at the same time regulate it through the use of its purchasing power to advance conceptions of sustainable development and social justice.

Green procurement and purchasing is extremely important—recent studies have shown that public expenditures can account for 45 percent of Gross Domestic Product (GDP) among developed countries, and government-driven consumption of products and services can reach 25 percent of total GDP among Organisation for Economic Co-operation and Development (OECD) countries, although with considerable variation in the level and composition of expenditure, according to A. Afonso et al., and the OECD. Even though these figures could be overestimated by the applied methodology, it shows the importance and economic relevance of the public sector, and consequently public procurement and purchasing for national, regional, and global economies. That being said, the magnitude of government purchasing creates an enabling background for environmental policy.

In fact, portions of the public sector worldwide are applying environmental criteria in their purchasing decisions. Such greener procurement and purchasing policies are a reflection not only of an increasing concern about the effects that purchasing decisions have on the natural environment, but also part of a guiding principle that the public sector should pursue practices that are coherent with those that it recommends and enforces to other economy actors, such as private enterprises and households, according to the OECD. These measures and increased concern with the environment lead to the belief that governments have to take the lead by improving their own purchasing habits.

Implementing Green Procurement and Purchasing

In order to implement GPP policies, countries have been applying a number of different strategies and types of instruments. GPP policies have the power to correct institutional deficits and improve performance, leading to lower purchasing costs and higher environmental quality that improves government's overall efficiency and impact a broad range of environmental issues, according to the OECD. These instruments depend on the nature of the goods and services concerned, and the environmental impacts that are to be mitigated through the GPP strategy.

These practices can vary widely, and can either be voluntary or mandatory in nature. But many countries currently require that purchasing guidelines of particular products contain a minimum amount of recycled content or achieve specific levels of energy. Increasingly, these guidelines are promoting the use of biobased or organic products, biofuels, clean sources of energy, water conservation, and incorporate fewer polluting technologies, especially in developed countries.

Based on a review of these instruments, the OECD states that they usually include:

- information-based tools to provide environment-related information to procurement officers and agencies: catalogues, environmental criteria, databases, questionnaires, and life-cycle assessment methodologies;
- training and communication tools to increase awareness among procurement officers and agencies: courses for procurement officers, networks, conferences, websites, and newsletters;
- accounting and financial tools to better reflect environmental characteristics of products when choosing between goods and services: life-cycle costing or value-for-money methodologies and spreadsheets, third-party financing, methodologies to quantify external costs, price preferences; and
- standards and directives introducing performance-based or technology-based in goods and services: minimum recycled content or energy-efficiency standards.

GPP is not a stand-alone policy tool, and is considered a complement to other policies. They can also affect and be affected by other environmental policies that are already in place. This is particularly important if one considers that environmental policy objectives will affect operations that fall within the responsibility of other administrations. Therefore, there is the need for policies and strategies aimed at integrating agencies and departments by modifying their competences and institutional missions. This means that various government organizations responsible for the implementation of different policy objectives must internalize environmental priorities, and that environmental authorities must ensure maximum efforts and assistance to facilitate the implementation of environmental management and environmental procurement initiatives.

In many cases, GPP is implemented in the context of reforms to public expenditure management systems in which environmental concern can coexist with an improved public expenditure management. With respect to implementation, there are two common approaches:

- Decentralized and bottom-up, targeting a broad range of products for a wide number of purchasers
- Strategic and top-down, for targeted products that strive to not only advance environmental goals, but also to create strong market incentives

Objectives of Green Procurement and Purchasing

Although one of the primary objectives of GPP is to improve the environmental performance of the public sector. It is sometimes true that these policies can also have indirect impacts on the environmental aspect of the economy as a whole. These policies provide the link to the significant indirect effects of changes in the purchasing decisions made by the private sector. Governments often intend that GPP policies not only change their own purchasing behavior, but also influence decisions of other economic actors by sending clear signals to the marketplace, according to the OECD. The private sector may follow a government's example—government purchasing of green products may influence the perception of practicability or acceptability of the green product. In this case, government purchasing acts as a kind of certification for the product.

It may also set a "moral" example that enterprises and private purchasers may choose to follow. This can foster an environment that encourages private firms to develop new, greener products that will not only satisfy government purchasers but also attract private costumers. Thus, GPP can result in the development of products that are environmentally preferable to existing alternatives, although the OECD has highlighted that these mechanisms and feedback effects are still poorly understood.

In general, the private sector responds to changes in government purchasing. It may become greener if the government policy reduces costs of purchasing green products by encouraging innovation that creates new, greener products, or enables green suppliers to realize significant economies of scale, or increases the market acceptance of green products by demonstrating their commercial feasibility. On the other hand, if the government policy results in higher prices for green products (or low prices for "brown" products) private green purchases may not increase, or may even become "browner."

Approaches and Strategies

As a result, GPP has to be viewed in the broader context of public administration, including public budget systems, procurement law, trade commitments, and other relevant areas. That is why coordination between GPP and policy instruments in other areas is key to the implementation of successful GPP programs and policies. Several approaches can support management and ensure that desired diffusion of this coordination really occurs. The general rule is that all of these policies, instruments, and management systems should be made as simple as possible, in order to not be perceived as a new burden. On the contrary, it should lead to easy and rational solutions.

The European Union, for example, offers a comprehensive GPP Training Toolkit in 22 languages, available for download on their website. The toolkit offers an action plan, and legal and practical modules, designed for purchasing officers, in a broad range of product and service groups, advising on purchasing from furniture to construction, to office supplies and textiles. The U.S. government operates a Federal Facilities Environmental Stewardship and Compliance Asistance Center that provides links and information on the Environmentally Preferable Purchasing (EPP) program, Comprehensive Procurement Guidelines (CPG), ENERGY STAR, the Green Procurement Program (GPP), and the Federal Energy Management Program (FEMP). Their website also maintains links to environmental regulations and policy, and provides training and examples.

Many recommendations can be made for effective and optimal GPP strategies to achieving environmental, economic, and market-oriented goals. These usually include the following:

- Commitment of senior management officials, as their leadership is essential in emphasizing the importance of GPP and in making it a priority for busy managers and procurement officers
- Mandatory GPP requirements are also very effective, when coupled with other measures such as accountability, monitoring and reporting requirements
- Simple and concrete GPP tools are essential to provide guidance to procurement officers
- Measurable targets, indicators and public reporting are key, to enable comparison across different public institutions
- Pilot projects foster creativity, due to their limited scope, and serve to demonstrate the effectiveness of cutting-edge technologies

These elements are important in understanding the extent to which governments can incorporate environmental standards and criteria in their procurement laws and procedures. Although there are still institutional barriers to overcome in many countries, GPP is a reality worldwide and has the potential to be further increased.

See Also: Agenda 21; Sustainability Indicators; Sustainable Development.

Further Readings

Afonso, A., L. Schuknecht and V. Tanzi. "Public Sector Efficiency: An International Comparison." *Public Choice*, 123/321–47 (2005).

Arrowsmith, Sue and Peter Kunzlik. *Social and Environmental Policies in EC Procurement Law: New Directives and New Directions*. New York: Cambridge University Press, 2009.

European Union. "Toolkit on Green Public Procurement." http://ec.europa.eu/environment/gpp/toolkit_en.htm (Accessed December 2009).

Hens, Luc and Bhaskar Nath. *The World Summit on Sustainable Development: The Johannesburg Conference*. New York: Springer, 2005.

McCrudden, C. "Using Public Procurement to Achieve Social Outcomes." *Natural Resources Forum*, 28/4:257–67 (2004).

Organisation for Economic Co-operation and Development (OECD). *The Environmental Performance of Public Procurement*. Paris: OECD Publishing, 2003.

Organisation for Economic Co-operation and Development (OECD). *Greener Public Purchasing: Issues and Practical Solutions*. Paris: OECD Publishing, 2000.

United Nations Environment Programme (UNEP). "Agenda 21." http://www.unep.org/Documents.Multilingual/Default.asp?DocumentID=52 (Accessed December 2009).

U.S. Government Federal Facilities Environmental Stewardship and Compliance Asistance Center. http://www.fedcenter.gov/programs/buygreen (Accessed December 2009).

World Summit on Sustainable Development. "Plan of Implementation on Sustainable Development. New York: United Nations," 2002. http://www.un-documents.net/jburgpln.htm (Accessed December 2009).

Rafael D´Almeida Martins
University of Campinas

GREEN ROOFS

A green roof is the roof of a building that is covered with vegetation and soil or another growing medium. Other layers such as a waterproofing membrane, a root barrier, or drainage and irrigation systems are often included as well. Sometimes referred to as rooftop gardens, ecoroofs, or vegetated roofs, green roofs are becoming increasingly popular for their environmental, economic, and aesthetic benefits.

The benefits attributed to green roofs are numerous. The first is energy efficiency. By providing a layer of insulation and absorbing the heat of the summer sun, green roofs have been found to lower energy requirements for cooling and heating buildings significantly. Green roofs can reduce the urban heat island effect and can reduce air pollution, as the vegetation filters both particulate matter and carbon dioxide. Another benefit of green roofs is that they reduce a city's impervious surface area. By absorbing rainfall, green roofs can help control storm water runoff, which in combined sewer systems is typically discharged with untreated

The green roof of Chicago's city hall was planted in 2000 to test the effects of such roofs on air quality and temperature. It contains over 100 species of shrubs and vines, mostly prairie plants native to the Chicago region.

Source: TonyTheTiger/Wikipedia

wastewater into receiving water bodies. A green roof can also protect the manufactured roof of a building from the damaging effects of sun and harsh weather, increasing its life span significantly. Aesthetically, green roofs increase green amenity space in urban areas and beautify city buildings. Finally, green roofs can provide habitat for urban wildlife, thereby maintaining and improving biodiversity within cities.

Types of Green Roofs

There are two main kinds of green roofs: intensive and extensive. The difference depends largely on the amount of maintenance they need and the depth of the planting medium used. Intensive green roofs tend to have deeper planting media as well as contain complex irrigation systems. Intensive green roofs generally include a broad range of ornamental plant species and large plants such as trees and shrubs, and they may have ponds and other decorative features. Intensive roofs are the most expensive kind of green roof. The roofs they are built on must be engineered to accept the weight of deep soil and large plants. Their need for ongoing maintenance also adds significantly to the long-term labor costs of the roof. Intensive green roofs have a greater emphasis on aesthetics than extensive roofs and are usually designed for public access and recreational use.

Extensive green roofs, which have lower construction and maintenance costs than intensive roofs, are the more common kind of green roof. They generally have a thin, lightweight growing medium, which lowers the weight load that the building must handle, allowing this type of green roof to be built on a larger variety of structures. The lightweight planting media, along with a lack of irrigation other than rainfall, requires that the plants on these roofs be chosen for their hardiness and drought-resistant properties, such as sedum. These features, along with ease of maintenance, are generally considered more important than the plants' aesthetic appeal. Extensive green roofs are designed to be virtually self-sustaining and to require only a minimum of maintenance. They are generally accessed only for maintenance and not for recreational purposes.

A third kind of roof that is sometimes added to typologies of green roofs is the containerized roof. Containerized roofs are made from numerous shallow containers filled with planting medium and placed side by side to cover some or most of a roof. The containers can be filled with hardy plants similar to those found on extensive green roofs, but they are often oriented toward growing vegetables or herbs in an urban setting. This type of green roof is popular for its relative ease and affordability of installation. However, as these roofs do not offer many of the benefits of intensive or extensive green roofs, such as building insulation, many experts do not consider containerized roofs to be true green roofs.

The initial cost of installing an intensive or extensive green roof can be several times more than installing a conventional roof. However, the long-term savings of green roofs can compensate for this higher upfront cost. Green roofs can save money through reduced energy use and by doubling or tripling the life span of the roof on which it is built. Green roofs can save cities money by reducing the pressure of storm water runoff on their often-aging infrastructure, thereby reducing replacement and building costs.

Not a New Idea

Although they have gained in popularity over the last several years, planted roofs are not new. The Hanging Gardens of Babylon, the most famous ancient green roof, were planted in the 7th century B.C.E. In Scandinavia and Iceland, people have used sod roofs to insulate homes from the extreme cold for centuries. In modern times, Germany began promoting green roofs in the 1960s and is often credited with developing many of today's advanced green roof techniques. Germany has more green roofs than anywhere else, with over 10 percent of the roofs in the country covered by vegetation. In Germany, a combination of tax incentives and regulations is used to encourage, and sometimes require, that new and existing roofs be green. Likewise, Switzerland has a national law requiring green roofs that also gives cities significant room for interpretation and flexibility. Many city governments across Europe offer combinations of tax incentives and regulations that are increasing the popularity of green roofs, and numerous European countries have active nongovernmental organizations working to promote green roofs. As such, green roofs have become much more common in Europe than in North America.

Although North America continues to lag behind Europe, this may be starting to change. In Canada, green roofs are beginning to grow in popularity and can now receive government money from funds dedicated to energy efficiency. In the United States, efforts to promote green roofs remain primarily local. The green roof on top of Chicago's City Hall is probably the most famous in the country. This semiextensive roof, which the city calls a rooftop garden, was planted in 2000 as a city demonstration project to test the effects of green roofs on air quality and temperature. It consists of over 100 species of shrubs and vines, most of which are prairie plants native to the Chicago region. Chicago is also home to what is generally considered the world's largest green roof. The city's Millennium Park, completed in 2004, is a 24.5-acre intensive green roof built on top of a parking lot and an underused train yard. Numerous other cities across the United States are also actively working to promote green roofs.

The increasing popularity of green roofs, as well as the growing willingness of local and national governments to provide the combinations of regulation and incentives necessary to encourage property owners to install them, are raising the visibility and the acreage of green roofs. It remains to be seen, however, whether green roofs can gain broad-enough acceptance and popularity to have a significant effect on energy use and water pollution in the world's cities.

See Also: Air Quality; Biodiversity; Heat Island Effect; Water Pollution.

Further Readings

Carter, Timothy and Andrew Keele. "Life-Cycle Cost-Benefit Analysis of Extensive Vegetated Roof Systems." *Journal of Environmental Management,* 87/3 (2007).

Dunnett, Nigel and Noel Kingsbury. *Planting Green Roofs and Living Walls.* Portland, OR: Oregon Timber, 2004.

Getter, Kristin L. and D. Bradley Rowe. "The Role of Extensive Green Roofs in Sustainable Development." *HortScience,* 41/5 (2006).

Lawlor, Gail, et al. *Green Roofs: A Resource Manual for Municipal Policy Makers.* Canada Mortgage and Housing Corporation, May 2006.

Corina McKendry
University of California, Santa Cruz

GREYFIELD DEVELOPMENT

Greyfield developments are adaptive reuses of former commercial or light-industrial properties in which remediation of hazardous wastes is not a financial barrier (or perceived
barrier) to redevelopment. Such developments can contribute positively to sustainable
communities because they reuse properties and existing infrastructure, and thereby reduce
the consumption of virgin land. The genesis of greyfields, along with an analysis of the
benefits and challenges of redevelopment, is outlined below.

The neologism *greyfield* parallels the terms *brownfields* and *greenfields* and is thought
to derive from the color of the vast asphalt parking lots surrounding obsolete strip malls,
declining regional malls, or vacated freestanding retail stores. Although the term is most
often used to refer to former retail properties, it may also be used more broadly to refer to
any parcel with nonresidential vacant or underused buildings unencumbered by the threat
of toxic substances and their remediation.

Greyfields originate as a consequence of suburban sprawl, when new commercial
development at the ever-expanding fringe of metropolitan areas supersedes older, smaller
developments in first- and second-ring suburbs, leaving behind a swath of outdated and
less visually attractive retail space. As population shifts to areas farther from the central
city, and as consumer tastes change, such retail spaces become progressively less
competitive and vulnerable to decline and abandonment. The Congress for the New
Urbanism estimated that in 1999, approximately 7 percent of the 2,700 regional or super
regional malls in the United States could be considered greyfield malls, with an additional 12 percent of all U.S. malls likely to fall into greyfield status by 2004. These
greyfield malls generate less income, have lower occupancy rates, and are smaller and
older (on average, 32 years old, with their last renovations on average 13 years ago)
than their counterparts in more affluent, newer areas farther from the center city. For
their communities, they represent potential gaps in the urban fabric: swaths of underused
land with declining tax revenues, visual blight, and diminution of services for nearby
residents.

Because of their large size and readily accessible locations (often near highway inter-
changes), greyfields present attractive opportunities for developers. Greyfields are also
attractive because infrastructure (water, sewer, electrical, communications services) is
already in place and because site work—grading, parking lots, and other site amenities—
has already been completed. Adjacent neighborhoods present a potential built-in market
for new uses.

For communities, greyfield development can also be appealing. It makes more efficient
use of infrastructure and municipal services, it increases the tax base, and it creates new
jobs. Greyfield development thus also has the benefit of reducing single-occupant vehicle
trips by clustering residential and other uses. It can revitalize surrounding neighborhoods
if care is taken to integrate the project into the existing urban fabric.

However, greyfield development is not without challenges. Cosmetic improvements to
facades or mall interiors, or even changes to the retail mix, are not sufficient to reverse the
decline of such properties, because newer and better retail opportunities already out-
compete these sites. Rather, developers must find a mix of new uses that are economically
viable and that will garner local support. Existing buildings must be able to be retrofitted

for these new uses while meeting current standards for the particular use, or must be demolished and replaced. "Big box" buildings, in particular, are not easy to retrofit because of their large footprints, lack of windows, and concrete block walls. Last, community support for redevelopment must not be taken for granted: Residents of surrounding areas may be apprehensive about congestion, overdevelopment, and changes in the character of their neighborhoods. For project developers, it is important to be in touch with community needs and desires and to put in place a planning process that respects neighborhood concerns and input.

Although much depends on unique site characteristics and market conditions, many successful greyfield developments have focused on a mix of uses including multifamily residential, commercial, medical office, entertainment, and often institutional uses at densities sufficient to create a critical mass that can support commercial uses and create a "sense of place" through pedestrian-friendly layouts and careful attention to building materials and design details. Redeveloping at higher densities than the original use also helps support public transit, and thus can align with community desires for more environmentally sustainable growth. In some projects, there is a desire to create a town center (in suburbs where one had not existed previously), and space is allocated for a town hall, library, community center, or even a school. Often, greyfield development relies on public–private partnerships to make projects economically viable; these may range from local support for or even local initiatives for rezoning, to density bonuses, to public financing of updated infrastructure, or even tax-increment financing.

Greyfield developments are often found in the U.S. south and southwest, where populations are growing and, consequently, suburban sprawl is increasing. Many of these developments show the influence of New Urbanist principles. Although space does not permit a list of more than a few examples, Milzner Park in Boca Raton, Florida, is considered to be an exemplary model of greyfield redevelopment. Other examples include City Center in Englewood, Colorado, and City Heights Community Center near San Diego, California.

See Also: Brownfields; Greenfield Sites; Infill Development; Smart Growth; Sustainable Development.

Further Readings

Bucher, David C. "Case Study: Greyfields as an Emerging Smart Growth Opportunity With the Potential for Added Synergies Through a Unique Mix of Uses." *Real Estate Issues* (Summer 2002).

Chilton, Kenneth M. "Greyfields: The New Horizon for Infill and Higher Density Regeneration." Louisville, KY: Center for Environmental Policy and Management, 2005.

Dunham-Jones, Ellen and June Williamson. *Retrofitting Suburbia: Urban Design Solutions for Redesigning Suburbs.* Hoboken, NJ: John Wiley & Sons, 2008.

Sobel, Lee S. and Steve Bodzin. *Greyfields Into Goldfields: Dead Malls Become Living Neighborhoods.* Chicago: Congress for the New Urbanism, 2002.

Judith Otto
Framingham State College

GREYWATER

In the average British or American home, as much as 60 percent of total wastewater produced is so-called greywater; that is, wastewater not of drinking quality, but not too heavily polluted with human or animal waste or industrial chemicals (i.e., not "blackwater"). With growing pressures on water supply caused by demographic change (urbanization and the rise in single-person households) and by climate change (which threatens a significant reduction in available freshwater resources for many parts of the globe), there is renewed interest in either reducing greywater production or finding new social and economic uses for it.

Before considering the technological issues attendant on wastewater reduction or reuse, a little historical context is important. Though it may seem entirely natural to us now, the use of large quantities of water for the removal of human wastes has not been with us forever. Indeed, the decision to flush away human waste with water was another Victorian contribution to world civilization. Following the cholera epidemics of the 1850s, politicians and health authorities made the connection between contact with human waste and contaminated drinking water supplies and embarked on a series of major public works that changed the face of London and, subsequently, other world cities. There is, of course, the famous story of Dr. John Snow cutting off a contaminated pump handle in London's Soho, with immediate reductions in cholera infection rates. Slightly later, Chief Engineer of the London Metropolitan Board of Works Joseph Bazalgette responded to crises such as the 1854 cholera outbreak and the "Great Stink" of 1858 to win approval for the construction of more than 1,000 miles of sewers to collect and transport London's "nightsoil" away from London. At about the same time, businessman Thomas Crapper began patenting and marketing indoor water closets for the convenient removal of human waste. The first public toilets had been installed at the Great Exhibition of London in 1851, but the idea caught on rapidly, and by 1900, there were more than 100 in the capital alone.

Together these and other urban innovators of the mid-19th century ushered in a "great sanitary awakening" that saw increasing volumes of water used not just for bodily waste removal but also for greater and greater levels of personal and domestic hygiene. Government authorities across the rapidly industrialized world even hired public health specialists whose job involved educating the poor, urban underclasses about washing their homes and themselves more often and to a higher standard. From this decade forward, human health was to become interlinked with the use of increasing volumes of water for washing bodies, homes, and (more recently) cars, and for removing our bodily wastes.

We have thus reached a situation whereby each and every European is consuming 60–70 liters of water and each and every American 200–250 liters each and every day for uses that do not necessarily make that water unusable for other purposes. It is also undeniable that our use of water for personal and domestic hygiene is grossly inefficient, with the average domestic shower unit consuming approximately 12–15 liters of water per minute—more if a shower pump is fitted. Similarly, modern clothes and dishwashing technologies often do not include water efficiency as a design criterion, consuming as much as 120 liters per cycle. The large differential between European and U.S. water consumption is the subject of continuing controversy, but it seems to be largely related to a higher propensity to have water-intensive sprinklers, high-capacity washing, power washers, and other technologies.

So we return, in the 21st century, to the problem of greywater. What can be done to reduce our production of greywater, and what can be done to possibly get more/better use out of the greywater that is unavoidable? It is an oddity of modern domestic architecture that we use the same drinking-water-standard water supply for all uses, whether this is efficient or not. Thus there is a rapidly growing trend, especially in Asian cities and parts of the U.S. Southwest, toward the installation of "purple pipe" systems, designed to carry water treated to a lower standard, including greywater, to appropriate points of use. Such systems mark a recognition that it is increasingly irrational to put the same water we drink into toilets and onto our gardens. The explosive growth of the bottled water market can also been seen in this light as part of the trend toward "twinning" our water supplies.

At the domestic level (and the relative contribution of domestic water consumption to overall water consumption varies widely in Europe and North America), attention is now turning to a twofold strategy of encouraging the installation of water-efficient personal and domestic hygiene and gardening appliances and the reuse or recycling of unavoidable greywater. Most such systems involve the collection of greywater from kitchens and bathrooms (but not toilets, which produce "blackwater") into a central reservoir or cistern for local reuse. Rainwater may also be incorporated into such systems, which can produce a considerable amount of water—as much as a 1,000 liters (1 cubic meter) each month. However, even greywater may contain harmful pathogens from washing or hygiene applications, plant nutrients from detergents (especially phosphates), and heat, as much of it is discharged warm or hot. Thus, greywater must either be used immediately or be treated and stored for later use. Agricultural or garden applications of greywater are possible but require careful management, as soil quality can be compromised and pathogens introduced into the garden where they may flourish. Similarly, there are now so-called intelligent toilet systems that can safely and efficiently use greywater, but again they require a step-change in thinking about domestic water management. The complexities and costs of the latter explain why greywater recycling is not terribly popular at the household level. Moreover, many jurisdictions and professional plumbing codes prohibit greywater reuse, considering it in the same treatment class as blackwater.

More promising are moves to address greywater reduction and reuse at the scale of larger developments or even local councils. Thus, for example, Global Water of Maricopa, Arizona, has been distributing treated "reclaimed" waters (black as well as grey) via purple pipe systems for applications such as golf courses, public landscaping, and even fire departments, and the city of Tucson, Arizona, is using treated wastewater to supply its popular (and aptly named) Sweetwater Wetlands. In the United Kingdom, complete water cycle studies, such as the Corby Water Cycle Study, offer a potential platform for the articulation of larger greywater treatment systems that might also be integrated with natural landscapes or ecosystem restoration of groundwater recharge schemes.

See Also: City Politics; Water Conservation; Water Treatment; Water Sources and Delivery.

Further Readings

Brix, H. "Use of Constructed Wetlands in Water Pollution Control: Historical Development, Present Status, and Future Perspectives." *Water Science and Technology*, 30/8:209–23 (1994).

Crettaz, P., et al. "Life Cycle of Assessment of Drinking Water and Rain Water for Toilet Flushing." *Aqua*, 48/3:73–83 (1999).

Environment Agency for England and Wales. "A Study of Domestic Greywater Recycling."
    London: Her Majesty's Stationery Office (HMSO), 2000.
Mcilwaine, Stephen and Mark Redwood. *Greywater Use in the Middle East: Technical,
    Social, Economic and Policy Issues*. Warwickshire, UK: Practical Action Publishing,
    Forthcoming.
Staddon, C. *Managing Europe's Water: 21st Century Challenges*. London: Ashgate, 2009.

Chad Staddon
University of the West of England, Bristol

GRIDLOCK

The term *gridlock,* coined by the Traffic Department in New York City in the 1970s, originally referred to a specific effect of traffic congestion in cities with a grid layout: Drivers with a green light could not fully pass through the intersection before their signal turned red, thereby blocking traffic in the cross direction from proceeding when the light turned green in the cross direction. The use of the term *gridlock* has subsequently been broadened to refer to any traffic congestion (including highways) in which stop-and-go movements or periodic complete stops are the norm rather than the exception, whether or not a gridded roadway network is involved. Gridlock imposes costs on human lives and the environment in many ways: It wastes fuel, adds to commuting times, and increases ambient air and noise pollution.

Cities like New York delineate crosshatched, "don't block the box" zones (visible here in the upper left corner) at traffic-prone intersections to battle gridlock.

Source: iStockphoto

Solutions for reducing gridlock fall into three general classes: increasing the capacity of the roadway network, reducing the number of vehicles in the network at any given time through transportation demand management strategies, and more broadly, working to reconfigure the relationship between land uses to reduce the number and length of vehicle trips. All of these classes of solutions have multiple approaches. The toolbox for increasing the capacity of the roadway network includes new vehicle lanes, double-decker lanes or tunnels in high-congestion areas, improvements at traffic signals to increase queuing capacity, automated toll collection systems, segregation of vehicles by type, and Intelligent Transportation Systems that automatically direct vehicles away from congested roadways. Because the first two of these are costly, complicated, and often politically controversial, transportation planners and engineers have more recently focused attention on the remaining solutions. Small-scale improvements to intersection layouts and adjustments to signal timing can result in improved traffic flow, especially if nearby signals are coordinated. According to U.S. Federal Highway Administration models, correctly timed signals may reduce fuel consumption by up to 13 percent. Another approach, automating toll collection systems, has the added benefit of reducing labor costs for transit authorities, in addition to reducing gridlock and travel time. Segregating vehicles by type can increase roadway capacity. Narrower lanes for automobiles, combined with use of center medians and shoulders, allow new lanes to be added without buying additional rights-of-way. This technique was used in Los Angeles in the early 1990s, giving rise to the term *L.A. squeeze.*

Intelligent transportation systems use information and communications technologies to manage vehicular traffic. Surveillance and data-collection devices such as pole-mounted cameras or pavement sensors collect real-time data on traffic speeds, traffic volumes, accident locations, and weather. This information can be processed and fed back to drivers, with continuously updated data for traffic conditions, travel time, and alternate routing. In its most futuristic application, intelligent transportation systems research groups are working to develop "driverless car systems," in which onboard computers can respond to changing conditions within microseconds, allowing the distance between vehicles traveling at highway speeds to be greatly reduced.

A second class of solutions for reducing congestion is managing roadway demand by reducing the number of vehicles in the roadway network at any particular moment. This can be achieved by shifting travelers out of automobiles into other modes of transit (buses, commuter rail, light rail, rapid transit, and bikeways), by reducing the number of single-occupant vehicles (through carpools, vanpools, or incentives like high-occupancy vehicle lanes), by shifting work hours or necessary trips to work (telecommuting, flextime arrangements), or by using market mechanisms to restructure the use patterns on roadways. This last option is termed *congestion pricing* and includes such options as higher tolls during peak hours, increases in parking fees, or automatic fees for driving within certain areas in center cities. Congestion pricing has been more widely adopted outside the United States than within it; examples include London, Stockholm, Ontario, and Singapore. One concern with congestion pricing is the question of socioeconomic equity: Those who are least wealthy, and therefore least able to afford congestion fees, are the least likely to have access to flexible scheduling to avoid rush-hour surcharges or high parking fees.

The third class of solutions begins with the premise that neither additional roadway capacity nor better utilization of it will ultimately eliminate gridlock as long as the predominant pattern of growth and development is suburban sprawl created by the segregation of sprawling land uses—single-family residential developments, campus-style industrial and office parks, and large-scale shopping centers and entertainment venues—and the need for personal vehicles (and their fossil fuels) to travel between them. Solutions focus on mixing small-scale land uses at medium-to-high densities to encourage pedestrian trips, using infill development as a way to take advantage of existing transportation infrastructure, and promoting even higher densities near transit stops and multimodal nodes in the transportation system. Some jurisdictions (Greater Ontario, Oregon) have instituted growth boundaries to force growth to occur at higher densities in center cities and to produce more integrated, efficient public transit systems. However, such smart growth initiatives are still the exception rather than the rule.

As can be seen from the foregoing, much of the debate around solving gridlock problems is predicated on the assumption that the prevailing pattern of suburban sprawl enabled by the automobile is here to stay. However, the emerging consensus seems to be that solutions to congestion are best developed at smaller, more localized scales, in response to particular local (and political) conditions. The one-size-fits-all, macroscale congestion solutions of past decades are no longer politically feasible or financially possible.

See Also: Air Quality; Carpooling; Congestion Pricing; Transit-Oriented Development; Transport Demand Management.

Further Readings

Arnott, Richard, et al. *Alleviating Urban Traffic Congestion.* Cambridge, MA: MIT Press, 2005.
Dunphy, Robert. *Moving Beyond Gridlock: Traffic.* Washington, D.C.: Urban Land Institute, 1997.
Federal Highway Administration (FHA). *Traffic Congestion and Reliability: Linking Solutions to Problems.* Washington, D.C.: FHA, 2004. http://www.ops.fhwa.dot.gov/congestion_report_04/congestion_report.pdf (Accessed August 2009).

Judith Otto
Framingham State College

HABITAT CONSERVATION AND RESTORATION

Habitat conservation and restoration are the processes and outcomes of actions taken to mitigate the detrimental ecological effects of urban development. They have roots in the conservation movement, but differ greatly from their predecessor. The conservation movement of the late 19th and early 20th centuries focused on preserving wild and scenic lands as refuges from the intensity of industrial urban areas. When nature was considered at all in relation to cities, it was generally as a recreational amenity, and a method of providing relief from the stress of urban life. Few visionaries saw the wider implications of the declining environmental quality in cities and the stress exerted by the resource demands of expanding urban populations and its effect on the health of natural systems in the surrounding areas. Fewer still made the connections to sustainable development or, ultimately, human habitation. The conservation movement did not shift its focus to cities until the middle of the 20th century, and even then the tendency was to create nature reserves outside of cities.

Despite the emergence of the new scientific field of ecology, the city was still viewed as somehow independent of nature, and laws adopted to protect the environment reflected that highly anthropogenic orientation. As conservation science has continued to advance and the pace of urbanization has continued, the necessity of habitat conservation and restoration has became increasingly apparent. Cities are reconsidering their relationship with their natural environments, returning to historical practices, and pioneering new ones.

A Matter of Scale

Habitat can be defined at various scales. Many of the earliest efforts at habitat conservation arose from concerns about declining bird populations in cities, and as such they defined habitat quite narrowly, often focusing on a single element (e.g., type of vegetation) in a single part of the species' range. A broader definition of habitat is that it is a physical space comprising the living and nonliving constituents upon which a particular organism, population of organisms, or interrelated groups or populations of organisms depend for survival. This space may be small or it may spread across continents. In practice, habitats are still often narrowly defined for a specific species (e.g., the Tennessee purple cornflower, or the golden capped fruit bat) or for a community of naturally occurring species (e.g., intertidal zones, or the blue gum high forest), but as our scientific understanding has grown, it has become increasing clear that no habitat is discrete, but is embedded in a highly complex hierarchy of systems. Our past gross misunderstanding of those relationships and of the role that biodiversity plays in maintaining the health of ecosystems has lead to habitat degradation or loss, either directly, or through the process of fragmentation.

Habitat fragmentation occurs when development introduces breaks in the physical fabric of a habitat. Where development is widespread, as in large urban agglomerations, only remnants of the original habitats may remain. Even in less intensely developed areas, man-made barriers, such as leap-frog (discontiguous) urban development, agricultural operations, and infrastructure projects (e.g., highways, dams, and pipelines), can effectively isolate habitat areas from one another and from the hinterlands. Fragmentation can occur with or without incurring habitat loss and, depending on the species or communities present, fragmentation may be minimal, or it may be devastating. Whether through environmental degradation or fragmentation, habitat is considered lost when it no longer meets the physical requirements of an organism or species (e.g., providing ambient climate or nutritional needs) or has so greatly disrupted the reproductive cycle, migratory routes, or social systems of a species that it no longer supports the long-term survival of the organism, species, or community of species.

Habitat restoration is the process of repairing a damaged ecosystem. The success of any such effort depends on a good understanding of the constituent elements of the system or, at least, of the causes of its degradation. In urban settings, habitat restoration often involves removal of nonnative species that have been introduced to or somehow invaded an area to the detriment of native species. It may also involve pinpointing sources of pollution, cleaning up existing pollution, or preventing further discharges to allow the habitat to recover on its own. In other instances, it may involve reforestation or reintroducing flora and fauna that were historically present.

Habitat conservation and restoration efforts have generally been myopic, undertaken only after marked declines in populations of species of local interest have been identified, and often only in response to legal requirements, such as the U.S. Endangered Species Act (ESA). Proactive efforts were rare until the 1990s. Technological advances that facilitated a combination of satellite imagery, powerful data management tools, and vastly improved mapping techniques allowed conservation biologists to significantly advance the science of habitat conservation. The ability to model the spatial distribution of species at the ecoregional scale vastly improved methods of identifying systemic problems and identifying local hot spots, and provided the opportunity to respond to both earlier.

Regulatory Approaches to Habitat Conservation and Restoration

At the time it was adopted, the 1973 ESA was among the strongest environmental protection laws. It was also fraught with loopholes and hampered by the state of environmental science. Furthermore, it cast habitat conservation as a reactive process, set in motion by the listing of a species or community of species as threatened or endangered rather than as a strategic process aimed at maintaining a healthy ecosystem. The Habitat Conservation Plan (HCP) process added to the ESA in 1982 encouraged planning for larger areas, but still allowed habitat to be defined narrowly, and as a result early HCPs often were arguably ineffective in protecting the targeted species. Those early experiences eventually led to the preparation of HCPs of much wider scope. State and local governments now routinely prepare HCPs that cover hundreds of thousands of acres and scores of species. Since 2006, the U.S. Fish and Wildlife Service has been promoting, though not requiring, the development of strategic habitat conservation plans to cover entire bioregions. This may eventually lead to a national conservation strategy.

The European Union has already devised a conservation strategy for its member nations. The Habitat Conservation Directive was drafted subsequent to the Convention on Biological Diversity issued by the United Nations conference in Rio de Janeiro in 1992. It recognizes biodiversity as essential to sustaining the health of the world's ecosystems, and acknowledges the EU's responsibility in maintaining biodiversity. The directive is similar to ESA in that it focuses attention on threatened species. Unlike ESA, however, it establishes a network of protected areas that exist to support the ecological health of the continent by promoting and safeguarding biodiversity. The directive also addresses habitat restoration as an issue independent of development, which the ESA does not do. Member countries of the EU have subsequently adopted complementary strategies to promote biodiversity within their own borders.

Urban Habitat Conservation

Cities, either directly or through the demands they place on natural resources, have been largely responsible for the destruction of habitats and the local extinction of many species. Improvements in spatial distribution modeling clearly show these impacts at the global scale, and, together with a better understanding of how cities rely on ecological processes to clean the air, filter water, and support agriculture, this has caused a shift in the way urban planners think about habitat conservation and restoration in their own localities.

Many are reconsidering historical development patterns that incorporated large, open and natural spaces, and provided interfaces with adjacent cultivated lands and hinterlands, and are including these design elements in plans for new urban areas. They are also devising strategies to retrofit or rehabilitate these elements as cities redevelop. Strategic placement of parks, golf courses, and gardens is increasingly recognized as a way to improve their functionality in sustaining biodiversity. Correctly sited and sized, they can substitute for lost or degraded habitats (such as wetlands disturbed or supplanted by agriculture), or they can act as buffers to defend sensitive habitats from urban impacts. Similarly, landscaping requirements are being rewritten to take ecological considerations into account rather than merely reflecting aesthetic tastes. Growth management boundaries are being adopted, and development standards amended to limit the intrusion of urban uses into habitat areas. Many cities limit the use of impervious paving surfaces as part of their efforts to protect watersheds and reduce heat-island effects. Restoration ecologists would push the envelope further. They advocate designing cities to work synergistically with adjoining natural spaces. Promoters of the "ecological land-use complementation" approach see it as a more effective way of promoting biodiversity, ecosystem resilience, and sustainable development than the current means of protecting habitats in urban settings.

Trends in Habitat Restoration

Ecologists are also rethinking approaches to restoring degraded habitats, given an incomplete understanding of the complexities of these systems. Habitats often do not respond to restoration efforts as anticipated, and changes in local climates due to global climate change, often preclude returning a habitat to a pristine state. The current trend is to try to restore resilience to a habitat. Such efforts entail quantifying and monitoring conditions, discerning the feedback loops, and responding to manageable threats before they destroy the ecosystem. Restoration is no longer an end-state, but an on-going process responding to evitable changes while acknowledging that some changes are inevitable.

Conservation Advocacy

Advocacy groups have been instrumental in habitat conservation and restoration. They have often forced developers to conserve habitat by threatening legal action, but in other instances, they have protected habitats through the outright purchase of land or by establishing conservation easements, which allow some land use (e.g., grazing or farming), but prohibit intensive development of the land. Their efforts have been critical in habitat restoration, as they heighten public awareness, campaign for government funding for these efforts, or fund them themselves through private contributions.

Among the earliest conservation organizations to form in the United States were the Sierra Club, founded in 1892, and the Audubon Society, which formed in 1905. Like many groups that formed later, both started with rather narrow foci and later expanded their scope to the international level.

Conservation organizations are nearly as abundant as the species and habitats they seek to protect. Many were formed in the 1960s when the modern environmental movement was born. Among the larger and more influential of these groups are the Environmental Defense Fund, the World Wildlife Fund for Nature, and Friends of the Earth, which is a network of 77 environmental organizations.

See Also: Biodiversity; Bioregion; Ecosystem Restoration; Environmental Planning.

Further Readings

Beatley, Timothy. *Habitat Conservation Planning: Endangered Species and Urban Growth.* Washington, D.C.: Island Press, 1994.

Colding, Johan. "'Ecological Land-Use Complementation' for Building Resilience in Urban Ecosystems." *Landscape and Urban Planning*, 81 (2007).

Diamond, Jared. *Collapse: How Societies Choose to Fail or Succeed.* New York: Penguin Books, 2005.

Platt, Rutherford H., Rowan A. Rowntree and Pamela C. Murick, eds. *The Ecological City: Preserving and Restoring Urban Biodiversity.* Amherst, MA: University of Massachusetts Press, 1997.

Susan H. Weaver
Independent Scholar

Hamburg, Germany

Hamburg is Germany's second largest city, with approximately 1.8 million residents in its municipality and over 4 million residents in the greater metropolitan area. Hamburg is also home to Europe's second largest port, which for much of the 20th century was the driving force behind Hamburg's economy. In 2009 the European Commission selected Hamburg to act in 2011 as the second European Green Capital because of the city's strict commitment to addressing environmental issues and its farsighted plans for urban renewal. By 2005, Hamburg had reduced its per capita CO_2 emissions by 15 percent (in comparison to 1990 emission levels), and the city council set the ambitious target of reducing emissions by 80 percent by 2050. Hamburg is often considered to be a stellar example for best practices in combining economic and environmental sustainability and pursuing green urban development.

Hamburg's history dates back to the 9th century, and from its beginnings, Hamburg developed a strong seafaring economy due to its sheltered location along the river Elbe, with access to the North Sea. Hamburg was a major player in the medieval Hanseatic League before developing into a metropolitan port city in the 19th and 20th centuries. Due to its status as a free city within federal Germany, Hamburg remains a major trade and transport hub.

However, increased competitive pressure on the global and European market, as well as sociopolitical and environmental problems caused by rising unemployment and flood problems, present Hamburg with a host of challenges. First, the city has to decide whether it wants to proceed with a port extension, the socioeconomic cost-benefit ratio of which remains unclear and which will necessitate a further deepening of the Elbe River. Prior river construction measures led to a significant loss of habitats and natural flooding areas, and negatively affected the river's oxygen content and self-cleaning capacity. Second, the city is attempting to tackle the social problems of unemployment, violence, and social fragmentation by engaging in major urban renewal projects. The biggest of these projects is the Hafencity, which aims to extend the inner city by 40 percent through redeveloping 1.5 km of former port area into a metropolitan business and recreational area, offering space for up to 5,000 residential areas and office space for over 20,000 jobs.

In 2009 the European Commission elected Hamburg to act as European Green Capital in 2011 because of the city's exemplary efforts in environmental protection and sustainable urban development. The European Commission was particularly impressed with the city's clear goals, financial commitment, and public outreach strategy. Hamburg developed in the process of Local Agenda 21 a comprehensive sustainability strategy, called "Environmental Roadmap—Goals for a Sustainable Hamburg," which sets a goal of reducing per capita CO_2 emission (compared to 1990 levels) by 40 percent by 2020 and 80 percent by 2050. By 2005 Hamburg had already cut its per capita emissions by 15 percent.

The city's sustainability strategy allocates 25 million Euros per year for over 200 environment-friendly measures and initiatives, especially in the areas of energy consumption and transportation. Hamburg already offers an excellent public transport system and cycling network, however, ample room for improvements exists, as thus far a congestion fee has not been levied, and a significant amount of container traffic to and from the port occurs on the road, not on the rail. Other environmental measures taken by the city council include replacing old boiler and heating systems in public buildings with new eco-friendly versions (at an estimated cost of 18 million Euros), converting traffic lights to LED technology, and installing energy saving lamps and bulbs in the majority of city-owned buildings. Also, Hamburg city is committed to generating energy from renewable sources. Hamburg installed over 36,000 square meters of solar collectors, and the city provides a subsidy for bioenergy plants. Moreover, Hamburg runs successful waste collection, heat recovery, and water treatment schemes.

The city is also engaged in several major urban redevelopment programs, notably in the Hafencity project and the quarter (Stadtteil) of Wilhelmsburg. Wilhelmsburg is an old residential area, located on an island within the Elbe River, that suffers from a host of social problems (e.g., high crime rate) and is in dire need of urban redevelopment. All new building developments in these areas must follow strict ecostandards, as laid out in the city's Climate Protection Act, which, for example, forbids the installation of electric storage heating systems.

Another important feature of Hamburg is that it offers its citizens a fair amount of green spaces. Of the city's 755 square kilometers, 68 square kilometers are public green areas (such as parks), 34 square kilometers are city-owned wooded areas, and another 61 square kilometers are nature reserve areas, meaning that in total about 22 percent of the city is greenspace. Due to this high percentage, almost 90 percent of Hamburg's inhabitants have a green area accessible within 300 meters of their home. Additionally, Hamburg covers its water needs by using its own high-quality groundwater.

However, despite Hamburg's laudable efforts to sustainable urban development, the city's selection as European Green Capital also received a fair amount of criticism. While Hamburg did, for instance, reduce its per capita CO_2 emissions by 15 percent, the city's 2005 per capita emissions of 9.69t of CO_2 per annum is still 50 percent higher than Stockholm's per capita emissions in 1990. Moreover, the bulk of Hamburg's proposals and initiatives seem to be concerned with cutting public spending on heating and fuel costs, while neglecting issues connected with port traffic and river construction. Port traffic and river construction measures, however, are chiefly responsible for the loss of habitats and natural flooding areas, both of which present a serious environmental concern. Also—like most major cities—Hamburg accumulates a significant ecological debt, as the ecosystem services the city requires cannot be sustained through the city's own resources.

Overall, Hamburg allocates time and resources to addressing the challenge of environmental sustainability, and spends a significant amount of money on disseminating its programs and initiative, hoping that other cities will follow its example and invest in green urban (re)development. Hamburg's push for environmental sustainability is also a push for a socioeconomic restructure of the city, which means that it will reach its environmental political and economic goals only if it follows an integrated approach of sustainable urban management.

See Also: Green Infrastructure; Healthy Cities; Renewable Energy; Stockholm, Sweden.

Further Readings

Deutsche Welle. "Hamburg and Stockholm Named European Green Capitals." http://www .dw-world.de/dw/article/0,,4051170,00.html (Accessed January 2010).

European Green Capital Award. "Hamburg Application." http://ec.europa.eu/environment/ europeangreencapital/docs/cities/2010-2011/hamburg_application_collect.pdf (Accessed January 2010).

Grossmann, Iris. "Three Scenarios for the Greater Hamburg Region." *Futures*, 38 (2006).

Fabian Schuppert
Queen's University Belfast

HEALTHY CITIES

A balance between urbanization and social, public spaces is crucial to the viability of a city. The term *healthy city* is rooted in a vision of what a healthy city can become. A city is viewed as a complex organism that is living, breathing, growing, and constantly changing. A healthy city is one that improves its environments and expands its resources so that people can support each other in achieving their highest potential. It is essentially created through the design of healthy public spaces and urban environments that encourage sociability, vitality, efficiency, and sustainability. Though the term *healthy city* is relatively new in urban literature, it is based upon concepts that have been around for a long time. It does not stand alone but is situated in a complex array of relevant issues including improved physical infrastructure and service provision for much of the city population. This means ideally each house should be provided with piped water and connection to a sewer (or other good-quality provision for sanitation) and each residential area with paved pathways and access roads, electricity, storm water drainage, and open spaces. There should be regular collection of household wastes, nutrition programs for vulnerable groups, health promotion and comprehensive healthcare, and emergency life-saving services. There is also need for considerable sophistication and sensibility to detect major health problems and to design and implement appropriate interventions. Thus, the definition of "healthy city" encompasses the process of enabling people to take control over and to improve their health.

The concept of a healthy city, in broader terms, is the ability of the urban area and its region to continue to function at levels of quality of life desired by the community, without causing adverse impacts inside and outside the urban boundary and without restricting the options available to the future generations. It is more about improving people's quality of life, maintaining environmental sustainability, and providing social justice. The concept implies that all sectors of urban planning, such as housing, energy, water supply, sanitation, transport, industry, education, healthcare, and so on, must take into account the concerns for maintaining the urban ecology and to oversee that the indicators for overall development have intersectoral linkages that make way for the success of healthy cities. Therefore, the key issue is not about planning and building "healthy cities," but to understand how the existing form, government structure, production systems, consumption patterns, waste generation, health, and management systems are compatible with present and future development goals for the city.

The creation of healthy cities within an urban ecology has a number of implications for the structure and functioning of government along with a shift in our values to understand that economic growth and development is no longer the overriding social and political objective, but merely one objective that has to be balanced with other objectives such as sustainability, equity, livability, social cohesion, and environmental quality. Most action to achieve a healthy city has to be formulated and implemented locally. Local action needs political support and commitments from city authorities to reorient policies toward equity, health promotion, and disease prevention. In particular, five elements make up the strategic framework for achieving healthy cities:

- Promoting healthy public policy
- Creating supportive environments

- Strengthening community participation
- Improving personal skills
- Reorienting health services

Currently, there is a worldwide movement of bottom-up, community-driven "healthy cities program" in which local authorities in more than 1,000 cities have sought new ways to work with the many different actors and interests within their boundaries in the promotion of health and prevention of diseases. This general principle is expressed more specifically in terms of the explicit qualities that a healthy city should strive to achieve:

- A clean and safe physical environment of high quality
- An ecosystem that is stable in the present context and sustainable in the long term
- A strong, mutually supportive and non-exploitive community
- A high degree of participation and control by the public over the decisions affecting their lives, health, and well-being
- The meeting of basic needs (for food, water, shelter, sanitation, education, income, employment, and safety) for all segments of the city population
- A well-integrated structure of city government having coordination between municipal authorities and community-based organizations for microlevel planning and management
- Access to a wide variety of skills, experiences, and resources, with extensive contacts, interactions, and communication
- A diverse, vital, and innovative city economy
- The encouragement of a connectedness with the past, with the cultural and biological heritage of city dwellers, and with other groups and individuals
- A built-in form that is compatible with and enhances the preceding characteristics
- An optimum level of appropriate public health and sick care services accessible to all
- High health status

Healthy Alternatives for Cities

Cities, with few exceptions, stimulate a worldview that is based solely on the realities of those built environments and not of the larger ecological systems within which the cities themselves are sustained. Achieving progress to build healthy cities depends on a number of issues that must be taken into consideration within the aegis of the planning mechanism. Although planning in the specific sense is applied to individual developments, it also plays an integral role in merging individual developments into a cohesive whole intended to achieve any given policy goal, in this case, adoption of healthy cities programs. Such programs promote innovation and change in local health policy by advocating new approaches to public health. They explore effective ways to translate the principles and targets of the national health strategy into practice in urban settings. They also provide public health leadership through mechanisms that recognize and mediate the interests of different groups in the community. Therefore, the planks of such a precise agenda for making the transition to future healthy cities need urgent consideration of the following.

Maintaining Urban Environment Through Sustainable Consumption

From an environmental point of view, cities are the net consumers of natural resources and producers of all sorts of wastes. However, cities also have great capacity to be more resourceful. Urban design integrated with nature can both save resources and allow city dwellers to enjoy a more natural environment. It is estimated that a serious commitment to conservation could reduce our consumption by 30–40 percent while maintaining or even raising our standard of living and health status.

Meeting Economic Needs

The successful transition to healthy cities can be ensured only when the populace has access to adequate livelihood or productive assets. Most urban authorities have limited funds for capital investments at their disposal to improve their environmental performance. Privatizing public services can be a source of capital for investment by which long-term urban strategies and programs can be successfully implemented.

Commitment to Health and Healthy Public Policy

This means realizing healthy cities are based on a serious commitment to health and devising a healthy public policy. The success of healthy cities is reflected in the degree to which policies that create settings for health are in effect throughout the city administration. Efforts to achieve such goals only work when homes, schools, workplaces, and other parts of the urban environment become healthier settings to live in.

New Approach to Migrants and Slum Dwellers

It is neither feasible nor desirable to stop the process of migration. Predetermined sites with basic civic amenities should be earmarked and developed in advance for these sections of the community. The more planned and purposive the process of absorption, the better it would be for the migrants, their life, health, and the city.

Political Decision Making

Building healthy cities requires political decision making for public health. Housing, environment, education, social services, and other programs of city government have a major effect on the state of health in cities.

New Framework for Urban Administration

Urban centers wrestling with new administrative, environmental, and health challenges must be supported by an appropriate institutional framework for translating both vision and will into the reality of practical action. It should be community-led and consensus-based and work inter-sectorally and collaboratively.

Public–Community Partnerships

The alternative to conventional urban administration is an emerging form of what has been called "environmental administration." It can be characterized as noncompartmentalized, open, decentralized, antitechnocratic, and flexible. For effective implementation of healthy community development, a great effort will be needed to turn the system of local government into a paragon of environmental administration. The most viable option for this is to build a "public–community partnership" between government and decentralized, flexible, locally controlled, politically popular, democratic, and open community-based organizations to implement measures of improvement at the community level.

Promoting Local Autonomy and Initiative

The first step in facing the challenge of healthy city growth is to establish meaningful and workable mechanisms for region-wide planning and coordination and control of development. Both autonomous and planned deconcentration within a city can create higher levels of interdependency and intraregional mobility. Local initiatives and control in a spatially and administratively decentralized settlement system would enhance self-reliance and sustainability for many urban functions. This would reduce the burden on the city government and thus enable it to cope with the increasingly complex systems of interaction and regional functions.

Empowerment of the Community

No planning methodology can work without the involvement of the communities. Neighborhood committees of those formally housed, as well as of those living in informal settlements should be empowered. It is possible to create such committees within the framework of the municipalities and ward or local-level committees who should not only look after neighborhood services but should also assist in delivery of healthcare services and medical emergencies. People participate in health through their lifestyle choices, their use of health services, their views on health issues, and their work in community groups.

Intersectoral Action and Nongovernmental Organization Initiative

The term *intersectoral action* describes the process through which organizations working outside the health sector change their activities so that they contribute more to health. Urban planning that supports physical fitness by providing ample green space for recreation in the city is an example of intersectoral action. For successful implementation of healthy cities programs, it is necessary to create organizational mechanisms through which city departments and other bodies come together to negotiate their contribution to such action. With the trend toward democratic governance and decentralization, there is an increasing emphasis on the role of civil society in governance and management, providing tremendous opportunities for organizations to support the voice of the civil society. Nongovernmental organizations need to significantly expand their scope of activities, strategically access and use the media, and in general, develop capacities for communication that are more effective, along with negotiations.

Planning, Precaution, and Innovation

Healthy cities, by definition, are focused on achieving immediate and long-term goals related to urban health. For cities this means adopting a precautionary approach to future developments to avoid the buildup of potential liabilities, such as contaminated land, polluted air, and so on, and to ensure that technological innovation is geared to healthy requirements. Promoting health and preventing disease through intersectoral action requires a constant search for new ideas and methods and their ability to create opportunities for innovation within a climate that supports change.

Conclusion

The future is difficult to predict, particularly when social, economic, political, and international issues of far-reaching consequences are involved in giving new shape to our cities. The vision of healthy cities should link ecological sustainability with social justice and the pursuit of healthy livelihoods. It is a vision that must acknowledge the ecological limits to growth, promote ecological and cultural diversity and a vibrant community life, and support a community-based, sustainable economy that is directed toward fulfilling human health needs, rather than just simply expanding its periphery. Building healthy cities also requires access to decision-making processes to ensure that economic and political institutions promote activities that are ecologically healthy, sustainable, and socially just. It requires that these institutions respect our needs as human beings and citizens, and not just as producers, consumers, and voters. Effective and acceptable local solutions require local decisions, which in turn require the extensive knowledge and participation of the people most affected by those decisions in their homes, in their workplaces, and in their communities. Most important, planners should have a special obligation and ability to frame issues, assume leadership, champion initiatives, and demonstrate healthy alternatives in their everyday practice.

See Also: Citizen Participation; Civic Space; Compact Development (New Urbanism); Environmental Planning; Green Communities and Neighborhood Planning; Infrastructure; Sustainable Development.

Further Readings

Campbell, Scott. "Planning: Green Cities, Growing Cities, Just Cities?" In *The Earthscan Reader in Sustainable Cities,* David Satterthwaite, ed. London: Earthscan, 1999.

Middleton, Charles. "The New Politics of Environmental Governance, Environmental Sustainability and Life in the City: A Challenge for Design Professionals." In *Sustainable Development and the Future of Cities,* Bernd Hamm and Pandurang K. Muttagi, eds. London: Intermediate Technology Publications, 1998.

Rees, William E. "Achieving Sustainability: Reform or Transformation?" *Journal of Planning Literature,* 9/4 (1995).

Roseland, M. "Dimensions of the Future." In *Eco-City Dimensions—Healthy Communities, Healthy Planet,* Mark Roseland, ed. Gabriola Island, British Columbia, Canada: New Society Publishers, 1997.

Satterthwaite, David. "Cities and Sustainable Development: What Progress Since *Our Common Future.*" In *The Brundtland Commission's Report—10 Years,* Guri Softing, et al., eds. Oslo: Scandinavian University Press, 1998.

Satterthwaite, David. "The Scale and Nature of Urban Change in the South." International Institute for Environment and Development (IIED) Working Paper. London: IIED, 1996.

Wacker, Corinne, Alain Viaro and Markus Wolf. "Partnerships for Urban Environmental Management: Roles of Urban Authorities, Researchers and Civil Society." *Environment and Urbanization*, 11/2 (1999).

World Health Organization (WHO). *Twenty Steps for Developing a Healthy Cities Project*, 3rd Ed. Copenhagen: WHO Regional Office for Europe, 1997.

Mohua Guha
Aparajita Chattopadhyay
International Institute for
Population Sciences, Mumbai

HEAT ISLAND EFFECT

Urban heat islands (UHIs) are regions of characteristic warmth associated with cities. They are perhaps the clearest expressions of unintentional climate modification by humans. Weather patterns, geographic setting, and urban configuration cause UHIs to vary in space and time; the heat island effect is typically most pronounced in city centers where building and population densities are highest, and during calm, cloudless evenings that promote maximum thermal differences between the city and surrounding countryside. UHIs have several economic, ecological, and social effects, notably on energy consumption, dispersal of air pollutants, and human health and comfort. Climate-sensitive urban design demonstrates some potential for mitigation of these effects, where appropriate.

The UHI effect traditionally denotes a near-surface air temperature difference between the urban canopy-layer (the layer of atmosphere below the building tops) and a nearby rural area. Pedestrians directly experience this canopy-layer UHI, which ranges from a few degrees Celsius in moderately sized cities to approximately 10–12 degrees Celsius in the downtown cores of the largest cities. Daytime urban–rural temperature differences are smaller and some portions of urban areas may even exhibit "cool islands." Heat islands vary across cities depending on the local degree of urbanization; urban parks are relatively cool, for example, whereas greater nighttime warmth tends to occur at nodes of more intense development. Nighttime temperature increases rapidly near the urban–rural boundary, and then rises more slowly toward the city center. Hence, UHIs are so named because their contours of equal temperature resemble those of elevation on a relief map.

Urban–rural temperature differences have been observed worldwide for centuries. The English chemist and amateur meteorologist Luke Howard was the first person to conduct formal observations of urban and rural thermal climates, as documented in his 1833 book *The Climate of London*. In distinguishing city and country temperatures around London, Howard observed that the city "partakes much of an artificial warmth, induced by its structure, by a crowded population, and the consumption of great quantities of fuel."

Howard is generally considered to be the founder of urban climatology; modern pioneers include T. J. Chandler, Helmut E. Landsberg, and Timothy R. Oke. Published literature on observational canopy-layer UHIs is vast, most of it descriptive and most of it originating from large, midlatitude European and North American cities. Although UHIs were initially linked to city size or population, the main physical processes causing heat islands were clarified in the latter part of the 20th century.

The genesis of nighttime canopy-layer UHIs can be traced to the unique materials and structure, or form, of cities. Urban materials such as asphalt, concrete, and brick are typically better than natural materials at storing daytime heat and releasing it at night; furthermore, vertical surfaces (building walls) in the city provide a greater surface area for this heat storage and release. Building walls also obstruct the sky and prevent nighttime radiant heat losses from streets. These processes slow the cooling of streets and the canopy-layer air during the late afternoon and evening, leading to a positive urban–rural temperature difference that persists through the night. Waste heat released from transportation, heating/cooling, and industrial processes also plays a role in UHI formation, especially in cities with cold climates, high densities, or major industrial facilities. Additional factors related to wind sheltering, the removal of green space, and increased radiant heating from the polluted atmosphere above the city may contribute to UHI formation. The relative importance of these factors varies with city form, function, and location.

Heat island magnitude, as measured by an urban–rural temperature difference, depends on season, latitude, weather, and the surface character and state of the surrounding countryside. Clouds and wind in particular decrease heat island strength by reducing differences in heating and cooling rates between city and countryside. Seasonal variation of rural properties, such as soil wetness, foliage amount, and snow cover, significantly affects rural air temperatures and therefore UHI magnitudes. Corresponding seasonal variations in cities are blunted by the predominance of paved surfaces, the scarcity of vegetation, and the removal of snow. Canopy-layer UHIs are typically maximized in the summer season in midlatitudes; however, they may also be significant in winter in climates with high space heating requirements. In tropical latitudes, UHI intensity tends to be smaller than in midlatitudes and is greatest during the dry season.

In addition to the canopy-layer UHI, two other heat islands are commonly distinguished: The term *boundary-layer UHIs* refers to differences in air temperature in the lowest 0.1–1.0 kilometers (km) of atmosphere above the city and are smaller than canopy-layer UHIs (usually 1–2 degrees Celsius), and the term *surface UHIs* refers to differences in temperature between urban surfaces, such as buildings and roads, and rural surfaces, typically soil and vegetation. Surface UHIs are maximized during the day and vary dramatically in magnitude. Although linked, each of these heat island types involves different physical processes and different human consequences.

Surface and canopy-layer UHIs affect building energy use for heating and cooling—warm urban temperatures reduce energy requirements for space heating during the cool season and increase demand for air conditioning in the warm season. Heat islands influence the thermal comfort of pedestrians and building occupants, intensify heat waves, increase heat mortality rates, extend the growing season for urban gardeners, and reduce the likelihood of road ice and snow accumulation. Local weather patterns are also affected by heat islands, notably in the genesis of urban–rural breezes and the enhancement of cloud and precipitation. Boundary-layer UHIs enhance smog formation by speeding up photochemical reactions. Overall, the UHI has negative effects in warmer climates but may present a mix of negative and positive effects in colder climates.

As understanding of heat islands and urban climates continues to mature, effective communication of the associated impacts and potential solutions to planners and other decision makers remains an ongoing challenge. One hurdle is the reconciliation of climate-derived objectives with other urban planning goals, principles, and realities (e.g., cost, aesthetics, transportation, etc.). A further challenge, given the diversity of urban meteorological and geographical contexts worldwide, is to distill a practical and widely applicable set of design recommendations from a growing urban-climate knowledge base. Generic solutions may exist; for example, the addition of reflective surfaces and vegetation on streets and roofs (e.g., green roofs) for daytime urban air temperature reduction shows promise in some settings. Other features of cities, such as their physical structure, are not easily modified. The availability of urban climate resources and guidelines to planners is inadequate at present, which may largely explain the deficiency of climate principles, and UHI mitigation strategies more specifically, in urban planning and design.

See Also: Air Quality; Density; Green Communities and Neighborhood Planning; Green Roofs; Parks, Greenways, and Open Space.

Further Readings

Landsberg, H. E. *The Urban Climate.* New York: Academic Press, 1981.
Oke, T. R. "The Energetic Basis of the Urban Heat Island." *Quarterly Journal of the Royal Meteorological Society,* 108/455 (1982).
Voogt, J. A. "Urban Heat Island." In *Causes and Consequences of Global Environmental Change*, I. Douglas, ed. *Encyclopedia of Global Environmental Change*, Vol. 3, R. E. Munn, ed. Chichester, UK: John Wiley & Sons, 2002.

E. Scott Krayenhoff
Iain D. Stewart
University of British Columbia

Historic Preservation

Historic preservation helps cities achieve sustainability in three interrelated ways. First, by definition, preservation is ecologically sustainable. The adaptive reuse of old buildings conserves the resources used in their production. Second, historic buildings and landscapes provide an economic benefit to communities. Preserving historic districts is a reminder of the past as well as an investment in the future of the community, which can enhance property values and improve the local tax base while also being used as a marketing tool. Finally, preservation is socially sustainable because it preserves traditional cultural values. Preservation can act as a tool to unite a community and promote citizen participation. Although generally seen as a positive planning activity, historic preservation is not without thoughtful criticism. Critics of historic preservation identify three main critiques: authenticity, consumerism, and power.

The History of Preservation

Uneasy with the pace of change facing the rapidly industrializing late-19th- and early-20th-century city, citizens became sentimental and nostalgic for earlier, simpler times. People yearned for traditions and values in the face of cultural change. Historic preservation served as a tool to preserve and protect those values and traditions by conserving buildings and landscapes as material manifestations of cultural ideals. Private individuals and groups carried out most of the early preservation projects to foster traditional values. In the United States, the patriotic fervor of Ann Pamela Cunningham and the Mount Vernon Ladies' Association led to the preservation of George Washington's home, and John D. Rockefeller preserved Williamsburg, Virginia. Each act of preservation used the built environment to cultivate nationalistic traditions, myths, and beliefs.

By the mid-20th century, the process of preservation in the United States became more organized with the creation of the National Trust for Historic Preservation in 1949 and the passage of the National Preservation Act in 1966. According to its website, the goal of the National Trust is to provide "leadership, education, advocacy, and resources to save America's diverse historic places and revitalize our communities." The passing of the National Preservation Act created the National Register of Historic Places and placed its oversight in the hands of the National Park Service and the secretary of the interior. The preservation of buildings and places listed on the National Register necessitates the application of the secretary of the interior's Standards for the Treatment of Historic Places. Developed in 1976 and updated in 1995, the standards apply to historic buildings as well as historic landscapes. The four treatments emphasized in the standards aid in making sound preservation decisions. The four treatments are defined here:

- *Preservation:* The act or process of applying measures necessary to sustain the existing form, integrity, and materials of an historic property. Work generally focuses on the ongoing maintenance and repair of historic materials and features.
- *Restoration:* The act or process of accurately depicting the form, features, and character of a property as it appeared at a particular period of time by means of the removal of features from other periods in its history and reconstruction of missing features from the restoration period.
- *Rehabilitation:* The act or process of making possible a compatible use for a property through repair, alterations, and additions while preserving those portions or features that convey its historical, cultural, or architectural values.
- *Reconstruction:* The act or process of depicting, by means of new construction, the form, features, and detailing of a nonsurviving site, landscape, building, structure, or object for the purpose of replicating its appearance at a specific period of time and in its historic location.

Bolstered by the National Preservation Act, historic preservation became a source of community power. As a symptom of the cultural changes occurring in the 1960s and 1970s, preservation became a tool for the organization of local residents resisting changes forced on them by corporations, developers, and municipalities in the name of urban renewal. The late 20th and early 21st centuries witnessed the use of historic preservation as a means to green the city through adaptive reuse of buildings and the creation of historic districts in older, centrally located communities. This period also witnessed the postmodern critique of preservation based primarily on issues of authenticity, consumerism, and power.

Adaptive Reuse

Adaptive reuse is the conversion of old buildings to uses other than that which were originally designed: for example, the conversion of factories to residential developments. The 19th century witnessed the explosion of the textile industry in New England. The Boston Manufacturing Company, the first fully integrated textile mill in which all phases of cotton cloth production occurred in the same factory, was built in 1813 along the banks of the Charles River in Waltham, Massachusetts. In the 1820s and 1830s, this mode of production was copied at Lowell, Massachusetts; Lawrence, Massachusetts; Manchester, New Hampshire; and several other industrial New England cities. By the 1920s, facing competition from southern textile mills, mills in New England entered a period of decline and many closed. Left to deteriorate, vacant mill buildings became targets for vandalism, arson, and other unsavory activities. As a blight on communities, unsightly mill structures promoted a negative community image. By the 1980s, adaptive reuse of these vacant buildings reversed this trend and fostered a culture of urban sustainability.

In general, there are two main types of adaptive reuse of mills in New England. The first is a conversion from mill to museum. Examples of this type of adaptive reuse include Lowell National Historic Park, Lawrence Heritage State Park, and the Massachusetts Museum of Contemporary Art in North Adams. Not only were the original structures, beams, brickwork, and machinery preserved and used in the reuse, but in the case of the Massachusetts Museum of Contemporary Art, brownfield remediation was necessary in order to remove polychlorinated biphenyls, volatile organic compounds, and heavy metals before the site could be used by visitors. The second type of mill conversion is from mill to residential use. Examples of mill-to-residence conversion include trendy loft-style condominiums that feature industrial elements (e.g., high ceilings, exposed brick, exposed beams) that highlight the mill theme. This type of residential reuse can be found in many formerly industrial communities throughout New England. Adaptive reuse conserves natural resources by recycling the existing structures, but increasingly these developments take sustainability farther, promoting green living by featuring green roofs, renewable energy sources, the use of greywater, or geothermal exchange to provide heating and cooling.

Historic Districts

By the 1920s and 1930s, Americans uneasy with the pace of change taking place in the urban landscape began to change zoning laws to preserve small portions of their local communities as historic districts. In 1931, influenced by a small group of citizens, Charleston, South Carolina, became the first city to pass a zoning ordinance establishing historic districts to enhance the sense of place. According to the Architectural Heritage Center, a historic district is defined as "a geographically definable area, urban or rural, possessing a significant concentration, linkage, or continuity of sites, landscapes, structures, or objects, united by past events or aesthetically by plan or physical developments." Similar to other periods of preservation, this preservation movement was grounded in nostalgia, collective memory, and sense of place. In the face of drastic changes to the landscape, local citizens nostalgic for simpler times set about to preserve districts full of landmark buildings and homes with distinctive architecture or history. These districts symbolized the local sense of place and collective memory.

The preservation of historic districts helps cities move toward sustainability by preserving older, centrally located communities. Preserving these older communities conserves the natural resources used in their construction. For example, adaptive reuse in Denver's Lower Downtown Historic District turned abandoned warehouses into locally owned book shops, restaurants and brewpubs, loft-style condominiums, and art studios. Historic districts preserve a dense and walkable community that is reminiscent of urban life before to the ubiquity of the automobile. Typically, these districts are located along commuter transit lines and reduce the community's carbon footprint. The preservation of historic districts has economic benefits as well. Because being located in a historic district provides unique qualities, historic districts help to improve property values. These districts also use signs, markers, banners, and flags to highlight the community's new vibrancy. Typically, communities create a new identity marked by a hip-sounding name for the newly preserved district. For example, Denver's Lower Downtown Historic District was transformed into *LoDo*, a term that was then used to market and sell the area. The preservation of historic districts is also socially sustainable. Historic districts serve to unite communities by further developing a sense of place and collective memory. The postmodernism of the late 20th century redirected the focus to include citizen participation of those living in forgotten ethnic enclaves.

Critiques of Preservation

The three main areas of preservation criticism are authenticity, consumerism, and power relations. The debate surrounding authenticity centers on the differences between history and heritage. In examining the rise of the heritage industry, David Lowenthal defined history as "the past that actually happened." As critical analysis, history tries to deal in the testable interpretation of facts—what happened, where, when, why, and how. According to Lowenthal, heritage, in contrast, "is not history . . . it uses historical traces and tells historical tales, but these tales are stitched into fables that are open neither to critical analysis nor to comparative scrutiny." Heritage distorts the past so that it conforms to an idealized representation of the past; it manipulates the past to suit the needs of the present—namely, using the past to sell the city.

In selling the city, the "authentic" past, in the form of a selective imaginary, is used to create historic districts, which help to brand the city and create the city as a theme park. Preservation privileges one historical narrative and erases others, cleansing history of any unsightly aspects of the past and making history more watered down and pleasant. This Disneyfication of the city results in fantasy environments based on the myths associated with nostalgia, sense of place, identity, and permanence. The original character and sense of place is stripped away, replaced with the tourist-friendly, aesthetically pleasing veneer of consumption. Therefore, preservation is an aesthetic means for economic development based on the illusion of authenticity. The creation of historic districts and the conversion of old buildings into museums and residences are part of a broader effort to reinvest and redevelop central cities, and in this way, historic preservation is an agent of gentrification.

Historic preservation was used as a tool to preserve older run-down, aesthetically unpleasing neighborhoods during the urban renewal craze of the 1950s–1970s. This resulted in the displacement of people of color and lower-income families. In their desire for tradition and community in changing times, urban populations of white-collar workers, upper-middle-class residents, and tourists seek out authentic representations of the past. They seek out older, run-down communities, where they can purchase properties at lower rates, improve the housing stock, and reap the rewards of property enhancements. As discussed earlier, the creation of historic districts and the conversion of old buildings to new uses can improve the aesthetic qualities of a neighborhood, which then helps to enhance property values. These increases in rents and land values result in the displacement of the original owners—largely people of color and lower-income families. It should be noted the benefits of preservation are not equitably distributed among all social classes. The preservation of "authentic" heritage, the use of history to sell the city, and urban redevelopment benefit the urban elite. Although some cities do represent diversity through preservation, the narrative told through preservation is still fixed largely on wealthy white men.

Conclusion

The rapid urbanization of the world necessitates a new approach to sustainable urban development. Ironically one approach is to look back to the past. Preservation helps cities become more ecologically, economically, and socially sustainable in three ways. First, adaptive reuse utilizes the existing built environment to conserve natural resources. Second, the preservation of historic districts provides a marketing tool to help the local economy by providing appealing images that enhance property values. Third, preservation fosters a sense of place and collective memory and unites communities. However, it should be noted that critical examinations of preservation should address the authenticity, consumerism, and power associated with historic preservation.

See Also: Adaptive Reuse; Citizen Participation; Green Housing.

Further Readings

Architectural Heritage Center. "Preservation Glossary." http://www.visitahc.org/node/47 (Accessed April 2009).
Lowenthal, David. *The Heritage Crusades and the Spoils of History*. Cambridge, MA: Cambridge University Press, 1998.
Murtagh, William J. *Keeping Time: The History and Theory of Preservation in America*, 3rd Ed. New York: John Wiley & Sons, Inc., 2006.
National Trust for Historic Preservation. "About the National Trust for Historic Preservation." http://www.preservationnation.org/about-us (Accessed April 2009).
Page, Max and Randall Mason. *Giving Preservation a History: Histories of Historic Preservation in the United States*. New York: Routledge, 2004.

Phil Birge-Liberman
Syracuse University

I

Indoor Air Quality

Though often taken for granted, indoor air quality is an important concern. Most people spend more time indoors than out, consequently the health risks posed by pollutants trapped within buildings can be significantly greater than those associated with outdoor air pollution. The main causes of poor indoor air quality fall into two broad categories: (1) emissions from building materials and furnishings, dirt and molds, chemical cleaning supplies, heating and air conditioning system fuels and filters, and site characteristics; and (2) poor ventilation that concentrates interior pollutants by preventing exchange with the outside air.

The heavy use of asbestos as a building material has resulted in major health problems, leading to extensive litigation against manufacturers over its use.

Source: iStockphoto

Emissions from building materials and furnishings emerged as a problem with the increased use of processed wood products, foams and plastics. Virtually every modern building contains some of these products, for example, particleboard, plywood, fiberboard, and paneling. The primary pollutant associated with this group is formaldehyde, a component of many adhesives. Problems with formaldehyde rose to prominence during the 1970s when urea-formaldehyde foam insulation was installed in many U.S. homes as an energy conservation measure. Soon after installation, occupants of these structures began to suffer from eye, nose, and throat irritations and upper-respiratory difficulties. The cause was traced to high interior concentrations of formaldehyde.

A similar situation occurred when manufactured homes containing formaldehyde-based components were provided to families displaced by the flooding in and around New Orleans as a result of Hurricane Katrina in 2005. Since 1985, the Department of Housing and Urban Development and many U.S. states have established specific emissions limits for building products containing formaldehyde, and limits on volatile organic compound (VOC) emissions from paint have been set by many air pollution control districts.

Radon is an example of natural pollutant characteristic of a building's location. An odorless, tasteless gas, radon emanates from the natural radioactive decay of uranium in the soil. Radon ions have a static charge that binds them to minute dust particles, which can easily be inhaled, a trait that links radon to lung cancer. Four conditions must exist for radon to pose a problem in a building. Two are geological: There has to be a source of uranium in the soil, and the soil must be permeable enough to allow radon to move through it. The two other requisites relate to building construction: There must be a pathway into the structure—a hole, crack, or gap, and a negative air pressure differential must exist between the building's interior and surrounding soils. Radon can be found throughout the United States, and since the 1980s, homeowners have been advised to test for its presence in their homes.

Before emissions from building materials and furnishings were identified as health threats, buildings were designed to be as air tight as possible in order to conserve energy and reduce heating and cooling costs. Consequently, indoor pollutants were trapped and concentrations tended to increase over time. Today, green construction techniques allow buildings to breathe, striking a more healthful balance between energy conservation and indoor air quality. Indoor air quality is enhanced by the use of nontoxic materials, those emitting few or no VOCs, and those resistant to mold and mildew problems. However, much of the current U.S. housing stock predates these design and product improvements, and poor indoor air quality still places an estimated 40 percent of the U.S. population at risk for allergies, respiratory disease, cancer, and developmental problems.

See Also: Air Quality; Green Design, Construction, and Operations; Green Housing.

Further Readings

Air Quality Services, Inc. "Energy Conservation and Indoor Air Quality: Benefits of Achieving Both in Homes." Marietta, GA: Air Quality Services, Inc., 2009. http://www .aerias.org/uploads/IAQ%20Energy%20Homes%20FINAL.pdf (Accessed January 2010).

Burroughs, Barney and Shirley J. Hansen. *Managing Indoor Air Quality*. Lilburn, GA: Fairmont Press, 2008.

Harris, Janie L. "Indoor Air Quality—Radon." http://fcs.tamu.edu/housing/healthy_homes/ indoor_air_quality/radon/radon.php (Accessed January 2010).

Marsik, Tom and Ron Johnson. *Effect of Outdoor Air Quality on Indoor Air: Modeling, Measurement and Control.* Saarbrücken, Germany: VDM Verlag, 2008.

Samet, Jonathan M. and John D. Spengler, eds. *Indoor Air Pollution: A Health Perspective* (The Johns Hopkins Series in Environmental Toxicology). Baltimore, MD: Johns Hopkins University Press, 1991.

Spengler, John Daniel and Richard Wilson, eds. *Particles in Our Air: Exposures and Health Effects.* Cambridge, MA: Harvard School of Public Health, 1996.

Spiegel, Ross and Dru Meadows. *Green Building Materials: A Guide to Product Selection and Specification.* Hoboken, NJ: John Wiley & Sons, Inc., 1999.

U.S. Environmental Protection Agency. "An Introduction to Indoor Air Quality." http://www.epa.gov/iaq/index.html (Accessed January 2010).

Justin Corfield
Geelong Grammar School, Australia

INFILL DEVELOPMENT

Infill development can best be described as the placement of residential or commercial property on land that is currently underused, vacant, or abandoned. As such, it is commonly used in urban areas as an example of a strategy to curb sprawl and is advocated by proponents of "smart growth" policies. Infill development has become a relevant contemporary issue as urban neighborhoods become more economically obsolete over time. Part of the impetus behind the provision of infill housing arose from the Supreme Court case of *Berman v. Parker* (1954), which allowed for the elimination of blighted properties from urban neighborhoods as the first step in urban renewal revitalization efforts. Although infill development is usually associated with urban areas, suburban cities have their own opportunities to provide such development once land parcels become underutilized, and thus obsolete.

Reasons for Advocating Infill Development

From the consumer side, infill housing provides more families with the opportunity to enjoy urban living, thus alleviating the need for significant commutes from outer suburban areas. Several cities are experiencing an increase in the number of their downtown residents, reversing a decades-long trend. This urban relocation is a reflection of the status symbol that households place on such prime locations. Lower commute time also provides an opportunity for more community involvement and greater access to cultural and other resources provided by the urban environment, as opposed to the suburbs and exurbs. Demographic shifts also suggest the growth of populations that would be most amenable to residing in infill housing. As housing markets experience shifts toward smaller families, a more active retiree population, households without children, and single households, infill development becomes a more attractive residential choice. In addition, a number of community government employees and workers from vital employment centers such as hospitals and universities are increasingly given incentives to live closer to their jobs, in essence furthering the demand for infill housing. Meanwhile, several households have indicated their interest in, through either official surveys or ballot initiatives, the displeasures associated with sprawl-type conditions within their communities, such as increased commute times, declining air quality, environmental degradation, and other quality-of-life issues. Although some consumers may not initially be attracted to infill development, there is evidence that prospective home buyers will purchase within these communities if they receive some form of community amenity in return, such as parks or access to bicycle routes.

Communities also gain from infill development strategies. Infill within urban settings provides homeownership opportunities for all income categories, which is one of the main drivers behind families fleeing to the suburbs. Infill development also provides more tax revenue by making obsolete land uses productive once again. On the cost side, local governments benefit from infill development, in that public expenditures for infrastructure will decrease as development moves away from the urban fringe. Infill development also provides one of the best methods for implementing a mixed-income housing environment—one of the key ingredients in contemporary community development efforts.

Obstacles and Barriers to Infill Development

One of the primary challenges of infill is community resistance to new development in already-dense areas. Such anxiety is based on perceptions that infill will likely lead to higher property taxes and more undesirable populations moving nearby. Because infill requires higher-density residential and commercial patterns, there is also concern about access to and availability of such services as street parking. Some communities also bemoan the decline in privacy that ultimately accompanies an increase in density. In older neighborhoods, residents are particularly concerned about the detrimental effects that poorly planned infill projects have on the community, particularly in ignoring its historical and sociological aspects.

There are other concerns about infill housing that more directly affect the average consumer and his or her willingness or capability to gain access to infill housing. The American ideal of homeownership has helped to fuel Americans' need for private spaces and access to a yard—housing amenities that are more closely associated with the suburbs. In addition, many potential infill sites may not have high levels of support services or facilities, such as quality schools, grocery stores, and retail shopping outlets, which hampers prospects for consumer demand for such development. Valuation of infill properties also becomes problematic, in that there are not enough adequate comparable sales to provide an accurate appraisal of the market value of the house. Lenders may be reluctant to provide financing for such a property, given the perceived riskiness involved with both the project itself and the underdeveloped neighborhood in which it is located. Obtaining reasonable property insurance rates and coverage may also be a concern.

Developers and building professionals face their own obstacles to infill development, no matter how well-intentioned their efforts are to provide such a product to the community. The cost of land assembly (which includes such things as acquiring the site, demolishing any existing improvements on the land, environmental cleanup, and site preparation) is significantly—sometimes prohibitively—higher than the cost of raw land in suburban or rural environments. To compound the issue, many economists and urban planners believe that an effective infill strategy is not one that is piecemeal but, instead, one that requires the assemblage of several acres of land covering hundreds of housing units. Meanwhile, the approval and regulatory process for infill developers is more cumbersome, and thus more costly, than for traditional developers. Although urban activists tout the presence of already-existing infrastructure such as streets and utility hook-ups as advantages of infill housing over suburban development, some developers still point out the need for infrastructure upgrades once infill is completed. Finally, there has been some concern about the availability of developers who specialize in this type of niche housing within some housing markets.

See Also: Green Communities and Neighborhood Planning; Smart Growth; Sustainable Development.

Further Readings

Danielsen, Karen A., et al. "What Does Smart Growth Mean For Housing?" *Housing Facts and Findings,* 1/3 (1999).

Farris, J. Terrence. "The Barriers to Using Urban Infill Development to Achieve Smart Growth." *Housing Policy Debate,* 12/1 (2001).

Suchman, Diane R. and Margaret B. Sowell. *Developing Infill Housing in Inner-City Neighborhoods: Opportunities and Strategies.* Washington, D.C.: Urban Land Institute, 1997.

Vallance, Suzanne, et al. "The Results of Making a City More Compact: Neighbours' Interpretation of Urban Infill." *Environment and Planning B: Planning and Design,* 32 (2005).

Andrew T. Carswell
University of Georgia

INFRASTRUCTURE

Infrastructures are network-bound large technological systems through which utility services such as drinking water and electricity or waste and sewerage services are supplied, distributed, and consumed. Infrastructures are relevant for green cities, as they enable and regulate the efficient supply, distribution, and consumption of environmentally relevant flows of energy, water, and materials. This article covers the past and present organization of infrastructures and explores the social and technical adaptations of infrastructures that are needed to make up a green city.

Evolution of Infrastructures

The provision and consumption of water, energy, and waste services in cities, and the development of infrastructures to accommodate this, has followed similar patterns in all Western societies. In preindustrial times, the services were provided autonomously by individual traders and by consumers themselves. Water was obtained from individual wells, from canals, or from private water sellers; energy supplies like wood or charcoal were provided individually or by private sellers; and waste was dumped in canals or collected by farmers to apply as fertilizer on their lands.

Following this long era of autonomous provision and consumption, in the 19th century, piecemeal modes of provision emerged in which independent suppliers became involved in providing local services through small technological networks. An example is electricity provision by local supplier firms who built the first local grids for small firms or affluent households. These consumers were no longer managers of their own energy supply, as the balance of management was shifted toward the local electricity provider. In a similar vein, the first piped water supply systems for individual users were constructed by private firms to serve citizens who could afford the new water-flushed toilets.

Around the turn of the 19th to the 20th century, local and regional governments took up the responsibility of larger infrastructure development and management as they realized that such investments would be beneficial for economic development and public health at large, but also that such investments were too big for private parties to bear. Such an integrated mode of organization of infrastructures meant that supply and demand of water, electricity, and waste services over larger, integrated infrastructures were optimized to make maximum use of the capacity installed (power stations, waste facilities, pumping stations, cables, and pipes). The effort of local and regional governments was to interconnect smaller networks to give all city dwellers access to the modern services of water, heat, electricity supply, and sewerage and waste collection services. However, infrastructure capacity remained a limiting factor, and demand was to be controlled and scheduled. The public was urged to conserve water and energy not for environmental reasons, but because of peak loads that could not be met by system capacities.

A next stage of infrastructure provision, mostly starting after World War II in most Western countries, is the "universal" mode of organization, in which demand was not regarded as something to be differentiated, promoted, or controlled, but as a nonnegotiable need that had to be met. This is the era of planning and building large-scale and nationwide uniform infrastructure, resulting in national electricity grids, water supply stations and mains, centralized sewer treatment, landfills, and waste incineration plants processing the domestic waste of millions of householders. As electricity, water, sewer, and waste networks were extended up to regional and national levels, demand had to be generated and sustained to keep these systems in operation. The exploration of newly discovered huge natural gas reserves in the Netherlands in the beginning of the 1960s serves as a case example for a universal mode of provision. Within a decade, a nationwide network of piped gas had been rolled out, and household consumers were being educated and subsidized to make the change to central heating and cooking based on gas. Demand for gas was heavily promoted to pay back the infrastructure investments in the short term, before the expectations of the takeover of nuclear energy in the national energy balance would become reality.

Although the huge public infrastructures that resulted from this phase could benefit from economies of scale, from the 1970s on, growing concerns over the environment and depletion of sources had emerged and had led to various experiments with alternative, down-scaled, autonomous forms of provision. Many technological innovations that have now been incorporated in modern electricity, water, and waste sectors had their start in this period: solar heaters, rainwater harvesting for domestic use, greywater recycling, home composting, and sustainable building were all invented and experimented with within local communities during the 1970s.

By the end of the decade, as a result of political concerns over inefficient public utility management and lack of public funding for reinvestment in infrastructures, the first calls for privatization of utility assets and liberalization of utility markets were made. Energy and water supply, along with railways and postal services, were gradually transferred from state companies to private companies, although not without extensive political controversies. It was the start of what has now become a "marketized" mode of infrastructure organization. Many infrastructures have since been privatized, new competitors for services have entered the thus-far monopolist utility markets, and the uniform networks have been diversified in terms of technology, scale, and market. It is this marketized mode of provision that sets the scene for exploring the role of infrastructures in greening the city.

Splintering Utility Services, Distributed Generation, and "Smart Grids"'

A marketized mode of provision involves the provision of diversified utility services by multiple providers through partially fragmented grids. Marketized provision evolved from a paradigm shift from universal modes of provision, which means that market principles were introduced in situations in which state companies had a natural monopoly, with captive consumers having no choice but to use and pay for the utility services provided. The logic of infrastructure management, in which fixed units of power and water were delivered at standardized cost as part of a wider universal service obligation, has been challenged by a new logic in which infrastructure resources are becoming commoditized—gradually differentiated in terms of cost, availability, and quality over space and time.

Much of the public and political debates around this issue have been devoted to the responsibility of the state versus private businesses to provide universal services to the population about tariff setting and investments in hardware and issues of equity and rights of access to basic services. However, the splintering of infrastructures also offers major opportunities for greening the city.

The opening of utility markets has led to the emergence of multiple providers who try to compete with differentiated services. The splintered, partly fragmented grids are being served by more localized resources and providers: Grey- and rainwater systems replace parts of the water supply, and combined heat and power stations and solar panels produce electricity on a localized scale. As an alternative to centralized electricity supply, models of distributed generation flourish; for example, diverse flexible local and regional power generation units supply to the main grid next to a single power plant at the center. It makes the grid much more resilient and offers opportunities to use diverse renewable energy resources. Recently, so-called smart grids have entered the scene, referring to self-regulating local grids, matching demand, storage, and supply from various energy sources throughout the grid. Consumption units, such as households, become production and storage nods in the network as well, regulated by a next generation of meter systems that not only display but also manage demand and supply. In these scenarios electric cars play a dominant role not only as a new green mode of transport but also as an aggregated battery to store abundant electricity loads during the night.

Green Consumption

Environmental issues have been used to differentiate among infrastructure providers since the 1990s. Green electricity was the first differentiated service in European electricity markets, as green providers were the first to differentiate from the normal ones. Consumers—who were so far considered passive connections—were being asked to consider switching to another electricity provider on the basis that they were greener than their competitors. The possibility for consumers to make a choice between different providers, their services, and even the sources by which they produce electricity signals the end of their role as captive consumer and the start of a new identity—that of "customer." However, the opening of utility markets has led to more than that. With the increased possibilities created by new legislation and subsidy programs for solar or wind electricity production, everyone, including domestic consumers, could become providers of electricity to the common grid. With this birth of the "co-providing consumer" in utility markets, the strict boundaries between the domains of consumption and provision of utility services have become blurred. More roles for consumers as utility service providers can now be defined. In addition to the roles mentioned earlier, the role of "citizen-consumer" should be addressed: consumers who base their new choices in utility services on societal or environmental considerations and who become local beneficiaries of a windmill association, supporters of a local waste recycling scheme, or protesters against prospective nonsustainable developments in wastewater treatment in the neighborhood.

Another implication of greening the consumption of infrastructure services is the need to make these services more visible and tangible for end consumers. When captive consumers become coproviders, services differentiated, and environmentalists plead to make consumption patterns transparent to encourage rational use, then visibility is a key issue in a utility world that was until recently at the back of consumers' minds, and literally underground. Visibility may refer to the infrastructure hardware (pipes and cables, water resources, solar panels); to the flows of water, energy, or waste; and to the quantity and quality of production and consumption. Smart meters for energy and water may reveal individual consumption patterns to providers, but they may also make providers visible to consumers. In addition, now that they have become more visible, infrastructure services may also become objects of conspicuous consumption and distinction with social and, likewise, environmental implications: solar panels, or rainwater devices, or low consumption patterns for energy, water, or waste services may well become objects of social distinction.

Infrastructures and Green Cities

Also, on a higher, institutional level, a further liberalization and differentiation of utility markets will show a blurring of boundaries not only between provision and consumption but also between the different utility sectors. New alignments between utility companies and sectors emerge as energy providers become involved in the supply of hot tap water to households, waste managers start to produce biogas from organic waste processing, and water suppliers get involved in geothermal heat projects and underground heat/cold storage technology. Green city development in physical terms is about making efficient use of renewable resources and closing the loops of material, energy, water, and nutrient flows. Infrastructures of energy, water, sewage, and waste need therefore to be planned and developed in an integrated way. The challenge is thus to develop urban public–private governance arrangements that allow for integrated planning of infrastructures for the greening of cities in a time when integrated and universal modes of utility organization have long passed and the former utility sectors have been splintered in social, economic, and technical terms. The role of city governments is likely to grow again in the initiation and planning of new, localized energy, water, and waste services provision, but mostly in public–private relationships with the privatized providers of infrastructure services and technologies, as well as the housing sector.

See Also: Ecovillages; Green Housing; Greywater; Power Grids; Renewable Energy; Waste Disposal; Water Sources and Delivery.

Further Readings

Cowan, R. S. *More Work for Mother: The Ironies of Household Technology From the Open Hearth to the Microwave*. New York: Basic Books, 1983.

Guy, S., et al., eds. *Urban Infrastructure in Transition: Networks, Buildings, Plans.* London: Earthscan, 2001.

Hughes, T. P. *Networks of Power: Electrification in Western Society, 1880–1930.* Baltimore, MD: Johns Hopkins University Press, 1983.

Southerton, D., et al., eds. *Sustainable Consumption: The Implications of Changing Infrastructures of Provision.* Cheltenham, UK: Edward Elgar, 2004.

van Vliet, B., et al. *Infrastructures of Consumption: Environmental Innovation in the Utility Industries.* London: Earthscan, 2005.

Bas van Vliet
Wageningen University

INTERMODAL TRANSPORTATION

Intermodal transportation is the management and operational integration between different transport modes and means in sequence by loading units (containers), avoiding manipulation of the goods contained in them. It is fair to say that intermodal transportation has existed since ancient times, as people and cargos have always traveled and made transfers from one means of transport to another. Intermodal transport, however, is a modern development resulting from the progress of transport technologies that have made the train, steamers, motor vehicles, and the aircraft available.

Intermodal Transport of Goods

Beginning from the 1940s, intermodal transportation has established itself chiefly in the goods transport sector by virtue of two technological innovations: the pallet and the container. The pallet is a platform made of wood, synthetic material, or metal and moved by lifting trolleys, on which packages ready for shipping are piled up. The modern pallet was first employed by the U.S. Army during World War II, and within a short time it became the indispensable logistic device for speeding up the shipping of large amounts of goods. Beginning from the 1960s on, the size of pallets has progressively been standardized by the Organization for Standardization to enable transport by various means: In Europe, where goods are generally carried by road, the most commonly used pallet measures 800 × 1,200 mm; in the United States, the standard dimension is 1,000 × 1,200 mm; in Asia, pallets measure 1,100 × 1,100 mm.

The container consists of a metal box designed for carrying goods by land, sea, and air. In the United States, the earliest containers, known as "lift vans," came into use in the 1910s. In Europe, the earliest systematic use of containers goes back to the 1920s. Traditionally, however, the age of containers begins in 1956, the year in which an American entrepreneur had the idea of converting a tanker into a container-carrying ship. As with the pallet, containers' sizes have been progressively standardized, first by the U.S. Department of Defense and later by the International Organization for Standardization (ISO) . At present, the two most common standards are the ISO 20' (6,058 × 2,438 × 2,591 mm) and ISO 40' (12,192 × 2,438 × 2,591 mm).

Pallets and containers have brought about the establishment of two prevailing forms of transport. The first, chiefly employed for medium distances, contemplates the carrying of containers on a road semitrailer to the nearest railway station. From here the container will travel on a special train carriage to the station nearest to its destination, where it will be again loaded onto a road semitrailer and taken to its final destination. The second, chiefly used for long distances, consists of the transport of a container (by road or rail) up to a shipping port, where it is loaded onto a ship and taken to a port near its destination, from which (still by road or rail) the container will be transferred to its final destination.

In both cases, thanks to pallets and containers, intermodal transportation enables the shipping of goods without handling them again after they have been loaded onto the container at the place of production or shipping. The absence of intermediate handling reduces damage risks, lowers shipping costs, and increases the speed of delivery.

Intermodal Transport of People

We can also speak of intermodal transport with reference to the movements of people. In general, travelers use various means of transport: For long distances the most common combination is car/train–airplane or car/train–ship. For short to medium distances, the most common combination is bus/tram–train or car–train. Airports, for example, are nowadays "intermodal terminals": within their precincts they are, in fact, provided with parking lots, high-speed railway stations, terminals for various urban transports (i.e., metropolitan railways, trams, buses), and locations for the principal car rental companies.

In the course of the 19th century and of the first half of the 20th, an integration of transports emerged. Generally speaking, stagecoach services organized their timetables to create connections with the arrivals and departures of trains, which then monopolized medium- to long-distance transport.

Beginning from 1950s, the transport businesses started to operate separately: Railway and road transport companies began to compete in organizing trains and buses, offering services on the same routes with the same timetables. In this way, the system of public transport has become inefficient, encouraging a constant increase in the use of private transport. This caused pressing road congestion problems in urban areas. For this reason, beginning from the second half of the 1980s, policy makers, urban planners, and traffic engineers have seen in intermodal passenger transport the best way of transferring part of urban traffic from the private car to public transport. The main solutions adopted for favoring the development of intermodal passenger transport have been park-and-ride, integrated ticketing, and bus rapid transit.

Park-and-ride is probably the most widely adopted measure by local administrations in the way of sustainable intermodal passenger transport (park-and-ride facilities, with dedicated parking lots and bus services, began during the 1960s in the United Kingdom). Car drivers coming from outside the urban perimeter can park in special lots that also serve as the terminals for urban public transport and transfer to a bus or rail system for the rest of their trip. In this way, the transfer from a car to a public means of transport is fast and economically advantageous, as there is usually a single ticket to pay for both the parking and the bus ride.

Integrated ticketing offers travelers the possibility of using different means of transport (usually the bus, tramway, subway, or train) by paying a single comprehensive fee. This tariff system—whose adoption is today made much easier thanks to electronic ticketing—had initially been adopted in the most extended metropolitan areas; it has since been extended to regional contexts and, in some cases, is used nationally.

Bus rapid transit indicates a series of systems of rapid transport—usually with buses running along dedicated lanes that operate separately from other traffic modes—whose function is to enable commercially viable speed in public transport to make it more competitive (say, compared with the private car).

Over the course of the last few years, new systems of demand-responsive transport have emerged. These are shared transport services—by car or minibus—which are planned on routes that cannot be served by other means of public transport. This new service presents some limitations, as those who book a trip must do it in advance and must adapt their needs to those of others using the same service. Demand-responsive transport services are active in the United States, in Australia, and in the United Kingdom.

Furthermore, in an ever-increasing number of cities, shared taxi services are beginning to appear; these offer the possibility of sharing a taxi trip in the company of other travelers along the same route.

Green Intermodal Transportation

Recently, a growing number of cities have adopted new measures aimed at promoting a more sustainable form of intermodal transportation (also called "green intermodal transportation") through carpooling, car sharing, and bike sharing.

Carpooling is mainly used in the United States and in Northern European countries: it consists of sharing a private car with a number of people who travel the same route to reach their individual places of work or study. In countries where this mode of transport is the most developed, there are designated gathering points for users that can be reached on foot, by bike, or by bus. The chief advantage of carpooling is that it reduces the number of cars in circulation.

Car sharing is a service offered by special car-hire companies. Generally speaking, car sharing enables users to book a car that they pick up at a special parking location. This may be defined as an intermodal form of transport, as users reach the parking points on foot or by public means (this depends on the distance between the individual home and the car park). Car sharing users pay an annual subscription fee in addition to the costs of use, based on the distances actually covered and on booking time. Notwithstanding the advantages for the environment created by car sharing (decrease in the number of cars in use, more efficient use of the car, and less pollution, as cars are replaced not according to age but according to mileage), this service has difficulties in growing because it involves limitations for the users; for example, distance from the car park, the need to book the times of use, and the impossibility of only one-way trips because the car must be returned.

One of the most effective systems of sustainable intermodal mobility is bike sharing. This is a bike-hire service. In general, bikes are owned by local authorities or by private companies and are put at the disposal of users at special sites, usually located near railway stations, bus terminals, and in the main squares of city centers. Bike sharing is probably the most ecologically agreeable means of transport, as it involves the use of both bike and public transport. In fact, bike-sharing users reach bike pick-up sites by public means and use the bikes to travel within city centers.

In an effort to encourage the diffusion of the above means of sustainable intermodal passenger transport, some public administrations have adopted policies of park pricing and road pricing. Park pricing policies are now fairly well distributed: It means applying varying parking fees according to area and time. Usually, parking costs increase as one nears the city center. Policies of road pricing envisage a payment of special tariffs for entering the urban perimeter of a city by private car. The reason for road pricing is to deter the citizen from using his or her private car by making its use expensive compared with the cost of public transport.

See Also: Bicycling; Bus Rapid Transit; Carpooling; Traffic Calming; Transportation Demand Management.

Further Readings

Brian, Solomon. *Intermodal Railroading*. St. Paul, MN: Voyager/MBI, 2007.

Leinbach, Thomas R. and Cristina Capineri, eds. *Globalized Freight Transport: Intermodality, E-Commerce, Logistics and Sustainability*. Cheltenham, UK: Edward Elgar, 2007.

Lowe, David. *Intermodal Freight Transport*. Amsterdam: Elsevier Butterworth-Heinemann, 2005.

Organisation for Economic Co-operation and Development (OECD). *Benchmarking Intermodal Freight Transport*. Paris: OECD, 2002.

Rodrigue, Jean Paul, et al. *The Geography of Transport Systems*. London: Routledge, 2006.

Federico Paolini
University of Siena

KABWE, ZAMBIA

Kabwe is a town in the Copperbelt province of Zambia and has approximately 210,000 residents. Kabwe is most commonly known for human fossils found in a cave just outside the city. However, Kabwe is sadly one of the 10 most polluted places on Earth, with severe amounts of lead, cadmium, and zinc pollution resulting from decades of unregulated industrial mining and smelting operations. It is estimated that altogether almost 250,000 people living in the area of and around Kabwe are suffering from varying degrees of lead poisoning. Kabwe, hence, is considered to be a paradigmatic example of the health hazards that uncontrolled industrial mining and smelting activities can pose to humans and to the environment.

In 1902, when Zambia was still under British colonial rule, rich deposits of metal were discovered in a hill outside the village of Kabwe, which immediately led to the beginning of mining activities and the development of a town (named Broken Hill from 1912 until 1967). Mining operations (and later also smelting) were initially controlled by the Rhodesian Broken Hill Development Company and were later taken over by Zambia Consolidated Copper Mines. Until the mid-1930s, Kabwe was the biggest and richest lead mine in the world. Industrial mining continued until 1994, when the Zambian government decided to close the mine.

Under British colonial rule, as well as under Zambian rule, mining and smelting operations in Kabwe went practically unregulated. Over 90 years of heavy mining and smelting activity without proper concern for the human and natural environment, thus, have led in Kabwe to severe pollution by toxic metal dust precipitation across a radius of at least 20 kilometers. Local levels of lead, zinc, copper, and cadmium particles in soils, vegetation, and water are significantly higher than normal.

Consequences of Industrial Waste

In 2006, the Blacksmith Institute, a New York-based nonprofit organization committed to fighting the health-threatening consequences of industrial wastes, published a list of the 10 most polluted places on Earth. The list featured Kabwe, La Oroya (Peru), and Rudnaya Pristan (Russia)—three places in which the local population is highly affected by lead poisoning from (former and ongoing) industrial mining activities. Although Rudnaya Pristan fell out of the top 10 in 2007, La Oroya and Kabwe remain among the 10 most polluted places.

In Kabwe, zinc, copper, lead, and cadmium concentrations in soils are significantly higher than those recommended by the World Health Organization. As all of these metals are nonbiodegradable, pollution in Kabwe is long lasting and cumulative. The consequences for the local population are immense because especially lead and cadmium are highly toxic and extremely dangerous even in small amounts.

Studies have recorded that local children's blood lead levels average between 50 and 100 mcg/dL, which is up to 10 times the maximum allowed by the U.S. Environmental Protection Agency and much higher than even in La Oroya. Children and young men are at especially high risk of lead poisoning; children in the Kabwe area already suffer at birth from high blood lead levels (as a result of placental lead transfer before birth), and children are also very likely to play in contaminated dust. Many young men are exposed to greater risk because many locals use the inoperative mine to scavenge for scrap metals. However, inhaling contaminated dust is a serious issue for all members of the local community.

A further major health risk is posed by a small waterway running from the former smelter right through the town center. The creek was used as a wastewater canal from the then-active smelter, resulting in severe contamination of the water, which even today exhibits major levels of metal pollution. Creek water is used by local families and children for bathing and washing, thus exposing a high number of people directly to dangerous amounts of lead, cadmium, and zinc particles.

The effects of metal poisoning in the human body range from vomiting and diarrhea to kidney failure, debilitating diseases of the central nervous system, and death. Commonly, a blood lead level above 120 mcg/dL is considered to be life threatening, but in certain townships of Kabwe blood lead levels as high as 250 mcg/dL have been recorded. Within the Kabwe area, an increased likelihood of kidney problems, slow mental development of children, and other metal poisoning–related diseases, as well as related premature deaths, has been observed. As a result of insufficient documentation by local authorities and the fact that the biggest part of the local community was for a long time unaware of the pollution problem, however, exact numbers of lead poisoning–related deaths and illnesses are not available.

One of the main problems in Kabwe is that the nonbiodegradable nature of the metals means that cleanup programs will be both protracted and costly. Thanks to increased scientific and popular interest in Kabwe and its pollution problems, however, at least the first steps toward improving the situation have been undertaken.

Cleaning Up the Mess

The World Bank approved a $40 million budget for cleanup measures under the framework of the Zambian Copperbelt Environment Project, roughly $20 million of which will be used for measures directly related to cleaning up Kabwe. In addition, in 2007, the Copperbelt Environment Project received a $10 million grant from the Nordic Development Fund.

The first step toward improving conditions is to educate the local population and raise their awareness of the dangers of lead poisoning and the possible sources of contamination. This includes the distribution of simple advice on how to avoid contact with lead and cadmium particles; for instance, through increased hygiene (e.g., washing hands before eating) and taking precautionary measures (e.g., keeping children from playing in dust and bathing in the wastewater creek). A large part of the local educational work is provided by the Kabwe Environmental Rehabilitation Foundation, a local nongovernmental organization committed to educating the local community.

However, current remediation plans also include relocation and resettlement programs, as especially those townships situated in the corridor of prevailing windfall suffer from life-threatening levels of lead pollution. Therefore, entire neighborhoods of Kabwe might need to relocate. Remediation measures will also address the problem of slag heaps and other contaminated waste relicts of Kabwe's mining and smelting operations.

As the Zambian government has finally started to acknowledge the severe nature of the pollution in Kabwe and its danger to the local population, first signs of progress should be seen shortly. Overall, Kabwe should serve as a warning about the potentially devastating consequences of industrial mining and smelting. Even 15 years after mining operations in Kabwe were discontinued, contamination levels are extremely high, and their effects on people's health are grave. Remediation in Kabwe will take years, during which the local community will continue to suffer.

See Also: La Oroya, Peru; Linfen, China; Sumgayit, Azerbaijan; Vapi, India; Waste Disposal; Water Pollution.

Further Readings

Backsion, Tembo, et al. "Distribution of Copper, Lead, Cadmium and Zinc Concentrations in Soils around Kabwe Town in Zambia." *Chemosphere,* 63/3 (2006).
Blacksmith Institute. "The World's Worst Polluted Places: The Top Ten of the Dirty Thirty." http://www.blacksmithinstitute.org/wwpp2007/finalReport2007.pdf (Accessed April 2008).
Leteinturier, B., et al. "Reclamation of Lead/Zinc Processing Wastes at Kabwe, Zambia: A Phytogeochemical Approach." *South African Journal of Sciences,* 97/11–12 (2001).

Fabian Schuppert
Queen's University Belfast

KAMPALA, UGANDA

Kampala City is known for its rolling hills and green landscapes. Having evolved in the Buganda Kingdom, Kampala has a mixture of traditional and modern character, exhibiting economic, social, and political hegemony in Uganda. Its sprawling neighborhoods, extensive road network, and infrastructure are, however, characterized by an informal process of development. Although the city provides opportunities to the population through administrative and economic roles, its management remains a huge challenge for sustainability. As the major industrial hub in Uganda, it absorbs over 40 percent of the total labor force and contributes greatly to the national gross domestic product. As an economic center, Kampala provides most of the country's opportunities for business, from physical and communication infrastructure to labor and financial services. Despite these great opportunities, the city continues to grow and develop unsustainably, degrading the ecosystems within it and around it, with a large ecological footprint. Ecological degradation is manifested in high per capita waste generation, energy use, land use change, and pollution. These impacts are explained by the underlying demographics. With a high population growth rate of 5.71 percent, the demand for energy, transportation, consumables, and housing is unsustainable.

Waste Generation and Lack of Disposal

Demand for consumables drives business in the city but also leads to waste generation. When coupled with inadequate waste management systems, the result is an accumulation of waste scattered around the city, posing environmental risks. The generation of methane related to waste management, in addition to carbon emissions, raises issues of sustainability. With one landfill already declared unsanitary, individuals are largely responsible for the unsustainable waste management practices of indiscriminate dumping, burning, and compositing—the biggest environmental challenges in the city. An estimated 1,000 tons of wastes are generated per day in the city, but only about 40 percent is collected and disposed of. Seventy-five percent of the waste is organic, and three-fourths of the organic wastes are peelings from bananas, potatoes, cassava, and sweet potatoes. This implies that nutrients transfer from the rural hinterland to the city. Making the city greener would require social, technological, and economic innovation to establish systems for recycling the nutrients. Various innovations at the neighborhood and city levels are being piloted, including nutrient recycling. An example is "food peelings," which has evolved as a technological and economic innovation. The system comprises four elements: collection for livestock feeding, composting the peelings for urban farmers, collecting and retransporting them to rural areas for manure, and a collection-drying-milling process used for poultry feed. The later innovation is the newest and can reduce the amount of landfill waste by 50 percent. This has multiple effects: a reduction in costs for waste management, reduction in energy for such transportation, and reduction in emissions. These neighborhood-level activities are turning the tide of over two decades' reduction in ecosystem services, making Kampala City greener.

Impact of the Transportation System

An important aspect of Kampala City is its transportation system. Urbanism has been described by various attributes, among which mobility is a key characteristic. Urban population mobility is driven by three broad factors: economic, characterized by movement between home and work place; social, in response to social need through visits; and political, in terms of the power relations among people at household, community, and city levels, as explained by the "political economy" theory. Associated with population mobility is demand for energy for the transportation sector. Although domestic energy demand can surpass that of transportation if the wood biomass is considered, recent studies show a high mobility rate among the residents for economic reasons. With this level of energy use, the toll on the environment is through greenhouse gas emissions. The transportation sector is largely inefficient and unsustainable because of high costs, travel time, and carbon dioxide (CO_2) emissions of from 0.11 kg CO_2/L to 3.54 kg CO_2/L of diesel fuel. In addition to the transportation-related energy demand, there is a demand to generate power for different uses, following a reduction in national hydropower generation. A government tax waiver on diesel products and generators triggered a surge in diesel fuel use across the country. This has increased CO_2 emissions, a situation requiring reversal to achieve the Intergovernmental Panel on Climate Change emission target of 2 tons per capita.

Impact of Urban Agriculture

Sustainable urban development discourse has extensively covered the issue of urban agriculture in cities of sub-Saharan Africa, including Kampala. With the many definitions and dimensions of urban agriculture, including the cross-social class engagement in urban agriculture, the practice can be traced back for decades in Kampala. Formerly an illegal form of urban land use, urban agriculture has the potential to regenerate the lush green ecosystem of the city. This is achieved by the coupling of urban agriculture with livelihood systems, and, more recently, as a means of adapting to climate change effects. Technologies could increase greenery in the city and provide the urban poor with alternative livelihoods. Thus fruit trees, vegetable production, and other perennial crops have been promoted as strategies for sequestering carbon into soil and terrestrial systems and for enabling plot-level infiltration with a reduction in runoff, which can translate into a reduction in flooding associated with climate change. Although studies have shown that dangers related to uptake of toxic metals by plants exist, the potential for urban agriculture lies partly with national-level changes in transportation and energy policies. This would imply reduction in the use of leaded fuel and responsive urban planning that would ensure energy-efficient systems.

Conclusion

In the future, Kampala should remain green in terms of terrestrial ecosystem, although it is under threat of degradation. Despite this threat, which is also associated with a reduction in ecosystem services, there are efforts to regenerate the ecosystems. Kampala's ecological footprint is currently too large, but it can be reduced through innovative waste management practices, nutrient recycling, and a move toward green energy, among other practices.

See Also: Green Energy; Sustainable Development; Urban Agriculture; Waste Disposal.

Further Readings

Byandala, A. J. "Traffic Congestion of Kampala City Roads—Solutions." Paper presented at Seminar on Traffic Congestion in Kampala City: Solutions, Uganda Institution of Professional Engineers, 1996.

Environmental Resources Ltd. "Solid Waste Disposal: Kampala Final Report." Ministry of Water Lands and Environment, World Resources Institute, 1990.

Kendall, H. *Town Planning in Uganda: A Brief Description of the Efforts Made by Government to Control Development of Urban Areas From 1915 to 1955.* Kampala: The Crown Agents, 1955.

Mubiru, F. "Solid Waste Management in Kampala." Paper Presented at Workshop on Solid Waste Management in Kampala. 1997.

Mugabi, S. D. *Domestic Solid Waste Management in Uganda; Attitudes, Practices and Policy Recommendations.* MUIENR. Kampala: Makerere University, 1998.

National Environment Management Authority. "State of the Environment Report for Uganda." Kampala: National Environment Management Authority, 2001.

Nostrand, V. "Kampala First Urban Study." Kampala: Kampala City Council, 1993.

Nyakaana, J. B. "Solid Waste Management in Urban Centers: The Case of Kampala City in Uganda." *East African Geographical Review*, 19/4 (2000).

Songsore, J. and G. McGranahan. "The Political Economy of Household Environmental Management: Gender, Environment and Epidemiology in the Greater Accra Metropolitan Area." *Journal of Political Economy*, 26/3:395–412 (1998).

Shuaib Lwasa
Makerere University

L

LANDFILLS

In the United States, 228 million metric tons per year of trash/solid waste materials are generated at an astonishing rate of 4.6 pounds (2.1 kilograms) per day per person. It is twice the amount generated by other major developed countries in the world. According to the U.S. Environmental Protection Agency, about 32.5 percent of solid waste is recycled, 12.5 percent is burned, and the rest—55 percent—is buried in landfills.

Landfills used by a single producer to bury their own waste at the place of production are known as internal waste disposal landfills. Landfills in urban settings are used by many producers and are known as municipal solid waste landfills (MSWLFs). They receive solid wastes from households or commercial operations and nonhazardous sludge, industrial solid waste, and construction and demolition debris. MSWLFs are carefully designed structures built into or on top of the ground for solid waste disposal, so that the wastes can be isolated from the surrounding environment including groundwater,

Compaction of waste at landfills like this one in Perth, Western Australia, is organized by cells. Once one cell is covered with compacted solid waste and topped with soil or other materials, another cell is dug close to it.

Source: Ashley Felton/Wikipedia

air, rain, scavenging animals, and so on. This isolation is accomplished by the provision of an impervious bottom liner and daily layering of soil on top of the wastes.

Landfill Construction Criteria

The MSWLFs are built in compliance with federal regulations because landfills are ticking environmental hazards. In the United States, taking care of trash and building landfills are local government responsibilities. An environmental impact assessment is conducted on the proposed site by the agency responsible for the landfill to make it commercially and environmentally viable. The important requirements for building a landfill include location, capacity, and stability: composition of the underlying soil and bedrock at the proposed location, surface water flow at the proposed site, the effect of the proposed landfill on the local environment and wildlife, the historical or archaeological value of the proposed site, and most important, cost.

Location is one of the major criteria for siting a landfill. It must be constructed far from densely populated area and should have easy transportation access for trash producers, as well as for transporting compost from the site. Several studies have shown that in the United States, landfills are being constructed close to poor neighborhoods or settlements of socially weaker or underprivileged populations. The capacity of the landfill must be determined on the basis of the density of the wastes, the number of people to be served by it, and the construction of lining, drainage layers, and daily and intermediate covering availability. Underlying geology, nearby earthquake fault lines, groundwater table, and location of nearby flood plains and water bodies such as rivers, streams, and lakes also help determine the construction of landfills. Landfills can easily pollute groundwater; therefore, special care should be taken to confirm the stability of a landfill. Surface water should not flow from a landfill to nearby properties, and landfills should not be constructed in locations where there are possibilities of interference with local or migrating birds, wildlife, and fisheries. Landfills should not be built on locations that contain historical and archeological artifacts. The final consideration for building a landfill is its cost. In the United States, the cost of landfills is mostly borne by the local public. A feasibility study must be conducted to determine the cost of building a landfill.

In general, landfills are constructed to fill existing pits in the ground created by mining activities, to fill naturally occurring valleys or canyons, or are piled above ground. However, mines or quarries frequently contact the groundwater supply. If the landfills are constructed to fill mines or other natural depressions, extreme care must be taken to observe that the groundwater is not getting contaminated. Therefore, groundwater monitoring wells should be dug at various points around the site to monitor water quality, so that escaping wastes can be captured.

Construction Process of a Nonhazardous Landfill

A typical landfill is constructed with a bottom liner system, cells to store solid compacted wastes, a stormwater drainage system to collect rainwater that falls into the landfill, a leachate collection system to collect percolated contaminated water through the landfill to stop its flow to groundwater, a methane collection system, and a covering or cap to seal the top of the landfill.

The bottom liner of a landfill prevents the waste contaminates from coming in contact with the outside soil, especially the groundwater and/or aquifer. In nonhazardous landfills, the liners are usually some type of durable and puncture-resistant synthetic plastic. They are made of polyethylene, high-density polyethylene, or polyvinylchloride materials and are usually 30–100 mm thick. Compacted clay soils are often times used as an additional liner over the plastic liner. The plastic liner often is surrounded on either side by a fabric mat to help keep it from tearing or puncturing from the nearby rock and gravel layers.

The landfill capacity is directly related to the cells' volume. The cells are the free space to store old or new compacted or fresh solid wastes. The cells are of sufficient volume to store at minimum a day's worth of waste. Compaction of waste is done in the cells, which are constructed mostly by bulldozers. Once one cell is filled with compacted solid waste and covered with soil or covering materials, another cell is dug close to it.

Waterways are built in the landfill site to discharge any rainwater before coming into contact with the landfill's decomposed wastes. Plastic drainage pipes and storm liners are used in the landfill to collect water from areas of the landfill and channel it to drainage ditches constructed around the landfill. These drainage ditches are either concrete or gravel lined and help carry water to collection ponds. In the collection ponds, the drainage water is left to settle and tested for contaminates before releasing it to outside land.

This water that comes from the landfill can never be contaminant-free. The dissolved contaminants in the water from landfills are called leachate. Leachates are typically acidic. Perforated pipes run throughout the landfill to collect leachate. These pipes then drain into a leachate pipe to carry the leachate to a collection pond constructed beside the landfill. Leachates are either pumped to the collection pond or allowed to flow to it by gravity.

A methane collection system is another optional facet of a typical nonhazardous landfill. Because of the anaerobic reaction, methane is formed from landfills on a regular basis. Series of pipes are embedded in the landfill to capture the produced methane and are either flared up or used to produce electricity.

Operating Procedure of Landfills

Three important operating procedures are applied to nonhazardous waste landfills. They include confining the wastes to as small an area possible for ease of operation, compacting the trash or solid waste through the use of mechanical compacters to reduce its volume, and covering the compacted waste with a soil layer on a daily basis. The wastes coming to the landfills are generally screened, or the waste producers are educated about the type of products that should be sent to landfills. In many communities, waste collection vehicles are employed to collect solid wastes curbside. These collectors are well trained to screen waste materials before dumping them into the collection truck. In landfills, compacters or bulldozers are used to spread and compact the waste. In addition to soil, sprayed-on foam products, temporary blankets, chipped wood, and degradable biosolids are used daily to cover the compacted solid wastes. To conserve space in a landfill, a tarpaulin is used as a cap or covering for the cells after each day of compaction and storing, giving the landfills more longevity.

Environmental Impacts of Landfills and Their Advantages

As mentioned earlier, landfills are potential sources of environmental hazards. Pollution is the main adverse impact of a landfill. Decomposed wastes produce huge amounts of contaminants that pollute the groundwater and/or aquifers if they leach into it. It is a huge problem in areas where the groundwater table is very high, such as in coastal areas. Air pollution is another common feature of a landfill. Methane gas is formed as a result of waste decomposition. Methane is a greenhouse gas that enhances global warming and subsequent climate change. Rats and flies and other harmful insects flourish in landfills and can spread diseases. Dust, foul odor, and noise pollutions are other adverse impacts of a landfill.

Once a landfill site completes its operational requirement, it can be reclaimed for other purposes. Parks, golf courses, and sports complexes and fields can be built on the reclaimed land. Many cities, including parts of New Orleans; Washington, D.C.; Mexico City; Helsinki; the Chicago shoreline; the city-state of Singapore; Manila; and Hong Kong are built over reclaimed land created over unusable land such as wetlands, natural depressions, and mining lands.

Fresh Kills landfill in Staten Island, New York, was the world's largest landfill or man-made human structure. It is now slated to become a park for local residents. Sydney Olympic Park (Australia) is the best example of the use of landfill-reclaimed land. The Olympic Park was the primary venue for the 2000 Summer Olympic Games. However, buildings constructed on landfill-reclaimed land must have a methane monitoring option inside the building to alert for the presence of a buildup of gases to a harmful level. Methane produced in landfills can be collected and used to generate electricity in a gas-fired power plant. In developing countries, landfills are a source of food and income for many poor residents.

Water and air are circulated into specially designed landfills, known as "bioreactor landfills," to accelerate the biological decomposition of waste material and maximize the production of biogas. These biogases are collected through perforated pipes fitted inside the landfills and can be burned to produce electricity.

Superfund sites are advanced landfills controlled by federal government to clean up uncontrolled hazardous waste materials that cannot be managed by MSWLFs. Landfills are tens of times more than the current available superfund sites in United States. In Hall County, Georgia, there are 13 landfills to cater to the need of the county's inhabitants.

See Also: Environmental Impact Assessment; Garbage; Recycling in Cities; Waste Disposal.

Further Readings

Freudenrich, C. "How Landfill Works." 2009. http://science.howstuffworks.com/landfill6 .htm (Accessed July 2009).

Neal, R. and A. Allen. "Environmental Justice: An Annotated Bibliography." Report Prepared for the Environmental Justice Resource Center. Atlanta, GA: Clark Atlanta University, 1998.

Schroeder, P. R., et al. *The Hydrologic Evaluation of Landfill Performance (HELP) Model: Engineering Documentation for Version 3*. Washington, D.C.: U.S. Environmental Protection Agency, Office of Research and Development, 1994.

Scorecard. "Pollution Locator: Superfund." 2009. http://www.scorecard.org/env-releases/land (Accessed July 2009).

U.S. Environmental Protection Agency. "Waste—Non-Hazardous Waste—Municipal Solid Waste." November 13, 2008. http://www.epa.gov/epawaste/nonhaz/municipal/index.htm (Accessed July 2009).

Sudhanshu Sekhar Panda
Gainesville State College

LA OROYA, PERU

The city of La Oroya, located in central Peru, is 112 miles (187 kilometers) northeast of Lima, the country's capital. It suffers from extremely high levels of pollution because of the smelting operation that is located there. At one time, it had the highest smelting chimney in the world.

Over many years there have been a high number of premature deaths around La Oroya, Peru, some of which have been closely linked to the city's copper smelters. The area suffers from extremely high levels of pollution because of the smelting operation.

Source: Maurice Chédel/Wikipedia

There has been human habitation at La Oroya from about 8000 B.C.E. The Spanish started a mine there in 1533, taking over workings by the Incas. The mine was used to extract copper and lead. However, because it was difficult to get to and from La Oroya, there was not much interest in the region. There was much focus on the town during the War of Independence, however, when the forces of Simón Bolívar were fighting the Spanish Imperial troops, most of whom were based in Peru. The crucial battle of Chacamarca was fought just north of La Oroya, known as San Jeronimo de Callapampa—the city assumed its present name in 1893, the year in which the railway connecting it to Lima was completed. The railway was built by American engineer Henry Meiggs, and it became the highest standard-gauge railway in the world. In 1904, the railway line was extended to Cerro de Pasco, and the silver from the area, as well as more zinc, lead, and copper, was transported through La Oroya. Most of these mines were owned by the Cerro de Pasco Corporation, which in 1922 decided to establish a copper smelter at La Oroya. Three years later, the La Oroya became the capital of Yauli Province. In 1942, it was elevated to the status of a city. By then it was hailed as the "metallurgical capital of Peru." Largely denuded of plant life, even by the 1950s, there were no birds or other fauna.

The copper smelter was so successful that in 1928 a lead smelter was built. In 1952, a zinc refinery was added. These factories dealt not only with the products of local mines but also with those from mines in Cerro de Pasco. The need for timber to run the smelter led to deforestation of the region. Indeed, this had been happening for so many years that when Meiggs built the railway, sleepers had to be brought from Oregon in the United States. Over much of La Oroya's history, effluent from the mines and the smelters was dumped in the nearby river. By this time, 6,500 of the total population of 26,000 were working at the various refineries and smelters. Many of the local elite made their money renting houses to these workers and to others, usually coming from other parts of Peru to operate restaurants, garages, and other service industries. This division has led to many issues over fixing problems because the elite have refused to pay for the cleanup, often clashing with the local trade union, Sindicato de Trabajadores Metalurgicos de La Oroya.

In 1974, the Peruvian government nationalized the Cerro de Pasco Corporation, which formed a part of Empresa Minera del Centro del Peru S.A. (Centromin), owned by the government. The smelters continued to pollute the environment, and the slag heaps around the town continued to grow in size. In 1987, the first edition of the Lonely Planet guide to La Oroya stated that it "is a place to travel through and not to." It sticks by this advice in more recent editions of its guidebook, and its rival, Rough Guide, noted that "La Oroya is a not particularly inviting place."

The lack of investment in the smelters meant that pollution had started to seriously affect the health of many people living in the city. This was made worse by the presence of copper, zinc, and lead facilities, which led to production of many by-products including gold and silver, as well as antimony, arsenic, bismuth, cadmium, indium, selenium, sulfuric acid, and tellurium. The facilities were profitable, providing the government with considerable revenue, but they also resulted in major pollution, helping to create acid rain that destroyed any surviving vegetation.

In 1993, the Peruvian government of Alberto Fujimori, elected president three years earlier, embarked on a major privatization campaign and split up and sold off parts of Centromin. Doe Run Peru—a part of the Missouri-based Doe Run Corporation—bought 99.97 percent of the La Oroya smelter in 1997 for $120.5 million, as well as a capital contribution to Metaloroya, another part of Centromin, of $126.5 million. At the same time, Doe Run was indemnified against any actions that might be brought against them for environmental matters that arose before their purchase. Doe Run Peru, now a part of the Renco Group, tried to modernize production facilities at La Oroya. One of their aims was to reduce emissions and improve air quality. In 1999, it was discovered that arsenic in the air was 85 times more than considered safe, and there were 41 times more cadmium and 13 times more lead than regarded to be safe.

Doe Run did invest heavily in the operations in La Oroya, which has produced an improvement in domestic water supply, better domestic waste disposal, and better facilities at the smelter. Much of this was because Doe Run had agreed, when it bought the smelters, to embark on an environmental remediation management program (Programa de Adecuación y Manejo Ambiental). Initially, Doe Run expected that the cost of sorting this out would be $107 million, but it later discovered that the environmental problems were far worse than they had realized and that $244 million would be required to make good the damage. As a result, it made a request to extend the deadline for implementing the program from 2006 on, and the Peruvian government accepted this request.

Over many years there have been a very high number of premature deaths in La Oroya, some from malaria and other diseases, as well as those closely linked to the smelters. Draining of marsh areas and aerial spraying have removed malaria as a problem, and as a result pollution remains a major threat to life, with lung ailments being common. Tests on nearly all children in the area have shown very high levels of lead in their blood. This led to the Blacksmith Institute naming La Oroya as one of the 10 most polluted places in the world.

See Also: Air Quality; Healthy Cities; Kabwe, Zambia.

Further Readings

Deustua, José R. *The Bewitchment of Silver: The Social Economy of Mining in Nineteenth-Century Peru*. Athens: Ohio University Center for International Studies, 2000.

Laite, Julian and Norman Long. "Fiestas and Uneven Capitalist Development in Central Peru." *Bulletin of Latin American Research*, 6/1:27–53 (1987).

WorstPolluted.org. "La Oroya, Peru." http://www.worstpolluted.org/projects_reports/display/41 (Accessed December 2009).

Young, Everild. *The Land of Three Worlds*. London: Andrew Melrose, 1953.

Justin Corfield
Geelong Grammar School, Australia

LEED (Leadership in Energy and Environmental Design)

Leadership in Energy and Environmental Design (LEED) is a third-party voluntary certification system that rates buildings based on the extent to which they reduce their indoor and outdoor environmental impacts. The system was developed by the U.S. Green Building Council (USGBC) and is currently being used in over 100 different countries. It has become the de facto standard for defining "green buildings" in the United States and elsewhere. LEED consists of nine versions for different building types and designates buildings at one of four levels—Certified, Silver, Gold, and Platinum—based on the number of credits or points earned. As of April 2009, there were 22,000 buildings registered with LEED and over 2,400 completed, certified buildings, with the numbers of both doubling every year.

LEED and the USGBC

The overall goal of LEED is to reduce the effect of the built environment on natural processes. Buildings in the United States account for 70 percent of electricity consumption, 39 percent of energy usage, and 12 percent of potable water usage, while generating 39 percent of greenhouse gas emissions, exclusive of any goods or services produced inside. Reducing these figures cannot happen only through the practices of users such as turning off lights or using less water; there must also be a change in the way buildings are designed and sited. LEED serves as neutral, third-party verification that a building has been properly designed to reduce its environmental impact. With this in mind, the USGBC has the goal of not simply producing greener buildings but also transforming the building industry. There are two aspects to this: ratcheting up the standards over time so that achieving certification is always something noteworthy, and also developing and disseminating green building practices so that they become part of every building, not just those that are LEED-certified.

The USGBC was founded in 1993 by three environmental consultants as a nonprofit organization with the express purpose of creating a system of nationwide standards to stimulate interest in building sustainably. The LEED program began in 1998 and has since gone through several major revisions. At this time, the actual certification process is run by the Green Building Certification Institute (GBCI), and USGBC handles revisions to the standards as well as their promulgation.

The voluntary nature of LEED is one of the reasons it has been so successful (although as discussed later, many governmental jurisdictions are now requiring its use). Building owners can choose the degree of "greenness" they want to achieve, and beyond a few required points, they can choose which credits they want to earn to achieve that level. The scope and flexibility of the system means that even though it is not possible for any given building to earn all available credits (e.g., redeveloping a brownfield site is incompatible with preserving agricultural land), it is not very difficult to earn the minimum points required for certification for any given project. A building owner registers his or her project with the GBCI before or during construction. Once the building has been completed, the owner submits documentation to the GBCI to verify the number of LEED credits that were earned. Unlike other systems such as the United Kingdom's Building Research Establishment Environmental Assessment Method (BREEAM), on-site inspection is not part of the certification process—everything is done through online submission of documentation. Once that documentation has been approved, the building is certified as green and a plaque is awarded for display.

The standards themselves, including their scope and strictness, are developed by consensus-based, volunteer committees with drafts open to public comment. For example, the first draft of the Neighborhood Development rating system (see following) garnered over 5,000 public comments and has since been revised and opened to a second round of comments. However, only member organizations, not individual members, are allowed to vote on the actual standards. This tips the power balance in terms of organizations rather than individuals, although that does include state and local governments, nonprofits, and educational institutions, in addition to private firms.

The LEED Rating System

The USGBC describes LEED as the equivalent to nutritional information on a food package, laying out the points or credits earned in each of the available categories and enabling comparison of one building to another. The 2009 revision of the standards, LEED version 3, simplified the rating system so that buildings are rated out of 100 points at one of four levels: Certified buildings must achieve 40 points; Silver, 50 points; Gold, 60 points; and Platinum, 80 points. These standards are meant to reduce the overall environmental impact of a building, including energy, water, materials, and other components, by a certain percentage based on the level: 30 percent for Certified buildings, 40 percent for Silver, 50 percent for Gold, and 70 percent or more for Platinum. (Because the specific credits that are earned vary from project to project, these figures are only approximations.) Of the buildings certified as of April 2009, 29 percent were at the minimum level of Certified, 34 percent were Silver, and 31 percent were Gold, with the remaining 6 percent achieving Platinum. It is clear that for the most part, building owners who choose to certify with LEED are not going for the minimum requirements but are aiming for higher levels.

Points or credits are earned in one of five categories. Sustainable Sites credits involve locating the building so as to minimize the effects on ecosystems and resources; for example, ensuring access to public transportation, minimizing stormwater runoff, or reducing the heat island effect. Water Efficiency credits include reducing water usage by at least 20 percent and minimizing wastewater. Energy and Atmosphere credits include reducing energy usage and emissions, as well as reducing the use of ozone-destroying refrigerants, using renewable energy, and having the building commissioned. The Materials and

Resources category incorporates the use of materials that are sourced within 500 miles of the construction site or are sustainably harvested or recycled materials, and minimizing and recycling construction waste. Indoor Environmental Quality involves reducing indoor emissions from carpet, paint, and so on, and increasing indoor daylight and ventilation. There are also bonus points possible for Innovation in Design, which includes going above and beyond the credits in the above categories (e.g., a higher-than-required reduction in water usage) or making use of new technologies not specifically referenced in the standards (such as hybrid vehicles). In response to user concerns about the geographical specificity of buildings' environmental impact, the Regional Priority category adds extra weight to credits that are targeted for a specific bioregion. For example, buildings in the Southwest receive more points for reducing water usage than do buildings in the Midwest.

Because buildings vary tremendously in terms of size, function, and user requirements, there are multiple rating systems set up under LEED. New Construction dominates the list, with 65 percent of registered buildings falling under that category (which also includes major renovations). The other categories are Core and Shell (building exteriors only, 12 percent of projects), Existing Buildings (focusing on operations and maintenance, 11 percent of projects), Commercial Interiors (where the tenant does not own the building, 8 percent of projects), Schools (3 percent of projects), and Retail (2 percent of projects). New systems under development and testing include Healthcare (because of the special 24-hour energy needs of hospitals and other healthcare facilities) and Neighborhood Development (which blends smart growth and green buildings to go beyond a building-by-building approach).

One additional category, LEED for Homes, is slightly different from the others. Because of the higher volume and mass-production style of housing, as well as the fact that many regions had a green home certification process in place already, LEED for Homes is implemented through regional providers who carry out the certification. Version 3 of this rating system is being developed separately from the others. There are nearly 2,000 certified homes across the United States as of April 2009 that are not included in the totals above. About half of these are at the Silver level, with the rest roughly equally split between Certified, Gold, and Platinum.

The LEED system applies to the buildings, not to the firms that design and build them. An architecture or engineering firm cannot be designated as a "green builder," no matter how many LEED projects they are part of. However, individuals can become Green Associates or Accredited Professionals; there are over 82,000 such individuals as of May 2009. Both designations indicate that the recipient has passed a test covering knowledge of the LEED process, as tested by the GBCI. However, only individuals who have actually worked on a green building project can qualify as Accredited Professionals, and they have to specialize in one of five categories based on the type of project. The advantage of gaining accreditation is twofold: it makes the registration and certification process easier if an accredited professional is involved, and it also earns the project one credit. The split into multiple categories of accreditation indicates the rapid growth of the program and its value to the private sector.

Implementation of LEED

Since the initial nine pilot projects in 1999, LEED has taken off. One of the reasons for this exponential growth in the use of LEED is the introduction of policies at various levels of government encouraging or mandating its use. As of 2003, the U.S. Government Services Administration, which leases and maintains federal buildings, requires that all new federal building projects or renovations meet LEED standards. The U.S. Army has implemented a similar requirement for its facilities, as have the departments of Agriculture, Energy, and State. Thirty-one state governments have some kind of policy in place encouraging or mandating LEED, as well as 131 municipalities and 54 public school districts, colleges, and universities.

These policies range widely in terms of their requirements and are in many cases as flexible as the LEED system itself. For example, policies can be implemented via city ordinance, state legislation, or agency regulation. Requirements can be based on square footage, the type of building, or whether a building receives any public funding. Policies may require certification at a specific level, such as Silver, or offer incentives such as expedited permitting for developers who pledge to achieve a certain level. Major U.S. cities with LEED-based policies include Chicago, Los Angeles, Boston, and Dallas. Cities such as Chicago or Portland, Oregon, that were early adopters of LEED policies have been able to expand the influence of green building beyond the early public sector projects to private developments across the city, and later adopters are hoping to follow.

The private sector has also been highly involved in integrating LEED with their building projects. Ford, Sprint, Steelcase, PNC Bank, and Toyota are among the firms that have achieved LEED certification for some of all of their new facilities over the last few years. LEED for Multiple Buildings and the Portfolio Program both allow organizations with either campus-style facilities or multiple copies of the same building (such as branches of a bank) to streamline the certification process, rather than having to take each structure into account separately. For example, Citibank has taken advantage of the Portfolio system by having a prototype building certified that will be constructed in over 100 branch locations.

As might be expected, the distribution of LEED-certified buildings across the United States is not spatially even. The highest concentrations are in the Pacific Northwest, university towns, and government centers. To some extent, this corresponds with policies at local, state, and federal levels. However, there are also significant clusters that have come about from nonprofit or private actions. For example, the metropolitan area with the highest number of LEED-certified buildings per capita is Grand Rapids, Michigan, thanks to the efforts of furniture manufacturers Herman Miller, Steelcase, Allsteel, and Haworth to have their headquarters and manufacturing facilities certified (giving them a marketing advantage for their products that can be used to achieve LEED credits for reduced indoor emissions). Pennsylvania and New Jersey have relatively high numbers of green buildings because of the efforts of PNC Bank to have all of their branches certified. Finally, Gainesville, Florida, has the highest number of LEED-certified buildings of any city in the state, despite the city's relatively small population, because of the University of Florida.

In many cases, firms or organizations that are committed to using the LEED system are taking advantage of the marketing benefits it provides. Demonstrating environmental concern through LEED certification has become an important way for firms and organizations to show their green credentials. Although there are concerns about spending extra money to obtain certification, recent studies show that the additional cost of building green is only 2 to 9 percent of the total project cost. Furthermore, increased energy efficiency and reduced employee absenteeism have been shown to make up for those increased costs within only a few years, and leasing rates may actually be higher for buildings that are certified as green. LEED is now marketed in terms of being good not only for the environment but for the bottom line as well.

Worldwide, programs such as BREEAM in the United Kingdom, Comprehensive Assessment System for Building Environmental Efficiency in Japan, and Green Star in New Zealand and Australia serve a similar function as LEED. The LEED standards themselves have been adopted at project sites in over 100 countries, including Brazil, China, and India, and BREEAM International has been employed across Europe and Asia. The major differences between the systems are in terms of how certification is approved (on-site inspection for BREEAM, online documentation for LEED) and in percentage reductions from a baseline (LEED) compared with meeting quantitative targets (BREEAM). The Green Globes rating system, developed in Canada and based on BREEAM, has become a competitor for LEED within the United States. It incorporates more life-cycle analysis than does LEED and is aimed at smaller buildings, but it focuses more on the building itself than on its site and surroundings. Many green building policies do not specify LEED but any third-party rating system, which includes Green Globes.

There are five main criticisms of LEED as it currently stands. First is the added cost of construction, although this cost is usually less than developers fear and is usually recouped faster than they expect. Second, the standards generally do not mandate the tracking of performance over time, only on completion of the building. For example, a credit can be earned for purchasing green power, but that only has to be maintained for two years. Third, although there is significant attention paid to the sourcing of materials, there are no credits concerning life cycle or cradle-to-cradle analysis, which is increasingly important to ecologically minded architects and builders. Fourth, standards are based on percentage reductions in consumption or emissions, rather than on achieving absolute targets that are ecologically meaningful. Although this has advantages in terms of flexibility and matches the USGBC's goal to transform the industry one step at a time, it means that the criteria can be seen as somewhat arbitrary. Finally, there is concern that LEED allows firms to continue with "business as usual," achieving certification and claiming that this makes them environmentally friendly even if most of their practices are still environmentally harmful. Each of these concerns is being addressed by the USGBC through its ongoing revision of the standards and through making people aware of the overall costs and benefits of using LEED to build green.

See Also: Bioregion; Construction and Demolition Waste; Green Communities and Neighborhood Planning; Green Design, Construction, and Operations; Green Housing; Indoor Air Quality.

Further Readings

Cassidy, R., et al. "White Paper on Sustainability: A Report of the Green Building Movement. Building Design and Construction." http://www.usgbc.org/Docs/Resources/BDCWhite PaperR2.pdf (Accessed August 2009).

Cidell, J. "Building Green: The Geography of LEED-Certified Buildings and Professionals." *Professional Geographer,* 61: 200–15 (2009).

Kibert, C. *Sustainable Construction: Green Building Design and Delivery.* Hoboken, NJ: John Wiley & Sons, 2008.

Turner, C. and M. Frankel. "Energy Performance of LEED for New Construction Buildings." Prepared by New Buildings Institute for U.S. Green Building Council. http://www.usgbc .org/ShowFile.aspx?DocumentID=3930 (Accessed August 2009).

U.S. Green Building Council (USGBC). http://www.usgbc.org/aspx?DocumentID=3340 (Accessed August 2009).

Julie Cidell
University of Illinois

LIGHT POLLUTION

One-fifth of the world's population cannot see the Milky Way at night because of so-called sky glow. This NASA composite image of Earth at night shows particularly impacted areas like Europe, Japan, and North America.

Source: Image by Craig Mayhew and Robert Simmon/ NASA GSFC

Light pollution generally refers to human-designed, artificial lighting that illuminates areas other than those the light was designed to illuminate. A classic example of light pollution would be a streetlight that is constructed to illuminate a sidewalk or street to allow walkers and motorists to travel with greater ease and safety at night. Depending on the design and construction of the light fixture, the light can either be directed directly to the ground beneath it or, alternatively, can release light in every direction. The light that escapes above the light fixture and is not helpful in illuminating the ground below contributes to what is called light pollution.

What Is Light Pollution?

Although light pollution may intuitively seem to be in a different category from other forms of pollution such as air or water pollution, it is rightly considered a form of pollution, if pollution is understood to mean the presence of something in the environment that is either harmful and/or toxic. Light pollution affects our environments aesthetically, and it also has been demonstrated to have harmful effects on the health of human and nonhuman species and to alter human and nonhuman animal behavior.

What Light Pollution Does

Light pollution has various effects. The most visible effect of light pollution is the reduction of star visibility. This is the result of what has been termed *sky glow*. Sky glow is what hangs over cities and urban centers. It is said to be the result of artificial light interacting with clouds and other elements in the atmosphere that causes the bright glow common to urban areas that presents a dome-like barrier between the inhabitants of urban areas and the larger cosmos beyond the glow. There are far fewer stars visible to residents of urban, light-polluted areas than to inhabitants of rural, less light-polluted areas. Most of the world's population currently lives under sky glow of greater or lesser severity. Sky glow prevents one-fifth of the world's population from seeing the Milky Way at night; this number is much higher in the most affected areas, such as Europe, Japan, and North America, where the number is set as high as two-thirds.

Light pollution can have severe cultural and physical effects. In terms of health and behavioral effects of light pollution, many species, including our own, are sensitive in our daily routines to the presence and absence of light. Our species is fundamentally diurnal (active by day) as opposed to nocturnal (active by night). To facilitate our activity by night, we need to light up our way. Unfortunately, excess artificial light can severely and detrimentally affect the health and behavior of other species. Light pollution has been cited as disrupting feeding patterns, migration patterns, mating, and so on. There are reported cases of birds being attracted by artificial light from buildings only to collide with them, or to circle the light source until dying of exhaustion. A famous example used in discussions of light pollution is that of the sea turtles that lay eggs on dark beaches; when the hatchlings emerge they gravitate toward the brightness of the sea. With beachfront development and its attendant light pollution, however, the light from the sea is often eclipsed by the light from the beach, thus disorienting the hatchlings, leaving many of them vulnerable and unlikely to ever reach the sea. Another example of the effect of light pollution on animals is the opportunity it provides for creatures to forage for food longer; however, this simultaneously leaves them more vulnerable to predation. The general point is that many species' health and behavior are closely connected to the cycles of light and day, and as we engineer the darkness into light and obstruct these rhythms, animals and ecosystems are being affected in ways we do not yet fully understand.

Our species, too, is subject to the natural cycles of light and darkness, day and night. There are two areas of emphasis open for further research regarding the effects of light pollution on human health. The first regards the effect of light pollution on our circadian rhythms—exposure to excess artificial light could disrupt our circadian rhythm and alter our sleep schedules. A second concern that draws attention studies a possible connection between light pollution and higher instances of breast cancer.

On a cultural level, the stars have figured largely in humanity's imaginings, stories, and understanding of our place within the universe for thousands of years. It can be seen as a cultural tragedy and injustice that a profound source of our own self-understanding and cultural identity is blocked from view for many of us because of where we live, unless we have the means to venture out into unpolluted areas to see the stars beyond in all their splendor. Losing the stars from sight may have as a disorienting effect on human beings in our own way as it does for the sea turtles mentioned above.

How Light Pollution Can Be Rectified

Although light pollution was likened to other forms of pollution in terms of its effects, light pollution is arguably much more easily addressed and ameliorated than other forms of pollution. Nuclear waste as a type of pollutant, for example, remains radioactive and poses great risks for an incredible length of time. Many forms of cleanup for various pollutants require great investments of time and money. Light pollution, in principle at least, is much easier to address. Stories are told of people who experienced profound moments when confronted with power outages and were able to see the night sky in a way never seen before from their homes. The effects of light pollution do not linger in the same way other pollutants do once the source of pollution is remedied.

One solution suggested and encouraged by many is better design in light fixtures. Rather than designs that allow light to disperse in every direction, pollution-conscious designers, individuals, businesses, institutions, and city planners could opt for light fixtures that contain the light to the desired area of illumination. To this end, the International Dark-Sky Association has a "fixture seal of approval" program and maintains a list on its website of approved light fixtures that are "dark-sky friendly." This is a situation in which dark-sky-friendly fixtures, and antipollution efforts in general, also make good economic sense.

Products and programs that attempt to be more dark-sky friendly are more efficient and cost less money in the long run. This is not a completely simple situation, however, because what constitutes pollution is not universally agreed on. Although one citizen may claim to be concerned for his or her health and the effect of light on their own circadian rhythm, alternatively, a company may want to defend its ability to advertise its product in a visible space as brightly as possible. There are difficult discussions to be had and conflicting ends. However, the promising situation of light pollution is that should communities local or global come to an agreement to remedy light pollution, there is good reason to believe we could be successful.

See Also: Daylighting; Energy Efficiency; Environmental Planning; Green Communities and Neighborhood Planning; Green Design, Construction, and Operations.

Further Readings

International Dark-Sky Association. http://www.darksky.org (Accessed April 2009).
Klinkenborg, Verlyn. "Our Vanishing Night." *National Geographic* (November 2008).
Scribner, Brad. "Light Pollution." http://ngm.nationalgeographic.com/geopedia/Light_Pollution (Accessed April 2009).

Jonathan Parker
University of North Texas

LINFEN, CHINA

Many Chinese cities have pollution levels that exceed World Health Organization standards—some are the world's most polluted cities. Linfen is an example. In the past the city experienced chronic air pollution for around 163 days of the year as a result of the operation of surrounding industries and coal mines. This pollution has led to widespread disease and increased mortality and, in some instances, affected the structural integrity of some buildings. Recently, the Chinese government has taken action to improve the quality of Linfen's air. This article considers the origins of the pollution problems that have beset Linfen, actions that have been taken to remedy the pollution, and prospects that Linfen's residents will enjoy a healthier future.

Sources of Pollution and Impacts

Located in China's Shanxi Province, inland from Qingdao, Linfen has a population of approximately 4.2 million. Pollution in Linfen comes from multiple sources and has affected the air, water, and soil quality. Coal mines in the nearby hills are a major source, as are iron foundries, steelworks, coking plants, and coal-fired power stations. Pollution also comes from cement works, tar refineries, various smelters, and a variety of other industries, as well as car and truck traffic. The main pollutants include polycyclic aromatic hydrocarbons, sulfur dioxide, carbon monoxide and dioxide, nitrogen dioxide, fly ash, and suspended particulates (PM2.5, PM10), as well as arsenic, lead, ammonia, and volatile organic compounds (e.g., phenols and benzene). Recent studies have also found that what little agricultural land remains in the area is affected by soil contamination, seriously affecting food safety. Crop yields are down as well, because pollution has stunted growth and reduced disease resistance.

The high levels of pollution have taken a serious toll on the health of Linfen's inhabitants. Many residents have suffered direct effects (e.g., eye, nose, and throat irritation; headaches; nausea; fatigue; and dizziness). Growing numbers of residents suffer from respiratory diseases as well. High lead levels have also affected local children, who have exhibited elevated rates of lead poisoning. In addition, a marked decline in sunlight, by as much as 10 to 30 percent, may have placed some children at risk of rickets. Many water supplies are also contaminated with arsenic, leading to epidemic levels of arsenicosis. Sufferers have symptoms including skin lesions, circulatory problems, hypertension, and in severe cases, cancer. Not surprisingly, the city's mortality rate has climbed in recent years, alarming local health authorities.

Government Action

Many of the problems described here are not unique to Linfen. The Chinese State Environmental Protection Administration was created in 1998 to begin to remedy these issues, and China now has a National Environment and Health Action Plan (2007–2015), which has established targets for cleaner production, energy efficiency, and reducing disease prevalence. The Ministry for Environmental Protection (which superseded the State Environmental Protection Administration in 2008) now assesses the environmental quality of 617 cities annually; it reported in 2008 that nationally, 65 percent of sewerage was being treated, 82 percent of garbage was being recycled, and 84 percent of medical waste was being appropriately managed.

China also appears to be slowly adopting ecocity principles, at least on paper. The Central Party has switched from a philosophy of "develop now and remediate later"—the model that was followed when developed countries like the United States industrialized—to now promoting the need for "sustainable development." All future development in China must exhibit ecologically sound construction and environmental protection practices. The Central Party has committed 1,000 billion Renmimbi (Chinese Yuan) in recent years to promote this cause. The Ministry for Environmental Protection also has also promulgated various legal documents such as "Total Emissions of Major Pollutants Assessment and Control" and "Environmental Protection Violations Law Action" to guide local governments in reducing pollution and protecting the environment. Remediating environmental contamination, however, has yet to emerge as a major goal. It is often the case that the central government's edicts take some time to filter down to local officials, and many local officials still pursue a "growth at all costs" philosophy. Real disparities also exist between chronic pollution levels in the western part of China compared to relatively better air and water quality in the country's southeastern cities.

Emerging Results

At the local level, some progress appears to have been made in Linfen. On January 15, 2007, Linfen government officials announced the "Blue-Sky-Green-Water Program," which contained energy-efficiency and emission-reduction actions. According to some Western nonprofit organizations, Linfen city officials planned to shut down 160 of 196 iron foundries and 57 of 153 coking plants by the end of that year. Pollution was to be dramatically improved through 30 steps, which can be summarized as five broad responses: updating industrial infrastructure, strengthening the control of industrial pollutants, improving city infrastructure, implementing ecologically sound construction, and establishing effective environmental management systems. This included more stringent entry criteria for industries, curbing the development of high-energy-consuming industries, and strengthening energy-saving incentives by improving the city's energy-saving tax system.

The city also embarked on four "addition and subtraction" strategies: supporting high-tech, efficient, and clean industries and eliminating inferior ones; enhancing the level of industrial development; heightening energy efficiency and upgrading corporate management; and increasing green space and decreasing contaminated land to improve livability in both urban and rural areas. According to Chinese sources, by the end of 2008 Linfen had shut down 265 heavily polluting factories and added an additional 686,700 m^2 of greenspace, reduced industrial accidents by 257 compared to the previous year, and increased the number of cleaner air days from 15 in 2006 to 310 days of better air quality by the end of 2008. The latest official report (July 2009) asserts that Linfen is now ranked 49th among the 113 key cities requiring environmental protection.

See Also: Air Quality; Beijing, China; Cities for Climate Protection; Environmental Planning; Green Communities and Neighborhood Planning; Healthy Cities; Renewable Energy; Sustainable Development; Tianying, China.

Further Readings

Blacksmith Institute. *The World's Most Polluted Places: The Top Ten.* New York: Blacksmith Institute, 2006.

Economy, Elizabeth C. "'The Great Leap Backward?' *Foreign Affairs,* 86 (2007).

Fagin, D. "Is China's Pollution Poisoning Its Children?" *Scientific American Magazine* (August 2008).

Fu, S., et al. "Composition, Distribution, and Characterization of Polycyclic Aromatic Hydrocarbons in Soil in Linfen, China." *Bulletin of Environmental Contamination and Toxicology,* 82 (2009).

Liu, L. "Urban Environmental Performance in China: A Sustainability Divide?" *Sustainable Development,* 17 (2009).

Shan, Y. "Investigation of Pb Pollution of Road's Green Land and Control Measures in Linfen." *Urban Environment & Urban Ecology,* 20 (2007).

Watts, Jonathon. "Fighting for Air: Frontline of War on Global Warming." *The Guardian* (March 26, 2007). http://www.guardian.co.uk/environment/2007/mar/26/globalwarming .china#article_continue (Accessed September 2009).

Zhang, Z. X. "China Is Moving Away From the Pattern of 'Develop First and Then Treat the Pollution.'" *Energy Policy,* 35 (2007).

Jason Byrne
Griffith University

Liu Li Can
International Centre of Communication Development

Zeng Li Qun
Beihai College, Beihang University

LOCATION-EFFICIENT MORTGAGE

Location-efficient mortgages (LEMs) are a financial instrument that allows homebuyers to shift a portion of the savings gained through reduced transportation spending over to their housing expenses. They are intended to reward homebuyers for choosing to live in walkable, transit-oriented communities, rather than electing to live in automobile-dependent outlying suburbs. As such, they are seen as a means to simultaneously address issues of sprawl, automobile overuse, unaffordable housing, and resource overconsumption. Rather than simply focusing on the price of the housing unit itself, LEMs approach housing affordability in a more holistic fashion, taking into account the overall affordability associated with living in particular regions of a city over others, based on likely transportation needs and associated expenses.

In the spring of 1996, three nonprofit organizations—the Center for Neighborhood Technology in Chicago, the Natural Resources Defense Council in San Francisco, and the Surface Transportation Policy Project in Washington, D.C.—formed a research team to investigate the possibility of establishing an innovative new mortgage product called the LEM. The LEM Partnership (now known as the Institute for Location Efficiency) was formed to conduct the related research, formulate the mortgage product itself, and seek ways to make it available in the primary and secondary mortgage marketplaces. The project is also partially funded by the U.S. Department of Energy, the U.S. Environmental Protection Agency, the Federal Transit Administration, and several private foundations. As a part of a Federal National Mortgage Association (Fannie Mae)–sponsored Alternative Underwriting Experiment, the LEM Partnership conducted negotiations with Fannie Mae to conduct market tests of the LEM. The first such test market was Seattle. At this time, there are only four housing markets in the United States in which the LEM is offered: Seattle, Chicago, Los Angeles, and the San Francisco Bay Area.

The rationale behind the LEM is that households in more-automobile-dependent communities have been found to spend on average more than 20 percent of their budgets on transportation compared with less than 17 percent in households located in communities with more transportation options available. The monetary costs of transport have a significant effect on standards of living, as the cost of private automobile transport is a substantial fraction of most family budgets. In fact, studies show that most American families spend more on driving than on healthcare, education, or food.

A 1994 paper by John Holtzclaw evaluated the relationship between four neighborhood characteristics on housing affordability—residential density, transit accessibility, neighborhood shopping, and pedestrian accessibility—as well as their effects on motor vehicle usage per household and total vehicle miles traveled annually per household. The study established that residential density and transit accessibility were the best statistical correlations with household transportation costs. In addition, higher residential density was found to be the most important factor in increasing location efficiency, as better transit accessibility, more local shopping, and a more pedestrian-friendly environment generally accompanied it. The study concluded that residential density and transit accessibility variables could be used to quantify household transportation costs with a high degree of reliability.

Similar products available include the energy-efficient mortgage, in which the costs of upgrading a home's energy efficiency are rolled into the cost of the mortgage, and the smart commute mortgage, which is similar to the LEM but with less emphasis on neighborhood characteristics.

Mortgages and Auto Dependency

The central premise of the LEM is that the demand for homes in suburban, auto-dependant locations is encouraged by conventional mortgage lending. Because conventional lending does not consider location efficiency, households qualify for the same mortgage no matter what the related cost of transportation is. In places where outer suburban housing is less expensive than inner city housing, conventional mortgage underwriting forces lower-to-moderate-income households to purchase their homes in locations that are overly dependent on automobile commuting. By placing a dollar value on desirable urban characteristics, the LEM can make such areas more affordable for lower-income households.

As a market-based tool, LEMs are intended as a "carrot" to encourage "smart growth," as opposed to more conventional and prescriptive antisprawl policies. As such, they can fill a role by augmenting the public sector's limited capacity to shape growth.

There are three steps involved in the calculation of so-called location efficiency savings for a given neighborhood. First, the model addresses homebuyer information (number of persons in the household and household income) and information on the area in which the home is located (households per total acre, pedestrian factor, households per residential acre, and transit access). The model then calculates auto expenses using econometrically estimated vehicle miles traveled and auto ownership, along with Federal Highway Administration data on the cost of owning and operating an automobile. Finally, the applicant's household automobile expenses are subtracted from the "base case"; that is, the automobile expenses for a household of similar wealth and size in a neighborhood with poor transit access, low pedestrian friendliness, and relatively low density.

Traditional lending guidelines require a minimum down payment of 5 to 20 percent of the appraised value, with a maximum housing-expense-to-income ratio of 28 percent and a maximum debt-to-income ratio of 36 percent. In contrast, a LEM requires a minimum down payment of only 3 percent of the appraised property value. Initially, the LEM allows a maximum housing-expense-to-income ratio of 35 percent and a maximum debt-to-income ratio of 45 percent. When the savings available from location efficiency are added on, the maximum housing-expense-to-income ratio becomes 39 percent and the maximum debt-to-income ratio becomes 50 percent. The mortgage term is from 15 to 30 years, with a fixed interest rate competitive with prevailing market rates.

To qualify for the LEM, homebuyers must do more than buy into a location-efficient area. Before purchasing their property, they must take part in counseling about location efficiency. Borrowers are also encouraged to use public transit, shop locally, and participate in an annual survey. Borrowers have also been able to take advantage of other benefits, such as bus passes at reduced rates.

Critiques of LEMs

LEMs have been criticized on a number of grounds, including the possibility that they are unlikely to create significant new demand for homes in targeted areas, as, presumably, many of the people who would be attracted to a location-efficient lifestyle already live in such areas.

There has also been concern expressed over the potential for higher default risks, as other products offering higher loan-to-value ratios have been consistently found to raise the probability of default. A 2001 study by Allen Blackman and Alan Krupnick showed that the absence of a negative correlation between the probability of mortgage default and location efficiency in effect amounts to making preferential loan terms available to a random sample of people.

Others point out that not all households are adaptable to the ideal LEM "lifestyle." In addition, despite the importance of lower levels of car ownership and use for those who do choose the LEM lifestyle, there is no way to prevent homeowners from buying or depending heavily on a car. Finally, although a high degree of availability to transit is assumed when a borrower is able to make the most of the LEM, it does not take into account proximity to employment.

Although LEMs have failed over the past decade to spread beyond their original four cities, numerous jurisdictions including Minneapolis, Philadelphia, Salt Lake City, and Pittsburgh have adopted the smart commute mortgage, which offers incentives to homebuyers to purchase houses accessible by public transit. To meet this requirement, borrowers must purchase a home within 400 meters of a bus route or 800 meters of a public transit light rail stop. Similar to the LEM, qualification guidelines are more flexible than those offered by a traditional 30-year fixed-rate mortgage. Incentives can include a down payment as low as 3 percent, two years of free public transit fares, and a $200 per month on paper income "increase" when calculating loan qualification.

Provisions for the expansion of the LEM program were included in the draft Green Resources for Energy Efficient Neighborhoods Act of 2008, but the bill never became law. However, given the severity of the home financing crisis that emerged in 2007, the prospects for the wider adoption of such "creative" home financing mechanisms may be in doubt.

See Also: Transit; Transit-Oriented Development; Walkability (Pedestrian-Friendly Streets).

Further Readings

Blackman, Allen and Alan J. Krupnick. "Location-Efficient Mortgages: Is the Rationale Sound?" *Journal of Policy Analysis and Management,* 20/4:633–49 (2001).

Holtzclaw, J. *Using Residential Patterns and Transit to Decrease Auto Dependence and Costs.* San Francisco, CA: Natural Resources Defense Council, 1994.

Krizek, K. "Transit Supportive Home Loans: Theory, Application, and Prospects for Smart Growth." *Housing Policy Debate,* 14/4:657–77 (2003).

McCann, B. *Driven to Spend: The Impact of Sprawl on Household Transportation Expenses.* Washington, D.C.: Surface Transportation Policy Project, 2000.

Michael Quinn Dudley
University of Winnipeg

LONDON, ENGLAND

Thanks in part to these floating, garbage-catching devices, the Thames is one of the cleanest metropolitan rivers in the world. London's growth, however, burdens a sewer system that dumps 52 million cubic meters of untreated sewage and rainwater each year into the Thames and Lee rivers.

Source: iStockphoto

London is a world city that faces many challenges in the context of climate change and a growing urban population. Although there is evidence that London's environmental quality is increasing, a growing urban population is putting pressure on land and resources. The mayor of London has a statutory duty to develop a strategic vision of how to address issues such as energy, water, and waste, and London's general approach to facing its challenges has been to involve key stakeholders from the business and community sectors in the environmental management of the city. A major focus of (re)development over the coming decades is East London, especially the Olympic Park and Thames Gateway developments.

London's Profile

London is the largest city in Europe, with approximately 7.5 million inhabitants, contributing around 20 percent to the United Kingdom's total gross domestic product. The administration is made up of a two-tier structure, in which the Greater London Authority (GLA), the citywide government, is in charge of strategic planning, economic development, transport, and police and fire services, and 33 local councils are in charge of local services such as schools, social services, local planning issues, and waste. The spatial development strategy set out by the mayor of London seeks to accommodate London's growth without intruding on the city's open spaces by increasing the city's currently relatively low density (4,730 inhabitants per square kilometer).

Climate Change and Energy

Climate change has become a top priority on the city's policy agenda since 2000, when the GLA (and the role of mayor of London) was established. In 2006, London's carbon dioxide emissions amounted to 44 million tons, comparable to that of countries such as Greece and Portugal and 8 percent of total U.K. emissions. A part of London's emissions are from landfill and waste incineration, but energy is the largest source of emissions, with the domestic sector accounting for 38 percent, followed by the commercial sector (33 percent), ground-based transport (22 percent), and industry (7 percent). Although emissions in the period 1990–2006 actually decreased by 1.5 percent, this is attributed to a decline in industrial activities in London and a more general shift in the United Kingdom's electricity generation mix from coal toward natural gas (and does not constitute a long-term trend).

The mayor set an ambitious target for London of a 60 percent reduction in carbon dioxide emissions below 1990 levels by 2025. As the bulk (75 percent) of emissions are caused by London's electricity and gas consumption, climate change and energy policy have become closely interlinked in London. The almost simultaneous publication of Energy Strategy and the London Plan in 2004 set up an unprecedented framework for tackling climate change in London (followed in 2007 by the Climate Change Action Plan). Policies have focused on three key areas: first, the planning system (by introducing a mandatory requirement of 10 percent on-site generation of renewable energy in new developments over a certain size); second, energy efficiency in the built environment by changing practices (schemes such as the Green Housing Programme and the Green Concierge Services for domestic users and Better Buildings Partnership, Green 500, and the Buildings Retrofit Programme for the commercial and public sectors); third, reducing London's carbon intensity by moving a quarter of London's energy supply to decentralized energy sources (mainly combined heat and power). To deliver these and involve key stakeholders from the business and third sectors, the mayor set up several strategic partnerships (such as the London Energy Partnership, London Hydrogen Partnership, and London Climate Change Agency). As a result there has been a steady increase in the installation of biomass boilers, heat pumps, solar thermal, and photovoltaic cells. London has also been a vocal player on the international climate policy stage, engaged in initiatives such as the C40 and the Clinton Climate Initiative. Every year the GLA releases the London Energy and Greenhouse Gas Emissions Inventory to monitor the capital's progress.

Air Quality and Transport

Road transport is the largest source of air pollution in London, followed by industrial processes, air traffic, and rail. Although trends over the last decade show reductions in the concentrations of many pollutants, air quality in the city still breaches European Union and national health targets. The trend of decreasing nitrous oxide and particulate matter smaller than 10 µg (attributed to engine and fuel improvements) in particular is expected to slow down considerably and eventually be offset by increasing traffic levels.

A key activity in this field is monitoring concentrations by automatic air quality monitoring (there are over 100 sites across the city). London has seen an increase in Air Quality Management Areas, which are areas where air quality is not expected to meet the national objectives. GLA strategies to reduce emissions from road transport include making the taxi fleet meet stringent Euro III emission standards and the bus fleet Euro II standard, introducing hybrid diesel–electric buses, discouraging the biggest polluters (through the London Low Emission Zone), and introducing an emissions-weighted congestion charging (a differentiated pricing system aimed at encouraging lower-emission cars) on top of the existing congestion charge.

Waste Management

London produces over 18 million tons of waste every year; however, municipal waste (where the GLA's statutory duty lies) accounts only for 4 million tons, of which over three-quarters is household waste. Although household recycling is increasing, it stills fails to meet the national household recycling target. One key policy area is to reduce landfilled waste (64 percent). Recovering energy from waste is another important area, especially a move away from incineration for electricity production (18 percent), perceived as inefficient and controversial because of emissions from burning and transportation, and the crowding-out of recycling. New technologies such as anaerobic digestion, pyrolosis, and gasification are being experimented with to produce hydrogen in the future. In the mayor's 2003 Waste Strategy, a proposal was made to set a single waste authority for London (currently London boroughs collect waste in their area, and only 21 boroughs dispose of waste collectively), which would have significant effects on the way waste is handled in the capital. In addition, a strategy to deal with so-called business waste (commercial, industrial, construction, and hazardous waste) is currently under consultation.

Water

Water consumption per capita in London is higher than the national and European urban averages but has remained stable at an average 156 liters a day (2004–2005) since 1990. At present, there is mostly enough water available to satisfy London's demand (the majority comes from the rivers Thames and Lee); however, the combined effects of climate change and increased supply leakages (attributed to an aging infrastructure) may reduce the amount of water available in the future. Thames Water is implementing an extensive mains replacement program to renew much of the Victorian network over the next three years. Progress will be monitored by the GLA. The mayor's Draft Water strategy proposes to improve water efficiency domestically, encourage water conservation, and use reclaimed water for nonpotable needs.

Water Quality

The Thames is one of the cleanest metropolitan rivers in the world, hosting 120 species of fish. However, climate change and a growing city put pressure on London's sewer system (which collects sewage and rainwater runoff together), and 52 million cubic meters of untreated sewage and rainwater pollute the rivers Thames and Lee each year. Improvements in London's sewage treatment works over the coming years will improve the quality of rivers in the medium term. The lower Lee River, which has been affected by urban and industrial pollution, sewer misconnections, and sewer overflow problems, will benefit from the regeneration associated with the Olympic Games. Also, there are plans for a 30-kilometer-long tunnel to intercept sewage and rainwater discharges along the length of the Thames and transport the wastewater for treatment in East London.

Flooding

A significant part of London lies on a tidal floodplain, so the city is vulnerable to flooding from the tidal Thames and its tributaries, surface water flooding from heavy rainfalls, and overflowing sewers. Climate change increases the probability of all these forms of floods, and a growing population increases their consequences. The Thames Barrier, a large flood control structure, is situated downstream from London and can be closed to protect the city from high tides and storm surges—something that has occurred about 70 times since 1982. However, increases in winter rainfall and tidal surges mean that flood risk management options other than the Thames Barrier will be necessary in the long-term future. The Environment Agency is currently working on the Thames Estuary 2100 project, which they will report on to the government in late 2009.

See Also: Air Quality; Distributed Generation; Environmental Planning.

Further Readings

Greater London Authority. "Action Today to Protect Tomorrow: The Mayor's Climate Change Action Plan." 2007. http://www.london.gov.uk/mayor/environment/climate-change/ccap/index.jsp (Accessed April 2009).
Greater London Authority. "Green Light to Clean Power: The Mayor's Energy Strategy." 2004. http://www.london.gov.uk/mayor/strategies/energy (Accessed April 2009).

Anne Maassen
University of Durham

Los Angeles, California

The City of Los Angeles has become a leader in the green cities movement through progressive legislation and innovative policy makers. Green policy is supported through a variety of initiatives, including state-owned alternate fuel vehicles, landfill diversions, and large-scale tree planting projects. Such actions have permitted Los Angeles to set high goals and achieve great successes in the environmental protection and renewal movement. Other major U.S. cities have copied many of Los Angeles's projects in an attempt to improve the quality of life not just for individual municipalities, but for the nation and planet as a whole.

Los Angeles's major green efforts stem from a plan set forth in Mayor Antonio Villaraigosa's Green LA Program. The Green LA Program seeks to reduce greenhouse gas emissions by the year 2030 to 35 percent below the levels recorded in 1990. The program goes further than the Kyoto Protocol and rests on the plan to increase use of renewable energy by 35 percent by 2020. The Green LA Program has undertaken major projects such as the "greening" of urban alleys, increasing urban energy efficiency, allowing utility rebates for solar power, and the reforestation of the Los Angeles metropolitan area, among other things.

Los Angeles possesses over 900 miles of alleyway, with much of this space underutilized and unsafe. Alleyways have historically resulted in a variety of dangerous and uneconomic activities, including criminal acts, the dumping of garbage, and underutilized space. In an effort to solve these problems, Los Angeles has sought to use these public spaces to benefit the public. Los Angeles has reclaimed its alleys, using some as pedestrian markets as well as public gardens. These actions have assisted with the beautification of the city, as well as reduced carbon dioxide in the city's atmosphere. This general reclaiming of alleys for public space has caught on and spread to other cities such as Seattle. A more central component of the movement that has been taken up in cities such as Chicago is the lining of alleyways with porous materials that will absorb polluted water and allow the clean water, after being filtered through the groundfill, to refill underwater basins. Urban runoff is the primary pollutant to the oceans, and as the alley water reclamation process reduces runoff into lakes and oceans, it has been very successful. In Los Angeles, the tourism dollars that the ocean brings are protected and augmented by such measures. The removal of trash bins and dumpsters from alleys has allowed Los Angeles's citizens a greater sense of steward-ship of their city, as well as providing an economically viable way to continue the green movement that has swept the city.

Urban efficiency extends to areas outside a municipality's direct control. Los Angeles has been very successful in encouraging the building of and conversion to of Energy Star buildings. Energy Star status is accorded to buildings that use at least 35 percent less energy than normal buildings and emit 35 percent less carbon dioxide into the atmosphere. According to the Environmental Protection Agency, Los Angeles has the most Energy Star buildings in the United States—at this time, over 260 Los Angeles buildings have earned the agency's Energy Star designation. Some buildings have achieved the Energy Star stan-dard through an upgrade to more energy-efficient lighting or by installation of more effi-cient motors for elevators, fans, or other mechanical systems. Such a conversion can have significant economic benefits for a building's owners. For example, one area building that was renovated to Energy Star standards enjoyed a 2.4 percent reduction in energy use, which resulted in annual savings of $90,000 in energy bills.

Energy Efficiency

Nationally, streetlights generally account for between 10 and 38 percent of a city's annual energy bill. In the United States alone, over 35 million streetlights exist. These streetlights use up to 1 percent of all electricity consumed annually in the United States. Los Angeles continues to seek energy efficiency, and the resulting cost-savings this efficiency engenders, through a variety of innovations. One such program seeks to replace over 140,000 existing Los Angeles streetlights with light-emitting diode (LED) lamps. This change to LED lamps will drastically decrease the city's energy use and reduce carbon dioxide emissions by 40,500 tons. This switch will also save Los Angeles approximately $10 million annually. Los Angeles's move to LED lights will thus reduce energy costs, cut carbon dioxide emis-sions significantly, and result in a 40 percent reduction in energy expenditure. Switching from the current streetlights to LED lamps is equivalent to removing 6,700 cars from Los Angeles's roads. The streetlight replacement project is being funded through a series of energy rebates, street light assessment fees, and a series of loans. The loans will be repaid over a seven-year period solely through savings in energy and maintenance costs. Follow-ing this seven-year repayment period, continued savings from the LED lamp replacement will result in profits to Los Angeles. This undertaking was a crucial initiative for Los Angeles and has drawn great interest from other cities because of the potential for energy savings. By using LED lights and other alternative energy sources, Los Angeles is finding ways to reduce emissions and to cut energy expenditures. The Green LA Program has been successful because of this dual approach, continuing to emphasize energy efficiency's ben-efits for the environment, but also recognizing the economic and business advantages. Los Angeles plans continued investments for efficient energy using the monetary benefits reaped from earlier energy-efficiency projects.

Los Angeles's move toward using more alternative energy sources has focused chiefly on solar energy. In support of this, Los Angeles has offered two rebates to reduce the cost of solar energy for Los Angeles residents. These rebates are from two sources: Common-wealth Edison (the primary utility provider in Los Angeles) and a 30 percent tax credit from the federal government. These rebates have allowed consumers and businesses to shift to greater use of solar energy for up to 60 percent reduction in start-up costs. The already generous rebates are increased for government buildings, nonprofit organizations, schools, and those offering affordable housing. Through this combination of rebates, Los Angeles has managed to make solar electricity less expensive than regular utility power. Because solar electricity has no continuing costs, barring maintenance and the eventual cost of replacement panels, the cost savings over time of this program are significant. Over 13 megawatts of energy have been created by such programs since 1999. The Los Angeles Department of Water and Power has structured a 10-step declining incentive plan to further encourage participation in a solar project that is already one of the largest in the nation. Customers receive a lump sum payment to purchase or lease photovoltaic systems to offset utility energy consumption at the installation site.

Urban Reforestation

The Los Angeles Green Movement also has focused on a large system of urban reforestation—one that has concentrated on the greening of both public and private property. The Los Angeles reforestation project has used economic benefits related to urban green spaces to garner support from citizens. Homeowners, for example, can see a 5–20 percent increase in property values for homes that have state-of-the-art landscaping compared with those that do not. The value and desirability of whole neighborhoods also increases because of the proximity of abundant green space. The placement of trees near houses and other occupied buildings also results in lowered utility bills because of increased shade. Los Angeles, similar to many cities, suffers from what is known as the "heat island effect," which results from the abundance of paved surfaces and buildings absorbing heat from the sun. The heat island effect contributes to 12 percent of air pollution in cities, making the reforestation project an attack on air pollution, as well as on the high temperatures of the city. By focusing on the economic benefits of reforestation, the City of Los Angeles thus found private landowners a ready and willing constituency to support the planting of additional trees on private lands.

Los Angeles has also been proactive about reforesting publicly owned lands that it controls, including parks, municipal buildings, and other recreational facilities. The City of Los Angeles Department of Recreation and Parks oversees more than 15,000 acres of parkland—space that provides recreational opportunities for city residents and other visitors. As part of the reforestation program, the Department of Recreation and Parks seeks to use the urban forest it manages to improve the economic, physical, and social health of the community while augmenting the environmental quality of the area. All of the park sites were evaluated by the department to determine which locations would support the planting of young trees. A detailed tree inventory was begun in 2004, using satellite-based global positioning system technology that allowed park officials to record the location of park trees to the accuracy of 12 inches. A reforestation program for Los Angeles's parks was initiated that concentrated on long-range plans (those taking 10 or more years), intermediate-range plans (those taking 5 to 10 years), and short-range goals (those taking less than 5 years).

Los Angeles's reforestation program is viewed as important to quality of life for members of the region, as well as for the community itself. The reforestation of public property, including parks, provides energy benefits, improves air and water quality, and enhances the overall experience of park goers. Reforested public spaces, for example, increase opportunities for bicycling, hiking, jogging, and other healthy pursuits. Reforestation of public spaces also offers many of the same social benefits as reforestation of private property, including noise abatement, strengthening wildlife habitats, reducing ultraviolet light exposure, increasing opportunities to experience nature, improving health, and creating jobs and educational opportunities. Park reforestation seeks the ideal park tree—one that provides shade, resists wind damage, seldom requires pruning, produces minimal debris, is deep rooted, has few pest and disease problems, and thrives in a wide range of soil conditions, irrigation regimens, and air pollutants. Because no tree has all of these attributes, tree species are matched to the planting site as much as is practicable. The reforestation project has also had a beneficial effect on helping the Department of Parks and Recreation provide a broader variety of tree species, so that parks do not experience devastating loss as a result of an unexpected outbreak that impinges on one particular species. Native California species are preferred for reforestation, and hardy nonnative species are allowed if they achieve a ratio of no more than 30 percent of one genera, 20 percent of one species, and 10 percent of one cultivar. The reforestation effort also has assisted with water pollution. Removing paved areas and replacing them with trees increases the amount of water that fills underground reservoirs and decreases surface runoff. Surface runoff is a major problem for the city—after rainstorms, pollutants from city streets flow directly into the ocean, contaminating a major part of the area's attraction.

Los Angeles is a sprawling city that is more heavily reliant on the automobile than compact, walkable cities with mass transit systems. Nevertheless, using a variety of public and private initiatives, and building citizen participation into the planning process, Los Angeles has been highly successful in joining the green cities movement. Although the Green LA Program set ambitious goals, a dual emphasis on its economic and environmental benefits has garnered wide support. The accomplishments Los Angeles has enjoyed to date have caused it to be emulated by leaders of other cities seeking to initiate sustainability movements in their own municipalities.

See Also: Air Quality; Bicycling; Citizen Participation; City Politics; Ecosystem Restoration; Heat Island Effect; Urban Forests.

Further Readings

Berg, N. "LA's Green Alleys." *Planning*, 75/6:24–5 (2009).
Kennedy, M. "Green as the New Norm." *American School & University*, 81/6:16–23 (2009).
Swidler, A. and S. C. Watkins. "'Teach a Man to Fish': The Sustainability Doctrine and Its Social Consequences." *World Development*, 37/7:1182–96 (2009).
Wilcox, K. "Greenest Skylines." *Scientific American Earth 3.0*, 19/2:10 (2009).

Stephen T. Schroth
Jason A. Helfer
Jordan K. Lanfair
Knox College

MALMÖ, SWEDEN

Determined to ward off any challengers to its self-proclaimed moniker, "The Sustainable City" (*den hållbara staden*), Malmö has widely publicized that 2020 is the year that it will become "climate neutral" by relying on renewable energy for 50 percent of its energy usage. The extent to which the gem of Sweden's southern region is committed to attaining these ambitious goals is manifested in its master plan and by two official publications of the municipality; namely, *Energy Strategy for Malmö 2008* and *Environmental Program for Malmö 2009–2020*. Although its population of 290,000 makes it Sweden's third city in terms of population, Malmö is not content to take a back seat to either Stockholm or Gothenburg on the world stage. Indeed, it has boldly proclaimed its "acceptance of the challenge to become the world's best at sustainable city development." Therefore, it is no surprise that the holding of the 15th Conference of Parties to the 1992 UN Framework Convention on Climate Change (COP-15) in Copenhagen was viewed as a propitious occasion to showcase Malmö's most creative and groundbreaking projects in sustainable urban development.

With Copenhagen just a 30-minute jaunt across the sound (Öresund), it was natural for Malmö to want to entice the 15,000 delegates attending COP-15 to take a short ferry ride to Sweden's southernmost metropolis for the purpose of becoming acquainted with a municipality that is "in the forefront of taking an ecological approach to planning, building, and construction in the urban environment." Hence, the Environment Department created "Climate Study Tours in Malmö" to coincide with the two-week period in December 2009 that COP-15 convened. Because the purpose of the tours was to showcase the projects that Malmö considered to be its most stellar achievements in terms of "mitigation and adaptation" and "increasing sustainability" in the urban setting, it is worthwhile to examine the four tours that were offered.

Malmö, Sweden: Tours From the Sustainable Center

Bo01 (Habitat 2001): City of Tomorrow

Bo01 is a tour of the Western Harbor, an area that has been transformed from an industrial park into an area with parks, swimming, schools, and housing. The reclamation of the area started in 1998 with the opening of Malmö University at the site. Then, in 2001, the European Home Fair was held there, with newly constructed environmentally sustainable apartments available for purchase on the spot. Bo01 is an area that relies on 100 percent locally produced renewable energy and showcases functioning local storm water management. A distinctive architectural landmark in this area, "The Turning Torso," is an apartment building in which all 147 apartments have garbage disposals that grind food waste into an organic sludge, which is then used to create biogas. The area demonstrates the feasibility of transforming an area that was "typical of urban redundant industrial land" into a center for knowledge and sustainable living.

Augustenborg

This is a 1950s neighborhood that was redeveloped in the late 1990s with the collaboration of local residents. The homes in the area now have green roofs, and the neighborhood has many green areas with multiple uses, as well as open storm water management.

Sege Park

Sege Park is a former hospital area that has been converted into a residential area, with an abundance of green areas for outside activities. It is accessible to the center of Malmö by public transportation. Sege Park has the largest photovoltaic panel installation in northern Europe, as well as an urban wind power plant that generates electricity.

Malmö Museum

Tours begin on the way to the museum, showcasing improvements in public transport and innovative bicycle lanes, as well as examples of locally produced solar energy. The museum contains exhibitions showing the impact of this green lifestyle on Earth's climate.

In June 1992, Agenda 21 was adopted at the United Nations Conference on Environment and Development. It is a plan of action to be undertaken globally, nationally, and locally to progress toward sustainable development in the 21st century. Agenda 21 was reaffirmed at the World Summit on Sustainable Development, held in August 2002 in Johannesburg. An important implication of Agenda 21 is the need to measure the effect of urban living on sustainable development. Over 50 percent of the world's population lives in cities: These cities represent just 2 percent of the world's landmass, but they consume 75 percent of the world's resources. In Europe, where 80 percent of the population lives in urban areas, tools for measuring the effect of urban living drew much attention. The European Common Indicators (ECI) initiative was launched in May 1999 and resulted in the development of 10 common indicators for sustainability at the local level; that is, areas in which local action can be taken toward achieving sustainable development.

In 2001, Malmö participated in a pilot ECI project that used the ECIs to measure the ecological footprints of municipalities in the United Kingdom, Finland, Spain, Norway, the Netherlands, Italy, and Sweden. In the final report on the pilot project, Malmö had distinguished itself as an ECI associated with average residential emissions (at 0.86 tons, it measured lower than Europe's southern urban areas), accessibility to public transport (with 31.3 percent of its population relying on public transport), and by having the highest percentage of environmentally certified enterprises of the 28 municipalities that provided information for this ECI. Malmö was also noted for the predominance of nonmotorized transport (with 23.2 percent of its inhabitants using bicycles and 20.5 percent walking for transportation).

Keeping up with its favorable showing in the 2001 pilot project, Malmö's city buses are engineered to use gaseous energy sources. Since late 2008, 42 percent of city buses have used biogas extracted from food waste. This percentage is expected to rise as a result of a planned increase in the production of biogas—an increase that should not be difficult to achieve, as 96 percent of household waste is collected for recycling, with food waste also being used to create biogas for vehicle fuel. Waste sources are also used to generate heat energy, and 25 percent of Malmö's heating is provided by the city incinerator via waste-to-energy sources. Another 16 percent of heating is provided by hot water that has been warmed by excess heat released from daily industrial activities in the city. Malmö made an impressive showing in the first European Green Capital Award competition, announced in May 2008 and established in 2010 as the inaugural year for qualified cities in the European Union to compete for the award. The competition uses 10 "indicator areas" that are similar to the ECIs, except that unlike the ECIs, the indicator areas do not measure socioeconomic factors in the urban environment. Malmö received a perfect score of 15 with respect to the "environmental management" indicator area (a distinction shared by only 3 of the 35 applicants for the award). Moreover, Malmö scored higher than Hamburg—the winner of the 2010 Green Capital Award—in the four indicator areas of local transport, noise pollution, water consumption, and environmental management. Malmö also tied with Hamburg in two indicator areas: local contribution to climate change and green urban areas.

By 2030, Malmö aspires to be a city that is run 100 percent on renewable energy.

See Also: Agenda 21; Copenhagen, Denmark; Hamburg, Germany; Renewable Energy; Stockholm, Sweden; Sustainable Development.

Further Readings

"European Common Indicator." Final Project Report. Milan, Italy: Ambiente Italia Research Institute, May 2003). December 9, 2009. http://ec.europa.eu/environment/urban/pdf/eci_final_report.pdf (Accessed July 2009).

European Green Capital. "Environment." September 9, 2009. http://ec.europa.eu/environment/europeangreencapital/about_sumenus/background. html (Accessed July 2009).

Malmö Stad. "Malmö—A City in Transition." http://www.malmo.se/english (Accessed July 2009).

Gwendolyn Yvonne Alexis
Monmouth University

MASDAR ECOCITY

In the city of Abu Dhabi, the initiative known as "Masdar," which means "the source" in Arabic, may just be the most innovative strategy yet to surface toward a sustainable future. Masdar is hoping to be the "green" envy of all, with new initiatives and partnerships for a green ecocity that will far surpass green technologies and innovations of the past.

Masdar's goals are the following:

- Masdar will be a zero-carbon, zero-waste city that explores innovative ideas regarding greenhouse gas emissions and the effect of our carbon footprint.

- Masdar will be a car-free area.
- Masdar hopes to drive the commercialization and adoption of sustainable energy technology.
- Masdar desires the honor of being an example for the rest of the world and wishes to be a leader in green ecocities and green technologies.
- Masdar aspires to become a research and development consultant for the world energy market.
- Masdar searches for solutions to securing energy sources, dealing with climate change, and educating society in regard to sustainability.
- Masdar wants to provide economic growth for its city through the use and invention of new energy technology, positioning itself for the future.
- Masdar wishes to create partnerships with companies on the level of an international platform.

The city's vision was designed by Foster and Partners and will definitely affect the architecture of tomorrow. Innovations that the city will use include a solar-photovoltaic power plant that will supply energy for the city. Designers also foresee using solar canopies shaped like sunflowers that provide shade as well as power. These sunflower-shaped umbrellas will close at night. No cars will be allowed in the city—people will travel from point A to point B by an electric light rail system that will link back to the center of Abu Dhabi, which is the capital of the United Arab Emirates. Wastewater will be purified and recycled and used as a water source to grow plants. The plants, in turn, will be used to supply biofuel. Desalination processes, much more efficient than current facilities, in Abu Dhabi will also be implemented within the city.

The Masdar project could not come at a better time. The human population on Earth has grown exponentially, and with this rise in population, the concerns and requirements for resources and fossil fuels increase at a tremendous rate. Concerns surrounding energy and energy security and climate change have pushed humans to be more conscious of greenhouse gases that have been rising over the decades. The increase in the need for energy has made the terms *sustainable* and *green* common household terms. Many companies and individuals are rushing to bring green ideas and energy-conserving measures to fruition. The diversity of green ideas are many and range from small-scale energy ideas to ideas that are hatched for a carbon-neutral zero-waste city like the Masdar Ecocity. However, such a project being in a city near you soon will be impeded by the initial up-front cost, estimated to be approximately $22 billion. Many cities will not be able to finance the costs of upgrading their city to a Masdar-type green city. In addition, each Masdar will require a diverse set-up geared toward various topography and climate variations. Also, for as many Masdars as may be built for many years, there will still exist carbon-producing cities that could possibly offset the carbon-neutral areas, especially if they are side by side. The advantages to Masdar are that it is the first to implement such a project, so many investors are interested in this cutting-edge technology and wish to be in on the ground-breaking new technology. Masdar also wishes to partner with other companies, and many investors have already contacted the Masdar executives. Abu Dhabi is also already known as one of the richest cities in the world, so money does not seem to be an issue for this area, which is known for its oil production.

Many have already partnered with the Masdar city initiative including the Massachusetts Institute of Technology (MIT), which has launched an energy initiative with the Masdar project. The institute hopes to attract some of the best academics from around the world to collaborate on the Masdar project and push energy technologies. The project wishes to be multinational and to use cutting-edge research and development. The initiative welcomes not only established companies but also startups. In January 2009, the World Future Energy Summit in Abu Dhabi attracted more than 16,000 visitors, including General Motors as well as small-scale start-up companies. The partnership with MIT is the first of its kind to concentrate on energy and sustainable technologies. MIT, along with the Masdar project, wishes to address the world's pressing energy demands. Other partners to Masdar include Boeing, which has announced that Abu Dhabi's Masdar Institute of Science and Technology will study a possible source of jet fuel that would be produced from a family of saltwater plants. Masdar has also signed an agreement with Bahrain's National Oil and Gas Authority to develop innovations to slash carbon emissions in the oil and gas industries. This initiative will use the Kyoto Protocol's schematic, which enables companies to benefit financially by trading carbon credits when they reduce emissions. These examples are only a few of the partnerships that have already been launched in light of the Masdar project. Many more partnerships will be sure to follow.

As the Masdar Ecocity increases its efforts and constructs its green utopia, many partners and affiliations will create and fuel innovative technologies that can eventually reach a city near everyone. However, Masdar has already been deemed the world's biggest petri dish or test tube. As with any project, naysayers exist. If Masdar does prove to be successful and affects the world as much as the ideologies suggest, the effect will be at the international level and will definitely have a trickle-down effect into other countries. Masdar may be the answer to future energy concerns, but how Masdar look-alikes will be implemented is still a matter for debate and will not be decided until after Masdar has had a trial run. If all goes well, Masdar could be the future energy technology of your hometown.

See Also: Carbon Footprint; Carbon Neutral; Ecological Footprint; Ecovillages; Energy Efficiency; Environmental Impact Assessment; Environmental Planning; Green Communities and Neighborhood Planning; Green Design, Construction, and Operations; Green Infrastructure; Green Jobs; Green Procurement and Purchasing; LEED (Leadership in Energy and Environmental Design); Power Grids; Sustainability Indicators; Sustainable Development.

Further Readings

Arabian Business. http://www.arabianbusiness.com (Accessed October 2009).

Carbon Offsets Daily. "Carbon Hero." http://www.carbonoffsetsdaily.com/news-channels/asia/carbon-hero-4332.htm (Accessed October 2009).

ESRI (UK). "The Challenge of the 21st Century." http://www.esri.com/industries/facilities-management/pdfs/masdar.pdf (Accessed October 2009).

Gordon, David. *Green Cities: Ecologically Sound Approaches to Urban Space.* Montreal: Black Rose Books, 1990.

Morris, Loveday. "Masdar's Seeds Start to Sprout Across the World." *The National* (September 18, 2009). http://www.thenational.ae/apps/pbcs.dll/article?AID=/20090919/WEEKENDER/709189804/1299/weekenderlisttemplate (Accessed October 2009).

Walsh, Bryan. "Abu Dhabi: An Oil Giant Dreams Green." *Time* (February 12, 2009). http://www.time.com/time/magazine/article/0,9171,1879168,00.html (Accessed October 2009).

Deanna Spraker
Virginia Polytechnic Institute

MAYORS CLIMATE PROTECTION AGREEMENT

The U.S. Mayors Climate Protection Agreement is a voluntary initiative in which participating mayors commit their cities to taking action to reduce greenhouse gas emissions. Signatory mayors adopt the goals of the Kyoto Protocol; namely, to reduce their cities' greenhouse gas emissions to 7 percent below 1990 levels by 2012. The Mayors Climate Protection Agreement is an initiative of the U.S. Conference of Mayors and was unanimously endorsed by the conference at its annual meeting in June 2005. As of August 2009, 969 mayors from all 50 states; Washington, D.C.; and Puerto Rico had signed the agreement, representing nearly 85 million people.

The Mayors Climate Protection Agreement has three main aspects. First, it urges state governments and the federal government to enact programs to meet Kyoto climate change goals, particularly through the development of renewable energy and fuel-efficiency technologies. Second, the initiative calls on Congress to adopt legislation to reduce greenhouse gas emissions. It urges that this legislation include specific timetables and emission limits and that it create a market-based national emissions-trading system. Third, the Mayors Climate Protection Agreement lists a number of specific recommended actions that mayors may take to meet or exceed Kyoto Protocol goals in their cities. The first suggested action is taking an inventory of city emissions and coming up with a clear plan to achieve reduction targets. The other actions include those that focus on city operations, those that address public policy, and those that attempt to influence the behavior of individuals in the city. Suggested changes in city operations include retrofitting city buildings for energy efficiency, purchasing Energy Star appliances, and converting city fleets to more fuel-efficient vehicles. Public policy changes include land use and zoning policies that make the city more walkable, increase green space and urban forests, and promote energy-efficient building practices. City efforts to influence individual behavior include climate change education programs, campaigns to encourage increased recycling, and incentive programs for carpooling, bicycle riding, and alternative transportation. Though listing a number of actions that participating cities may take, the agreement is clear that it is strictly voluntary, that the listed action items are just suggestions, and that they are not exhaustive of the steps cities may take to reduce greenhouse gas emissions.

The U.S. Mayors Climate Protection Agreement was launched by Seattle mayor Greg Nickels on February 16, 2005, the day the Kyoto Protocol went into effect for the 141 countries that had ratified it. The United States, having declined to ratify the treaty, was not a party to the Koto Protocol, and the initiative was promoted by Mayor Nickels as a response to federal inaction on the issue of climate change. The original goal was to have at least 141 mayors sign the agreement by the U.S. Conference of Mayors annual meeting in June of that year, symbolizing the 141 countries that had ratified the Kyoto Protocol. This goal was achieved, and at the annual meeting the Mayors Climate Protection Agreement was unanimously endorsed by the U.S. Conference of Mayors. In addition to endorsing the agreement, the U.S. Conference of Mayors stated its commitment to encouraging cities to join the agreement and to work with appropriate civil society organizations to track progress and implementation of the initiative. Though somewhat more concentrated in Democratic-leaning states, the Mayors Climate Protection Agreement has significant bipartisan support, being adopted by Republican and Democratic mayors alike in cities of all sizes and in every state in the country.

Benefits of the Agreement

Analyses of city initiatives to address climate change such as the Mayors Climate Protection Agreement cite a number of benefits of these programs. First of all, city governments often have jurisdiction over a number of key policy areas that can significantly influence local greenhouse gas emissions, the most important of which include land use planning, waste management, and building codes. Second, local climate change actions have a flexibility that may not exist at the state or national level. The smaller scale of citywide climate-mitigation projects enables more experimentation by policy makers and allows policies to be tailored to the specific conditions and preferences of each locality. Furthermore, this space for experimentation and innovation, it is argued, can help discover effective ways to reduce greenhouse gas emissions that can then be translated to other cities and to larger political scales. Finally, the widespread adoption of the Mayors Climate Protection Agreement shows that there is broad political support across the United States for addressing climate change, despite national political rhetoric to the contrary. As such, these local efforts may push the federal government to take action on this issue, both domestically and internationally.

Opinions vary as to what motivates mayors to sign onto the Mayors Climate Protection Agreement. Some mayors may be inspired by the global consequences of climate change; others, particularly those in coastal areas, by the long-term effect of climate change on their cities. Others may be convinced to join the agreement by the immediate, localized benefits that can emerge from programs to reduce a city's greenhouse gas emissions. These benefits include reducing air pollution and traffic congestion; saving money on municipal, business, and household energy bills; improving quality of life, with more green spaces and a more walkable city; and the availability of improved transportation choices. A more cynical perspective, however, argues that these motivations are probably not as significant as the political benefits of being seen as a "green" mayor. According to this argument, mayors may not be as concerned with climate change as an environmental or social issue as they are with the political capital that can be gained by signing on to the voluntary greenhouse gas reduction goals of the initiative. Whatever combination of factors influence a mayor's decision to join the Mayors Climate Protection Agreement, what does seem to be the case is that the wide variety of cities represented by the nearly 1,000 participating mayors is indicative of the growing consensus that climate change needs to be addressed and signifies that there is substantial political support for doing so.

Critiques of the Agreement

Despite the widespread embrace of the U.S. Mayors Climate Protection Agreement and its potential benefits to participating cities, a number of critiques of the initiative have also been offered. First of all, critics raise the concern that despite the promising rhetoric or the best intentions, it appears unlikely that any but a small handful of cities will actual reach the greenhouse gas reduction goals outlined by the Mayors Climate Protection Agreement. There are a number of reasons for this. First, though city governments may have jurisdiction over land use and waste management and may use this jurisdiction to reduce carbon emissions, very few cities in the United States have control over local utilities, a major source of greenhouse gas emissions. Although city governments can promote conservation, without control over utilities, it is not possible for cities to change utility operations to reduce emissions or to transition to less-carbon-intensive forms of energy production. Furthermore, most cities share control over transportation with numerous other municipal governments in the region. To be effective, efforts to reduce greenhouse gas emissions by improving public transportation or reducing traffic congestion must therefore be coordinated on a regional scale. Finally, many greenhouse gas–reduction policies such as investments in energy conservation technologies require significant up-front funds that cities often are unable or unwilling to spend, even if these programs would save them money in the long run. For all these reasons, even the mayor with the best intentions may have difficulty dramatically reducing city contributions to climate change.

The second major critique of city-based climate initiatives is that the city is an inappropriate or ineffective scale at which to address this issue. One reason, as discussed above, is that cities often do not have the institutional capacity to make the changes necessary to reduce carbon emissions significantly. Second, as many states that have participating cities also have climate change policies, there is potential for problematic policy interaction. For example, the patchwork of city and state programs may put an unnecessary burden on industry, and duplication of enforcement efforts for similar policies enacted by overlapping jurisdictions may cause significant confusion and waste regulatory resources. Third, some studies indicate that the majority of greenhouse gas reductions that have occurred in cities are a result not of city policies but of actions taken at the state or national level, such as the adoption of a state renewable portfolio standard or the recent increase in vehicle fuel-efficiency standards. As few cities appear to be slowing their emissions faster than their state average, some critics assert that it is these higher levels of government where changes are most effective and, therefore, where carbon reduction programs need to be enacted.

Despite these critiques, many continue to see cities as having a crucial role to play in any successful effort to address climate change, and participation in the U.S. Mayors Climate Protection Agreement continues to grow. Though in the early years of the agreement it was frequently noted that the initiative posed a challenge to federal inaction on climate change, the federal government is now beginning to support city climate change mitigation efforts through programs such as those that help fund energy-efficiency initiatives. It remains to be seen, however, what long-term effect the Mayors Climate Protection Agreement will have on federal climate change policy and whether it is able to significantly contribute to the reduction of greenhouse gas emissions in the United States.

See Also: Agenda 21; Cities for Climate Protection; City Politics; Seattle, Washington.

Further Readings

Bailey, John. "Lessons From the Pioneers: Tackling Global Warming at the Local Level." *Institute for Local Self-Reliance.* (2007). http://www.newrules.org/de/pioneers.pdf (Accessed September 2009).

Byrne, John, et al. "American Policy Conflict in the Greenhouse: Divergent Trends in Federal, Regional, State, and Local Green Energy and Climate Change Policy." *Energy Policy,* 35 (2007).

Droege, Peter. *Urban Energy Transition: From Fossil Fuels to Renewable Power.* Oxford: Elsevier, 2008.

Lutsey, Nicholas and Daniel Sperling. "America's Bottom-Up Climate Change Mitigation Policy." *Energy Policy,* 36 (2008).

Corina McKendry
University of California, Santa Cruz

Mexico City, Mexico

Mexico City is going green. Particularly after Mexico decided to participate in Agenda 21, government agencies, businesses, citizens' groups, academic programs, and national and international nonprofit organizations have been working to make the city greener. Programs

Among Mexico City's most pressing challenges are increasing energy demands, a water shortage, ineffective waste management, pollution, and urban sprawl, which is apparent in this view of the city.

Source: iStockphoto

changing the ways people live, work, and think range from legislative reforms and the greening of government agency practices to sustainable building and comprehensive waste treatment. Green is in, but significant challenges must be overcome to make the sustainable city movement more powerful, while making ongoing socioeconomic development more equitable.

Mexico City is the metropolitan area that encompasses the country's capital, Mexico D.F. (the Federal District), and surrounding areas in the state of Mexico. Mexico City is among the 10 wealthiest cities of the world by gross domestic product. With a population of more than 20 million, the metro-

politan area is also one of the world's largest. Mexico City's population growth exploded when industrialization engulfed the country in the 1930s. In 1930, the city's population was just over 1 million. Population growth has slowed since the 1970s, but settlement has continued to advance into surrounding areas of the state of Mexico, with immigrants to the city establishing settlements on the fringes of the metropolitan area. The urban sprawl that accompanies population growth, coupled with the inability of efforts to keep up with worsening environmental problems overwhelm attempts by the city's governmental authorities to make the city greener.

Although many political leaders and policy makers are beginning to address challenges to sustainable development in Mexico City, greening the city is not yet a comprehensive administrative endeavor. The report published by the Ministry of Finance and Public Credit on the country's 2010 Economic Program illustrates that Mexican policy makers are still compelled to focus on increasing the growth of the economy, productivity (especially in sectors that promise to help alleviate unemployment), and more immediate means of combating poverty. For Mexico City's policy makers, addressing such pressing concerns is also the primary agenda. Comprehensive planning, effective governance, and reconciling individual and collective interests in the megacity have always been major challenges, so dealing with the complexity of interrelated causes of environmental problems will be a daunting task for the municipal government. Nonetheless, some municipal government programs are already making progress in solving environmental problems as the movement to make the city greener gains momentum. Policy makers and planners recognize that demands are on the rise for solutions to environmental problems;

green products and services; energy-efficient housing; more cost-effective, safe, and environmentally sustainable public transport systems; green jobs; and pleasant and enjoyable public spaces.

An Environmental Agenda

Policy initiatives are spearheaded by city government administrations elected every six years. The current administration of Marcelo Luis Ebrard Casaubón of the Party of the Democratic Revolution has set an environmental agenda for 2007–2012 to be coordinated and carried out by the Federal District Government's Environment Secretariat (SMA). The other key government agency responsible for implementing green city initiatives is the Secretariat of Urban Development and Housing. Aware that their city's worsening environmental problems and growing population necessitate significant economic reform and that municipal government alone cannot handle the gargantuan task, Mexico City policy makers are organizing educational outreach and supporting cooperative initiatives. For example, the SMA coordinates a new effort called the Green Plan. Its council is made up of 30 representatives of other governing bodies, educational institutions, consulting firms, industry groups, and organizations representing citizens and businesses. The Green Plan addresses urban mobility, land conservation, green spaces, the city's drinking water shortage, air quality, solid waste management, and greenhouse gas emissions. The SMA has also established a Mexico City Center for Environmental Information and a Directorate of Environmental Education to provide easy access to information about SMA programs and initiatives of national and international organizations dealing with environmental problems, as well as to facilitate educational outreach.

Federal government agencies in Mexico that coordinate and sponsor programs in Mexico City include the Ministry of Environment and Natural Resources, the National Institute of Ecology, the National Council of Science and Technology, the Energy Secretariat and its National Commission for Energy Conservation, and the National Commission for the Knowledge and Use of Biodiversity. Representative projects include coordinating and funding technological innovation to make commerce and industry more sustainable, providing technical assistance to public sector enterprises adopting better resource-consumption practices, establishing energy performance standards, and even administering appliance swap programs that encourage consumers and businesses to retire older appliances for more energy-efficient models. Such efforts establish new protocols and simultaneously support the creation of markets for green products and services.

The Ecologist Green Party of Mexico calls public attention to climate change, unsustainable resource consumption, socioeconomic inequality, and lack of education and information that exacerbate environmental problems in the city and beyond. It also promotes a policy agenda that emphasizes government reform, restructuring of public financing, and promotion of public dialogue.

Alongside policy makers and politicians, academic institutions and nonprofit organizations are working to green the city by organizing academic and informational programs to make sustainable living more possible. The Autonomous University of Mexico offers respected academic programs that prepare future scientists, professionals, and policy makers to address the city's environmental and socioeconomic challenges. In addition to its academic programs, the Ibo American University has implemented a "Green Campus" program. The Mexican Fund for the Conservation of Nature, in cooperation with the

World Resources Institute, created New Ventures México in 2004 as a green-business incubator and consulting group. In 2007, New Ventures launched the Green Pages (Páginas Verdes), a high-profile directory of sustainable products and services.

Efforts to green the city are also supported by intergovernmental organizations. Examples include the United Nations Division for Sustainable Development, the U.S. Agency for International Development, the Commission for Environmental Cooperation created under the North American Agreement on Environmental Cooperation, and the British Foreign and Commonwealth Office. They provide policy models, promote green city initiatives, and support partnerships and educational programming that transform incentives and awareness of political leaders and citizens.

Private sector endeavors are also important. Leaders in commerce and industry are positioning their firms to take advantage of new approaches to development and emergent changes in the city's regulatory environment. Consulting, research and development, and credentialing firms are providing services to venture capitalists, both transnational and domestic, positioning themselves to market their ideas, products, and services. The manufacturing and service sectors can use the city's environmental problems as opportunities to establish viable new enterprises and markets. Industry groups like the International Union of Architects make sharing ideas and plans for transforming Mexico City possible. Public-sector agencies, institutions, and organizations that pioneer programs to make the city greener are often backed by banks and corporate sponsors. Many of the most ambitious green initiatives are codeveloped and supported by specialized consulting firms that provide advising and financial services, including establishment of trusts and identification of loans and grants from intergovernmental government aid agencies and international organizations such as the World Bank.

Challenges

Among the city's most pressing challenges are urban sprawl, an ever-increasing demand for energy, a serious water shortage, pollution, ineffective waste management, and an urban mobility crisis. Architects and engineering firms are addressing the need for more sustainable urban planning, green spaces, and sustainable buildings. City planners hope to convert up to 10 percent of the city's rooftops to "green roofs." Innovative building designs can make buildings into carbon traps and enable more urban farming, water reclamation, and solar energy collection. Relatively simple construction plans can take advantage of prevailing wind patterns for ventilation and orientation that enables passive solar heating and cooling.

Tremendous quantities of waste are produced by commerce, industry, and the activities of daily life in the city. In an effort to deal with these problems, relatively simple municipal regulations have been implemented. A partial ban on circulation of vehicles that emit high levels of carbon dioxide, more regulation of informal transportation and garbage collection services, and a ban on plastic grocery bags that will take effect in 2010 are representative. Businesses are offering products and services to help citizens and consumers adapt to policy changes.

A Mexico City Waste Commission has undertaken the daunting task of greening the city's solid waste disposal. The city produces over 12,000 tons of garbage daily, and its main landfill is leaching contaminants and poses other dangers to public health and the safety of hundreds of workers and scavengers who make a living there. Working with assistance from the International Solid Waste Association, agencies and investors are building new processing facilities to recycle, compost, and burn waste for energy and make the city's garbage collection and recycling efforts more efficient and safe. Methane gas collection systems that generate electrical energy have been installed at city landfills.

The desiccation of the aquifer under Mexico City has been ongoing for over 500 years. Today the city is subsiding into the depleted hydrologic basin that contains it and is plagued by a water shortage that has been addressed in recent years by importing water. Scientists, development firms, government agencies, and international organizations are exploring possible solutions, including the restoration of part of the basin as a smaller lake fed by runoff and treated wastewater that would provide more water for the city and induce microclimatic change to help mitigate air pollution. The myriad of other initiatives to conserve water and improve water resource management include improving wastewater treatment and improving irrigation systems in agricultural zones around the city.

Making Mexico City greener will depend on the capacity of political leaders, industry and commerce groups, entrepreneurs, organizations, and increasing numbers of politically active citizens to keep the momentum going. Especially important will be comprehensive and progressive public policies, enforcement of regulations, innovative conservation and development projects, business and industry strategies that take advantage of the changing regulatory environment, and shifting values. Education, democratic decision making, and alleviation of poverty are also fundamental prerequisites to greening this city that is home to many millions.

See Also: City Politics; Energy Efficiency; Environmental Planning; Garbage; Green Roofs.

Further Readings

Associated Press. "Mexico City Vows to Green Garbage Dumps." http://www.msnbc.msn .com/id/28777897 (Accessed October 2009).

Centro de Información Ambiental de la Ciudad de México. http://www.sma.df.gob.mx/ceina (Accessed October 2009).

Joint Academies Committee on the Mexico City Water Supply, Commission on Geosciences, Environment, and Resources, National Research Council, Academia Nacional de la Investigación Científica, A.C., Academia Nacional de Ingeniería, A.C. *Mexico City's Water Supply: Improving the Outlook for Sustainability*. Washington, D.C.: National Academies, 1995.

Las Páginas Verdes. http//:www.laspaginasverdes.com (Accessed October 2009).

Mexico City Government Environment Secretariat. "Plan Verde." http://www.planverde.df .gob.mx (Accessed October 2009).

Muñetón Pérez, Patricia. "Urbanismo Sustentable: Retos, Acciones y Beneficios para una Ciudad en Crecimiento." *Revista Digital Universitaria*, 10/7 (2009). http://www.revista .unam.mx/vol.10/num7/art47/int47.htm (Accessed October 2009).

New Ventures México. http://www.nvm.org.mx/newventures.html (Accessed October 2009).

Organisation for Economic Co-operation and Development (OECD). "Eco-Innovation Policies in Mexico." Environment Directorate, OECD, 2008. http://www.oecd.org/ dataoecd/27/17/42876980.pdf (Accessed October 2009).

Secretaría de Desarrollo Urbano y Vivienda. http://www.seduvi.df.gob.mx/seduvi (Accessed October 2009).

Secretaría del Medio Ambiente. http://www.sma.df.gob.mx/sma/index.php (Accessed October 2009).

Secretaría del Medio Ambiente. "Agenda Ambiental de la Ciudad de México: Programa de Medio Ambiente 2007–2012." http://www.sma.df.gob.mx/sma/links/download/archivos/agendambiental2008/15completo.pdf (Accessed October 2009).

Stefanie Wickstrom
Independent Scholar

MILLENNIUM DEVELOPMENT GOALS

The Millennium Development Goals (MDGs) were adopted by the United Nations (UN) in September 2000. Their creation came after more than a decade of UN conferences and summits on major issues linked to poverty and health in developing countries, and it set deadlines on a range of issues. The set of eight issues are referred to as the MDGs, each with specific objectives; they have been accepted by 189 countries and are designed to meet the world's most significant development priorities

The introduction of these goals signaled an unprecedented move to reduce extreme poverty through cooperation between all countries and development organizations. The MDGs are assessed for the world as a whole, as well as for groupings of individual countries. Groupings of developing regions are designated and are further broken down into subregions; this grouping is done primarily for statistical analysis. Different regions have different responsibilities for meeting the MDGs. Although developing countries have committed to improving governance and to making healthcare and education priorities, wealthy, developed countries have committed to supporting poorer countries through aid and better and fairer trade, as well as debt relief.

A host of UN-based organizations are directly involved in meeting the MDGs, including the UN Development Programme, UN Economic and Social Affairs, UN Environment Programme, World Bank, UN Children's Fund, World Health Organization, International Monetary Fund, UN-HABITAT, and the Food and Agriculture Organization of the UN (FAO), to name a few.

Millennium Development Goals

Goal One: End Poverty and Hunger

Reduce by Half the Proportion of People Living on Less Than $1 a Day

The first of the MDG goals is to reduce by 50 percent the number of people living below the poverty line. Originally, this was under the premise that extreme poverty consisted of people living on under $1 per day; however, this has been adjusted to $1.25, which leads to the conclusion that the number of people living in extreme poverty was and is higher than originally estimated/assumed.

Achieve Full and Productive Employment and Decent Work for All, Including Women and Young People

In many developing regions, jobs do not necessarily translate into living above the extreme poverty line. Many people are paid less than $1 per day. Women are also prohibited from being part of the formal workforce in many parts of the world. Contributing to the spread of poverty is the trend of increased urbanization, leading to higher populations in the world's slums, in which 1 billion people are exposed to increased risk of illness, hunger, and lower quality of life. Successful initiatives for addressing poverty include the spread of microfinance projects, which provide small loans for the establishment of micro-enterprises; new types of more productive crops; and capacity for emergency response to meet needs during and following disasters.

Reduce by Half the Proportion of People Who Suffer From Hunger

Although the proportion of people who are living without access to sufficient food has dropped since the early 1990s because of population growth, the actual number of people without enough food has risen. This trend has been reinforced with the rise in food prices that occurred during 2008. Much remains to be done to address the first MDG goal. In particular, work is needed to provide food availability, food aid, school feeding programs, promote urban development of slums, and facilitate better multilateral trading for the least-developed countries.

Goal Two: Achieve Universal Primary Education

Goal two has only one target: ensure that, by 2015, children everywhere will be able to complete a full course of primary schooling. In 2006, there were 73 million children worldwide who were not enrolled in primary education; although this represents a decrease from more than 100 million children in 1999. The goals for education, however, are not limited only to the number of children enrolled; they also encompass the quality of education received. Secondary school enrollments are significantly lower than for primary schools; at least 55 percent of children who are of age to attend secondary school remain unenrolled. Successful strategies to address low enrollment rates have included the abolishment of school fees and provision of materials to schools in need.

Goal Three: Promote Gender Equality and Empower Women

Goal three also has just one target: eliminate gender disparity in primary and secondary education, preferably by 2005, and in all levels of education no later than 2015. Gender equality is an important challenge in many of the developing countries, where historically girls have often been prevented from attending primary or secondary school. Worldwide, girls make up more than half of the children who are not enrolled in school.

Goal Four: Reduce Child Mortality

The goal is to reduce by two-thirds, between 1990 and 2015, the under-five-years mortality rate, the infant mortality rate, and to increase the proportion of 1-year-old children immunized against measles.

Goal Five: Maternal Health

The two targets to be achieved under this goal include reducing the maternal mortality ratio by three-quarters, and achieving universal access to reproductive health.

Goal Six: Combat HIV/AIDS, Malaria, and Other Diseases

The three targets to be met under this goal by 2015 include, first, that the spread of HIV/AIDS is halted; second, that there is universal access to treatment for HIV/AIDS for all those who need it; and third, to halt and begin to reverse the incidence of malaria and other major diseases.

Goal Seven: Environmental Sustainability

There are four targets to be met under this goal: to integrate the principles of sustainable development into country policies and programs and reverse the loss of environmental resources; to reduce biodiversity loss; to halve the proportion of the population without sustainable access to safe drinking water and basic sanitation; and to achieve a significant improvement in the lives of slum dwellers.

Goal Eight: Global Partnership

The five targets to be achieved under this goal include addressing special needs of least-developed countries, landlocked countries, and small-island developing states; developing further an open, rule-based, predictable, nondiscriminatory trading and financial system; dealing with debt of developing countries; providing access to affordable essential drugs in developing countries; and making available benefits of new technologies in developing countries.

Progress on the Goals

Since signing the convention, the achievement of the MDGs has become an increasingly difficult challenge. Some of the difficulties include a lack of capacity to implement them, lack of financial resources, and at times, lack of political will. It has been argued that the recession has taken a major toll on efforts, and the UN believes that the recession has hit those earning less than $1.25 particularly hard; the number of people earning those wages is significantly higher than was projected before the economic challenges in 2008 and 2009.

A major issue obstacle to urban areas achieving these goals is reaching the "urban poor," who generally live in slums, squatter settlements, or are marginalized. This group constitutes hundreds of millions of people, and to meet these needs, a focus on change in local governance is needed. This is important because "in urban areas, it is locally applied government rules and procedures that determine whether low-income households can send their children to school and afford to keep them there; whether they can get treatment when ill or injured; whether they are connected to water, sanitation and drainage networks." One of the major policy challenges in urban poor areas will be to legalize land tenure and provide the area with basic infrastructure, including safe drinking water and sanitation facilities, educational institutions, and hospitals. Various UN agencies, for example, UN-Habitat, are working in partnership with the agency United Cities and Local Governments to develop a program to localize MDGs.

Achieving goal seven, supporting environmental sustainability, has its own difficulties. The poor often exploit local resources when faced with hunger and the survival of their livelihoods. In this regard, the FAO "supports the integrated management of land, fisheries, forest and genetic resources, including through conservation agriculture, integrated pest management, water conservation and responsible water-use practices, and the protection of biodiversity." The MDGs strive to "create a vibrant, equitable and productive urban environment" at local and national levels, and stress that efforts to reduce poverty and achieve sustainable development should be enacted simultaneously to maintain a healthy planet and ecosystem.

See Also: Agenda 21; Sustainability Indicators; Sustainable Development.

Further Readings

McGillivray, Mark. *Achieving the Millennium Development Goals* (Studies in Development Economics and Policy). New York: Palgrave Macmillan, 2008.
"Millennium Development Goals: End Poverty 2015." http://www.un.org/millenniumgoals (Accessed October 2009).
Werhane, Patricia H. *Alleviating Poverty Through Profitable Partnerships: Globalization, Markets, and Economic Well-Being.* London: Routledge, 2009.

Velma I. Grover
United Nations University–Institute for Water,
Environment and Health (UNU–IWEH)

MITIGATION

Mitigation refers to a wide variety of policies, programs, and practices intended to compensate for natural resource impacts. Commonly, these measures have been developed to offset impacts caused by urban development. For instance, an active compensatory wetland mitigation program currently exists in the United States in which wetland impacts are mitigated by restoring or creating wetlands elsewhere. The concept of mitigation can be a somewhat controversial because it implies that the resources (in this case, wetlands) are being adequately compensated. It remains debatable whether human measures to compensate for wetland impacts truly replace the lost functions and values provided by natural wetlands. Other examples of mitigation practices include programs for listed species. Mitigation can also refer to the variety of practices intended to compensate for future calamities, such as floods, droughts, or global warming.

Wetland mitigation has developed with wetland policy in the United States. The passing of Section 404 of the Clean Water Act in 1972 and amendments in 1979 lead to the regulation of wetlands at the federal level. Through Section 404, wetlands are protected as "waters of the United States," which normally require approval by the U.S. Army of Corps of Engineers (COE) before most dredging or filling can occur.

Wetland regulations associated with Section 404 are most often applied in urban settings where land development occurs. Other federal laws and policies apply to wetlands influenced by other land practices (e.g., agriculture, forestry). Permission to impact wetlands can be received in several forms, but when impacts are substantial, an applicant usually seeks an Individual Permit. During review of an Individual Permit application, the COE evaluates proposed wetland impacts to see if the applicant has first demonstrated avoidance of the impacts to the greatest extent possible. If impacts are deemed unavoidable, they then evaluate the project to determine if impacts have been minimized to the greatest extent possible. Only after avoidance and minimization are demonstrated does the COE examine an applicant's plan to mitigate wetland impacts. Wetland mitigation has been proposed in many forms, but most often occurs as wetland restoration (restoring a previous wetland back to wetland conditions) or creation (creating a wetland where there wasn't one before). Other measures include wetland enhancement (improving an existing wetland), wetland preservation (providing additional legal protection, such as a conservation easement), and in-lieu fees (where applicants provide payment to a conservation program). These measures are generally discouraged, however, because they do not directly replace wetlands and result in a net loss of wetlands.

There is still considerable controversy regarding the use of wetland mitigation. Research evaluating created and restored wetlands generally show that these wetlands are not equivalent to natural wetlands in terms of their ecological functions and values (habitat, water quality improvement, and biogeochemistry). Many of these wetlands are still relatively new, and it has been suggested that it may take decades for them to develop conditions comparable to natural wetlands. Nevertheless, the use of wetland mitigation to compensate for wetland impacts seems to be established as part of the regulation of wetlands by the COE and other agencies.

One element regarding mitigation that has developed recently is the use of mitigation banks. Previously, permit applicants requiring mitigation would often establish their own area for wetland compensation. However, private companies have increasingly emerged to meet the needs of those requiring mitigation. Mitigation banks are started by separate companies to generate large amounts of mitigation that can then be sold as mitigation credit to those who need it. Mitigation banks have been used extensively to mitigate for impacts associated with wetlands and some listed species habitat.

These arrangements have been praised by some because they are privately operated businesses that provide greater flexibility to urban developers who need mitigation to meet the stipulations of environmental law. Regulatory agencies also stand to benefit. The COE has increasingly encouraged the use of wetland mitigation banks because of their higher success rates. Also, when wetland mitigation is aggregated into single, larger areas it becomes more feasible for agency personnel to monitor and ensure permit compliance. The use of mitigation banks has contributed to sometimes elaborate methods of calculating wetland "debits" (the value of the impact) and the "credits" (the value of the mitigation) of a natural resource. Once the bank sells all of its credit, it can no longer be used, and the area reverts to a long-term managed conservation area, as agreed upon by the regulatory agencies during the banks approval. One criticism of this approach is that it often redistributes wetlands on the landscape. For instance, wetlands impacted along the urban fringe are often mitigated in rural areas with lower land costs (thus making mitigation banks more profitable).

Mitigation programs are also in place to meet the laws and regulations for listed (imperiled, threatened, or endangered) species. State and federal wildlife agencies often require habitat conservation plans that propose compensation whenever an impact to listed species habitat is proposed. Listed species mitigation may include managing existing land for a listed species, relocating listed species, or providing funds for purchasing and protecting habitat (often through a state sponsored program). Mitigation banks are also used to purchase credit and meet the requirements of the Endangered Species Act. Individuals needing to impact listed species habitat may buy mitigation credit (depending upon the extent of the impact) from a bank that manages habitat for their species. From an ecological perspective, this may be desirable when the habitat being regulated is already fragmented and the long-term viability of a population is questionable. Many conservationists, however, criticize the practice, and contend that these practices are often deficient and circumvent the intent of the Endangered Species Act by contributing to important habitat loss.

The term *mitigation* can also describe measures for reducing the impact of future calamities, such as floods, droughts, or climate change. Flood mitigation measures may include nonstructural measures, such as urban planning practices, pertinent policy (such as restrictions on floodplain development), municipal flood emergency plans (evacuation plans and contingencies), and insurance measures. Structural mitigation measures may include levees and designed floodwater storage. All of these measures are intended to increase preparedness and decrease the impact from a future flood event. Preventative policy measures like these have also been designed to mitigate for droughts. These measures identify at-risk groups or regions and make the necessary preparations to minimize water-shortage impacts and hardships when a drought does occur.

Similarly, mitigation policies have been developed to address the threat posed from global climate change. Scientific consensus has determined that increasing carbon dioxide and other greenhouse gas levels in the atmosphere have lead to a global warming trend that has the potential to cause substantial environmental changes. To counter these trends, an international effort has focused on identifying measures needed to slow and reverse global warming. Global mitigation plans emphasize the reduction of greenhouse gas emission and the increase in carbon sink programs. Strategies to reduce greenhouse gases usually include energy conservation and developing clean energy programs. The other side of the equation is to enhance natural carbon storage in oceans, soils, and vegetation while improving other man-made technologies to sequester carbon.

See Also: Carbon Neutral; Carbon Trading; Ecosystem Restoration; Habitat Conservation and Restoration; Wetlands.

Further Readings

Bonnie, Robert. "Endangered Species Mitigation Banking: Promoting Recovery Through Habitat Conservation Planning Under the Endangered Species Act." *The Science of the Total Environment*, 240 (1999).

Hildreth, Richard G., David R. Hodas, Nicholas A. Robinson and James Gustave Speth. *Climate Change Law: Mitigation and Adaptation*. Eagan, MN: West, 2009.

National Research Council. *Compensating for Wetland Losses Under the Clean Water Act*. Washington, D.C.: National Academies, 2001.

Christopher Anderson
Auburn University

NATURAL CAPITAL

Natural capital is the way economists refer to the natural environment, and in particular, aspects of nature that are of use to humans including minerals, biological yield potential, and pollution absorption capacity. The term is used to include the goods and services provided by the environment as components of the economic system.

According to the Organisation for Economic Co-operation and Development (OECD), natural capital consists of natural resource stocks, land, and ecosystems. In a city, natural capital that add to environmental amenity would include rivers, harbors and beaches, wild areas, sporting facilities, parks, community gardens, as well as the trees, flora, and fauna that live there. Urban natural capital helps improve the lives of residents, both mentally and physically, and can be instrumental in attracting tourism, a skilled workforce, and can help increase area property values.

The air, soil, and water supply are also part of urban natural capital that are often taken for granted until they become polluted. A related term is *cultivated capital,* which is natural capital that has been transformed or adapted by humans. Examples would include domesticated animals, plant varieties, and urban parks.

Traditionally, economists paid little attention to the natural environment, assuming it would always be there to provide its goods and services. However, as this notion was undermined during the 1970s and 1980s, the natural environment was included in economic models to the extent that it provides inputs to the economic system and takes back outputs—waste products from the economic system.

The notion of natural capital incorporates the idea that the natural environment should be treated in the same way as "human capital" (skills, knowledge, and technology) and "human-made capital" (buildings, machinery, etc.). Economists argue that this is necessary because the management of the environment should be seen as an economic problem of allocation of scarce resources.

National Accounts

Most nations measure their economic progress in terms of rising gross national product (GNP). However, this is a measure that takes no account of the depletion of natural capital that might be occurring. Many people have called for national accounts to be adjusted to take account of this loss of natural capital.

Various modifications to GNP have been proposed over the years as a way of incorporating social and environmental factors. In 1985, the Organisation for Economic Co-operation and Development made a commitment to develop "more accurate resource accounts," and in 1987, the Brundtland Commission recognized the need to take full account of the improvement or deterioration in the stock of natural resources in measuring a nation's economic growth.

However, for the environment to be integrated into national accounts, it has to be valued in monetary terms, which creates problems. It can be done more easily for minerals and resources that have a market value, but finding a monetary value for noncommercial wild species, for example, or ecosystems, is far more difficult.

The people who put together the United Nations' system of national accounts, based on GNP, have decided that there should not be any major changes to them. Rather, they suggest that a separate system of satellite accounts be worked out that would give measures of natural capital, and that, at some time in the distant future, these might be incorporated into the main GNP figures. Norway, Canada, and France have instituted systems of extensive resource accounts that are separate but supplementary to their national economic accounts. These are physical measures of the country's natural resources such as forests, fish, and minerals.

The desire to incorporate natural capital into national economic accounts is based on the need to take account of environmental degradation and loss of environmental resources. However, the underlying assumption behind integration is that it is the aggregate of natural, human, and man-made capital that matters.

Weak Sustainability

The weak sustainability view, promoted by many economists, businesspeople, and policy makers, is that the total amount of capital (human, man-made, and natural) should be maintained, but the exact proportion of natural capital does not matter. In other words, a community can use up natural resources and degrade the natural environment, so long as they compensate for the loss with human capital and human-made capital. The Business Council of Australia, for example, argued that sustainable development does not require natural capital to be kept constant because what is most important is maintenance of the capacity to generate resources into the future, and the form of capital used to do this is merely a matter of efficiency.

Advocates of weak sustainability point out that if the money obtained from exploiting an exhaustible resource, such as oil, is invested so that it yields a continuous flow of income, it is equivalent to holding the stock of oil constant. They therefore argue that not only is some substitution inevitable when it comes to the commercial exploitation of natural capital but it is also consistent with intergenerational equity, provided that the income from that exploitation is used to generate capital of some sort. Economist David Pearce says that this means that the Amazon forest can be removed so long as the proceeds from removing it "are reinvested to build up some other form of capital."

Weak sustainability provides a rationale for continuing to use nonrenewable resources at ever-increasing rates. If there are temporary shortages of resources, the prices will go up and new reserves will be found, substitutes discovered, and/or more efficient use encouraged.

Natural Limits

Although the economic value of natural capital can be easily replaced, its functions are less easily replaced. Most people, even economists, agree that there are limits on the extent to which natural capital can be replaced without changing some biological processes and putting ecological sustainability at risk. In this view, the proportion of natural to human-made capital does matter.

Pearce and his colleagues give several reasons for maintaining a minimal level of natural capital. The first is nonsubstitutability. There are many types of environmental assets that are essential for life-support systems for which there are no substitutes; for example, the climate-regulating functions of ocean phytoplankton, the watershed protection functions of tropical forests, and the pollution-cleaning and nutrient-trap functions of wetlands. Moreover, for those people who believe that animals and plants have an intrinsic value, there can be no substitute.

Second, there are other environmental assets for which we cannot be sure there will be substitutes in the future, nor what the consequences of continually degrading them will be. Scientists do not know enough about the functions of natural ecosystems and the possible consequences of depleting and degrading natural capital, and it would be irrational to destroy these parts of the environment if we do not know whether we will benefit from it in the long term.

Third, the depletion of natural capital can lead to irreversible losses, such as the loss of species and habitats, which once lost cannot be re-created through man-made capital. Other losses are not irreversible, but their repair may take centuries—for example, the ozone layer and soil degradation.

There is an equity issue involved in replacing natural resources and environmental assets—that are currently freely available to everyone—with human-made resources that have to be bought and may only be accessible to some people who can afford them. Moreover, a substitution of wealth for natural capital does not mean that those who suffer from the loss of environmental amenity are the same people as those who will benefit from the additional wealth.

Finally, human-made capital often lacks an important feature of natural capital—diversity. Diverse ecological systems are more resilient to shocks and stress. Biological diversity ensures that ecosystems are robust and more likely to survive disruption, disease, and natural disasters. Even in economic systems, diversity helps to spread risks and maintain options.

Strong Sustainability

Understandably, environmentalists generally reject the concept of weak sustainability, even if it incorporates the idea of maintaining minimal environmental functions. They argue that the environment should not be degraded for future generations, even if they are compensated with greater human-made capital. They claim that human welfare can only be maintained over generations if the environment is not degraded; in economists' terms, if natural capital is not declining. They point out that we do not know what the safe limits of environmental degradation are; yet if those safe limits are crossed, the options for future generations would be severely limited.

Second, many environmentalists and others reject the idea of natural capital—that the natural environment should be treated like other forms of capital that are interchangeable.

They believe that most losses of environmental quality cannot be adequately compensated by gains in human or human-made capital without loss of welfare. Therefore they argue that future generations should not inherit a degraded environment, no matter how many extra sources of wealth are available to them. This is referred to as *strong sustainability*.

The production and consumption values and absorption capacity provided by the environment may be able to be replaced or extended, particularly through technological innovation, and in this way it may make sense to speak of human capital compensating for natural capital, but this is not the case with other environmental values. To maintain recreational, entertainment, and aesthetic values, the environment must not be spoiled.

If an old-growth forest is cut down, a tree plantation may replace much of the economic value of the old forest; however, a plantation is unlikely to recreate the original ecosystem and support the biodiversity provided by the natural forest. Nor will it have the beauty or spiritual value of the original forest. The plantation will be an impoverished version of the original forest, with many of the values associated with forests gone.

Bryan Norton asks "[S]uppose that our generation converts all wilderness areas and natural communities into productive mines, farmland, production forests, or shopping centres, and suppose we do so efficiently, and that we are careful to save a portion of the profits, and invest them wisely leaving the future far more wealthy than we are. Does it not make sense to claim that, in doing so, we harmed future people, not economically, but in the sense that we seriously and irreversibly narrowed their range of choices and experiences? A whole range of human experience would have been obliterated."

Future people who have never experienced wilderness would not miss it and would make do with human-made landscapes. They would not know they were worse off. However, current generations would clearly have diminished the range of future choices and opportunities and impoverished future lives. In this way the environment is quite different from conventional forms of capital.

In addition, the relative value of natural capital to human-created capital can change over time because the demand for environmental resources is likely to increase as they become scarcer, as incomes increase and people have more leisure time, and as the tourism and recreation industries are developed. In times when human capital was short and natural capital plentiful, it was sensible to use up natural capital, but now that natural capital is in short supply and human capital plentiful, it no longer makes sense.

A professor of international and environmental law, Edith Brown Weiss, argues that not only can natural resource consumption increase the real prices of those resources for future generations but that resources also may be depleted before they are identified as useful or before their best use is discovered. Weiss concludes that intergenerational equity requires preserving options, environmental quality, and access for future generations.

When resources are depleted and species extinct, the options available to future generations are narrowed. Overdevelopment reduces options and reduces diversity. Retaining environmental quality requires passing on the environment in as good a condition as we found it. It does not preclude some tradeoffs and compromises, but it requires that those tradeoffs do not endanger the overall quality of the environment so that environmental functions are reduced and ecosystems are unable to recover. The principle of "conservation of access" implies that not only should current generations ensure equitable access to that which they have inherited from previous generations but also that they should ensure that future generations can also enjoy this access.

See Also: Green Infrastructure; Infrastructure; Sustainable Development.

Further Readings

Beder, Sharon. *Environmental Principles and Policies*. Sydney: UNSW Press; London: Earthscan, 2006.

Costanza, R. and H. E. Daly. "Natural Capital and Sustainable Development." *Conservation Biology*, 6 (1992).

Norton, Bryan. "Ecology and Opportunity: Intergenerational Equity and Sustainable Options." In *Fairness and Futurity: Essays on Environmental Sustainability and Social Justice*, Andrew Dobson, ed. Oxford: Oxford University Press, 1999.

Pearce, David, et al. *Blueprint for a Green Economy*. London: Earthscan, 1989.

Weiss, Edith Brown. "In Fairness to Future Generations." *Environment*, 32/3 (1990).

Sharon Beder
University of Wollongong

NEW YORK CITY, NEW YORK

Because of the density of its urban form and the extensiveness of its mass transit system, which allow for lower per capita carbon emissions, New York City consumes relatively less energy and produces less greenhouse emissions than most American metropolitan areas. In addition to such infrastructural advantages, New York City has received a considerable boost toward the goals of sustainability thanks to the Bloomberg administration's ambitious environmental efforts.

Several studies contend that a metropolitan area's overall density is inversely proportional to its carbon emissions: Densely populated cities such as New York and San Francisco tend to have relatively smaller transportation and residential carbon footprints than low-density metro areas, which tend to be larger emitters of carbon dioxide per capita. A recent report by the Brookings Institution, which quantifies carbon emissions from transportation and residential buildings for the 100 largest U.S. cities, concludes that large metro areas offer greater energy and carbon efficiency than their suburban or rural counterparts: The average metro area resident's partial carbon footprint in 2005 was 86 percent of the average American's partial footprint. The difference primarily results from shorter car trips, lower automobile ownership, and less residential electricity and oil use. Spatially compact mixed-use development patterns also allow for lower carbon emissions, and households in apartment buildings with several units consume considerably less energy than those in single-family homes. According to David Owen, residents of New York City walk more, drive less, and leave a significantly smaller carbon footprint than people living anywhere else in the United States. Data from the 2009 U.S. Census contend that New York has the highest ranking in transit ridership in the United States, with 195 yearly trips per capita—in 2005, 37.9 percent of all transit ridership in the United States occurred in the New York City area, as opposed to 7.3 percent in Los Angeles and 6.6 percent in Chicago.

Hence, just by virtue of its density, its compact urban form, and the extensiveness of its mass transit system, New York City is structurally well equipped to become a leading "green city." Nonetheless, densely populated metropolitan areas like New York face a wider array of environmental challenges than low-density urban areas, including a stronger concentration of pollutants and severe infrastructure overloads.

New York City's air quality is among the worst in the United States, according to a 2008 survey that ranks it as the eighth worst for ozone pollution (smog) and as the 13th worst for particle pollution (soot). Moreover, densely populated metropolitan areas are likely to endure the most severe effects of climate change as compared with their suburban or rural counterparts. In the past century, the average temperature in New York has risen by two degrees, and the trend shows no sign of slowing down: Six of the hottest summers ever recorded in the city have occurred since 1990, and the 2002–2009 winters all had snowfalls far above average. According to a recent study led by the New York City Panel on Climate Change, frequent heat waves, intense rainstorms, and rising sea levels may pose a serious threat to New York City's infrastructure in the years to come.

In the United States, "green" policies can benefit from legislation at the federal level, such as the U.S. Energy Bill approved on June 26, 2009, which grants hundreds of billions of dollars in energy-related spending as part of the economic stimulus, allowing opening the market to new technologies and services, and creating new jobs in the "green economy."

During the second Bloomberg administration, New York has not only taken up the environmental challenge but has also sought to cast itself as a model for green cities worldwide, and Mayor Michael Bloomberg is seeking to establish himself as a national environmental leader. Although the challenges ahead are multifaceted, New York City may be on its way to take on a leadership role among U.S. cities in environmental matters.

Guidelines for a "Green" New York City: Bloomberg's PlaNYC 2030

In 2007, Mayor Michael R. Bloomberg unveiled the first long-term sustainability plan for New York City, called "PlaNYC 2030: A Greener, Greater New York." On the basis of an expected population growth of about 1 million residents by 2030, the plan outlines a growth agenda aimed at managing future development in the city while addressing the challenges of a rising population, a deteriorating infrastructure, and environmental threats. The plan outlines 127 specific proposals on energy, land use, water supply, housing, and transportation that are expressly aimed at sustaining growth while reducing greenhouse emissions by 30 percent by 2030 (compared with 2005 levels). The plan lays out a wide array of strategies for "smart" development in different sectors—including, among others, the retrofitting of city buildings to improve their energy efficiency; the remediation of brownfield sites; the development of new land by building platforms over transportation infrastructure, such as rail yards, rail lines, and highways; the provision of energy-efficiency tax rebates; the development of underused or vacant waterfront land; and the conversion of unused schools, hospitals, and other municipal sites for new housing, parks, and public space. The plan also calls for the reforestation of 2,000 acres of parkland by planting 1 million new trees, with the goal of creating a green area within 10 minutes, walking distance, of every New Yorker's home. To improve air quality, the plan promotes hydrogen and hybrid vehicles, introduces biodiesel into the city's truck fleet, enforces anti-idling laws, and lowers the maximum sulfur content in heating fuels.

The plan has spurred a wave of legislation in a range of different sectors, from land use to water supply, from transportation to housing. The next section reports the major measures, articulated along thematic areas, that the city has set forth before and after the plan's introduction. The final section lays out a preliminary analysis of their progress and partial outcomes.

Green Buildings

According to the U.S. Government Official Energy Statistics, carbon dioxide accounted for over 80 percent of U.S. greenhouse gas emissions in 2007. The majority of anthropogenic carbon dioxide is released when carbon-based fuels are burned to create energy for heating, cooling, and lighting of buildings. Residential and commercial buildings alone account for 39 percent of carbon dioxide emissions in the United States; transportation accounts for one-third, and the industrial sector is responsible for 28 percent.

In New York City, residential, commercial, and government buildings are responsible for nearly 80 percent of the city's total carbon dioxide output: City officials have therefore become increasingly concerned with their efficiency and environmental performance. Local Law 86, also known as the Green Buildings Act, which became effective in January 2007, requires new municipal buildings, as well as additions and renovations to existing city-owned buildings (around 1,300), to achieve Leadership in Energy and Environmental Design standards of sustainability, meaning they must be designed and engineered to reduce energy cost by a minimum of 20 percent.

A new package of legislation approved in April 2009 extends mandatory energy-saving improvements to large commercial and residential buildings. It requires about 22,000 buildings larger than 50,000 square feet to have an annual benchmark analysis of their energy consumption and mandates upgrades once every decade. Property owners are required to retrofit their buildings only if the upgrade will be paid back through energy savings after five years. The package includes $16 million in stimulus funding to help landlords afford the upgrades. City officials maintain that by 2022, the package will result in $2.9 billion in private investment and generate 2,000 new jobs in energy auditing and related fields, as well as thousands of temporary construction jobs, while reducing the city's carbon dioxide emissions by around 5 percent compared with 2005 levels. However, the cost of such improvements may lead landlords to shift the burden onto their tenants by increasing rents, even if the improvements ensure long-term gains. The package makes no provision for such matters, posing a serious threat to the already troubled situation of affordable rental housing in the city.

Since the introduction of the Green Building Tax Credit program, which was signed into law in May 2000 and provides for tax breaks to developers who build according to energy-efficiency standards, green buildings have mushroomed in the last decade. The program, which will be in place until 2014, has received strong support from environmental organizations, as well as from the real estate industry. In 2004, the Solaire in Battery Park City was the first residential high-rise building to be certified by the U.S. Green Building Council. By the end of 2009, Battery Park City will have eight green residential buildings and a green Goldman Sachs headquarters featuring solar panels, green roofs, organic gardens, and other energy-saving measures. New costly projects, such as the Hearst Tower and One Bryant Park in Midtown, 1400 5th Avenue in Harlem, and Octagon Park at Roosevelt Island, all boast cutting-edge energy-efficient features, and all have benefited from the tax credit program.

Transportation

On the basis of the premise that regional transportation projects may benefit both city residents and commuters from the region, PlaNYC aims to establish an independent regional financing authority—a partnership between the city and New York State—to ensure a steady stream of funding for the projects undertaken. A primary focus of the transportation plan is to increase capacity on key congested commuter routes that may otherwise be overloaded beyond capacity by 2030. Projects along this line include the Second Avenue Subway, a third track on the Long Island Rail Road main line, a second Trans-Hudson Express Tunnel for New Jersey Transit, East Side access for Long Island Rail Road commuters, Metro North Service to Penn Station, a new privately operated ferry system along the East River, and an overall improvement of bus service throughout the five boroughs by implementing a bus rapid transit system, which uses dedicated bus lanes to speed up bus travel, making it more reliable and effective.

Another key goal of the plan is to finally bring the city roads and transit systems into a state of good repair. In 1981, the Metropolitan Transit Authority halted all expansion projects until the entire system could reach a state of proper maintenance. Bloomberg's proposal for a Sustainable Mobility and Regional Transportation Financing Authority would provide the Metropolitan Transit Authority with a grant to cover the unfunded requirements to finally achieve a full state of good repair, plus allowing for future extensions.

To relieve the concentration of pollutants, the city is also encouraging the elimination of city sales taxes on energy-efficient hybrid vehicles, and the replacement of the diesel-powered city fleet with green vehicles. By 2009, over 15 percent of the city's taxi fleet had become hybrid. Furthermore, the plan strongly fosters the implementation of the city's 1,800-mile bike lane master plan: As of April 2009, around 141 miles of bike lanes had been installed.

The plan also contains prospective locations for the creation of new public spaces and pedestrian plazas in heavily congested areas. The first ones have been created in Manhattan at Madison Square, along Broadway between West 35th and West 42nd Streets, and in the Bronx, and plans are laid out for over 20 additional plaza sites.

In an effort to develop extra funding sources for its transportation initiatives, the plan also called for an elaborate plan for "congestion pricing"—a fee for vehicles driving into Manhattan on weekdays. Congestion pricing has been the subject of heated public debates and has come to define Bloomberg's second term in office. Even though New York has the lowest number of cars per household of any city in the United States and its commuters rely primarily on public transit, the idea has run into stiff resistance and was rejected by the State Assembly in Albany in April 2008. The demise of Bloomberg's plan meant a loss of $354 million in federal transportation aid, in addition to $400–$500 million in projected annual revenue from the traffic fees.

Brownfield Redevelopment

PlaNYC includes strategies to speed the cleanup and redevelopment of brownfield sites—areas whose soil has been contaminated by chemicals or industrial discharge—and dedicates a $15 million fund to foster their redevelopment. The city has established an Office of Environmental Remediation, which is supervising the remediation of 7,600 acres of unused land in the metropolitan area.

Energy Provision and Alternative Energies

To ensure steady power supply to the city, PlaNYC aims to create extra capacity by repowering old plants, building new ones, and creating dedicated transmission lines. The city is also testing the feasibility of plans to place solar panels on city-owned buildings, to install green roofs that absorb stormwater runoff, and to place wind turbines atop the city's sky-scrapers and bridges, as well as off the coastline of Queens and Brooklyn. PlaNYC may also implement a water conservation program to reduce citywide consumption by 60 million gallons a day.

Reforestation

The One Million Trees Initiative, launched in October 2007, aims to expand the city's urban forest by 20 percent; this measure promises to help improve microclimate, reduce pollutants, save energy on cooling, raise property values, and foster neighborhood revitalization. The city Department of Parks and Recreation has been granted $400 million to plant 600,000 new trees. Private businesses, residents, and other organizations are to plant the remaining 400,000 trees. In line with its greening efforts, in April 2008 the city mandated the planting of one tree for every 25 feet of street frontage in new developments and conversions of existing buildings. As of April 2009, nearly 225,000 trees have been planted across the five boroughs.

Air Quality

As of this writing, the city still fails to meet the federal standards for particulate and ozone pollution. With the launch of PlaNYC 2030, the administration pledged to "achieve the cleanest air quality of any big city in America"; by retrofitting buildings, switching to more efficient power plants, and replacing the city's fleet with hybrid vehicles, the plan aims to reduce global warming emissions by more than 30 percent by 2030. In 2009, these emissions dropped by 3.5 percent compared with 2008 levels.

Health and Nutrition

The Bloomberg administration has taken a leadership role in educating citizens on healthy eating habits and has launched a series of initiatives aimed at improving New Yorkers' access to fresh food. In an effort to counteract the epidemic of obesity and diabetes, in 2006 the city banned the use of trans-fat-laden products in restaurants and bakeries. To make fresh food available to residents in low-income neighborhoods, in June 2009 the city engaged 1,000 street vendors to bring vegetables and fruit into areas that lack traditional supermarkets that offer fresh produce. Along the same line, in September 2009 the City Planning Commission approved a proposal that offers zoning and tax incentives to supermarkets and large grocery stores locating in low-income areas of Harlem, central Brooklyn, the South Bronx, and Queens.

Conclusions

In addition to its many infrastructural assets, New York City has seen a tremendous boost toward the goal of becoming a leading green city, thanks to the Bloomberg administration's bold environmental agenda. PlaNYC 2030 lays out an ambitious roadmap for a "greener" New York, while setting the premise for future growth and leading the way out of the 2008–2009 recession. Over the past two years, the city's environmental efforts have gained momentum: According to the 2009 PlaNYC Progress Report, which chronicles the progress made since the plan was introduced, 85 of the 127 proposed initiatives are either on time or ahead of schedule.

The Bloomberg administration's proactive approach to ecofriendly forms of development is becoming a model for big cities worldwide. New York's emerging "green economy" is the result of concerns over global warming as much as it is a policy response to overcome potential limits to capital accumulation by creating new markets and services. By going "green," New York is creating new industry, new jobs, and new wealth.

However, New York's "green" agenda has downsides as well as benefits. Although its environmental efforts are highly commendable, it does not explicitly address the goals of social and economic equity, and it does not ensure that new growth will generate equal opportunities for all New Yorkers. The preliminary process that led to PlaNYC engaged a variety of energy and planning experts, business leaders, environmentalists, and other stakeholders, but it did not effectively involve the many civic groups and community organizations that are active in the city. A lively public discussion, while fostering concrete support for the plan's goals, also would have raised attention to issues of major concern among citizens.

For instance, the burden of efficiency upgrades in residential buildings may become unsustainable for moderate- and low-income New Yorkers if specific forms of control over property owners are not enacted. Mandatory improvements could lead landlords to increase rents—this in a city where, in the third quarter of 2008, only 10.6 percent of housing was affordable to people earning the median area income. The costly upzonings of undeveloped land and the rise of luxury residential projects funded through the Green Building Program raise the issue of whether a green New York can be affordable to all. Moreover, although the plan provides incentives to landlords and developers who go green, it does not grant the same kind of support to civic groups and neighborhood organizations that are willing to do their part in the greening of the city. Although it engages experts, lawyers, and managers in the newly established environmental authorities and agencies, as well as construction workers in new constructions and conversions, the plan needs to more expressly commit to an overall living wage–workforce development initiative, ensuring that as many New Yorkers as possible may also contribute in the new "green economy."

See Also: Air Quality; Brownfields; Carbon Footprint; Citizen Participation; Congestion Pricing; Density; Energy Efficiency; Green Design, Construction, and Operations; Renewable Energy; Transit.

Further Readings

American Lung Association. *State of the Air 2008*. http://www.lungusa2.org/sota/SOTA2008 .pdf (Accessed October 2009).

Angotti, Tom. "Is the Long-Term Sustainability Plan Sustainable?" *Gotham Gazette* (April 21, 2008). http://www.gothamgazette.com/article/sustainabilitywatch/20080421/210/2495 (Accessed October 2009).

Angotti, Tom. "Too Many Consultants, Not Enough Community." In "Planning for a Bigger, Greener City." *Gotham Gazette* (April 30, 2007). http://www.gothamgazette.com/article/ issueoftheweek/20070430/200/2160 (Accessed October 2009).

Baron, Eve. "How to Build Support." In "Planning for a Bigger, Greener City." *Gotham Gazette* (April 30, 2007). http://www.gothamgazette.com/article/issueoftheweek/20070430/ 200/2160 (Accessed October 2009).

Brown, Marilyn A., et al. "Shrinking the Carbon Footprint of Metropolitan America." In *Blueprint for American Prosperity.* Brookings Institution Metropolitan Policy Program (May 2008).

Byron, Joan, et al. "Growing Equality." In "Planning for a Bigger, Greener City." *Gotham Gazette* (April 30, 2007). http://www.gothamgazette.com/article/issueoftheweek/20070430/200/2160 (Accessed October 2009).

City of New York. "PlaNYC." 2007. http://nytelecom.vo.llnwd.net/o15/agencies/planyc2030/pdf/full_report_2007.pdf (Accessed May 2011).

City of New York. "PlaNYC Progress Report 2008." http://nytelecom.vo.llnwd.net/o15/agencies/planyc2030/pdf/planyc_progress_report_2008.pdf (Accessed May 2011).

City of New York. "PlaNYC Progress Report 2009." http://nytelecom.vo.llnwd.net/o15/agencies/planyc2030/pdf/planyc_progress_report_2009.pdf (Accessed May 2011).

City of New York, New York City Panel on Climate Change Climate Risk Information. 2009. http://www.nyc.gov/html/om/pdf/2009/NPCC_CRI.pdf (Accessed October 2009).

Dodman, David. "Blaming Cities for Climate Change? An Analysis of Urban Greenhouse Gas Emissions Inventories." *Environment and Urbanization*, 21/1:185–201 (2009). http://eau.sagepub.com/cgi/content/abstract/21/1/185 (Accessed October 2009).

Owen, David. *Green Metropolis: Why Living Smaller, Living Closer, and Driving Less Are the Keys to Sustainability.* New York: Riverhead, 2009.

Revkin, Andrew C. "City Plans to Make Older Buildings Refit to Save Energy." *The New York Times* (April 22, 2009).

Satterthwaite, David. "Cities' Contribution to Global Warming: Notes on the Allocation of Greenhouse Gas Emissions." *Environment and Urbanization*, 20/2:539–49 (2008). http://eau.sagepub.com/cgi/content/abstract/20/2/539 (Accessed October 2009).

Shiffman, Ron. "Making a Great Plan Even Better." In "Planning for a Bigger, Greener City." *Gotham Gazette* (April 30, 2007). http://www.gothamgazette.com/article/issueoftheweek/20070430/200/2160 (Accessed October 2009).

U.S. Census Bureau. "Transit Ridership in Selected Urbanized Areas: 2005." In *2009 Statistical Abstract of the United States* (Table 1077). http://www.census.gov/compendia/statab/tables/09s1077.pdf (Accessed October 2009).

U.S. Energy Information Administration. "Report on Emission of Greenhouse Gases." 2008. http://www.eia.doe.gov/oiaf/1605/ggrpt (Accessed October 2009).

Alessandro Busà
Center for Metropolitan Studies, Berlin

NIMBY

NIMBY (Not In My Backyard) and its derivations, such as NIMBYism, refers to efforts by groups or communities to exclude important, but locally undesirable land uses (LULUs). The types of land uses excluded include environmental facilities, such as toxic waste sites and incinerators, or human services facilities, such as homeless shelters and low-income housing. As an approach to understanding conflicts over the location of various land uses, the term *NIMBY* is often used negatively to characterize exclusionary efforts by upper-class communities, and such class-based practices do occur. Much of the literature on

NIMBY describes such struggles and focuses on ways to overcome such opposition. Recent scholarship points to the complexity of NIMBY situations. NIMBY can also refer to efforts by poorer communities to prevent themselves from serving as dumping grounds, can reflect distrust of government decisions, and can be tied to urban social movements that aim to give communities input into decision making about where to site various types of facilities. Thus, NIMBY can be looked at in at least two ways. It may refer to attempts by wealthy, relatively powerful communities to resist the siting of certain kinds of facilities in their communities. However, in a residentially segregated society, NIMBY also can represent efforts by poor, less powerful communities to avoid having unpopular land use facilities located in their backyards. Scholars and policy analysts increasingly recognize the complexity of situations characterized as examples of NIMBYism and have started questioning the assumptions and usefulness of NIMBY analyses in favor of more nuanced approaches that take the distribution of power and effects of social structures into account.

Some scholars suggest NIMBY is often an oversimplified analysis of struggles over land uses, reflecting assumptions that opponents of such land uses are parochial in their concerns and focus on their own narrow interests. Those advocating the land uses are seen as representing the more general public or civic good. It is further assumed that experts have carefully analyzed the particular situation to arrive at the best possible decision for land use siting. More specifically, scholars and policy analysts often suggest that five specific elements characterize communities in NIMBY situations:

- Communities take narrow and localized attitudes toward the problem without focusing on broader implications
- Communities distrust those sponsoring the proposed project
- Communities usually have limited information about project siting, risks, and benefits
- Communities have high levels of concern about project risks
- Communities evince highly emotional responses to the conflict

The implications are that communities should focus on broader implications, should trust project sponsors, should obtain or rely on those with more complete information, should be less concerned about risks, and should approach such conflicts in a more rational manner.

Recent research has suggested that these five conditions are not endemic to all NIMBY situations and may be an oversimplification. For example, research on opposition to the location of proposed sites for nuclear waste repositories found that local opponents were concerned about risks and did distrust sponsors, but did not exhibit low information, localized attitudes, or emotional responses. This and other research finds that only concerns about risks and distrust of project sponsors are regularly present in NIMBY situations. Some research suggests that distrust of project sponsors is not a cause but, rather, an effect of the NIMBY situation and the proposed project. Opponents of the siting of LULUs often are aware of broader implications of projects and have concerns about project risks based on rational responses to situations and well-founded distrust of supposedly neutral experts.

"Civic Interest" or "Self-Interest"

Concerns about the NIMBY syndrome lead to different analytical and policy concerns. First, some worry that the NIMBY syndrome and other associated phenomena such as BANANAism (Build Absolutely Nothing Anywhere Near Anything) and NOMP (Not On

My Planet) make it harder to build necessary regional and national projects anywhere. This often leads to the juxtaposition of NIMBY to the civic or public good mentioned above. Some writers have suggested that these issues imperil the ability to build any new projects. Second, issues of environmental justice and racism are involved in some NIMBY situations. Efforts by more powerful communities to exclude LULUs can add to already powerful pressures to site those uses in communities that have less power and clout. This can lead to LULUs becoming even more disproportionately sited in poorer, less-powerful communities, which then bear the risks associated with them. The degree of residential segregation in the United States exacerbates the effect of such decisions. Finally, because some communities do have more power, they can often exclude important and necessary facilities from their areas, and those projects may not get built at all, thus depriving areas of vital resources. For example, some communities in the Boston metropolitan area have opposed extending commuter rail lines through them, thus leading to more traffic congestion for other communities and the metropolitan region.

Too Much Ado?

It is possible, though, that fears of the NIMBY syndrome making it increasingly difficult to site important facilities anywhere may be overblown. In the article in which he originated the term *LULU*, Frank J. Popper noted the wide range of facilities that have been publicly opposed and argued that governments and private businesses have been able to cope with opposition to the location of development projects in the past. Popper's list of LULUs is inclusive, including waste management facilities, low-income housing projects, power plants, airports, prisons, halfway houses, sewage treatment plants, strip mines, power lines, highways, dams, oil refineries, rail lines, military installations, junkyards, cemeteries, amusement parks, highways, and various other types of facilities. Because distrust of actors and experts who make siting decisions is widespread, those deciding on where to site LULUs ought to consider a wide range of issues when making location decisions. Environmental justice is one such issue.

Where various types of development and land use projects that may be seen as undesirable might stimulate opposition from well-off communities, planners and policy makers may not even discuss locating facilities in better-off areas so that they will not provoke—often effective—NIMBY responses. In response to or in anticipation of resistance from upper-class communities, decision makers will often opt for what they see as the path of least resistance. This often leads to the siting and agglomeration of LULUs, which often do entail real risks and threats to quality of life, in communities less able to resist. From this perspective NIMBY can be seen as an aspect of environmental racism, defined by Robert Bullard as "practices or policies that disparately impact (whether intended or unintended) people of color and exclude people of color from environmental decision-making boards and commissions." Such communities are more likely than better-off white communities to live near LULUs. Obviously, a broad array of practices contributes to this situation, but NIMBYism is seen by many scholars as one factor in it.

Gibson argues that some of the best known research on NIMBY situations, research generally seen as reflecting a progressive orientation, reflects untenable assumptions as well as other problems in deploying the conventional NIMBY perspective. For example, in an early and influential analysis of understanding and overcoming the NIMBY perspective, Michael Dear notes that local opposition has made it difficult to locate important, but unpopular, facilities. Although sensitive to nuances of race and class, Dear—while acknowledging that opposition can be useful in some cases—focuses on how to manage local opposition to the siting of human services facilities. Similarly, Deirdre Oakley, in studying opposition to locating services for the homeless suggests that power over siting such facilities may be shifting away from local governments. Based on his Seattle research, Gibson argues that more equitable planning should open discussion and take seriously the distribution of power in local communities and regions to give less powerful communities a voice in planning and decision making.

See Also: Brownfields; Citizen Participation; City Politics; Environmental Justice; Environmental Planning; Environmental Risk; Green Communities and Neighborhood Planning; Parks, Greenways, and Open Space.

Further Readings

Bullard, Robert. "Residential Segregation and the Urban Quality of Life." In *Environmental Justice: Issues, Policies, and Solutions*, B. Bryant, ed. Washington, D.C.: Island Press, 1995.
Gibson, Timothy. "NIMBY and the Civic Good." *City & Community*, 4 (December 2005).
Oakley, Deirdre. "Housing Homeless People: Local Mobilization of Federal Resources to Fight NIMBYism." *Journal of Urban Affairs*, 24 (2002).
Popper, Frank J. "Siting LULUs." *Planning Magazine* (April 1981).
Smith, Eric R. A. N. and Marisella Marquez. "The Other Side of the NIMBY Syndrome." *Society & Natural Resources*, 13 (2000).

Walter F. Carroll
Bridgewater State College

NONPOINT SOURCE POLLUTION

Nonpoint source pollution is the introduction of impurities into a surface-water body or groundwater from dispersed origins (e.g., fertilizer runoff from lawns) compared with point source pollution, which originates from a definitive location (e.g., wastewater treatment facility). Most nonpoint source inputs fall into six major categories: sediments, nutrients, road salts, heavy metals, toxic chemicals, and pathogens. The primary pollutant for any given city is largely determined by the location and development status of an urban area. For example, Atlanta, Georgia, receives less snowfall than cities in the northeastern United States, thus, road salts will be a minor pollutant in Atlanta compared with heavy metals and toxic chemicals. In addition, heavy metals and toxic chemicals from road and parking lot runoff will be more prevalent in downtown Atlanta compared with Atlanta's sprawling suburbs, where increased construction activity may cause sediments to be the primary nonpoint source pollutant.

In urban regions, nonpoint source pollution can pose a threat to aquatic life in rivers and lakes or encourage the growth of invasive species, thereby discouraging tourism and recreation. Polluting surface waters and groundwater also pose a health threat to those relying on drinking water, which receives nonpoint source inputs. Regulating and treating nonpoint source pollutants is challenging because of the dispersed nature of nonpoint source inputs and usually intermittent discharges, associated with a rainfall or snowmelt event, compared with point sources. Although point sources have been regulated through the National Pollutant Discharge Elimination System since the passing of the 1972 federal Clean Water Act, nonpoint source inputs were not regulated until the 1987 amendments were passed, referred to as the 1987 Water Quality Act. In these amendments, greater attention was focused on controlling nonpoint sources and required storm water regulation of municipal storm sewer systems, industrial activities, and construction activities. The 1987 Water Quality Act was instrumental in recognizing the threat of nonpoint source pollution to human health and aquatic life, thereby requiring cities in the United States to develop management plans to reduce nonpoint source inputs into rivers, lakes, and groundwater.

Nonpoint Sources: Sediments

Sediments are one of the most ubiquitous nonpoint source pollutants in urban areas and include suspended (clay- and silt-sized particles) and deposited (sand-sized particles or larger) sediments in surface waters. Excessive sediment loading to surface waters can impair aquatic life by burying habitat and breeding structures in stream channels or by clogging gills of aquatic insects and fish. Furthermore, a decrease in water clarity would inhibit algae growth, which provides the essential base for food webs in all surface waters. The negative effects for humans would include taste, color, and odor problems in drinking water and blocked water intake pipes at drinking-water treatment plants. In addition, metals, nutrients, and toxic chemicals bind to sediments during runoff of impervious surfaces, which causes hazardous chemicals to be transported farther downstream in rivers than would occur otherwise. Excessive sedimentation would also raise streambed elevations, causing greater occurrences of flooding along channels and/or altering the path of the channel.

A review of the economic consequences of sedimentation in surface waters of North America titled "Economic Considerations of Continental Sediment-Monitoring Program," by W. R. Osterkamp, estimated that the annual cost of damage resulting from sediment pollution is $16 billion. In urban regions, runoff, which carries sediments to surface waters, can originate from construction sites, roads, and parking lots. Rainfall or snowmelt can erode loose sediments from disturbed land or, in the case of roads and parking lots, flush particles from impervious surfaces into adjacent rivers or stormwater systems. Any particles from roads and parking lots are also likely to include toxic chemicals such as hydrocarbons, road salts, and heavy metals. These toxic chemicals can further increase the threat to aquatic life and humans relying on receiving the water supply farther downstream.

Nutrients

Nonpoint pollution from agricultural and urban areas is a major source of nutrients, such as nitrogen and phosphorus, to surface waters and groundwater. The most common impairment of surface waters in the United States is cultural eutrophication—the biological enrichment of a body of water caused by excessive nutrient or sediment loading by human activities. In a review of nonpoint pollution titled "Nonpoint Pollution of Surface Waters With Phosphorus and Nitrogen" by Stephen Carpenter et al., eutrophication was found to account for half of the impaired lake areas and 60 percent of the impaired rivers in the United States. Excessive nutrients added to surface waters can lead to unsightly or potentially toxic algae blooms in lakes during spring and summer. If left unmanaged, lake beds could become hypoxic (low oxygen concentrations) in late summer, as dying algae settle to the bottom of lakes to be decomposed by microbes—a process that consumes dissolved oxygen in water. If oxygen levels throughout a lake decline below 5 milligrams per liter, fish kills become an imminent concern. For humans and animals, phosphorus is not considered a direct toxin; however, the U.S. Environmental Protection Agency has established a maximum contaminant level for nitrate-nitrogen of 10 milligrams per liter in drinking water for infants younger than 3 to 6 months of age. At these concentrations, nitrates can convert into nitrites in the infant's gut and interfere with the oxygen-carrying ability of blood, causing the infants to suffocate.

In a 1995 study on cultural eutrophication in Lake Erie, the Ontario Ministry of the Environment identified the three most significant nonpoint sources to be atmospheric deposition and subsequent runoff from impervious sources, septic system effluent, and lawn fertilization. Poorly maintained personal septic systems and pet wastes are a source of phosphorous, and lawn fertilization is often a source of nitrogen in surface waters. The greatest source of nitrogen, however, is atmospheric deposition of industry and automobile emissions on roads and parking lots. Rainfall or snowmelt will collect and transport free nitrogen and sediment-bound nitrogen to surface waters via runoff or through storm drains.

Road Salts, Heavy Metals, and Toxic Chemicals

Road salts, primarily consisting of sodium and calcium chlorides, are extensively used throughout the United States for deicing roads in the winter. When dissolved salt runs off roads, parking lots, and sidewalks, it can increase salinity levels in surface and groundwaters. High salt concentrations can be toxic to fish and to other aquatic life, as well as a concern for treating drinking water downstream from cities and towns in the colder climates of the United States. In addition, surface waters containing high salt content can be corrosive, causing damage to bridges and underwater pipes, such as those that draw water for drinking purposes or discharge storm water.

Heavy metals and toxic chemicals are of concern because of their toxic effects on aquatic life and human health. Copper, lead, and zinc are the most prevalent heavy metals found in urban runoff. Heavy metals bioaccumulate in aquatic life, becoming more concentrated in predatory fish such as trout and bass, which are higher up in the food chain than mussels and aquatic insects. This can lead to human health warnings about consuming fish from lakes and rivers in which metals have accumulated in the sediments. Together with polycyclic aromatic hydrocarbons and other toxic chemicals, the leading source of heavy metals and toxic chemicals from urban areas is automobiles.

Pathogens

Pathogenic organisms include bacteria, viruses, and protozoa, all of which pose a threat to human health if concentrations are high enough to cause gastrointestinal illnesses or other conditions affecting the upper respiratory tract, ears, eyes, or skin. High concentrations of pathogens in surface waters can disrupt recreational activities or affect human health. Before 1986, the EPA recommended the use of fecal coliforms as an indicator of pathogenic bacteria in surface waters. Recently, they have found that *Escherichia coli* for fresh water and enterococci for fresh and marine water are better indicators of acute gastrointestinal illness, causing states in the United States to replace their fecal coliform criteria. Two species of protozoa that are of particular concern in urban areas are *Giardia lamblia* and *Cryptosporidium parvum*, following the 1993 outbreak in Milwaukee, Wisconsin, caused by these protozoa passing through a drinking-water treatment plant, which led to 400,000 people becoming ill and approximately 100 dying.

The three primary sources of pathogens in cities are human waste, pet waste, and domesticated wildlife. Poorly managed on-site wastewater treatment systems leak human waste into surface waters or groundwater. Another source of pathogens to surface waters is when heavy loads of human waste and rainfall events coincide, resulting in wastewater being sent into storm drains to avoid flooding wastewater treatment plants. Pet waste and feces from domesticated wildlife, which include raccoons, geese, pigeons, and rats, enter surface waters via runoff of lawns and parks. Because most cities do not connect storm drains with wastewater treatment plants, any runoff of wastes into storm drains would be directly discharged into surface waters.

Management of Nonpoint Sources

Best Management Practices (BMPs) are a common technique for protecting and improving the quality and quantity of surface waters and groundwater affected by nonpoint source pollutants. The application of BMPs should focus on managing inputs, rather than treating pollutants that have already entered a river or lake. This approach is more economical and requires a collaborative effort of cities along a common body of water, expanding Best Management Practices to an entire watershed.

Examples of BMPs include the use of erosion fences at construction sites to prevent runoff of disturbed soil into surface waters—a forested or vegetative buffer strip between lawns or community parks to impede runoff of pet wastes, fertilizers, and sediments and ensuring that all on-site wastewater treatment systems are functioning properly and placed away from surface waters or shallow groundwater. Stormwater management practices have become increasingly important as cities expand, causing greater volumes of surface runoff to be directed to surface waters via storm drains. Storm drains provide a direct transport of nonpoint source pollutants from roads, parking lots, and homes to surface waters without the natural filtration action that results when runoff passes through vegetative buffer strips or wetlands in rural areas. Management practices that retain and treat stormwater include retention ponds, constructed wetlands, and infiltration trenches.

The importance of managing nonpoint source pollution is to protect water resources so that they may be recycled for multiple uses. Cities have come to recognize the benefits of clean water, particularly for drinking water purposes, which has led to the control and treatment of nonpoint source pollutants. However, many stormwater systems are now reaching the end of their useful lives or are poorly maintained, which means that stronger regulations and more creative management schemes are required to ensure the future health of our freshwater.

See Also: Stormwater Management; Water Conservation; Water Pollution; Watershed Protection; Water Treatment.

Further Readings

Carpenter, Stephen, et al. "Nonpoint Pollution of Surface Waters With Phosphorus and Nitrogen." *Ecological Applications*, 8/3 (1998).

Leeds, Rob, et al. *Nonpoint Source Pollution: Water Primer*. Columbus: Ohio State University Extension, 1996.

Osterkamp, W. R., et al. "Economic Considerations of Continental Sediment-Monitoring Program." *International Journal of Sediment Research*, 13/4 (1998).

U.S. Environmental Protection Agency (EPA). "National Management Measures to Control Non-Point Source Pollution From Urban Areas." Document EPA-841-B-05-004. Washington, D.C.: EPA, 2005.

Matthew R. Opdyke
Point Park University

NORILSK, RUSSIA

With the discovery of massive deposits of nickel in the 1920s, slave labor camps to mine these reserves soon followed, and the city of Norilsk was born. Officially founded in 1935 as the Norilsk Combine, for decades the city was a key island in Stalin's industrial gulag archipelago. The People's Commissariat for Internal Affairs (or NKVD) was given responsibility for construction of Norilsk as a test of its ability to manage large projects that relied on prison labor. Today, with a population of approximately 135,000 people, Norilsk, which is located on the Taimyr Peninsula, is the world's second-largest city (after Murmansk) above the Arctic Circle.

Mining remains the primary industrial activity and source of employment in Norilsk, as the world's largest nickel deposits, almost all of Russia's platinum group metals (platinum, palladium, and rhodium), and half of the country's copper can be found within the city's environs. These reserves are exclusively controlled by the multinational firm Norilsk Nickel. Because of the decades of heavy metal mining and processing that took place at its outdated smelting plants, Norilsk has the dubious distinction of regularly being recognized as one of the world's most polluted cities.

At its peak in 1951, Norillag, or the Norilsk Corrective Labor Camp, had 72,500 prisoners. Norillag was tasked not only with mining the deposits but also with all labor-intensive spheres of economic activity: building bridges, roads, and settlements, and even fishing and hunting. The city these laborers built is in typical Soviet geometric style, with the geometric structure of long avenues punctuated by large symbolic squares. The harsh weather (gale-force winds, heavy snowfall, and permafrost) necessitated some architectural ingenuity, including driving steel pilings deep into the soil to ensure the structural integrity of buildings. Residential blocks also included closed courtyards to avoid wind-blown snowdrifts.

The Taimyr Peninsula remains the home of the Nenets, Enets, Dolgan, and Nganasan peoples, some of whom continue to herd reindeer along the vast stretches of tundra. However, forced Soviet-era collectivization policies, coupled with the lure of high wages in the mines and smelting factories, led many to abandon their nomadic lifestyles, as in many other regions of the Soviet Union. Nevertheless the region around Norilsk, at least on paper, enjoyed special autonomous status as the Dolgano-Nenetskii (Taimyrskii) Autono-mous Okrug from 1930 to 2007 because of the large percentage of "native" groups (approximately 20 percent of the total population by the 21st century). However, the administrative center was Dudinka, the port that serves Norilsk, whereas the city of Norilsk itself paid taxes to Krasnoyarskii Krai, not the Okrug, rendering it relatively pow-erless. In 2007, the Okrug was formally dissolved and all administrative and territorial duties were handed over to the Krai.

For much of the 20th century, Norilsk was only accessible by traveling up the Yenisei River. Residents speak of traveling to and from "continent," and many still view the city as a temporary location, where one makes money and then moves on. Nevertheless, some residents have developed a strong sense of place. Despite a program to resettle residents of the Russian north, many have resisted leaving, even those who no longer work. This poses problems for the Russian government, as the cost of providing services for a resident of the polar region is four times higher than the average in Russia.

Norilsk Nickel and Pollution

Once a state-owned conglomerate, Norilsk Nickel was privatized under the controversial loan-for-shares auction program rolled out under the Yeltsin administration. This meant the de facto privatization of the Norilsk "factory-state," in which workers, because of the high cost of relocation, had no alternative for themselves in Norilsk. Many ended up suf-fering from a "Stockholm Syndrome" of loyalty to the factory.

Privatization of Norilsk Nickel turned Vladimir Potanin and flamboyant business partner Mikhail Prokhorov into Russian oligarchs. The latter, no longer associated with the company, used some of his immense wealth to create the Prokhorov Foundation, which still supports social and environmental causes in Norilsk, and some to buy the NBA basketball team the New Jersey Nets. Fully privatized by 1997, Norilsk Nickel has production facilities in six countries, but the bulk of its production still comes from its Norilsk mines and, to a lesser degree, from mines in the Kola Peninsula. Although head-quartered in Moscow, Norilsk Nickel has a hand in virtually all facets of the city's operations.

The company's smelting operations are the largest source of air pollution in Russia and the Arctic as a whole. Norilsk Nickel's three massive smelters annually release millions of tons of pollutants into the air, primarily in the form of sulfur dioxide, which turns into acid rain. A 1992 assessment by the Russian government estimated that acid rain had destroyed 180,000 hectares (ha) of forests and compromised an additional 382,000 ha. The Soviet and Russian governments, however, have long been accused of underreporting emissions levels. Getting access to the region has been difficult, in part because the city has been largely closed to foreigners since 2001, ostensibly because of a nearby missile silo facility. Analysis of heavy metal pollution, for example, of the soils indicates that nickel and copper emission have been significantly underestimated.

A. S. Yakovlev et al., found in 2009, "irreversible, catastrophic impacts" within four kilometers of Norilsk, with a landscape characterized by high concentrations of heavy metals, absence of trees, and permanent soil damage. Significant soil and water pollution were measured 25 kilometers away. Greenpeace-Russia has declared a 30-kilometer radius around the city a "dead zone," claiming acid rain has affected an area the size of Germany. National Aeronautics and Space Administration (NASA) space images clearly show the geographic scope of the smelter's effect, largely visible in the form of lack of vegetation.

According to NASA, heavy metal pollution of the soils near Norilsk is so severe that it has become economically feasible to extract platinum and palladium from them.

Life expectancy for factory workers has been estimated as 45 years of age—10 years below the national average. Norilsk Nickel has taken some measures to reduce the toxicity of the air emissions and broadly improve health conditions. However, a 2007 report indicated that air pollution is responsible for 37 percent and 22 percent of children and adult morbidity, respectively.

Prospects

In many respects, Norilsk is emblematic of the path-dependent human settlement patterns that have their roots in Soviet Era development policy that continue to shape much of present-day Russia. Norilsk is an integral component of Russia's resource geography, a node in the "petrostate" economy. Yet ameliorating the ongoing toll on the environment and the health of local residents will likely necessitate a complete transformation of Norilsk's embedded mining infrastructure, including closing the most polluting plants. O. Vendina finds hope that a network of economically strong cities could diversify both political power and economic centrality away from Moscow and St. Petersburg, knitting together Russia's regions. This horizontality of Russian space would in turn catalyze the transformation of polluted cities such as Norilsk and its nearby regions.

Ironically, the greatest hope for a "greener" Norilsk may lie outside Russia. Although the Arctic industrial enclave and former gulag feels remote in the geographic imaginary, commodities that originate from Norilsk's mines and smelters can be found, in some version, in virtually every home, driveway, and office on the planet. Nickel is used to make stainless steel, rechargeable batteries, electric guitar strings, magnets, and coins. Platinum is used for catalytic converters, jewelry, anticancer drugs, spark plugs, and more. In this manner, we are all connected to Norilsk.

See Also: Carbon Footprint; Environmental Justice; La Oroya, Peru; Waste Disposal.

Further Readings

Blacksmith Institute. "The World's Worst Polluted Places. The Top Ten of the Dirty Thirty." September 2007 (updated 2009). http://www.blacksmithinstitute.org (Accessed November 2009).

Boyd, Rognvald, et al. "Emissions From the Copper-Nickel Industry on the Kola Peninsula and at Norilsk, Russia." *Atmosphere Environment,* 43:1474–80 (2009).

Ekspert Online. "Opasno Dlia Zhizni" (September 14, 2007). http://www.expert.ru/news/2007/09/14/statistika (Accessed November 2009).

Ertz, Simon. "Building Norilsk." In *The Economics of Forced Labor: The Soviet Gulag,* Paul R. Gregory and Valery V. Lazarev, eds. Stanford, CA: Hoover Institution, 2003.

Gelman, Vladimir. "Ot Mestnogo Samoupravleniia k Vertikali Vlasti." *Pro et Contra,* 1/35 (2007). Moscow Center Carnegie. http://www.polit.ru/research/2007/04/16/gelman.html (Accessed March 2008).

Goldman, M. *Petrostate: Putin, Power, and the New Russia.* Oxford: Oxford University Press, 2008.

Kotov, V. and E. Nikita. "Norilsk Nickel: Russia Wrestles With an Old Polluter." *Environment,* 38 (November 1996).

Newell, J. *The Russian Far East: A Reference Guide for Conservation and Development.* McKinleyville, CA: Daniel & Daniel, 2004.

"Norilsk Restricts Access for Foreigners, Who Are Drawn by Its High Living Standards." *The Current Digest of the Post-Soviet Press,* 53/45:14 (2001).

Panov, E. "Norilsk pri novoi komande." *Russia Today,* 2001. http://www.russia-today .ru/2001/no_23/23_hope_1.htm (Accessed November 2009).

Pederson, William D. "Norilsk Uprising of 1953." In *Modern Encyclopedia of Russian and Soviet History,* vol. 25. Gulf Breeze, FL: Academic International, 1976.

Solzhenitsyn, Aleksandr I. *The Gulag Archipelago: 1918–1956.* New York: Harper Perennial Modern Classics, 2002.

Vendina, O. "Moscow: Post-Soviet Developments and Challenges." *Eurasian Geography and Economics,* 43/3:161–9 (Apri–May 2002).

Yakovlev, A. S., et al. "Assessment and Regulation of the Ecological State of Soils in the Impact Zone of Mining and Metallurgical Enterprises of Norilsk Nickel Company." *Eurasian Soil Science,* 41/6:648–59 (2008),

Josh Newell
University of Southern California

Megan Dixon
College of Idaho

P

PARKS, GREENWAYS, AND OPEN SPACE

Parks, greenways, and open space make up the most prominent features of urban greenspace. Although the creation, conservation, and protection of urban greenspace has a long history in urban development, parks, greenways, and open space are becoming an increasingly important aspect of building sustainable cities. If used individually, parks, greenways, and open space would have an important effect on greening the urban landscape, but when these elements are effectively integrated, they can have great effects on a city's environmental integrity and economic stability, as well as have great benefits for nonhuman systems.

Parks have long served city residents as places of leisure, recreation, and respite from the density of the city. Here, early risers in New York relax near the lake in Central Park.

Source: iStockphoto

Broadly defined, open space is any undeveloped land that remains in either a natural state, such as a wetland or forested hillside, or is in agricultural land use. Because it is one of any city's scarcest, and therefore most precious, assets, preserving open space is often one of the core goals of any urban greening strategy. Throughout the 19th and 20th centuries, urban open space has been thought to have beneficial social effects. More recently, open space is being seen by many planners as a socially necessary and valuable addition to cities because it can protect urban watersheds, preserve wildlife habitat, maintain agricultural land, provide scenic amenities, and provide outdoor recreation opportunities for residents. Since the onset of rapid urban sprawl in the mid-20th century, open space conservation has been a central goal of many urban environmental movements.

The public desire to preserve open space continues to be one of the most powerful barriers to contemporary urban encroachment into previously undeveloped greenfields. Urban growth boundaries and green belts, or legally protected undeveloped land surrounding a city, are important tools that many cities have used to preserve open space for use by their residents and the preservation of the environment.

Urban parks are the most well-known forms of urban greenspace. Parks have long served city residents as places of leisure, recreation, beauty, and respite from the density of the city. Historically, parks were public spaces designed to respond to the perceived social needs of a community. Urban park researcher Galen Cranz identifies a useful set of characteristics of the historical uses of these important community spaces. Many of the most famous large urban park systems such as Central Park in New York City were designed as pleasure grounds for upper-middle-class city residents (1850–1900). Frederick Law Olmsted, the much-acclaimed father of landscape architecture, designed Central Park and other parks as pastoral settings that were designed to operate as the "lungs of the city." Later, other parks were designed as places of social reform for children and immigrants (1900–1930) or as places of recreation for suburban families (1930–1965). Still other parks were designed as a way to incorporate open space into the city to facilitate the mental and social health of city residents (since 1965). Although many of these historical parks were designed to respond to perceived social problems, rarely did they attempt to address environmental concerns. It has been only been recently, as city residents and officials have begun to recognize important connections between urban social and environmental problems, that parks have become critical assets to creating sustainable cities. Many new parks are situated on redeveloped industrial brownfield sites or on top of sealed landfills. In this way, the city is rid of a potentially toxic site and in its place has instead a vibrant urban environmental amenity. In addition, park planners are beginning to use parks to create more environmentally sensitive cities by using more native plant species in the design, educating park users in environmental stewardship, and linking parks through greenways that provide corridors for wildlife and human movement.

A greenway is a linear corridor of open space in a city that typically features a recreational trail. Greenways are often used to link a city's parks and open spaces to create an interconnected green network. Olmsted was a pioneer in greenway development. His design for the Emerald Necklace park system in Boston created a string of greenway-connected parks that still represents for many planners the ideal park system. Greenways such as the Emerald Necklace, and more recently the Midtown Greenway in Minneapolis and the High Line in New York City, are dedicated to improving the health of city residents and restoring the environmental integrity of the urban landscape, while also assisting the urban transportation infrastructure. Most recent linear parks follow natural features such as rivers, lakes, or streams. Many greenways are situated on abandoned and redeveloped railroad corridors. In the United States, the Rails-to-Trails Conservancy has been particularly effective at initially creating legislation that makes rail trail building possible, and then providing tools for cities interested in acquiring, cleaning, and transforming abandoned rail beds for use as recreational corridors. Increasingly, greenways such as the Beltline Emerald Necklace in Atlanta, Georgia, are being coupled with public transit lines to further improve public transportation options for residents. This project will incorporate 23 miles of trails, parks, and mixed-use park-centered developments alongside Atlanta's 20-mile transit loop. Effectively planned greenways can help create sustainable cities by providing a buffer zone that can protect rivers, streams, and lakes from pollution, creating wildlife movement corridors in dense urban areas, providing additional transportation choices for pedestrians and bicyclists, and offering city residents an accessible avenue for recreation and exercise.

Parks, greenways, and open space work together to contribute to a green infrastructure for a sustainable city. These greenspace options are environmentally progressive, as they help to preserve open space, protect critical environments, and provide important corridors and habitat for species, as well as serve as natural floodplains. Urban greening projects also increasingly provide economic stability to the city. In addition, urban environmental amenities such as parks, greenways, and open space have become important elements of urban development strategies that have helped to support growing economies. Parks, greenways, and open space also aim to achieve social equity goals by providing publicly accessible recreational amenities to benefit the health of all city residents.

See Also: Adaptive Reuse; Environmental Planning; Green Belt; Green Infrastructure; Intermodal Transportation.

Further Readings

Benedict, Mark A. and Edward T. McMahon. *Green Infrastructure: Linking Landscapes and Communities*. Washington, D.C.: Island Press, 2006.
Cranz, Galen and Michael Boland. "Defining the Sustainable Park: A Fifth Model for Urban Parks." *Landscape Journal,* 23 (2004).
Hellmund, Paul Cawood and Daniel Smith. *Designing Greenways: Sustainable Landscapes for Nature and People*. Washington, D.C.: Island Press, 2006.

Jeremy Bryson
Syracuse University

PERSONAL RAPID TRANSIT

The idea of personal rapid transit (PRT), in which there would be podcars or the like that would be able to transport people to and from work, as well as to and from shops, was first raised with the expansion of a number of European and North American cities in the 1890s. However, it was not until 1953 that Donn Fichter, a U.S. city transportation planner, came up with a viable means of implementing it, writing a book, *Individualized Automatic Transit and the City*, which was published in 1964. Two years later, the U.S. Department of Housing and Urban Development started collecting information to see whether a system of PRT could be developed that would be different from public transport. Essentially, its aim was to replace cars with people having the ability to travel in a form of public transport, but alone or in family groups, and so avoid some of the perceived problems with public transportation. This would combine the privacy of a car with the environmental benefits of public transport. In many ways it would resemble a taxi service.

The Aerospace Corporation in the United States, established by Congress, tried to come up with a system of PRT, and in 1969 their findings were published in *Scientific American*. Nine years later, members of their group—Jack Irving, Harry Bernstein, C. L. Olson, and Jon Buyan—wrote *Fundamentals of Personal Rapid Transit*, which was published by D. C. Heath and Company. It was in the same year, 1978, that the term *PRT* was formally coined by J. Edward Anderson from the University of Minnesota, who tried to develop PRT technology for defense contractor Raytheon.

Obviously, the benefits from the use of PRT would be large, mainly as they would remove the need for many people to drive, thereby reducing toxic gas emissions. However, they would also massively reduce the demand for parking spaces in cities and encourage more people to use public transport, albeit different from existing buses and trains.

The only viable method of delivering PRT has been using the system of light rail or tram lines, whereby people using the podcars could designate the places where they wanted to stop. The problem is essentially the high cost of the infrastructure involved. As most major cities have varying levels of public transport, the sheer cost of introducing a new one has meant that most places have decided not to use it because it would be economically unviable. Certainly interest in PRT increased with the rise in the price of oil in 1973, and again in 1979. However, most cities have used this as an opportunity to improve public transport, rather than to work on PRT, which was largely seen only in science fiction books and films. This, however, did not stop work on the computer-controlled vehicle system in Japan throughout the 1970s. However, the cost of running this system led to the idea being scrapped on safety grounds. In 1987, the French Aramis project was also shelved after 20 years of work on a "virtual train." There was also a German scheme designed by Mannes-mann Demag that was essentially based on the PRT replacing the taxi service in Hamburg.

The first place to introduce PRT—and indeed the only one to have actually used it for any length of time—has been Morgantown, West Virginia. There they have laid 13.2 kilometers of track since 1975. Morgantown became well known on account of the PRT, which has been found to be reliable. The PRT system there allows 20 passengers per vehicle, and some suggest that what is actually being used there is similar to trams or light rail with smaller carriages, rather than PRT. It had problems, although it carried 15,000 passengers a day in 2003 and still runs to West Virginia University.

The next place to introduce PRT is London's Heathrow Airport. Construction there was completed in 2009, and it is still being tested, with 3.8 kilometers of track connecting Terminal 5 with the long-term car park. This will be the first truly commercial PRT system anywhere in the world, and the airport hopes that, if it is successful, they will expand its use to other parts of the airport, and possibly also other airports.

In Masdar City, in Abu Dhabi in the United Arab Emirates, the idea of having PRT as an everyday method of travel is going to be tested; because automobiles will not be allowed into the city, it is expected that PRT will be the only powered transport for getting around the city, with an intercity light rail system to take residents to other areas. It is expected that the PRT at Masdar City, which is being constructed on the Persian Gulf, will be able to be used in 2011, according to current plans.

Abu Dhabi is also considering the introduction of the PRT schemes in Dubai, as well as Lulu Island at Abu Dhabi, both in the United Arab Emirates. There is also a proposal to introduce PRT in Bawadi, a new development being planned to have 51 hotels, including the largest in the world. The reason for the United Arab Emirates' introducing these measures seems to have been influenced by an aim to establish residential places that architects hope will improve the lifestyles of the people living there.

The next place expected to introduce PRT will be Daventry, in Northamptonshire, in Britain. There a network is being established that is hoped to cover 4.9 kilometers, but it is expected that this distance will be increased to 55.3 kilometers (34.4 miles). The planning envisages a major reduction in the use of automobiles and an improvement in the lives of the residents, with no need for expensive car parking facilities. This in itself aims to show a major change in thinking—the region around Daventry became well known as the center of the British automobile industry, although Daventry itself remained largely unaffected until it became regarded as an outer suburb of Birmingham. Based on this proposal, Santa Cruz in California is also hoping to introduce a PRT system.

See Also: Traffic Calming; Transit; Transit-Oriented Development; Transport Demand Management.

Further Readings

Anderson, J. E. "Optimization of Transit-System Characteristics." *Journal of Advanced Transportation,* 18/1:77–111 (1984).

Cole, Leone M. and Harold W. Merritt. *Tomorrow's Transportation: New Systems for the Urban Future.* Washington, D.C.: U.S. Department of Housing and Urban Development, Office of Metropolitan Development, 1968.

Fichter, Donn. *Individualized Automatic Transit and the City.* Chicago: B. H. Sikes, 1964.

Irving, Jack, et al. *Fundamentals of Personal Rapid Transit.* Lanham, MD: Lexington Books, 1978. http://www.advancedtransit.net/content/fundamentals-personal-rapid-transit-book (Accessed November 2009).

Justin Corfield
Geelong Grammar School, Australia

Portland, Oregon

A riverside pedestrian walk in Portland, Oregon. By some measures, Portland topped the list of sustainable U.S. cities in 2008.

Source: iStockphoto

Portland is the largest city in Oregon, with a municipal population of more than 530,000 people, and a metropolitan-area population of more than 1.5 million, according to the United States Census Bureau. Portland grew as an urban center because of its agricultural base and the nearby timber industry, and preservation of agriculture in the region remains a key goal for planners and other civic leaders.

Portland has been called the most environmentally friendly city in the United States and one of the most environmentally friendly cities in the world; it has also been listed as one of America's most livable cities in several national publications. These accolades may be attributed to the following:

- Comprehensive planning policies mandated by the state to curb sprawl and piecemeal development, and a generally strong tradition of land use planning;
- A state-of-the-art multimodal public transit network with extremely high ridership for a metropolitan region of its size;
- A network of urban open spaces connected to permanently protected wilderness areas;
- An elected regional government, the only one in the nation, that administers growth management policies;
- A young, ecologically conscious urban population committed to "green lifestyles"; and
- A tradition of grassroots civic and community involvement.

Oregon was an early adopter of laws promoting Earth-friendly policies, paralleling wide interest in environmental issues beginning in the late 1960s and responding to citizens' growing concerns since World War II about urban sprawl in the Willamette River Valley, land speculation in the eastern part of the state, and development pressures in the rugged landscapes of the Pacific coast. Although the passage of these environmental laws—including public access to beaches, public bonds for remediation of pollution, bans on billboards, funding for cycling paths, and mandated bottle return—was controversial, and ultimately required compromises between powerful opposing interests, the underlying concepts generally received broad support from most Oregonians, who were fiercely protective of the unique natural resources of their state and anxious to protect them from overdevelopment.

A statewide law mandating local planning was adopted in 1973, setting forth a lengthy set of planning guidelines that cities and counties were required to follow in order to meet state goals for responsible growth. This law also included the requirement for each municipality to define an urban growth boundary, which separates urban areas from rural or agricultural areas and heavily constrains nonagricultural uses in the latter. The effect of the law and the urban growth boundary is to push growth at higher densities to the central core of cities, while limiting suburban sprawl on the perimeter.

Portland adopted its urban growth boundary in 1979, which has helped to shape its urban forms and patterns: a mosaic of distinctive pedestrian-friendly, compact neighborhoods with shops, restaurants, and other services within walking distance; new development in accordance with strict design guidelines governing architectural character and detailing; and reclamation of obsolete industrial areas for new mixed-use neighborhoods. Since 1995, the state legislation has required cities to plan to accommodate 20 years of population growth within the urban growth boundary.

Portland's neighborhoods, center city, and suburbs are supported by a comprehensive, multimodal public transit system, which in turn makes transit-oriented development possible at relatively high densities. The regional public transit agency, established in 1969, is called TriMet (for the three counties which are included in its 570-square mile service area), and is governed by a board of directors appointed by the governor. Transit services include 52 miles of light rail (the Metropolitan Area Express, or MAX) on four lines; 81 bus lines; 15 miles of commuter rail; and an eight-mile streetcar loop, operated since 2001 by the City of Portland. The City of Portland also operates an aerial tram that connects the Oregon Health and Science University's main campus to a satellite campus at the South Waterfront.

There is no charge for rides on the MAX and streetcar within a portion of the downtown area in order to reduce vehicular congestion and to facilitate automobile-free short trips. The Portland International Airport (PDX) has been connected to the light rail system via a spur since 2001; further expansion of the light rail system is planned to the southeast side of the metropolitan area.

TriMet has calculated that public transit eliminates 74,000 automobile trips and 4.2 tons of pollutants each day. A ridership report states that TriMet serves more people than any other transit agency of a similar size and that growth in ridership has continuously increased over the last 21 years, while in the last decade, increases in ridership have outpaced both population growth and the increase in vehicle miles traveled. Reports from the agency also highlight its attention to the environment: the use of biodiesel and ultra-low sulfur diesel in trains and buses; state-of-the-art strategies for improving fuel efficiency; and environmentally-sensitive construction of new facilities, including bioswales and landscaping for attenuating and filtering stormwater runoff, and the use of sustainable materials wherever possible.

A key aspect of Portland's distinctive character is its emphasis on bicycling as a means to foster the good health of its citizens while improving air quality, reducing dependence on fossil fuels, and cutting emissions of greenhouse gases. Bicycle Master Plans were prepared in 1996 and 2009. As of 2007, Portland had 266 miles of bikeways, up from 144 miles in 1996, and an almost fourfold increase in ridership measured at the four main bridges across the Willamette River. Improvements in this 11-year period included new facilities for bike parking; arrangements for allowing bikes on light rail, streetcars, buses, and the aerial tram; public awareness programs for cycling, in order to encourage a larger percentage of the population to use bikes; and decreased accident rates. The 2009 Master Plan laid out steps to more than double the bikeway miles while taking steps to decrease the speed and volume of vehicular traffic on streets with bikeways.

Metropolitan Portland's planning process includes aggressive strategies to acquire open space and to develop an integrated network of parks and recreation areas throughout the region. In 1995, voters in the Portland metropolitan area passed a regional bond measure to acquire valuable natural areas for fish, wildlife, and people. By 2005, an additional 8,100 acres (about 8.7 percent of the total area of the city) of ecologically valuable natural areas had been purchased and permanently protected from development with these funds. Between the City of Portland and Metro (the regional government), 22,000 acres of parks and natural areas are managed as public open space. Highlights of the park system include the Governor Tom McColl Waterfront Park, a 29-acre swath of riverfront that was created in 1978 when a highway along the Willamette River was demolished; Washington Park, at 130 acres one of Portland's oldest parks; Forest Park, over 5,100 acres of woods and trails; and the 40-Mile Loop, a nearly-completed 140-mile trail system, based on a 1904 plan by the landscape architects the Olmsted brothers that will link 30 parks and open space areas together and to nodes of residential and commercial activity within the metropolitan region.

Metro, the only directly elected regional government in the United States, serves as the Metropolitan Planning Organization (MPO) for transportation planning and development, and oversees the regional public transit system. Metro was created in 1992 as the consolidation of and successor to other regional agencies, and is responsible for managing regional growth, infrastructure, and development issues that cross municipal boundaries. Metro also oversees wastewater management, solid waste management and recycling, open space preservation and management, and management for conventions and other large-scale public events.

Metro also actively manages development projects. Its Transit-Oriented Development Program selects developers to propose developments on sites that are owned by the agency near transit stops and provides financial incentives to enhance the economic feasibility of higher density mixed-use projects.

Redevelopment of specific parcels is also overseen by the Portland Development Commission, a quasi-public agency that oversees housing development, neighborhood revitalization, and economic development projects. The PDC also uses state-authorized powers of urban renewal to redevelop areas that have been designated as blighted or underused; in 2008, 11 such areas were under study or development by the Commission.

Despite the previous description of the various scales of government and their active role in promoting sustainable development within the City of Portland, it is also important to note that grassroots civic involvement and citizen participation is a strong tradition in Portland, and many of its most significant environmental achievements have been at the initiative of individual citizens or citizen groups. All planning agencies are required by law to consult with affected citizens and neighborhoods throughout the planning process, and the city's Office of Neighborhood Involvement works with and provides support for 95 formal neighborhood associations.

What distinguishes Portland from other places is the willingness of its citizens to use the powers of government at a variety of scales (state, regional, and municipal) to control the pace and shape of development. Further, Portland's integration of land use planning and transportation infrastructure is an essential feature in targeting growth to compact, walkable areas in and around the center of the city, while preserving a greenbelt of thousands of acres of forests and parks throughout and surrounding the city. While some have criticized the urban growth boundary concept and other growth controls as contributory to increased taxes, inflated real estate prices, and dislocations for the most vulnerable populations within the city, Portland's continuing ability to attract new residents and major corporations to this setting is proof that these regulations fit residents' own desires to live more lightly on the planet and to steward its natural resources.

See Also: Bicycling; Citizen Participation; Green Communities and Neighborhood Planning; Smart Growth; Transit; Walkability (Pedestrian-Friendly Streets).

Further Readings

City of Portland. "Portland Bicycle Master Plan for 2030." http://www.portlandonline.com/transportation (Accessed December 2009).

Girling, Cynthia and Ronald Kellett. *Skinny Streets and Green Neighborhoods: Design for Environment and Community*. Washington, D.C.: Island Press, 2005.

Stephenson, R. Bruce. "A Vision of Green: Lewis Mumford's Legacy in Portland, Oregon." *Journal of the American Planning Association*, 65/3:259–69 (1999).

Svoboda, Elizabeth. "America's 50 Greenest Cities." *Popular Science* (February 2008). http://www.popsci.com/environment/article/2008-02/americas-50-greenest-cities?page=1 (Accessed December 2009).

Judith Otto
Framingham State College

PORTS

Ports are powerful economic engines that play a major role in global production and distribution systems, as well as major sources of environmental problems and controversies. According to the United Nations, seven and a half billion metric tons of seaborne trade cargo flowed in and out the world's seaports in 2008. The growth in the amount of goods being shipped around the world, and the expansion of ports to accommodate ever-growing cargo volumes, has been followed by a corresponding growth in pollution that affects the natural environment and the health of communities that goods move through. These growing environmental threats are compounded by the fact that both ports and the shipping interests that rely on them are difficult to regulate. Although environmental externalities, given free rein in the global commons, ports' adverse health effect are experienced locally by those who work and live near or downwind from ports. Increasingly, throughout the world, ports are facing pressures to reduce their carbon footprints and to become more sustainable, which entails balancing economic development goals with concerns for social responsibility and the environment. The conflictual nature of such a goal has made urban ports and waterfronts contested spaces in which competing interests collide.

The growth in the volume of goods shipped worldwide, and the expansion of ports to accommodate this ever-growing cargo, has been followed by a corresponding growth in pollution. Here is an image of a busy commercial shipping port in Italy.

Source: iStockphoto

The shift toward containerization in the second half of the 20th century revolutionized the flow of goods, increased the volume of international trade, and brought about a separation between ports and their historic connection to inner city waterfronts. As a result of containerization, new port technologies, changes in the size and nature of ships, and transport systems that required deeper and wider channels, as well as vast amounts of waterfront space, inner city piers were abandoned and left to deteriorate as ports migrated to more geographically suitable areas.

The construction and expansion of container ports and the decline and subsequent revitalization of inner city ports has become a worldwide process that has exerted a heavy environmental toll on natural resources and public health.

Ports and Water Pollution

Waste from ships; contaminated bilge water; storm water runoff that contains pesticides, metals, and other pollutants; and routine and catastrophic oil spills all contribute to declining water quality around ports, which negatively affects ecosystems and human health. Unwanted harmful aquatic organisms and invasive species that travel in the ballast water of ships and wreak havoc with the local marine ecology have long been a problem around ports.

Dredging to create new channels and deepen existing ones necessary for larger ships and navigability has damaged coastal ecosystems by disturbing sediments contaminated with toxic chemicals, including polychlorinated biphenyls, mercury, other heavy metals, polycyclic aromatic hydrocarbons, and pesticides. Dredging also increases water turbidity, damages habitat and wetland areas, and disturbs threatened and endangered species around ports. Although preventing contamination is the key goal, dredging in an environmentally safe way that does not disturb submerged contaminants—and finding proper disposal sites for contaminated sediments and beneficial uses for noncontaminated dredge material—has emerged as a major environmental concern for port operators.

Air Pollution, Goods Movement, and Environmental Justice Struggles

Along with the ever-increasing numbers of goods that they bring to consumers, ships also bring about increasing levels of air pollution that unnecessarily threaten the health of people. In addition to burning fuel that contains excessive amounts of sulfur, many of the ships that flow in and out of ports lack the most basic pollution control devices. Their smokestacks spew tons of particulate matter, which triggers asthma attacks, cancers, and premature deaths. In 2009, the U.S. Environmental Protection Agency announced a new proposal to reduce ship pollution within 200 miles of U.S coasts by establishing emission control areas that would require ships to use cleaner fuel and to install more effective pollution control measures than those currently used. This action followed in the footsteps of an international agreement that adopted similar emission control standards and called for measures that would allow countries to petition the International Maritime Organization—an international agency of the United Nations that establishes and maintains regulatory frameworks for the global shipping industry—to create emission control areas off their coasts.

Air pollution is not limited to ships. The cranes and cargo-handling equipment, as well as trucks and trains that move containers in and out of ports, all rely on diesel-powered engines that emit harmful amounts of particulate matter, volatile organic compounds, nitrogen oxides, and sulfur oxides that are harmful to human health. Most are not equipped with pollution control devices.

Beginning in the late 1990s, combating pollution at major U.S. ports and the environmental health effects of "goods movement" emerged as a major environmental justice issue. As goods move from ships to trucks, trains, warehouses, and retail destinations, they pass through sensitive environmental areas, city streets, and densely packed neighborhoods that in many cases are populated by low-income and minority residents. People in communities adjacent to ports and goods movement distribution routes experience a disproportionate share of adverse environmental impacts from exposure to particulate matter and nitrogen oxides that include but are not limited to premature death, cancer risk, respiratory illness, and increased risk of heart disease. High noise levels, traffic congestion, and visual blight are also common in these neighborhoods, which environmental justice activists have come to refer to as "sacrifice zones" because they subsidize the economic flow of goods movements with people's diminished health. The Environmental Protection

Agency has called the movement of freight "a public health concern, at the national, regional and community level."

In many places, the public authorities that oversee ports are using their leverage to force their tenants, that is, shipping and trucking companies, to make their operations more sustainable. For example, the public authorities that manage the ports of Los Angeles and Long Beach, the two largest ports in the United States, have created a green truck program that would significantly reduce air pollution and carbon emissions by replacing old diesel-spewing trucks with newer ones and upgrading and retrofitting newer ones with pollution control devices, such as particulate filters, that would reduce harmful emissions. The program is supported by labor, community groups, and environmental organizations but opposed by truckers and trucking companies, who fear that such regulations would make them less competitive. Similar conflicts appear in ports throughout the world as community activists, environmental organizations, and government agencies strive to make them more sustainable.

Although the discourse of sustainability has become ubiquitous in port revitalization and expansion schemes throughout the world, it is elusive and contradictory. Its promise of win–win scenarios with regard to economic growth development, environmental health, and social justice and community concerns obscures the incompatibility and tensions between these goals. Greening ports is an uphill struggle, given that in many cases global pressures to be economically competitive overshadow environmental and equity concerns.

Revitalizing brownfields—sites where previous industrial uses have left behind contamination—is one port sustainability initiative that has the potential to be beneficial for the environment, the economy, and the community. Port expansion that relies on redeveloping brownfields is a smart growth strategy that improves water quality, restores wetlands and habitat, and improves air quality and public health. Brownfield redevelopment creates economically productive property and frees up other areas that can be used for recreational and community-oriented purposes.

Waterfront Revitalization of Redundant Ports

In recent years sustainability has become a mantra for public and private urban waterfront redevelopment and revitalization schemes throughout the world. Beginning in the 1970s, decaying and derelict waterfront spaces that were made redundant by the relocation of ports became new frontiers of economic opportunity. Throughout North America, residential, commercial, leisure, and tourist-related developments proliferated on waterfronts that were formerly occupied by port-related facilities. Gradually, urban waterfront regeneration became a worldwide phenomenon.

Urban waterfront development has been uneven—some areas became high-end residential and commercial gold coast areas, whereas others remained derelict waterfront wastelands. The worldwide waterfront renaissance has brought about a condition in which pockets of luxury coexist alongside vast spaces of dereliction, poverty, and environmental degradation. In many waterfront regeneration schemes there is an ongoing tension between pressures to redevelop derelict, potentially toxic spaces and environmental concerns. In many cases, economic interests downplay potential environmental risks.

Although modern ports have migrated to the urban fringe and become less visible to urban residents, both ports and the cities that house them are indissolubly linked. Both the working port and the revitalized waterfront are interconnected parts of a regional economic and ecological system. Worldwide efforts to make ports and urban waterfronts more sustainable might help to bring about an awareness of this connection.

See Also: Air Quality; Brownfields; Environmental Justice; Intermodal Transportation; Water Pollution.

Further Readings

Environmental Defense Fund, "Harboring Pollution: The Dirty Truth About U.S. Ports." 2004. http://www.nrdc.org/air/pollution/ports/contents.asp (Accessed July 2009).
Hall, Peter. "Seaports, Urban Sustainability and Paradigm Shift." *Journal of Urban Technology,* 14/2 (2007).
Hricko, Andrea. "Ships, Trucks, and Trains: Effects of Goods Movement on Environmental Health." *Environmental Health Perspectives,* 114/4 (2006).
Levinson, Marc. *The Box: How the Shipping Container Made the World Smaller and the World Economy Bigger.* Princeton, NJ: Princeton University Press, 2006.
U.S. Environmental Protection Agency. "Green Ports USA." http://www.epa.gov/otaq/diesel/ports (Accessed July 2009).

Steven Lang
LaGuardia Community College

Power Grids

Power grids are an interconnected network of transmission lines or buried cables that carry electricity from electrical utilities to businesses, organizations, and homes. Experts describe them as "the circulatory system of the electric utility sector" and as the "lifeblood of modern life." If a power grid fails, as happened in the blackout of August 2003, a large percentage of economic life grinds to a halt.

The current North American power grid is composed of two very large, interconnected systems that stretch across all of North America except Texas. Each grid is a gigantic web of power plants, transmission lines, transformers, control centers, and consumption sites.

Electricity is first generated in a large, regional power plant, usually through the burning of fossil fuels, but also through capturing energy from hydroelectric dams, nuclear power, or the sun or wind. It then flows freely, via a path of lowest resistance, across a network of high-voltage aluminum or copper transmission lines. In the first process of delivery to electrical consumers, electricity flows along transmission lines: large, high-voltage lines that connect a large power plant to smaller substations along the grid. Within these substations, the electricity passes through transformers that lower the voltage of the electrical current being transported. The lower-voltage current then flows to consumers through the second process of delivery, via a network of lower-voltage distribution lines that connect power substations to homes, businesses, and other organizations.

Upgrades Needed

By almost all accounts, the U.S. power grid system is in desperate need of an upgrade. Not only is the system overtaxed—running at carrying capacities that greatly increase the risk of power outages—but it is also severely underequipped to handle the demands of renewable energy as the United States moves to reduce its carbon dioxide emissions.

Frequently, the U.S. grid system carries energy far greater than the level required to maintain reliable delivery to consumers. This results in power outages and issues with power quality that cost consumers and the industry between $80 and $188 billion per year. This weak and overtaxed system also greatly increases the risk of major blackouts, similar to the one that occurred in August 2003, when the homes and businesses of nearly 50 million people lost power. The blackout rolled through 11 states in the eastern United States and Canada and cost $6 billion.

There are several reasons that the grid system has reached its current state. Between 1974 and 2004, electricity demand grew by 100 percent, yet investment in grid improvements declined by 50 percent. Given the lack of investment, a lot of grid equipment is simply wearing out.

Deregulation also plays a large role. When the U.S. grid system was being built, the electric power industry was composed of regulated local monopolies that provided electricity to consumers at a price controlled by state regulation. In this context, the grid system was designed for two purposes: to allow public and private utilities to generate large quantities of energy and deliver it to customers via vertically integrated (company-owned) transmission lines, and to create a system that ensures against electrical power outages by connecting a number of different power generators to the same grid. If electricity generation and distribution failed in one part of the grid, another generator and distributor could kick in and keep the power running smoothly to consumers.

Through a process of market deregulation that began in 1978, the function of power grids changed in a manner that even further diminished their reliability. Previously, an electric utility would own and operate both the power plants and the lines that connected them to their consumers. Power flowed through a relatively closed system of predictable demand. Reliability was secured by linking local, closed systems to neighboring grids that would supply power should an outage occur. Under deregulation, however, ownership of power generators and grid systems became decoupled. Electric power is now traded wholesale, transported through a separately owned grid to a customer that can be located thousands of miles from the generator. Predicting and, more important, regulating power flows across the grid becomes much more difficult. When power flows cannot be regulated, congestion is more likely to occur, and with congestion come power outages.

As the United States and Canada switch to renewable power sources, and as cities take steps to purchase more of their energy from renewable, distributed power sources and encourage residents to generate energy using solar and wind technologies, the demands placed on the grid will become even more complicated. By 2008, carbon reduction laws were in effect in 27 U.S. states and four Canadian provinces. To comply with these laws, the electric utilities industry will need to shunt aside some of the coal-burning plants currently connected to the power grid and substitute electricity from solar plants or wind turbines, which are usually located in areas that lack sufficient transmission capacities.

Investments and Innovations

In 1999, power grid investments began to improve; they were projected to grow to $11 billion in 2010. Despite this increase, the North American Electric Reliability Corporation predicts that even more transmission will be needed to maintain reliability. Companies are developing a number of innovations to make this happen, yet change is slow.

The widespread adoption of wind and solar energy will require investments not only in transmission lines but also in energy storage devices. One of the most well-developed innovations is the smart grid, a sophisticated computerized system for regulating energy flows across the grid. The smart grid is designed so that people at all phases of the energy delivery system—generators, consumers, and grid operators—have varying degrees of control over energy flow. Consumers can opt to turn off key appliances at peak consumption times when prices are highest, and grid operators can locate outages more rapidly and rely on computers to instigate hundreds of actions to contain the damage.

Other innovations are in the wings, awaiting investment sufficient to make them a marketplace reality. These include the development of massive batteries to store energy captured by wind farms and utility-scale solar generators. Because wind and solar power is generated only when the weather cooperates, adequate storage is essential to the development of renewable energy on a large scale. Another innovation relevant to the development of renewables is distributed generation, which relies on small-scale generators (often solar or wind) to produce power for consumer who live very near the generator site.

See Also: Distributed Generation; District Energy; Renewable Energy.

Further Readings

Borlase, Stuart. *Smart Grids: Infrastructure, Technology, and Solutions (Power Engineering)*. Boca Raton, FL: CRC Press, 2010.

Brain, Marshall. "How Power Grids Work." *How Stuff Works*. http://www.howstuffworks.com/power.htm (Accessed November 2009).

Gellings, Clark W. *The Smart Grid: Enabling Energy Efficiency and Demand Response*. Boca Raton, FL: CRC Press, 2009.

U.S. Energy Information Administration. "What Is the Electric Power Grid, and What Are Some of the Challenges It Faces." http://tonto.eia.doe.gov/energy_in_brief/power_grid.cfm (Accessed November 2009).

Maura Troester Nuñez
University of Colorado, Boulder

R

RECYCLING IN CITIES

The vast amount of waste created within city limits is a growing concern. Cities such as Naples, Italy, for example, have been facing serious concerns because of overfilling landfills. Recycling in cities is one answer to the problem of urban waste. Recycling is the recovery of discarded material for its reuse in new products. By reusing material destined for disposal, municipal recycling reduces an individual's ecological footprint by saving energy; avoiding the use of virgin resources, such as trees; and creating jobs.

Curbside recycling is the most common recycling service provided in urban areas. There are a variety of ways to handle curbside pickup. Some municipal governments offer recycling as a citywide service. In these instances, the city either contracts with a for-profit or a nonprofit hauling company or provides recycling services itself. If the city does not offer recycling services, which is becoming increasingly uncommon, curbside pickup may be available via independent companies. Residents also have the option of dropping off their recycling at the processing center themselves. These latter two options require more awareness and initiative on the part of individual residents.

Separate bins are often provided for trash and recycling, and if curbside pickup is offered, it generally occurs on the same day as trash pickup. Single-stream recycling offers an efficient means of collection. The term

New York City offers recycling as a citywide service and provides recycling bins in public spaces, like this one along the East River.

Source: iStockphoto

single stream means using one bin to collect all recycling, rather than the two or more bin system, usually mixed paper and commingled containers. This eliminates the need to run two or more recycling routes to collect the streams, provides an opportunity to standardize the collection vehicles, and thus decreases the energy, pollution, and economic costs associated with recycling pickup. In addition, the community participation rate often increases with the establishment of a single-stream recycling program. The downside to single-stream recycling is that there is a greater danger of contamination, the quality generally decreases, and there can be higher infrastructure costs up front. San Francisco, Toronto, Denver, Tucson, San Jose, Philadelphia, and Dallas are examples of cities that use single-stream recycling.

Both environmentalists and recyclers are concerned that the global economic downturn will spell disaster for municipal recycling programs. Where haulers were once paid by recycling centers, the bottom has dropped out of the market to such an extent that now haulers often have to pay the processing centers to accept their goods. As Western curbside recycling programs become less profitable, local governments will be forced to reevaluate the importance of recycling as they struggle with large deficits. The fear is that this may result in fewer municipally run recycling services.

There is no standard for what materials get recycled from city to city because different haulers pick up different items. They are limited by what their trucks are equipped to handle and what the processing facility they transport to has the capacity to process, as well as the market for the material. In addition, cities vary in their commitment to recycling, and therefore also vary in the stringency of their recycling standards. Although municipalities are generally responsible for the cost of waste disposal and recycling, throughout the European Union countries have adopted producer responsibility laws that require the producers of packaging material to either recover packaging material for recycling or pay into a system that ensures the packaging is collected and recycled.

Can Everything Be Recycled?

The resources available for recycling in cities are not limited to glass, plastic, aluminum, and paper. For instance, the city of Austin, Texas, has been collecting yard debris since 1989 and using it to create a product known as DilloDirt, which is sold locally. Many other cities collect yard waste, and more recently a handful of cities have begun to pick up food waste for composting. It has been estimated that composting can eliminate as much as 18 percent of disposal costs. Seattle, San Francisco, and Los Angeles offer curbside collection of household food waste. San Francisco even accepts meat and bones—materials not generally compostable at a household scale. Other cities encourage backyard composting. Philadelphia, for instance, offers residents a large bin in which they can combine household compostables and yard debris on their own property. Composting food waste and yard debris can play an active role in mitigating the effects of waste on climate change because it prevents the release of methane, a greenhouse gas significantly more potent than carbon dioxide.

Cities provide opportunities for large-scale water, energy, and land recycling as well. Recycling greywater—water draining from the shower, washing machine, dishwasher, or bathroom and kitchen sinks—reduces pollution, saves energy, and lessens a community's consumption of water. A simple rain barrel placed under household gutters can prevent a significant amount of water from entering the sewer. This water can then be used to water lawns and other landscaping.

Much energy is wasted transporting power from the plants where it is generated to the consumers that use it. To reduce this waste, it has been suggested that energy can be recycled within city limits through combined heat and power systems. Consolidated Edison in

New York City provides one example of such a program. The company recycles steam to heat thousands of buildings in Manhattan. Other cities in Europe have similar programs in place. Denmark recycles near 50 percent of its energy, and in Gothenburg, Sweden, a waste incinerator plant heats the water for a quarter of a million homes. The city of Boston, Massachusetts, is planning a unique indoor composting facility that would capture methane gas from rotting leaves and burn it to generate electricity for 1,500 homes, as well as to run on-site greenhouses.

Urban areas are littered with abundant empty lots and other brownfield lands. As defined by the U.S. Environmental Protection Agency, brownfields are real property, the expansion, redevelopment, or reuse of which may be complicated by the presence or potential presence of a hazardous substance, pollutant, or contaminant. By cleaning up, reinvesting in, and recycling these properties for new purposes, development pressures are removed from undeveloped, open land—often called greenfields. Brownfield land is an important urban resource that is often overlooked as waste.

Electronic waste, or e-waste, has become an increasing concern in cities, as computers and other electronics become an integral part of urban life. Many cities offer electronics pickup days where residents are invited to drop off old refrigerators, televisions, and other electronics at no cost. Generally speaking, it is costly to dispose of such items. New York City has taken this a step further and is the first major metropolitan area in the U.S. to pass a law requiring residents and manufacturers to recycle used electronic devices, including MP3 players, televisions, and computers. Other organizations such as Free Geek in Portland recycle computers on-site with the help of a mostly volunteer staff.

A Dirty Business

As much as there are many environmental benefits from recycling waste from cities, it is not always a clean business. A significant portion of the waste from U.S. cities gets shipped overseas for processing, creating both a large carbon footprint and concern over who is suffering the consequences of dirty recycling methods. In this sense, recycling is an environmental justice issue: Those who create the waste should take responsibility for its full life cycle. Such concerns are particularly poignant regarding e-waste, which is often harmful to human health throughout the recycling process.

Encouraging product reuse is another method for cities to keep material out of the disposal system. Cities often support materials exchange programs to facilitate the reuse and exchange of a wide range of items. For example, the Twin Cities Free Market, run by the nonprofit recycler Eureka Recycling in the Twin Cities metro area of Minnesota, and reuse centers, such as Bring Recycling in Eugene, Oregon, offer the opportunity to reduce the amount of durable goods entering the waste stream. The Internet has provided unique opportunities for grassroots recycling efforts along these lines. Websites such as the FreeCycle Network and Craigslist connect people with available resources by allowing them to post items they no longer need or items they are looking for. These sites enable city residents to reuse items that would otherwise end up in the landfill and reduce the energy costs of production.

Both cities and private organizations provide waste-reduction education and advocacy. For example, the Urban Corps of San Diego, a nonprofit agency that provides education and jobs to young adults, is building an Education and Community Outreach Center to serve as a resource for educating, training, and demonstrating how recycling and conservation preserves natural resources, reduces pollution, and reduces waste hauling costs/expenses.

Some cities are making commitments beyond basic recycling. For example, Boulder, Colorado, has signed a Zero Waste Resolution, and San Francisco has gone so far as to ban bottled water from city sponsored functions, in part to reduce the waste associated with disposable water bottles. Large-scale urban events provide templates for what can happen on a citywide scale. At the 2009 Superbowl, recycling and composting were encouraged, with the composting of biodegradeable plates, forks, napkins, and straws. The cooking grease and plastic drink bottles were also recycled.

Cities also facilitate recycling by supporting the construction and operation of facilities to separate mixed streams of recyclables into marketable components and by committing to purchase products made with postconsumer recycled content, thereby stimulating the market for recycled raw materials.

See Also: Adaptive Reuse; Brownfields; Combined Heat and Power (Cogeneration); Greywater; Landfills; Waste Disposal.

Further Readings

Kellogg, Scott and Stacy Pettigrew. *Toolbox for Sustainable City Living*. Cambridge, MA: South End, 2008.
Gandy, Matthew. *Recycling and the Politics of Urban Waste*. New York: Palgrave Macmillan, 1994.
Weinberg, Adam S., et al. *Urban Recycling and the Search for Sustainable Community Development*. Princeton, NJ: Princeton University Press, 2000.

Shannon Tyman
University of Oregon

RENEWABLE ENERGY

Geothermal plants usually drill deep into the Earth to reach the steam produced by the heat of the Earth. Here, steam pipes lead from a geothermal field toward a power-generation plant in New Zealand.

Source: iStockphoto

Renewable energy is the energy produced by using natural resources that will regenerate in time. Primary sources of renewable energy are water, geothermal heat, wind, and sunlight. The energy produced by using different organic materials, such as wood, agricultural rejects, forestry residues, and urban solid waste (called biomass) is also regarded as "renewable." According to authoritative international agencies like the United Nations Environment Programme and the Intergovernmental Panel on Climate Change, renewable energies represent the most realistic alternative to fossil fuels. At present, renewable sources are employed chiefly in the production of electricity, but also to produce biofuels and to provide heat for industrial and private users. Over the last 10 years, the development of renewable energies has been stimulated by concern over global warming. At this time, it is believed that renewable energy sources will play a key role in limiting greenhouse gases emissions.

Water

Hydroelectric energy has represented, and still represents today, the principal source of renewable energy: With reference to the global production of energy, hydroelectric represents 2.15 percent of the total. Concerning the construction of new large-scale hydroelectric plants today, development seems stalled because they require infrastructure (dams, reservoirs, catchment drains) that causes a considerable environmental impact. Although they produce no polluting emissions, they considerably alter the landscape, impair the balance of local ecosystems, and diminish the volume of water available for uses other than energy production. For these reasons, it seems more appropriate to develop hydroelectric plants with less than 10 megawatts (MW) of power production: These are small plants with a low environmental impact that produce energy by utilizing minor streams and river waterfalls. The realization of such small hydroelectric plants creates only modest environmental and technical problems, particularly where the construction of dams is unnecessary, and the cost of the electricity produced more or less equals that of larger plants.

Some countries today are experimenting with new technologies capable of producing power by using energy flows from seawater. Experiments are being conducted to try to exploit tidal power (in France), the energy potential of waves (in the United Kingdom, Norway, and Japan), and the temperature gradient—the difference in temperature that exists between deep and surface waters (in the United States).

Geothermal

Geothermic energy is generated by physical processes that occur in deepest layers of the Earth's crust (underground, the temperature rises by 30 degrees Celsius every 1,000 meters of depth). Geothermal heat rises to the Earth's surface, where it can generate a thermal current measuring about 0.065 watts per square meter. In general, geothermal plants use the following operational procedure: Through drilling, the steam produced by the heat of the Earth is brought to the surface and conveyed to pipes that carry it to a turbine, where, by means of a driving shaft, it is turned into mechanical energy, and into electrical power by an alternator. In cases where the geothermal fluid does not reach high temperatures, the required temperatures are reached by using hot water. The resulting heat is used, for example, in district heating plants or by agriculture for heating greenhouses.

With reference to energy production, geothermal energy today represents 0.41 percent of the global total. The principal users of geothermal energy are Iceland, Italy, the United States, Costa Rica, New Zealand, Japan, Kenya, and Ethiopia. From an environmental point of view, the exploitation of the geothermal resource generates carbon dioxide emissions (about 0.2 kg of carbon dioxide per kilowatt hour produced), hydrogen sulfide, ammonia, mercury, and radon.

Wind

Wind power—the kinetic energy produced by the wind or the masses of air generated by the uneven heating by the sun of the Earth's surface—is transformed into electrical power by a machine called a wind turbine or windmill. At present, the most common wind turbines are those with a horizontal rotor axis set parallel to the direction of the wind. The working of turbines with a horizontal axis is as follows: The blades that make up the rotor are fixed to a hub that connects to a main shaft, rotating at the same angular speed as the rotor. The main shaft is connected to a gearbox, to which a driveshaft is also connected. On the latter there is a brake, and lower down, a power generator, from which electrical cables carry the electricity produced to ground level. Vertical axis turbines also exist; these have a rotor spinning on top of a vertical axis perpendicular to wind direction. These machines have a low rotation speed and modest output, which makes them particularly suitable for mechanical use.

Today, as a rule, wind turbines are installed and connected to the main electricity network, chiefly in the form of multiple installations or "wind farms." Wind farms are usually installed in places where winds blow at a speed above 5.5 meters per second. Wind farms may be installed on land or offshore; offshore wind farms have important advantages, as offshore winds are more stable and reach higher speeds.

Despite the fact that between the mid-1990s and 2008 wind power represented only 0.2 percent of global energy production, wind power has undergone considerable development, gaining an important share of the market. Between 1996 and 2008, the global cumulative installed capacity has grown by 1,880 percent: In 1996, the global cumulative installed capacity reached 6,100 MW, and in 2008 it reached 120,791 MW. The country with the highest installed capacity is the United States, with 25,170 MW, followed by Germany with 23,903 MW, Spain with 16,754 MW, China with 12,210 MW, and India with 9,645 MW. These five countries possess 72.6 percent of the world's total installed capacity.

The success of wind power—despite the fixed costs of wind plants generally being higher than those of fossil fuels technologies—has been determined by the low variable costs, thanks to low operating costs and the fact that using wind eliminates the need to purchase fossil fuels to produce power. Furthermore, according to the G8 Task Force for Renewable Energy, wind is the energy source with the lowest externalities (costs resulting from the utilization of the plants). The costs of alternate power sources do not fall directly on the producer, but on the community; for example, in the emission of greenhouse gases by hydroelectric plants, or the risk of radioactive leaks in the case of nuclear power plants.

However, although wind power is regarded, along with biomass, as the chief alternative to fossil fuels, wind farms have become one of the main objects of NIMBY protests (the acronym for "Not In My Back Yard," normally used to indicate public protests aimed against construction of infrastructures of public interest). Among the most famous cases involving wind farms are the constructions in Nantucket Sound in Massachusetts and in County St. Lucie in Florida.

The main concerns with wind farms relate to land seizure, noise, electromagnetic wave diffusion, visual impact on the landscape, interference with birds (the most well-known case is that of birds of prey killed by wind turbines in Altamont Pass, California), and in the case of off-shore power plants, concerns over sea life. Studies carried out so far, however, have somewhat reduced alarm by some in the environmentalist movement.

Over the few last years, "small wind systems" have begun to take shape with the support of environmentalist associations. These are power-generating systems that use turbines less than 30 meters high. These plants are quite common in Great Britain, where, from the early 1990s, they have been supported by the British Wind Energy Association. Compared with the large-scale wind farms, the "small wind system" is easier to integrate into the landscape. These farms may make it possible to create a self-sufficient network: a string of bioenergetic firms and districts that use a portion of the energy produced and trade a portion on the network, with investment returns possible in just a few years.

Solar

Solar energy, the energy radiating from the sun, can be harnessed to obtain electrical power or heat. At present there are three main technologies employed in the exploitation of solar energy: photovoltaic cells, solar thermodynamic systems, and solar thermal systems. Photovoltaic cells exploit the photovoltaic effect; in other words, the property that some materials have to generate electrical power when directly hit by rays of the sun. The photovoltaic cell constitutes the base element of the process transforming solar radiation into electrical power: Assembled cells make a photovoltaic module, which is generally made up of 36 cells, each of them producing around 40 to 50 watts of power. When a number of modules are assembled as a single structure, it becomes a solar photovoltaic panel.

Solar thermodynamic systems instead exploit sunlight to produce electricity from the energy released by heated fluids at a high temperature (about 400 degrees Celsius). This method is the most competitive solar system because it can be put into operation quickly and is the most flexible. At this time, there are three types of solar thermodynamic systems: parabolic dish, parabolic trough, and solar tower. Parabolic dishes concentrate the energy of the sun on a tube placed in the focal line of the collector. Carrier fluid runs inside of the tube, which heats up and transfers its heat to a heat exchanger. The parabolic trough uses concave, dish-shaped reflecting parabolic mirrors to concentrate light on a receiving system placed at the focal point of the dish. Solar towers use a series of mirrors that follow the movement of the sun and reflect the sun's light onto a heat exchanger placed on top of a tower. Thermodynamic solar technologies have been developed chiefly in California, where between the 1980s and 1990s solar power plants were created with a capacity of 350 MW. Recently, a 64-MW thermodynamic solar power plant has been built in Nevada, and two additional solar power plants are being researched in the state of Florida and in Spain.

Solar thermal systems are employed for heating water destined for air-operating domestic heating systems. Solar thermal systems absorb heat through a solar manifold, transferring it to a collecting or using place by means of a fluid carrier (usually water or air). Nowadays, solar energy represents about 0.16 percent of global energy production.

Among the different solar technologies, the one that has scored real success in recent years has been the photovoltaic system. The global installed capacity of photovoltaic systems has risen from 1,200 MW in 2000 to 9,200 MW in 2007—an increase of 667 percent. In 2007, 85 percent of photovoltaic plants could be found in Europe, North America, and the Pacific. In particular, 73 percent of the market was concentrated in Europe, where the photovoltaic system has been promoted by German and Spanish investment. The country with the highest installed capacity is Germany, with 3,800 MW, followed by Japan with 1,935 MW, the United States with 814 MW, and Spain with 632 MW.

According to the European Photovoltaic Industry Association, in the next few decades the photovoltaic market will be dominated by Asia, Africa, and South-Central America; the developing countries, therefore, will be able not only to catch up but also to become the exemplary market for what concerns photovoltaic energy. China in particular will take up the role of leader, with investments that will double the European. Already today, there is a city in China—Rizhao—that can be defined as "a solar energy city." In Rizhao, in fact, solar thermal systems provide hot water to 99 percent of the families who reside in the central districts and to 30 percent of those residing in the outskirts. In total, there are 500,000 square meters of solar collectors in Rizhao.

From an environmental point of view, solar energy is, together with wind energy, the energy source with the least negative externalities. During their life cycle (25–30 years), solar plants do not cause risks to the natural environment or to human health. The only negative aspects concerning their use may be represented by the seizure of land (in the case of solar plants) and by the possible alteration of the urban landscape (photovoltaic plants being placed on roofs or on facades of residential buildings). The main environmental impact occurs during the production of panels when toxic substances such as silane, phosphine, diborane, and cadmium are employed. In addition, panels produce a special type of waste, the disposal of which necessitates the recovery of the above-mentioned toxic metals.

Biomass

Concerning biomass—organic materials employed to produce energy—wood, woodworking residue, agricultural discards, and the by-products of processing agricultural and industrial products (such as marc or oil-pressing residue) are mainly used to generate power. Also used as biomass is the biogas obtained through anaerobic fermentation of urban solid waste. At present, biomasses (including biogas that is produced by the anaerobic fermentation of organic substances contained in urban waste and waste from animal farms, essentially comprising methane and carbon dioxide) are the most used renewable source in the global production of energy (10.46 percent of the total). This figure is largely because in many developing countries, firewood is still the cheapest fuel and the easiest to get.

In developed countries, wood scrap constitutes 50 percent of the biomass employed in the production of energy. In the United States, for example, 10,000 MW of electricity are derived from the combustion of forest-clearing scrap, and the city of St. Paul, Minnesota, uses a teleheating plant that uses 250,000 tons of wood materials in place of coal, with a reduction of carbon dioxide emissions of 76,000 tons per year. In Europe—where electricity derived from biomass represents 2.4 percent of the total production of electricity—Finland and Sweden are the leading countries in the use of wood scrap, whereas France, Germany, and Austria are the leading countries in producing electricity from burning of urban solid waste.

The use of biomass for the production of energy is curtailed by problems. First, biomass is not always available: One only has to think of seasonal harvests (such as corn) that are only available at a certain time of the year. Second, to have the availability of sufficient quantities of biomass to fuel large plants for the production of electricity, vast agricultural areas would need to be diverted from cultivating food crops: To fuel an electrical plant producing 2,500 MW, it would be necessary to devote about 5,000 square miles of agricultural area to the cultivation of biomass. Third, a forced production of biomass on a large scale (short rotation forestry) would cause environmental problems that have yet to be fully assessed.

From an environmental point of view, the use of biomass would have a more serious impact than that of photovoltaic and wind plants because of emissions deriving from cultivation, cutting, transportation, and combustion activities that, albeit in reduced quantities, would produce nitrogen dioxide.

At present, the chief biofuels are biodiesel and bioethanol. The former is derived by the esterification of rapeseed and sunflower oils. The latter is obtained mainly from sugar cane, corn, and beetroot, through a process of fermentation and esterification. The use of edible vegetables to produce biofuels might upset food markets, as is already the case with corn. In addition, forests and grazing land converted to biofuel cultivation might contribute to carbon dioxide emission. (It is believed that the production of biofuels may generate an increase of from 17 to 420 times the annual quantity of carbon dioxide emissions than what might be saved by replacing fossil fuels.)

To avoid such problems, second-generation biofuels are being tested. An example is the production of bioethanol from a hybrid variety of *Miscanthus sinensis* (Japanese silver grass), *Miscanthus giganteus*. It is a grain grass that grows up to 13 feet tall with a very high potential return—equal to 60 tons of dry material per hectare, which is equal to around 60 barrels of oil.

See Also: Carbon Footprint; District Energy; Energy Efficiency; Green Energy; NIMBY.

Further Readings

European Photovoltaic Industry Association and Greenpeace. *Solar Generation V-2008. Solar Electricity for Over One Billion People and Two Million Jobs by 2020.* Brussels: EPIA-Greenpeace, 2008.

Global Wind Energy Council. *Global Wind 2008 Report.* Brussels: Global Wind Energy Council, 2008.

Hills, Richard L. *Power From Wind: A History of Windmill Technology.* Cambridge, MA: Cambridge University Press, 1994.

International Energy Agency (IEA). *World Energy Outlook 2008.* Paris: IEA, 2008.

Mallon, Karl. *Renewable Energy Policy and Politics: A Handbook for Decision-Making.* London: Earthscan, 2006.

Perlin, John. *From Space to Earth: The Story of Solar Electricity.* Cambridge, MA: Harvard University Press, 2002.

Rosillo-Calle, Frank, et al. *The Biomass Assessment Handbook: Bioenergy for a Sustainable Environment.* London: Earthscan, 2008.

U.S. Energy Information Administration (EIA). *International Energy Outlook 2008.* Washington, D.C.: EIA, 2008.

Federico Paolini
University of Siena

RESILIENCE

The concepts of resilience and the "resilient city" have gained considerable currency in recent years among urbanists, not only as they relate to ecological sustainability but also in terms of urban disaster planning and adaptations to the anticipated threats of the 21st century, among them peak oil and climate change. While resilience emerged from the ecological sciences in the 1970s, it is now widely discussed in many disciplines, and interest in its application to cities has surged since the September 11, 2001, terrorist attacks, the 2003 North American power blackout, and Hurricane Katrina. The term is sometime used interchangeably with *sustainability*, and while they may be closely related in practice, they are distinct, both semantically and theoretically. For example, while sustainability is always considered an ecological and social good, a social-ecological phenomenon may be undesirable yet still be quite resilient to change. Resilience is a function of the adaptive capacities found in nature, but when applied to human societies is highly normative and would demand an almost thorough departure from present structures and practices. At the very least, it requires an interventionist state to fund or subsidize infrastructure renewal, which may be politically controversial.

Generally, resilience is understood as the degree to which a complex system is flexible enough to respond and adapt to an externally-imposed force or change, and thus persist over time, while retaining its structure and functions. Conversely, a vulnerable system would be one in which conditions are inflexible, key resources comprise a monoculture, there is little learning capacity, and choices for addressing crises are constrained.

Resilience can be manifested in both ecologies and in human societies: Each are highly complex systems in which the interrelationships and synergies between elements are fundamentally important to their potential resiliency. Indeed, related literature explicitly recognizes social-ecological systems (SES) as the most appropriate unit of analysis, for human societies are indivisible from their biotic bases. Social-ecological systems such as cities cannot therefore be considered resilient unless these adaptive capacities are present not only in the natural environment but also within the full range of social, cultural, economic and political relationships.

The concept of resilience saw its origins during the early 1970s in the ecological sciences. Crawford S. Holling used mathematical models of natural systems to determine what makes them adaptive and resilient. Holling observed that forests have an adaptive cycle of growth, collapse, regeneration, and regrowth. In the growth stage, the ecosystem gathers biomass and becomes increasingly complex and interconnected. Eventually, self-regulation mechanisms kick in, developing efficiencies as specialized organisms fill a range of niches, and the system seeks to conserve these efficiencies. Eventually, however, the forest becomes so oriented to a particular and specific set of environmental circumstances that it can't absorb shocks, be they invasive species or changes to climatic conditions. The introduction of such elements—particularly if they are the result of violent or abrupt change—can therefore cause collapse of the ecosystem. Yet in the wake of this collapse comes the opportunity for new organisms to gain hold, which are not closely interconnected to others at first and can develop on their own. With these new opportunistic organisms come system regeneration and reorganization, as well as the beginning of a new growth stage. The ability of a system to regenerate also depends on the health of larger-scale complex systems in which they reside: If the climate is stable, the forest will regenerate.

Holling was careful to distinguish between system equilibrium and resiliency. The fact that an ecosystem is stable does not mean it can persist indefinitely; in fact, he argued that long-term homogenous conditions work against resilience by reducing diversity and flexibility, thereby discouraging novelty. By contrast, a resilient system may fluctuate greatly in terms of its condition and populations, but it nonetheless will demonstrate a greater ability to persist over time. Resilience is therefore not the result of any one element in the system, but the nature of the relationship between the elements. These elements need to be connected to others, but not so rigidly that they can't also operate independently. Implicit in this adaptive model is the ability of a system to self-organize, which requires that its various component parts have coevolved through the presence of flexible network connections that facilitate communication and other adaptive relationships.

Urban Resilience

In light of these concepts, cities, economies, and global ecosystems are increasingly being recognized for their vulnerabilities, ranging from the 2007 housing bubble, to the 2008 recession, to global climate change. Resilience theory tells us that failure in one part of a system can reinforce collapse in another, which can then result in a cascading series of failures. To address these possibilities, the resilience literature emphasizes self-organization, flexibility, and adaptation through redundancy, distribution of resources, and the development of learning capacity. Interconnections between elements should also be loosened, making the system in question (be it a forest or a city) better capable of bearing and absorbing shocks. However, the literature also recognizes that the application of these principles to human societies runs counter to most of the goals and norms of our present hyper-organized, centralized, efficient, and globalized economic and infrastructural arrangements.

The problem is that modern cities have few if any adaptive capacities. They are completely dependent on rapidly depleting energy sources, inefficiently distributed through an aging and highly centralized network of refineries and power plants. Flows of commodities are sourced from remote and often vulnerable locations and delivered via trucks on expensively maintained roadways. A heavy reliance on private automobiles often results in a transportation monoculture that is prone to disruption. Modern deindustrialized cities fill highly specialized functions, most of which are unrelated to the manufacture and distribution of necessary goods. Most seriously, for the past two centuries, cities have been built with an almost complete disregard for natural processes, with watercourses and prime farmland paved over and filled with buildings that rely on ever more energy-intensive technologies such as air conditioning and gas heating to compensate for external environmental conditions. The need to learn how to better conceive of human settlements from nature would seem to be obvious.

In the early years of the 21st century, several significant and devastating events contributed to a heightened interest in the use of biological models for understanding and repairing social-ecological systems. The terrorist attacks of September 11, 2001, in New York and Washington, D.C., traumatized the world and revealed not only the inherent risk in centralizing operations and resources, but that cities are vulnerable to volatile global geopolitics. Less than two years later, the blackout of August 2003 knocked out power to millions of people from Toronto to Detroit, demonstrating the dangerous overextension of North America's outdated and centralized energy technologies. The devastation wrought by Hurricane Katrina on New Orleans and the Gulf Coast in 2005 was exacerbated by the prior destruction of coastal wetlands for development, the poor state of infrastructure intended to protect New Orleans, and the many social inequities that left thousands of African American residents with no means of escape.

In light of these and other cases of disaster and recovery, the recent literature on hazard mitigation advocates strategies to promote urban resilience, such as ensuring reserves of key resources; equitably distributed and redundant infrastructure; and healthy social networks of trust to ensure that people can share information and come to one another's aid. Still, just as Holling showed that ecosystem resiliency depends on the health of larger global systems, so, too, did recovery from these events in each of these cases depend upon the health of larger complex systems, such as those in the political economy, and, indeed the biosphere. In other words, true resilience must include an "outside world" to come to the assistance.

Another body of literature considers urban resilience with a view to an uncertain future, and how human societies might cope with peak oil and climate change. Accordingly, the resilient city of the 21st century will need to include renewable energy; carbon neutrality; dispersed utilities; local agricultural and fiber production; closed-loop industries; local economies; and a transportation hierarchy built around compact urban environments supportive of walking, cycling, public transportation, and electric vehicles.

Key to all of these visions is an interventionist public sector. While a high degree of social capacity it also clearly desirable, much of what is required for the resilient city depends upon a wide range of investments on the part of governments in public utilities, renewable energy, and transportation systems, as well as incentives and subsidies for retooling economies and enabling publics to make more appropriately sustainable and resilient choices. Because such investments are generally politically objectionable to many fiscal conservatives and libertarians, they would certainly be the subject of considerable controversy.

This leads to another important consideration, which might be termed the *political economy* of urban resilience. According to Lawrence Vale and Thomas Campanella, the creation of politically palatable narratives of resilience and recovery in response to urban disasters is needed to reinforce the legitimacy of the governments involved. Because there is no such thing as a resilient city as such—resilience being dependent on the capacities and relative advantage among distinct urban populations—such narratives will always be contested. Furthermore, respective local narratives of resilience become ideological national narratives as well. Ultimately, however, these authors stress that resilience is always more than just the physical rebuilding of a city: It must include a demographically viable population to support community building and economic recovery.

Since 1999, Holling and colleagues from more than 20 organizations around the world have participated in the Resilience Alliance research consortium, which researches and promotes the principles of resilience and the theory of Panarchy, in which nested adaptive cycles operate at different scales. The Alliance also produces the journal *Ecology and Society*.

As the new century progresses, and ecological, social and economic conditions become ever more complex and threatened by the disruptions associated with energy depletion and climate change, principles of urban resiliency are sure to become an increasingly necessary part of the political landscape.

See also: Adaptation, Climate Change; Compact Development (New Urbanism); Distributed Generation; Environmental Risk; Green Infrastructure; Infrastructure; Power Grids; Renewable Energy; Sustainable Development.

Further Readings

Castree, N. and B. Braun. *Social Nature: Theory, Practice and Politics*. Oxford: Blackwell, 2001.

Demos. "Resilient Nation." 2009. http://www.demos.co.uk/projects/resilient nation (Accessed October 2009).

Derrida, J. and J. Caputo. *Deconstruction in a Nutshell*. New York: Fordham University Press, 1997.

Garmestani, A. S., C. R. Allen and L. Gunderson. "Panarchy: Discontinuities Reveal Similarities in the Dynamic System Structure of Ecological and Social Systems." *Ecology and Society*, 14/1:15 (2009).

Gunderson, Lance and Crawford S. Holling, eds. *Panarchy: Understanding Transformations in Human and Natural Systems*. Washington, D.C.: Island Press, 2002.

Holling, C. S. "Resilience and Stability of Ecological Systems." *Annual Review of Ecology and Systematics*, 4:1–23 (1973).

Homer-Dixon, Thomas. *The Upside of Down: Catastrophe, Creativity and the Renewal of Civilization*. Toronto, Knopf: 2006.

Hopkins, R. *The Transition Handbook: From Oil Dependency to Local Resilience*. Totnes, UK: Green Books, 2008.

Newman, Peter, Timothy Beatley and Heather Boyer. *Resilient Cities: Responding to Peak Oil and Climate Change*. Washington, D.C.: Island Press, 2008.

Vale, Lawrence J. and Thomas J. Campanella, eds. *The Resilient City: How Modern Cities Recover From Disaster*. New York: Oxford University Press, 2005.

Walker, B. H. and D. Salt. *Resilience Thinking: Sustaining Ecosystems and People in a Changing World*. Washington, D.C.: Island Press, 2006.

Michael Quinn Dudley
University of Winnipeg

REYKJAVIK, ICELAND

Located at the head of the Faxaflói Bay in southwestern Iceland, Reykjavik adopted its town charter in 1786 and became the capital of Iceland following the grant by Denmark of Home Rule in 1874. It was named Reykjavik, or Smoky Bay, soon after the first Norse settlers arrived in the late 9th century. The hot springs that give the city its name are heated by active volcanoes, and are in large part responsible for Reykjavik's acclaim in geothermal energy use and technologies. The city has been recognized as a Best Practice example for renewable energy by C40Cities, a network of cities that is affiliated with the Clinton Climate Initiative. Reykjavik has declared its intent to eliminate fossil fuel use within its jurisdiction altogether by 2040, and in January 2010 announced that it will seek the European Commission's European Green Capital designation for 2012 or 2013. The worldwide financial crisis of 2008 hit Iceland and Reykjavik with particular force, but the city's resolve to create a sustainable community is undiminished.

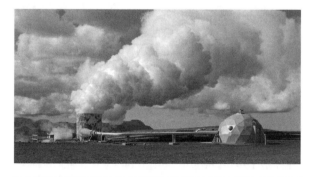

Most houses in Reykjavik are powered by the largest natural heating system in the world, with steam vents that tap deep into the Earth for geothermally heated water.

Source: iStockphoto

Reykjavik formally committed to planning for sustainability when it signed the Aarlbourg Charter in 1997. With Aarlbourg, the city adopted its Master Plan 1997–2016 as its Local Agenda 21 (LA21), and subsequently updated it in 2001 under the urban policy program titled "The Future City." That program included provisions for regular review, and by 2003 it had become clear that another revision was needed, because 90 percent of the projects listed in the 2001 LA21 had already been completed or were substantially underway. A new LA21 titled Shaping Reykjavik was adopted in 2006. It placed greater emphasis on urban form and transportation than the previous documents.

Energy management has played an enormous role in Reykjavik's progress toward becoming an exemplary green city. Long a leader in the development of geothermal energy technologies, Reykjavik has also stepped into the forefront of methane capture and conversion technology development. Geothermal resources comprised 53.2 percent of Reykjavik's energy profile in 2000. Oil accounted for the second highest proportion (27.3 percent), followed by hydropower (16.5 percent), and coal (2.9 percent).

Reykjavik's abundant geothermal resources have permitted the city an enormous head start in the race to become independent of fossil fuel. The city has been taking advantage and developing the potential of its underlying hot springs since the 1920s—though through the 1930s, coal furnaces remained the main source of heating. The geothermal system came on line in 1943, and use of geothermal heating became widespread in 1960. Over the next 40 years, the geothermal system expanded to serve 95 percent of the households in Reykjavik. Since 1944, the use of geothermal energy is estimated to have reduced total annual CO_2 emissions by 37 percent compared to what would otherwise have been the case. Geothermal heating costs are one-third the cost of heating with oil—a significant financial benefit in this arctic country.

Reykjavik Energy (RE), a municipal corporation, is responsible for geothermal resource development. It is an ISO 9001 compliant company working toward ISO 14001 certification. RE operates five geothermal plants in and around Reykjavik with combined capacity of 750 MW of thermal power from steam. Together they supply 26.5 percent of the country's electricity and 87 percent of structure heating. The largest of its facilities is the Nesjavellir plant. In addition to hot water used for space heating and other domestic and industrial uses, Nesjavellir has produced electricity since 1998 when it was fitted with two 30 MW steam turbines. Over the years its generating capacity has increased to 120 MW.

Oil consumption is rising in Reykjavik, as in all of Iceland. Since 1983, national per capita usage has risen by 56 percent, from 491 tons to 765 tons. Though some of the increase is attributable to increased industrial activity, the majority is due to the steady increase in vehicle miles traveled. In 1983 vehicles and equipment used 31.9 percent of petroleum imports. By 2007 their share had risen to 45.5 percent. LA21 programs aim to reduce demand for petroleum through transportation and land planning policies described below.

The LA21 Shaping Reykjavik set the reduction and reuse of waste as an objective, and many of the means to that end have already been implemented. Solid waste is handled by SORPA, an independent nonprofit formed in 1991, and jointly run by Reykjavik and surrounding communities. Prior to its founding, all but 8 percent of waste went into landfills. Today, 28 percent is recycled, and SORPA aims to meet the European Community 2016 goal of 65 percent.

While more than 70 percent of refuse still goes into landfills, the management technology is far advanced beyond that utilized at SORPA's inception. Facilities now capture methane and convert it to biogas, which fuels the company's fleet of collection trucks and administrative vehicles. SORPA also markets biogas through a related company, Metan Ltd.

Hazardous materials receive special handling and, as with general refuse, they are recycled or reused when possible. The majority of hazardous waste is exported to specialized handling facilities overseas. While there is a campaign to increase domestic composting, SORPA also collects green waste, which is converted to organic fertilizer distributed under the trade name MOLTA. The sewage system in Reykjavik, largely built in the 1960s, is in good repair. The 1997 LA21 set clearing the coastline of sewage related pollution as a goal, and by 2000 that goal had been accomplished, as had many of the other sewage-related provisions. To preserve the city's good water quality, Reykjavik conducts sewage receptor audits and has an urban surface runoff program that includes the use of bioswales and filtering ponds.

Land Use and Open Space Planning

A principal objective of Shaping Reykjavik is to "use land and resources...in an efficient manner, in harmony with the social and natural environment." The means by which it proposes to do this include smart growth strategies of urban infill, densification, and mixed-use development.

Though Reykjavik is a European city, it has developed in a low-density and autocentric pattern typical of North American cities. It did not achieve city size until the mid-20th century. It grew slowly, prior to World War II, as a town mainly comprising urban farms and with no bridges to link it to the surrounding countryside. It was during the occupation of Reykjavik by the allied forces that bridges were constructed and city population swelled. The transformation to a modern city had begun. Houses with suburban gardens replaced farms. In-town farming was formally banned in 1994.

Reykjavik, like the whole of Iceland, was largely treeless from the 9th century until the early 20th century, by which time approximately 1 percent of the land had been reforested. As a consequence, a uniquely Icelandic aesthetic has developed that values broad vistas. To the pragmatic people of Reykjavik planning for parks and open space seemed superfluous until the 1990s. A new attitude, reflected in Saving Reykjavik, recognizes the need for environmentally-friendly development, diverse recreational areas, and conservation of natural systems and spaces. Reykjavik's approach to this is called the Green Scarf, greenways wending between the city center and surrounding neighborhoods, connecting them with walking and cycling paths. Also as part of the "Green Steps" advanced in 2007 to support Shaping Reykjavik, city officials proposed planting 500,000 trees within the capital region.

Transportation Planning and Air Quality

While traffic volume and congestion contribute to several classes of urban problems, their effects on air quality are of particular concern. Between 1995 and 2007 the number of private vehicles per resident increased by nearly 40 percent in Reykjavik. The current ownership rate—in excess of 600 vehicles per 1,000 residents—is among the highest in Europe. An estimated 69 percent of the city's GHG emissions are attributable to transportation.

One of city's first responses to this increase was to mount a campaign to encourage public transportation ridership. Despite these efforts, increase in ridership has so far been minimal. Shaping Reykjavik sets a goal of doubling bus use over the next 20 years. Development of park-and-ride facilities for commuters and reduction or elimination of transit fares have been implanted to achieve this, but the car culture is well entrenched in Reykjavik. With the exception of the city center, few roads in Reykjavik form a connecting grid. This lack of connectivity poses an immediate challenge to reducing car use.

Though Shaping Reykjavik addresses density, it gives no guidance on the form of the road network. The document does propose to develop pedestrian and cycling facilities as integral alternative mode components of the transportation network, and to use the city's geothermal resources to heat pedestrian and cycling paths to make them safely accessible during winter months. Apart from these alternative mode facilities and public transportation, Shaping Reykjavik focuses on a fairly standard suite of traffic demand management strategies—staggered work schedules, carpool campaigns, replacement of traffic signals with roundabouts to improve traffic flow—none of which have successfully reduced vehicle miles traveled in other cities where they have been applied.

In general, air quality in Reykjavik is good. NO_2 and particulate matter levels have declined over the last decade, but hover around their respective limits set for healthful air. Greenhouse gas (GHG) emissions are rising, trending with increased metal production and vehicle use. Per capita, 2007 GHG emissions in Iceland were 14.8 tons CO_2-equivalent, up from 13.4 tons in 1990. In their 2010 Climate Change Performance Index, the NGO GermanWatch ranked Iceland behind 12 other countries. Though overall emissions levels and policies rated good, emission trends rated poorly.

In its quest to have the cleanest air of any city in the world, Reykjavik is considering stricter air quality controls on business and industry. Proactively setting an example, the city it has put an environmental management system in place to monitor its own practices and encourages private business to do the same and seek environmental certifications such as Green Flag and ISO 14001. The city also participates in the Nordic Cities Environmental Indicators Project (NCEIP), which collects data for 11 comparable data sets across 14 environmental parameters. The major air quality challenge for Reykjavik relates to vehicle use. Data collected for NCEIP revealed that GHG emissions from vehicles rose by 30 percent between 1999 and 2005. That increase correlates with increased ownership rates and vehicle miles traveled. Additionally, much of the city's airborne particulate matter arises from the use of asphalt paving and studded tires.

Given Reykjavik's urban form and entrenched car culture, one of the more promising approaches to reducing vehicle related emissions appears to be conversion of the private fleet—both automotive and marine—to alternative fuels. In 2007 the national government forecast that up to 50 percent of the cars on Icelandic roads will be powered by electricity by 2013. The annual operating cost advantage of electric cars may cause urbanites to make the switch since the expense is a mere 10 to 20 percent of the cost of gasoline powered cars. As an added incentive, RE currently provides free recharging at designated stations. Range and recharge issues, coupled with the need for four-wheel drive vehicles in suburban settings, may slow widespread acceptance.

Iceland is also actively promoting use of hydrogen to power vehicles on both land and sea. City and federal agencies are working with Icelandic New Energy (INE) to develop that market. INE operated the demonstration phase of the Ecological City Transport System (ECTOS), which introduced and ran hydrogen-powered buses in the city of Reykjavik for three years.

Two direct disincentives to gasoline powered automobile use were adopted in 2009. On January 1, 2010, the value added tax on gasoline increased by 25.5 percent, and a special carbon emissions tax will be applied to any fuel imports clearing customs after that date. Sharp decreases in vehicle miles traveled were seen in U.S. cities when gasoline prices reached $5 per gallon in 2008. The average price per liter of gas was 192 Iceland

Kronas in late December 2009—the equivalent of $5.53 per gallon. The price reportedly had increased 10 percent by mid-January 2010 as a result of increased taxes and the weakened purchasing power of Iceland's Krona. Additionally, Shaping Reykjavik proposes two more pricing disincentives: one to discontinue parking subsidies, and one to stop offering car related perquisites currently extended to both government and private sector employees.

Other transportation related measures to improve air quality listed in Shaping Reykjavik include shortening the periods during which studded tire use is permitted and eventually banning them, using geothermal resources to keep roadways clear of ice and snow, and replacing asphalt paving with concrete.

Challenging Opportunities

In 2008, the world financial crisis hit Iceland, and Reykjavik, particularly hard. The previous five years were boom years in which the wealth of the average family in Iceland had increased by 45 percent. At that point, the international market for mortgage backed securities and collateralized debt obligations imploded. Iceland's three major banks, holding over $160 billion of what came to be known as toxic debt, were failing. Their losses amounted to 11 times the nation's gross domestic product. The banks and the county faced bankruptcy. Personal fortunes evaporated. Borrowers saw not only their foreign currency-based debt increase, but interest rates soar, as the Krona's value plunged. To make bad matters worse, England and the Netherlands presented Iceland with bills, $3.7 billion and $1.7 billion respectively, as reimbursement for deposit insurance payments these governments had paid to their citizens who had invested in Icesave, an online subsidiary of Landsbanki, a Reykjavik-based bank. In January 2010, Iceland touched off an international furor by refusing to pay the Icesave bills pending a national referendum. The bills equate to $17,400 per citizen. To pay them would deepen the national recession. To refuse may delay Iceland's inclusion in the European Union, which is deemed necessary to keep its economic recovery program on track.

Some analysts believe Iceland's economic woes will set its quest to become fossil fuel free back by a decade, but Icelanders see the necessity to persevere. In November 2008, the Reykjavik City Council announced that despite decreased revenues and increased prices, there would be no layoffs or cuts in basic services. They proposed to instead tighten up procurement and decrease their costs by 15 percent in the near term. Rather than cancelling capital investment projects, they will continue to seek financing and will reprioritize projects based on urgency and the associated employment impacts. They will focus development where adequate infrastructure currently exists. Icelanders know that there is no viable option but to pursue the green agenda, which they expect will lead to long-term savings.

See also: Agenda 21; Air Quality; Carbon Footprint; Renewable Energy; Waste Disposal.

Further Readings

City of Reykjavik. "Local Agenda 21: The Environmental Action Plan of the City of Reykjavík." 2001. http://nepal.vefurinn.is/Nepal_Skrar/Skra_0000774.PDF (Accessed January 2010).

City of Reykjavik. "'Shaping Reykjavik' Reykjavik Local Agenda 21: Policy Formulation Towards a Sustainable Community in Reykjavik to 2015." 2006. http://www.rvk.is/Portaldata/1/Resources/umhverfissvid/umhverfisaaetlunrvk/shapingRVK.pdf (Accessed January 2010).

Reynarsson, Bjarni. "The Planning of Reykjavik, Iceland: Three Ideological Waves." *Planning Perspectives*, 14 (1999).

Susan H. Weaver
Independent Scholar

RIO DE JANEIRO, BRAZIL

Rio de Janeiro, on the Atlantic coast of southeastern Brazil, is one of Latin America's megacities. It is the second largest urban agglomeration in Brazil after metropolitan São Paulo. The municipality of Rio de Janeiro, measuring 1,182 square kilometers (km²) is the state capital of Rio de Janeiro State (43,696 km²). The estimated municipal population in 2008 was about 7.8 million inhabitants in a state of about 15 million people. About 11.8 million people occupy an area of 5,645 km² in the 20 municipalities of the Rio de Janeiro metropolitan region. As a megacity in one of the world's largest emerging economies, Rio faces tremendous challenges in forging a path toward achieving sustainable development.

Estácio de Sá founded São Sebastião do Rio de Janeiro for the Portuguese crown in 1565, ending a short-lived occupation by French forces. The administrative capital city of the Portuguese colony of Brazil was transferred from Salvador, Bahia, to Rio de Janeiro in 1763. The transit of enslaved Africans through the port of Rio continued through much of the 19th century. In the 19th century, a series of changes spurred economic modernization and industrialization in Brazil. The seat of the Portuguese empire was transferred from Lisbon to Rio de Janeiro in 1808. Brazil achieved independence from Portugal and established a monarchy in 1822. Brazil's coffee cycle started in the 1830s, leading to new waves of immigrants and giving rise to new agrarian and mercantile elites. Slavery was abolished in Brazil in 1888, and the monarchy ended in 1889 when the Federalist Revolution, backed by the coffee oligarchy, deposed emperor Dom Pedro II.

The deep social and economic inequalities evident in urban Brazil today are legacies of historical injustices. Rio's outer suburbs house the majority of its low-income population. Employment opportunities remain concentrated in the city's central areas. The number of *favelas* (shantytowns) in Rio increased from 26 in 1920, to 147 in 1960, to about 600 in the late 1990s. By 2000, there were more than a million *favela* dwellers in the city. The growth of *favelas* is part of broader processes of urban space production. Many affluent Brazilians—who previously self-segregated into exclusionary spaces, such as luxury residential high-rise buildings and gated communities within the country's largest cities—are now investing in and moving to medium-sized Brazilian cities.

Rio's *favelas* are notoriously associated with extreme poverty, illegal drug trafficking, and related violent crime. However, in many *favela* communities, residents have mobilized to create civil society organizations offering political empowerment, education, health services, income generation, social entrepreneurship, creative self-expression, and improved environmental quality.

Rio de Janeiro State is situated entirely within the biologically diverse Atlantic Forest region. Some of Brazil's earliest protected areas were established in Rio. Endorsement of sustainable urban development in Rio emerged in the early 1990s. In 1992 Rio hosted the United Nations Conference on Environment and Development and the Global Forum. Rio is preparing to host the final of the 2014 FIFA World Cup and the 2016 Olympic and Paralympic Games, which puts pressure on the city to upgrade facilities, infrastructure, and services.

Rio is preparing to cope with climate variability and the expected impacts of climate change including heavy rainfall, landslides, sea level rise, and storm surges. The city is involved in two international climate change initiatives. In 1998, the municipal government joined Cities for Climate Protection, a campaign of the International Council for Local Environmental Initiatives for cities interested in taking action to reduce energy consumption and greenhouse gas emissions. Researchers at the Center for Integrated Studies on Climate Change and the Environment at the Federal University of Rio de Janeiro prepared the Greenhouse Gas Emissions Inventory of the City of Rio de Janeiro published in 2000. The inventory analyzes emissions of carbon dioxide and methane from energy generation, energy consumption, waste management, transport, industry, land use, forests, and agriculture. In 2006, Rio joined the Clinton Climate Initiative's C40 Large Cities Climate Leadership Group, an association of 40 of the world's leading cities committed to confronting climate change cooperatively and taking measures to reduce greenhouse gas emissions.

See Also: Green Communities and Neighborhood Planning; Sea Level Rise; Urban Forests.

Further Readings

Baer, Werner. *The Brazilian Economy: Growth and Development*. Westport, CT: Praeger, 2001.

de Souza, Marcelo. "Metropolitan Deconcentration, Sociopolitical Fragmentation and Extended Suburbanization: Brazilian Urbanization in the 1980s and 1990s." *Geoforum*, 3 (2001).

O'Hare, Greg and Michael Barke. "The *Favelas* of Rio de Janeiro: A Temporal and Spatial Analysis." *GeoJournal*, 56 (2002).

Perlman, Janice. "The Metamorphosis of Marginality: Four Generations in the Favelas of Rio de Janeiro." *Annals of the American Academy of Political and Social Science*, 606 (2006).

Sandra Baptista
The Earth Institute at Columbia University

S

SAN FRANCISCO, CALIFORNIA

San Francisco, California, is often considered a green city on the cutting edge of environmental policies. It boasts an impressive natural setting—situated on the picturesque coast of the Pacific Ocean, nestled between striking hills—providing its residents a mild climate. On the cutting edge of environmental policies, San Francisco is a beacon for other cities to follow and emulate. San Francisco's official environmental mission is "to improve, enhance, and preserve the environment, and to promote San Francisco's long-term wellbeing."

San Francisco is home to many environmental policies and ordinances that have helped it become a leading green city. Some of the city's policies were the first of their kind in the nation.

San Francisco Government

San Francisco's Commission on the Environment was created in 1992 to advise the city's Board of Supervisors on legislation and policy related to the city's environment. The seven-member committee, appointed by the mayor, develops the city's environmental policies and programs on energy efficiency, environmental justice, toxics reductions, and recycling and waste.

In addition to the Commission on the Environment, SForward is current San Francisco Mayor Gavin Newsom's environmental agenda. It functions as the hub of environmental planning for the city. Included in SForward are such plans for a carbon-neutral city government, energy goals, and transportation alternatives.

San Francisco's Climate Action Plan is the commitment to reduce greenhouse gas emissions by 20 percent below 1990 levels by 2012. This ambitious goal is a direct response to the increasing threat of global climate change. The city has created numerous programs to make the Climate Action Plan a reality, including using renewable resources and the goal of becoming a zero-waste city by 2020. San Francisco was the first city in the nation to have its emissions certified by an external third party (the California Climate Registry), which gives it measurable and independent data with which to gauge its progress.

Recycling and Waste

In June 2009, San Francisco's Board of Supervisors amended the San Francisco Environment Code to include a "Mandatory Recycling and Composting Ordinance." This ordinance, effective October 21, 2009, requires all persons living within San Francisco to separate, recycle, and compost refuse. The most comprehensive ordinance of its kind in the country, the ordinance stands apart because it includes mandatory composting. Although cities such as Pittsburgh and San Diego have mandatory recycling laws, the laws do not include mandatory composting. Seattle passed a law in 2003 requiring residents to have a compost bin but did not require that all food waste go in it, unlike the San Francisco law. Once a grace period has passed, giving residents and businesses time to adjust to the new ordinance, those that fail to comply with the new ordinance of proper sorting can face fines from $100 to $500. The rationale behind the strict ordinance is directed toward the amount of space that food scraps and plant clippings take up in landfills, producing large amounts of methane gases. Monthly, 5,200 tons of compost from San Francisco and Oakland are delivered to Jepson Prairie Organics. The nutrient-rich compost is then sold back to San Francisco Bay Area farmers and vintners, helping the area keep its reputation for high-quality food and wine production. As of 2009, the city currently diverts 72 percent of its waste. By comparison, if the recyclables and compostables currently going into landfills were diverted, San Francisco's recycling rate would jump to 90 percent. The ordinance is touted as a major step in getting San Francisco to its plan of sending nothing to landfills or incinerators by 2020.

Energy

San Francisco boasts the largest city-owned solar panels in the country. A 675-killowatt solar system is attached to the Moscone Center, the largest convention and exhibition center in San Francisco. AT&T Park, home of the San Francisco Giants baseball team, houses solar-powered scoreboard panels. San Francisco has replaced incandescent traffic lights with light-emitting diode lights, reducing energy demand by 1 megawatt (MW), as well as saving the city over $7 million in energy and future maintenance costs. Over 2 MW of energy and a $20 million will have been saved over the next 15 years as a result of a major retrofit of hospitals, convention centers, and wastewater treatment plants. On the cutting edge of alternative energy sources, San Francisco is currently participating in a pilot program that is studying ocean wave and tidal energy. The program is conducting a study to find the best possible location for an offshore wave power plant. One of the potential sites is San Franciscos Ocean Beach.

Transportation

A current example of why San Francisco is truly a green city comes from the aftermath of the 1989 Loma Prieta earthquake. The powerful act of nature destroyed many structures in and around the San Francisco Bay Area, including freeways. After an outpouring of public support, the decision was made to demolish the waterfront Embarcadero Freeway, not to rebuild. The decision successfully reconnected the people to their historic waterfront and led to the redevelopment of the San Francisco Ferry Building located on the Embarcadero waterfront.

Along with being a green city comes commitment to environmentally friendly transportation. The city has converted all 1,500 municipal diesel vehicles to biodiesel, and Mayor Newsom has pledged to make all San Francisco taxis hybrid or run on alternative fuels by 2011.

San Francisco encourages alternative modes of transportation in many ways. The city has implemented or plans to implement such programs as expanding the bicycle network to make bicycle paths safer and more prominent, and over 500 parking spaces have been designated around San Francisco for car-sharing vehicles. Furthermore, in January 2009, all employers in San Francisco with 20 or more employees are required to offer a commuter benefit program. Ranging from discounted to free public transit passes, the goal is to encourage employees to take advantage of the many alternative options to get to work. To further encourage municipal employees to use alternative transportation, San Francisco offers an Emergency Ride Home service that offers a free or discounted mode of transit home if an unexpected emergency keeps the employee at the office later than expected. Commuter benefits also extend to the federally approved program that allows an employee pretax deduction of up to $230 per month to pay for transit and vanpool expenses. Other sectors in San Francisco are doing their part to encourage and reward those who use alternative modes of transportation. The California Academy of Sciences offers a discount on the price of admission to those who take nonautomobile transportation to the museum, located in Golden Gate Park. Other businesses offer a discount on purchases with proof of a bus fare.

Environmental Justice

Similar to most cities, San Francisco reflects an uneven distribution of environmental decision making based along class and racial lines. However, the City of San Francisco shows a commitment to environmental justice ideals. The effort to help each citizen enjoy his or her right to a clean and healthy environment drives many programs involving food security, as well as energy and air quality concerns.

The Bayview Hunters Point neighborhood in the southeastern part of the city is one example of an area on the receiving end of unjust environmental decisions in the past. One of San Francisco's poorest neighborhoods, Bayview Hunters Point is home to a decommissioned power plant, a sewage treatment plant, and over 300 toxic sites, including a U.S. Navy Shipyard, which is a federally designated Superfund site. The low-income residents of Bayview Hunters Point lack access to healthy produce and reliable transportation. In 2005, the California Department of Justice awarded a $150,000 grant to the Bayview Hunters Point Farmer's Market Collaborative. A collaborative effort between SF Environment, agricultural experts, and neighborhood groups, the collaborative provides healthy and fresh produce to the neighborhood in addition to nutritional education and cooking demonstrations. The farmers market accepts Special Supplemental Nutrition Program for Women, Infants, and Children and food stamps.

Bayview Hunters Point is also the location of a 2006 decommissioned power plant. In agreement with plant operator Pacific Gas & Electric, the city agreed in 1998 to shut down the power plant. The plant was deemed unreliable, but the decision also reflects a commitment to environmental justice goals. The southeastern part of the city had been unfairly housing San Francisco's power plants, overburdening the residents with poor air quality and high rate of children's asthma, among other health conditions.

Environmental justice can be reflected in other ways. Community gardens have a large presence in San Francisco, providing a respite from city life, as well as providing food for residents. Moreover, community gardens reflect small-scale grassroots democracy in action. Gardens flourish as a result of member participation, hard work, and organizing.

Green Building and Green Business

Construction and demolition debris accounts for 40 percent of the waste stream in San Francisco. Recognizing the problem, the City of San Francisco adopted mandatory green building standards in 1999. The Resource Efficient Building Ordinance pertains to all municipal building construction. In 2004 the ordinance was amended to make new municipal building construction adhere to the Leadership in Energy and Environmental Design Silver standards.

The newly renovated California Academy of Sciences in Golden Gate Park achieved the highest Leadership in Energy and Environmental Design certification of Platinum. It is the world's most visited green museum and also the world's largest public Platinum-rated building. The Academy of Sciences boasts a living roof that acts as a natural insulator and prevents 3.6 million gallons of rainwater from becoming storm water. Skylights on the roof act as both a natural lighting source as well as a natural cooling system for the building.

In an effort to get San Francisco businesses to go green, the City of San Francisco has implemented many programs designed to assist businesses. For instance, the San Francisco Green Business program is a joint venture between the San Francisco Department of Public Health, SF Environment, and San Francisco Public Utilities Commission. Its mission is to help San Francisco businesses adopt environmentally sustainable practices that are also profitable. The program sets criteria for businesses to follow, provides technical assistance, and promotes businesses that follow environmentally sound practices. San Francisco Green Business currently provides standards for hotels, restaurants, offices, dentists, and retailers. There are plans to expand the covered sectors to printers, garment cleaners, janitorial services, auto repair shops, and commercial buildings.

Effective June 1, 2007, the Food Service Waste Reduction Ordinance prohibits any establishment that serves food from using polystyrene foam (Styrofoam) to-go containers. Any to-go containers now in use must be either compostable or recyclable. By the spring of 2009, the city reached a 94 percent compliance rate.

San Francisco's cutting-edge environmental policies and programs make it a symbol for green politics and living for others to emulate. The city highlights what can be accomplished when a committed government and a citizen population work together to achieve a better environmental future.

See Also: Carbon Neutral; City Politics; Environmental Justice; Green Design, Construction, and Operations; Transit.

Further Readings

Coté, John. "S.F. OKs Toughest Recycling Law in U.S." *San Francisco Chronicle* (June 6, 2009).

Dreyfus, Philip J. *Our Better Nature: Environment and the Making of San Francisco.* Norman: University of Oklahoma Press, 2008.

Hatch, Alex. *Cracks in the Asphalt: Community Gardens of San Francisco*. San Francisco: Pasha, 2008.

San Francisco Green Business. http://www.sfgreenbusiness.org (Accessed August 2009).

SF Environment. http://www.sfenvironment.org/index.html (Accessed August 2009).

SForward. "Building a Bright Future: San Francisco's Environmental Plan 2008." http://www.sfenvironment.org/downloads/library/lisforward.pdf (Accessed August 2009).

Austin Elizabeth Scott
University of Florida

SEA LEVEL RISE

Within the next century—and possibly sooner—many of the Earth's coastal cities and agricultural lands will be at extreme risk because of the effects of the rising level of the planet's seas. Sea level rise, defined as the amount of increased volume of ocean, is caused by several factors and may eventually result in the flooding and erosion of coastlines and associated low-lying areas across the world. Some varied locations at high risk include cities such as New York City; Cape Lookout, North Carolina; London; Singapore; Galveston, Texas; New Orleans; and areas such as the Florida keys, Bangladesh, the Netherlands, the Maldives, and many other island nations, such as Kiribati, where evacuations have already begun. The causes of sea level rise include, but are not limited to, the following:

- *Thermal Expansion:* Caused by warming seas and water expansion upward, the primary cause of thermal expansion is global warming, the trapping of greenhouse gases within the atmosphere, and the resulting expansion in volume.
- *Ice Melt:* Also the result of warming, ice melt causes waters to rise as glaciers and ice caps melt, adding water to the seas.
- *Ocean Current Variations:* An example of ocean current variations is El Niño, which is a natural and seasonal surface occurrence that "stacks" seawater. This results in sea level rise because of the movement of warmer water flowing into higher latitudes, where cooler water lies.
- *Natural Variation in Sea Levels:* Variation can be up to two meters at different spots on the globe, as indicated by topographical studies and interpreted as an additional indicator of rising seas.

Scientists report a consensus in which they expect the rise in sea level to be 60 to 90 centimeters by 2100. Over the past 100 years along the U.S. Atlantic Coast, reported sea levels have risen 30 to 40 centimeters, which is higher than the average sea level rise around the globe. Because the sea is not at a consistent, even level and in its natural state, it rises at uneven rates around the globe. In some areas sea level has risen above the mean, and in other areas it has actually fallen below the mean sea level.

Scientists now agree that the Intergovernmental Panel on Climate Change (IPCC) report of 2007 underestimated the rate of sea level rise. Scientists believe that sea level is rising faster than the IPCC prediction because the panel failed to account for the melting of landed ice sheets in Greenland and the Antarctic. Since the IPCC report, scientists have documented an increased melting rate in those locations, with 2009 observations noting movement of 125 feet per day in one Greenland glacier. It has also been recorded that the rate of mountain glacier melting has itself increased, also affecting the rise in sea level around the globe. Sea level rise is also occurring at a faster rate over the years, with a rise of 3 or more millimeters a year since 1993. Historical records indicate a rise of close to 20 centimeters since 1880, and that the warmer the sea gets as a result of climate change, the faster it rises.

If sea levels continue upward at the current rate, a great many locations will be threatened by significant land loss, and some areas are expected to go completely under water. Human habitation in such areas will thus eventually be forced to relocate to higher ground. According to the U.S. Environmental Protection Agency, even a two-foot rise in sea level would inundate 17 to 43 percent of existing wetlands, most of which are in Louisiana. The IPCC reports conservatively that by 2080, fully as much as one-third of all coastal marshland worldwide could be completely submerged.

New York City is in extreme danger from sea level rise because of storm surges that hit the coastline there and multiply the effect. The city averages less than 16 feet above sea level, with some parts of the city lying even lower—measuring as little as 5 feet above the mean North Atlantic sea level. Predictions indicate that the sea level around the northeastern U.S. coast will increase about twice as fast as the world average, with more precise estimates for New York City given as about 8.3 inches higher than what is expected elsewhere around the world.

More modest predictions say that the rise in sea level will flood wetlands, exacerbate the erosion of beaches, increase flooding in developed areas, and contaminate freshwater sources via saltwater intrusion. Sea level rise also has great potential to kill off many species, particularly within ecosystems held in place by marshlands. Some locations already affected by encroaching sea levels include Thailand, Israel, China, and Vietnam. Coastal regions in these countries have already started to see saltwater intrusion affect their underground freshwater systems, which in turn, affects their everyday life.

Should the conventional prediction of a one-meter rise in sea level be correct, other low-lying countries that would be much affected by extreme flooding include Bangladesh, the Maldives, the Netherlands, and most of the inhabited portions of the U.S. state of Florida. Areas in Britain such as the Norfolk Broads and the Thames estuary are expected to be fully immersed. Along with locations already mentioned, the predicted one-meter rise by 2100 will submerge the Maldives, the city of Alexandria in Egypt, and many island nations.

The first island nation to go completely under water was Tuvalu in the Pacific Ocean—its population but a tiny portion of the estimated 10 percent (about 600 million individuals) of the world's population now living in low-lying areas who are continually threatened by the rising sea. Kiribati is another threatened Pacific island nation, with its land disappearing an inch at a time; some of its villages have been moved to higher ground, and citizens have already begun to abandon their homeland. Of Kiribati's 100,000 residents, 75 each year have been uprooted from their homes and moved to New Zealand because of a rising ocean incursion into their villages. Kiribati is expected to be completely submerged within the next 50 years.

Bangladesh, one of the countries most threatened by sea level rise, has many rivers and coastal deltas, so the people who live there are already accustomed to seasonal flooding. However, sea level encroachment on their communities is worsening those floods. Within the next 100 years, Bangladesh is expected to lose 15 to 17 percent of its coastal land, with further erosion inland possible. If sea level predictions are accurate, Bangladeshis expect eventually to be taken away from their low-lying homes and farmland, forever changing their way of life. Already long-threatened elsewhere in the world, mangrove swamps and coral reefs that have survived in Bangladesh will succumb, with the concomitant toll on fisheries and tourism.

Vietnam is home to another culture long identified with the sea and whose population resides mostly along its extensive coastline. Already seeing the early effects of predicted sea level rise, some of its regions will experience a rise larger than the mean. Assuming Vietnam experiences the predicted one-meter increase, 75 percent of the urban population will be uprooted. Depending on location, Vietnam's coastal regions could undergo complete submersion, flooding, erosion, and saltwater intrusion in surface and groundwater. The Camau Peninsula, site of the Mekong River delta, is expected to be fully inundated. The estimated land loss for just this peninsula is 8,000 square kilometers, with an estimated 1.5 million people to be relocated. For the urban-coastal society of Vietnam, the combination of flooding from sea level rise and normal flooding levels will cause extreme damage to many densely populated areas.

Wealthier nations, such as the United States and the United Kingdom, predicted to be hit by significant sea level rise, but better able to adapt, can afford to spend considerable amounts to protect their coastal lands. Urban natural capital—such as nearby wild areas and water resources, soccer fields and community gardens—is generally more available in developed nations, and offers an array of psychological, physical, and financial resources and dividends, which can greatly improve the ability of advanced nations to respond usefully to sea level rise. Proper exploitation of urban natural capital can help cities improve health and social cohesion in crises; sustain and expand participatory urban cultures; maintain clean air and water; and provide more recreational opportunities and better urban aesthetics. The United States is predicted to have to spend in the billions of dollars to protect its coast, and London has begun to explore raising the needed investments to provide its own defenses. Intelligent utilization of urban natural capital will be a key ingredient in this defense. The U.S. Environmental Protection Agency has published information concerning property rights protection for coastal landowners, creating a model for "rolling easements"—a legal arrangement that will allow owners to maintain rights when, endangered by encroaching shoreline, they remove and relocate buildings farther inland.

Two U.S. states, Florida and California, are preparing for the prospect of sea level rise attacking their coastlines. In an effort to save unique species, most of which are found only in the Florida Keys, Florida officials are moving them to higher ground. These animals and plants attract visitors, and produce economic prosperity for the state. Florida plans to incorporate building codes requiring residents to build on higher ground, and communities have started reducing their contributions to global warming. Arnold Schwarzenegger, the governor of California from 2003 to 2011, has implemented an executive order for the state to adapt and become resilient to sea level rise. This order includes the building of new transportation structures that incorporate the rise of sea level in their planning stages. Governor Schwarzenegger also requested a study to be conducted to determine which areas will be most affected by sea level rise.

Caused by thermal expansion and added water from ice melt, sea level rise will have severe effects on low-lying coastal regions all over the world in many different ways over the next several decades. Most of the areas currently affected are already experiencing threats to human populations, flora, and fauna. Many species—and human populations in many countries—can expect to be displaced. An integral effect of global warming, an estimated rise of one meter by the year 2100 is the most conservative prediction of scientists. Whether planned for or not, the effect of sea level rise on human civilization will be profound.

See Also: Coastal Zone Management; Wetlands.

Further Readings

Nash, Steve. "Wetlands, Icecaps, Unease: Sea-Level Rise and Mid-Atlantic Shorelines." *Bioscience*, 58/10 (2008).

"Rising Sea Levels Set to Have Major Impacts Around the World." *ScienceDaily*. http://www. sciencedaily.com/releases/2009 (Accessed April 2009).

"Sea-Level Rise Due to Global Warming Poses Threat to New York City." *ScienceDaily.* http://www.sciencedaily.com.releases (Accessed April 2009).

U.S. Environmental Protection Agency. "Coastal Zones and Sea-Level Rise." http://www.epa .gov/climatechange/effects/coastal (Accessed March 2009).

Anthony R. S. Chiaviello
Felicia Bates
University of Houston–Downtown

SEATTLE, WASHINGTON

Aptly named "The Emerald City" for its year-round lush green scenery, Seattle has been recognized as a leader in urban sustainability and environmental initiatives in nearly every green city index. Washington's largest city, Seattle has a city population of 586,000 and a greater metropolitan population of 3.5 million. Set between the Cascade and Olympic mountain ranges and adjacent to the Puget Sound in the Western quadrant of the state, Seattle is known for its picture-postcard natural beauty, outdoor recreational activities, and broad community support of sustainability efforts. Ubiquitous images of Mt. Rainier, salmon, and orcas reflect the city's strong identification with its natural surroundings. The city boasts numerous green spaces, recycling containers on nearly every street corner, plentiful biking and walking trails that enjoy year-round use, farmers markets and community-supported agriculture, and an environmentally conscious population proud of their city's reputation as a national leader in green living.

Seattle has appeared on all three of the major indices that measure the sustainability of U.S. cities, ranking first in political scientist Kent Portney's list of sustainable cities, third in SustainLane's index of sustainable cities, and 24th in the "Top 25 Green Cities" in *National Geographic*'s Green Guide. Seattle received a rating of 31 of 35 possible points in Portney's ranking system and was recognized for its sustainable indicators project (Sustainable Seattle) and supportive city governance. In SustainLane's index, Seattle ranked high in categories such as city innovation, energy and climate change, and knowledge and communications. The Green Guide recognized the city's high number of green buildings, large percentage of hydroelectric power use (roughly half of Seattle City Light's electricity comes from hydroelectric sources), and clean air and smoking ban, which went into effect statewide in 2005.

Key City and Grassroots Initiatives

The City of Seattle is involved in approximately 30 different specific initiatives and programs related to sustainability. The city government includes an Office of Sustainability and Environment, with a mission to collaborate with community organizations to put forward sustainable practices. Seattle Public Utilities launched one of the first curbside recycling programs in the nation, and Seattle City Light offers residents and businesses in Seattle the option of purchasing wind energy in the form of renewable energy credits through their "Green Up" program. In 2000 the city implemented a Sustainable Building Policy, the first city in the United States to do so, and it boasts the highest number of LEED (Leadership in Energy and Environmental Design)-certified buildings of any city worldwide, with 23 of 609 commercial buildings LEED-certified, as well as 17 percent of all new residential construction. The Department of Neighborhoods P-Patch Program provides 68 volunteer-run neighborhood gardens to Seattle residents, with an emphasis on Seattle neighborhoods with low-income and immigrant populations and youth. In June 2008, the city passed an initiative banning the use of Styrofoam and plastic food containers and utensils by food service vendors.

Seattle's 51st mayor, Greg Nickels, has established himself as a leader in environmental efforts since he took office in 2002. Nickels was named one of the United States' top mayors for sustainability by Portney's "Our Green Cities"—a web resource on sustainable cities in the United States, and in 2005 he was recognized as one of *Rolling Stone* magazine's 25 "Warriors & Heroes" for his efforts in climate protection in Seattle and for organizing mayors around the United States to take similar actions. On February 16, 2005, the day the Kyoto Protocol went into effect, Nichols established the U.S. Mayors Climate Change Agreement through the U.S. Conference of Mayors, putting forward 12 broad policy recommendations for U.S. cities that met or exceeded Kyoto Protocol targets. Policy recommendations included creating an action plan to reduce global warming emissions, promoting alternative transportation options, and increasing recycling. In November 2007, Nichols hosted the U.S. Conference of Mayors Climate Protection Summit in Seattle—a two-day event that attracted more than 100 mayors from across the United States and received widespread coverage in U.S. and international media. By April 2009, 935 mayors representing all 50 states, Washington, D.C., and Puerto Rico, had signed the agreement.

Seattle's reputation as an environmentally conscious city is in large part a result of the many grassroots efforts that the residents have organized over the decades. Seattle Tilth was started in 1978 by three friends and provides public education programs for urban gardening, including city chicken tours, where residents visit chicken coop owners in the City of Seattle. In the early 1990s, when many cities began launching sustainability plans, a group of Seattle residents launched Sustainable Seattle. Although many sustainability plans have been criticized for not providing consistency in measurement or types of data, Sustainable Seattle has been used as a model for at least 90 additional sustainable projects in cities through the United States. Since 1991, Sustainable Seattle has assessed 40 sustainability indicators in five categories: environment, population and resources, economy, youth and education, and health and community. The nonprofit group has been recognized for laying the groundwork for the city's comprehensive sustainability policies. SCALLOPS (Sustainable Communities All Over Puget Sound), supports 60 green neighborhood chapters across the Puget Sound region in their sustainability efforts, including more than a dozen neighborhood chapters in Seattle.

Challenges to Sustainability

Despite its glowing record as a leader in the current green movement, Seattle faces challenges to its sustainability, including traffic congestion, housing affordability, more inclusive environmental policies, and ongoing salmon habitat recovery. The U.S. Census Bureau report on commuting patterns ranked Seattle among the top 10 cities in the nation for percentage of residents who bike to work, walk to work, work from home, and use public transit. Seattle has also received low ratings for housing affordability, natural disaster risk, and metro street congestion. In particular, the Alaskan Way Viaduct, a portion of State Route 99 that spans the waterfront in downtown Seattle, has been judged as unsafe in the event of an earthquake, an eyesore, and a major contributor to air pollution in downtown Seattle, with more than 100,000 vehicles passing over it daily. Construction on a new tunnel replacing the viaduct is scheduled to begin in 2011. With Seattle real estate among the highest-priced in the nation, many people who work in Seattle live outside the city and commute by car.

Some of Seattle's environmental policies have been criticized as heavy-handed. A set of ordinances limiting rural development approved by the King County Council in 2004 highlights the tension between urban and rural residents over environmental policies. The ordinances were met with protests in downtown Seattle, and a series of efforts to reverse the ordinances were ultimately successful in 2008. The city's mismanagement of back-to-back snowstorms in December 2008, which shut down much of the city, caused some residents to complain that the city privileged environmental concerns over resident safety. The city had opted to use sand to de-ice the roads instead of salt, following a 1998 decision based on concerns that the use of salt on the roads would negatively affect Puget Sound. Mayor Nickels has also been criticized for what some residents suggest is a disproportionate focus on the South Lake Union neighborhood of Seattle—an area close to downtown Seattle that has become a major redevelopment project of local billionaire and Microsoft cofounder Paul Allen. The South Lake Union development features mixed-use buildings, an electric streetcar, walking and cycling trails, and the city's first biodiesel station. Although Seattle's history of development led to a devastating loss of salmon habitat that nearly rendered Chinook salmon an endangered species in 1999, salmon recovery efforts have been under way for the last decade, with varying degrees of success. As Seattle confronts the key challenge of a green city—combining urbanism with environmentalism—it will continue to be a model of sustainability for cities around the world.

See Also: Adaptation, Climate Change; Bicycling; Carpooling; Cities for Climate Protection; Citizen Participation; Community Gardens; Commuting; Environmental Risk; LEED (Leadership in Energy and Environmental Design); Recycling in Cities.

Further Readings

City of Seattle, Office of Sustainability and Environment. http://www.seattle.gov/environment (Accessed July 2009).

Karlenzig, Warren. "What Makes Today's Green City?" In *Growing Greener Cities: Urban Sustainability in the Twenty-First Century*, E. L. Birch and S. M. Wachter, eds. Philadelphia: University of Pennsylvania Press, 2008.

Klingle, Matthew. *Emerald City: An Environmental History of Seattle.* New Haven, CT: Yale University Press, 2007.

McRandle, Paul and Sara Smiley Smith. "Top 25 Green Cities in the U.S." http://www.thegreenguide.com/travel-transportation/top-25-green-cities (Accessed April 2009).

Our Green Cities. "The Top 12 Most Sustainable Cities." http://ourgreencities.com (Accessed April 2009).

Sustainable Communities ALL Over Puget Sound (SCALLOPS). http://scallopswa.org (Accessed July 2009).

Sustainable Seattle. http://www.sustainableseattle.org (Accessed July 2009).

SustainLane. "2008 U.S. City Rankings." http://www.sustainlane.com/us-city-rankings (Accessed April 2009).

"Warriors and Heroes: Twenty-Five Leaders Who Are Fighting to Stave Off the Planetwide Catastrophe." *Rolling Stone* (November 3, 2005).

Deborah R. Bassett
University of Washington

SMART GROWTH

Smart Growth is an approach to urban planning that emphasizes compact development patterns, with the aim of building more socially, economically, and environmentally sustainable communities. Smart Growth advocacy gained strength in the mid-1990s as evidence emerged tying undesirable environmental and social adverse effects to the suburban pattern of development that had predominated the post–World War II period. Development based on Smart Growth principles is relatively compact and engineered to limit environmental impacts. It typically features a mix of residential and commercial uses along streets laid out to encourage and accommodate pedestrian activity. Increased neighborhood walkability and provisions for cycling and transit are intended to reduce the necessity for private automobiles imposed by typical postwar suburban configurations. Because Smart Growth development exerts fewer detrimental environmental impacts than conventional suburban development accommodating similar populations, it has been endorsed by the U.S. Environmental Protection Agency, which has established a program to promote its adoption.

Principles of Smart Growth

Ten principles guide Smart Growth planning. These tenets are interrelated and intended to foster thriving communities. The preeminent principle advocates mixing land uses within neighborhoods. Smart Growth advocates consider the strict segregation of land uses widely adopted by U.S. cities in the postwar period as a major contributor to urban sprawl and the subsequent reliance on automobiles. This view is validated by data collected by the U.S. Department of Transportation's Federal Highway Administration for the 1990 Personal Transportation Survey, which attributed the majority of the growth in vehicle miles traveled between 1983 and 1990 to land use patterns, rather than to demographic changes.

A second principle calls for the use of compact building designs, which reduces both the spatial footprint of the community and the environmental footprint, as less land and infrastructure are required per structure. A third principle advocates developing neighborhoods in which homes, schools, parks, offices, services, and shops are located within walking distance of one another along streets that intersect at distances convenient for pedestrians. The primary objective is to reduce the need for automobiles by increasing walkability, but in the process it is also intended to correct for other social, public health, and environmental side effects associated with automobile-oriented development. Accessibility is further addressed by the fourth principle, planning for a full range of transportation modes, including sidewalks for pedestrians, corridors and facilities for cyclists, and enhanced public transportation options. Automobile access is also included, but the planning focus is shifted from accommodating private vehicles to providing better accessibility for people, whatever their chosen mode of travel. In highly urbanized areas and suburban bedroom communities, transit-oriented development is closely associated with Smart Growth.

A fifth principle advocates including a range of housing types at a variety of price points within each neighborhood as a way of building strong, cohesive communities in which people can remain even as their life stages and housing needs change. Apartments, smaller homes, and shared residences accommodate singles, newlyweds, and the elderly, and larger residences of various configurations house growing families or those preferring or requiring larger spaces. A sixth principle focuses on architectural design, calling for standards that ensure neighborhoods are at once attractive and distinctive—places in which residents feel "at home."

In addition to the six principles that emphasize neighborhood design, two specifically address resource conservation. The seventh principle is to preserve open space. This serves multiple ends: not only does it conserve locally important natural habitat areas and enhance community aesthetics by preserving landscape viewsheds but it also plays a vital role in airshed and watershed management. The eighth principle advocates directing development into existing communities, which supports resource conservation but also has public finance implications, as contiguous urban development has been shown to minimize public facilities construction, maintenance, and operations costs.

The last two principles focus on community involvement and equity. The ninth demands an inclusive planning process that facilitates and welcomes participation by all stakeholders. The last calls for a decision-making process that is transparent, fair, and cost-effective for all parties involved.

Environmental Benefits

The environmental benefit most frequently identified with Smart Growth is improved air quality as a result of reduced reliance on single-occupant vehicles, but Smart Growth addresses the full range of environmental concerns. It emphasizes conservation of natural resources, beginning with land. Smart Growth promotes developing infill lots and redeveloping parcels and obsolescent structures before extending the urban boundary into greenfields. Compact building designs are intended to diminish the spatial footprint of development per population, which helps to conserve agricultural land and preserve wild land and the natural ecosystems that are essential to clean air and water.

These policies fuel the charge most frequently fired by critics that Smart Growth promotes socially undesirable levels of density. Advocates respond that density is not promoted for its own sake and that good design alleviates the social problems that critics associate with density. Furthermore, they contend that Smart Growth corrects for social problems associated with isolation and economic segregation in suburban settings. In response to claims by critics that density increases environmental impacts, advocates point to evidence that well-designed, context-sensitive compact development reduces deleterious impacts.

Relative to standard suburban development serving similar populations, Smart Growth patterns require smaller investments in infrastructure as a result of their having fewer miles of streets, narrower streets, shorter runs of water and sewer mains, and fewer miles of electrical wires and telecommunications cables. Knock-on effects of these reductions include reduced operations and maintenance needs and lowered demand for both construction materials and raw materials. Reducing the amount of impervious surfaces reduces undesirable water runoff and flooding and contributes to maintaining local aquifers. The adaptive reuse of older structures reduces the demand both for new construction materials and for landfill space. Studies indicate that multiunit residential structures are more energy efficient than an equal number of detached residences. Evidence also suggests they require less water than detached buildings, partly because they use less for landscaping.

Advocates contend that each of these economies contributes to the larger goal of minimizing the environmental impact of urban growth. Given global climate change, it is apparent now that clean air and clean water are not infinitely available, nor are they merely local issues. Open spaces that support forests and serve as watersheds are vital components of a sustainable future. Smart Growth seeks to promote a more environmentally frugal approach going forward than what has prevailed in the recent past.

Social Benefits

Though Smart Growth was first promoted by urban planners and public administrators, it has many adherents in other social science disciplines. Sociologists cite its emphasis on creating attractive public places for residents to meet and interact as a way to reestablish the community cohesiveness that weakened in socially isolated and economically segregated postwar suburbs. The ability of many residents to stay within their neighborhood as their status or life-stage changes is limited in typical suburban developments where dwellings are segregated by type, size, and valuation. Smart Growth calls for a mix in residential styles within neighborhoods so that as finances, household composition, or other circumstances change, residents can move without leaving their community behind.

The mix of uses within a neighborhood and increased transportation mode options are intended to increase personal mobility by alleviating the dependence on automobiles that isolates nondrivers in suburban settings. For public health researchers and officials, this is not just a matter of individual convenience. Research ties the dependence on automobiles imposed by sprawl to increased health risks for the population in general, and to increased childhood obesity in particular. Hence there is wide support among public health agencies for Smart Growth's emphasis on street connectivity and neighborhood walkability as corrective measures that encourage incorporating physical activity naturally into everyday life.

Economic Benefits

For many citizens, the most persuasive arguments for Smart Growth deal with financial frugality. Studies suggest substantial public finance savings are associated with Smart Growth, starting with the reduced costs of building roads and water and sewer service systems for relatively more compact development, and extending to reduced maintenance and operations costs for this infrastructure. Lower costs to municipalities imply lower tax burdens for property owners. Smart Growth may also lower personal transportation costs when individuals are able to shift from using automobiles to using other modes of travel.

Other public agencies thought to benefit from compact development patterns include school districts, which may realize relatively lower costs for facilities and buses. Similarly, where police and fire departments' service areas are more compact, both operational costs and service response times may be reduced.

See Also: Compact Development (New Urbanism); Density; Infill Development; Sustainable Development; Transit-Oriented Design.

Further Readings

Burchell, Robert W. and Sahan Mukerji. "Conventional Development Versus Managed Growth: The Costs of Sprawl." *American Journal of Public Health,* 93/9 (2003).
Ewing, Reid, et al. "Relationship Between Urban Sprawl and Physical Activity, Obesity, and Morbidity." *American Journal of Health Promotion,* 18/1 (2003).

Smart Growth Network. *This Is Smart Growth*. Washington, D.C.: International City/ CountyManagement Association and the U.S. Environmental Protection Agency, 2006. http://www.smartgrowth.org/library/articles.asp?art=2367 (Accessed October 2009).

U.S. Environmental Protection Agency. "Our Build and Natural Environments." http://www .epa.gov/piedpage/pdf/built.pdf (Accessed October 2009).

Susan H. Weaver
Independent Scholar

STOCKHOLM, SWEDEN

Stockholm is the capital of Sweden, with approximately 825,000 inhabitants in its municipality and over 1.3 million inhabitants in its greater urban area. Stockholm is widely recognized for its natural and historic beauty (which led to its nickname "The Venice of the North") as well as for its political, economic and cultural importance for all of Scandinavia and Northern Europe. In 2009 the European Commission selected Stockholm to act in 2010 as the first ever European Green Capital because of the city's unique commitment to environmental issues and the protection of its natural resources. In 2005 Stockholm had reduced its per capita greenhouse gas emissions by 25 percent (in comparison to emission levels in 1990) and the city council set the target of becoming free of fossil fuel use by 2050. As a result, Stockholm is considered to be a stellar example for best practices in environmentally sustainable planning and pursuing green urban development.

Stockholm was founded in the 13th century, and from its beginnings, it occupied a strategically and economically important position due to its access to the Baltic Sea and lake Mälaren. Mälaren, a fresh water lake, is a unique natural resource for the city's urban ecology, while the brackish Baltic Sea is predominantly of aesthetic and economic value for the city's inhabitants. The most striking feature of Stockholm is that it is an extremely green city; of Stockholm's total area of 215 square kilometers, 56 square kilometers are parks and green spaces, while 28 square kilometers are water areas, meaning that almost 40 percent of the city's entire area consists of biologically productive land. This is a considerably higher percentage of green spaces and water than commonly found in cities of this size.

To date, 90 percent of Stockholm's population lives less than 300 meters from a green area, and the city boasts excellent swimming and fishing opportunities. However, Stockholm's green ambitions are not restricted to recreational and cultural aspects. In 2006 Stockholm city council introduced a congestion charge for cars entering or leaving the city center. Within the first year of the congestion charge, auto emission levels reduced by 10 percent, and air quality improved by approximately 5 percent. Moreover, Stockholm offers its citizens a highly efficient, reliable, and affordable public transport system, as well as over 700 kilometers of cycling lanes.

In addition, more than 65 percent of city households have access to district heating, which uses up to 70 percent of renewable energy. The city also has an ambitious biogas program, which focuses on using food waste and waste water for biogas production. Overall, Stockholm aims to be fossil fuel free in 2050. The city council laid out its ambitions for Stockholm in a document called Vision 2030, which states that Stockholm will foster innovation and growth.

According to the European Green City Index, presented during the United Nations Climate Change Conference in Copenhagen in 2009, Stockholm is the second greenest

European major city, a close second to Copenhagen. The study, which was carried out by the Economist Intelligence Unit of Siemens, evaluates cities in eight categories: CO_2 emissions; energy; buildings; transportation; water; air quality; waste and land use; and environmental governance.

The final report of the expert panel for the European Green Capital Award also rated Stockholm second, although behind Hamburg (which will become European Green Capital in 2011), while Copenhagen (also a short listed candidate) took eighth place. Stockholm scored best of the eight candidates in the categories of noise pollution, access to green areas, sustainable land use, and waste production and management.

Overall, Stockholm undertakes a range of measures in order to lower its emissions of greenhouse gases, to treat waste effectively, and to promote environment friendly urban living. Therefore, Stockholm managed to cut its per capita CO_2 emissions by 25 percent (in comparison to emission levels in 1990), which means that its per capita emission is half of the Swedish average. Stockholm has an efficient waste management system that makes use of innovative developments, such as vacuum controlled underground transportation of solid waste.

Although Stockholm presents a laudable commitment to addressing the environmental issues of urban life, production, and consumption, several problems have not been adequately addressed. One of the biggest problems is an excessive consumption of water: water is viewed as an abundant resource, and many water pipes are outdated and of poor quality, which leads to a huge loss of water through leakage. Current policies seem to neglect and underestimate the amount of freshwater that is required to uphold existing ecosystem services.

Ecosystem services are a range of processes and resources that are needed in order to maintain human welfare, such as air filtration, climate regulation, sewage treatment, noise reduction, crop pollination and energy sources. According to recent research, Stockholm County (which includes the capital area and its surrounding communities) requires about 40 times the amount of its inhabitants direct water consumption (blue water) in order to sustain the County's ecosystem services with fresh water (green water). Hence, underestimating the freshwater dependence of current lifestyles and practices through ecosystem services runs the risk of leading to shortages of a presumable abundant resource (i.e., freshwater).

A further problem of Stockholm, just like any other major European city, is that current CO_2 emissions still exceed the capacity of existing urban ecosystems to absorb CO_2. While Stockholm features an unusually high amount of urban carbon sinks, such as urban forests, lawns, parks, street trees and wetlands, currently only 60 percent of the city's CO_2 generated by traffic can be accumulated by these sinks. This does not yet include other anthropogenic greenhouse gas emissions. Stockholm thus depends on other regions' ability to accumulate its emissions and to provide it with ecosystem services that are not covered by its urban ecosystems, such as providing food.

Overall, Stockholm is one of the greenest cities in Europe, and many of the city council's projects are laudable, especially in regards to public transport, waste management and noise reduction. However, a metropolitan city like Stockholm heavily relies on ecosystem services outside the city, and making urban life truly sustainable is a gargantuan task that requires prolonged and serious effort on behalf of governments, industries, and citizens, inside and outside of metropolitan areas.

See Also: Copenhagen, Denmark; Hamburg, Germany; Healthy Cities; Greening Suburbia; Sustainable Development; Transportation Demand Management.

Further Readings

Bolund, Per and Sven Hunhammar. "Ecosystem Services in Urban Areas." *Ecological Economics,* 29 (1999).

European Green Capital. "The Expert Panel's Evaluation Work and Final Recommendations for the European Green Capital Award of 2010 and 2011." http://ec.europa.eu/environment/europeangreencapital/docs/apply/eval-report_2010_2011.pdf (Accessed January 2010).

Jansson, Asa and Peter Nohrstedt. "Carbon Sinks and Human Freshwater Dependence in Stockholm County." *Ecological Economics*, 39 (2001).

Sidenblath, Göran. "Stockholm: 300 Years of Planning." In *World Capitals: Towards Guided Urbanizations,* H. Wentworth Eldreage, ed. Garden City, NY: Anchor Press, Doubleday, 1975.

Fabian Schuppert
Queen's University Belfast

Stormwater Management

If the stormwater from rainfall or snowmelt does not soak into the ground, it will either flow to natural-surface waterways or into storm sewers, which ultimately direct the water back into natural watercourses.

Source: iStockphoto

The term *stormwater* refers to surface water that originates during rainfall or snowmelt. If stormwater does not soak into the ground, it will flow either to natural surface waterways or into storm sewers, which ultimately direct the water back into natural watercourses with or without treatment. In cities with combined sewers and storm drains, during rainfall, stormwater mixed with untreated sewage is designed to overflow into receiving bodies so that it does not inundate sewage treatment plants. This combined sewer overflow is a significant source of urban nonpoint source pollution.

In the United States, the Environmental Protection Agency regulates stormwater discharges through the National Pollutant Discharge Elimination System Stormwater Program from municipal separate storm sewer systems, construction, and industry. Within this overall framework, states manage their own stormwater control systems according to local conditions.

In the United Kingdom, responsibility for regulating stormwater lies with the Environment Agency for England and Wales and the Scottish Environmental Protection Agency. Here emphasis has shifted away from stormwater management systems and toward development control measures that reduce the cumulative effect of stormwater. For example, starting in April 2008 it became illegal for householders to pave over their front gardens (often done to provide extra off-street parking) without full planning permission (a complex process managed by district councils). The idea is to reduce the total land area covered with impermeable surfaces, which should in turn reduce the aggregate volume of stormwater finding its way into stormwater sewers (many of which are still combined sewer overflows). In the floods of June and July 2007, too-rapid runoff across impermeable surface is thought to have exacerbated the damage to natural and built environments.

In Scotland the main legislation covering the regulation of surface water discharges is the Water Environment (Controlled Activities) (Scotland) Regulations 2005. Through the Scottish Environmental Protection Agency's implementation of these regulations, certain categories of surface water discharge automatically require an application for a license; this includes housing developments of more than 1,000 houses, more than 1,000 car park spaces, industrial estates, and motorways. In most cases, licenses will either encourage or may even stipulate the installation of sustainable drainage systems (see following).

Mitigation

Until recently, water and environmental managers sought to deal with stormwater by providing ever-bigger, more capacious stormwater drainage systems. Cities are attempting to mitigate the effect of combined sewer overflow by building infrastructure to hold the stormwater and sewage during wet weather until it can be pumped through the sewage treatment plant after the precipitation ends.

In the last two decades, however, a revolution in attitudes has occurred, and "green infrastructure" (GI) now is seen as a cost-effective, sustainable, and environmentally friendly approach to stormwater management. GI approaches seek to implement modes of development or industrial practices that slow down or store or even reduce stormwaters locally. Thus, a variety of technologies is now available that mimic natural processes of infiltration, evapotranspiration, and attenuation. For example, encouragement (through the development control or planning gain processes) of permeable surface covers can permit greater infiltration, and the creation of natural swales or wetlands can boost evapotranspiration while also increasing local attenuation capacity.

The benefits of GI can include the following:

- Reduced and/or delayed stormwater runoff volumes, including the reduction of peak run-off volumes
- Enhanced groundwater recharge—an important benefit because groundwater can help maintain natural flow rates in rivers and streams, as well as augmenting local water supplies for domestic or other purposes
- Reduction of pollutant loads, especially through infiltration, because once runoff is infiltrated into soils, plants and microbes can naturally filter and break down many common pollutants found in stormwater
- The number of severity of combined sewer overflow events can also be reduced through the reduction of runoff volumes and by delaying stormwater discharges

There are also a number of indirect benefits of GI that follow from the creation of green seminatural spaces such as wetlands, swales, meanders, and so on. These benefits include improved animal and plant habitat, improved air quality as a function of more green space, greater carbon sequestration potential, and reduced impact of the urban heat island effect. Many studies have suggested that GI can achieve much greater and more cost-effective removal of pollutants than traditional hard-engineering approaches.

At the largest scale, the preservation and restoration of natural landscape features (such as forests, floodplains, and wetlands) are critical components of GI. By protecting these ecologically sensitive areas, communities can improve water quality while providing wildlife habitat and opportunities for outdoor recreation.

On a smaller scale, green infrastructure practices include rain gardens, porous (permeable) pavements, green roofs, infiltration planters, trees and tree boxes, and rainwater harvesting for nonpotable uses such as toilet flushing and landscape irrigation.

In the United Kingdom, GI measures are often referred to as *sustainable drainage systems* (SUDS) because they reduce flooding frequencies and intensities, encourage wildlife, and provide visually attractive and educational amenities in the form of wetland habitat, which is increasingly under threat in the United Kingdom.

At this time, the Environment Agency for England and Wales has two demonstration sites: one at the M40 Motorway Services Area in Oxford and another at the Hopwood Motorway Services Area on the M42. The latter sustainable drainage system incorporates an enhanced wetland feature that is designed to store runoff from the large car and bus parks. A similar scheme at the University of the West of England in Bristol involves a wetland and (attenuation) swale system, which both stores stormwater for slower release and waters an artificial wetland feature on the main campus.

Though entirely logical, uptake of sustainable drainage system and GI techniques for stormwater management has been relatively slow. In part this is a result of regulatory problems, including the legal definition of "sewer" and the delineation of responsibilities for stormwaters and for facilities for their management. Nonetheless "whole life" cost-benefit analyses have been strongly supportive of SUDS and suggest that well-designed and maintained SUDS are more cost effective to construct and cost less to maintain than traditional drainage solutions that are unable to meet the environmental requirements of current legislation.

See Also: Combined Sewer Overflow; Green Infrastructure; Water Pollution; Water Treatment.

Further Readings

Abbott, C. and B. W. Ballard. "Sustainable Urban Drainage—New Research Initiatives." *Water and Environment Manager*, 7/1:10–12 (2002).

Charlesworth, S. M., et al. "A Review of Sustainable Drainage Systems (SuDS): A Soft Option for Hard Drainage Questions?" *Geography*, 88/2:99–107 (2003).

Hendry, S. "Enabling the Framework—The Water Environment and Water Services (Scotland) Act 2003." *Journal of Water Law*, 13 (June 2003).

Jones, P. and N. MacDonald. "Making Space for Unruly Water: Sustainable Drainage Systems and the Disciplining of Surface Runoff." *Geoforum*, 38/3:534–44 (2007).

Staddon, C. *Managing Europe's Water Resources: 21st Century Challenges*. London: Ashgate Books, 2009.

Stephenson, A. "Closing the Floodgates With SUDS." *Sustain Magazine*, 10/2:57–58 (2009).

Chad Staddon
University of the West of England, Bristol

SUKINDA, INDIA

The town of Sukinda, in the Cuttack Division of the state of Orissa on the east coast of India, is the location of 97 percent of India's total deposits of chromite. This makes it a lucrative place, with 12 chromite mines currently in operation, as well as mines for other minerals also found in the area. These mines have operated unchecked for many decades, dumping much of their waste, with the result being that the town is now regarded as one of the 10 most polluted places in the world.

Essentially, much of Sukinda's problems stem from waste from the chromite mines. An oxide mineral—iron magnesium chromium oxide—chromite has a very high heat stability, which means that there is a great demand for it in China, South Africa, and Pakistan, where it is used to make ferrochromium, an important part of the process for making stainless steel and a number of other alloys. It can also be used for the production of magnesium. In spite of the need for chromite in the manufacturing process, there are relatively few places in the world where it exists. Some 44 percent comes from the Bushveld in South Africa; 18 percent from India—the vast majority from Sukinda; 16 percent from Kazakhstan; Zimbabwe, Finland, Iran, and Brazil account for most of the rest.

Even before work began on excavating for chromite, nickel, and other minerals, Sukinda and the region around it already had a reputation as being an area in which local officials rode roughshod over the local population. During the 1930s, this led to a loss of much of the woodland in the region, hastening the process of destroying the environmental infrastructure of the land. This resulted in a decline in farmland, and consequent problems between landlords and peasants. This conflict left the area open to exploitation by mining companies, and many of the landless laborers were desperate for work. The Orissa Estates Abolition Act tried to assist by providing land reform after independence, but many people still were unable to benefit from this major redistribution of agricultural land. Legislative moves have also been enacted to try to preserve the remaining forests in the state, with the Orissa Forest Corporation playing a major role in this.

One of the advantages of excavating chromite in the Sukinda Valley is that the mineral is close to the surface and can be excavated in open-cast ore mines. The 12 mines operating there have no environmental management or political oversight, and as a result, the waste, estimated at 30 million tons of rock, has been left on surrounding areas or dumped into the Brahmani River. This alone has had major effects on the waterways, and large amounts of untreated water have also been dumped into the river, causing major contamination. Many nearby settlements obtain untreated water from the river, and tests have shown that nearly three-quarters of surface water and 60 percent of drinking water contains very high levels of hexavalent chromium—more than double the level thought to be safe both internationally and locally. Some tests have shown the level of hexavalent chromium to be 20 times the international safety standards.

Although people downstream of the mine have been and are affected, it is the health of the workers in the mines and of their families that is of most concern, because of the very high level of infertility, stillbirths, and babies born with deformities. Some studies have shown that almost a quarter of the local people have suffered badly from pollution.

There have also been side effects from nickel mining. This mining is also done using open-cut mines because the nickel ore is some 19–20 meters below the surface. These areas were previously identified by low shrubbery, all of which were destroyed during mining.

There have been few attempts to deal with the ecological and physical problems caused by extensive mining, which has exacerbated medical conditions such as gastrointestinal bleeding and asthma. Furthermore, the mining operations have damaged the immune systems of many, and tuberculosis is widespread.

See Also: Environmental Impact Assessment; Environmental Risk; Kabwe, Zambia; La Oroya, Peru.

Further Readings

Panda, Ranjan K. "Orissa PCB's Report on Sukinda Mine Favours Industries." *Down to Earth*. http://www.downtoearth.org.in/full6.asp?foldername=20080630&filename=news&sec_id=4&sid=33 (Accessed December 2009).

Pati, Biswamoy. "Of Movements, Compromises and Retreats: Orissa 1936–1939." *Social Scientist*, 20/5–6:64–88 (May–June 1992).

Roy, S. "Geobotany in the Exploration for Nickel in the Ultramafics of the Sukinda Valley, Orissa." *Quarterly Journal of the Geological, Mining and Metallurgical Society of India*, 46:251–256 (1974).

Justin Corfield
Geelong Grammar School, Australia

Sumgayit, Azerbaijan

The city of Sumgayit in Azerbaijan, on the Aspheron Peninsula on the Caspian Sea, is only 19 miles (about 30 kilometers) from the country's capital, Baku, and is the center of the local chemical industry. It is regarded as one of the most polluted places in the world.

Sumgayit was officially proclaimed a city on November 22, 1949. In 1959, the city's population was 52,000—already the third-largest city in what was then the Azerbaijan Soviet Socialist Republic. In 1973, the National Steel Corporation of Pittsburgh was involved in consultancy work on the establishment of an alunite mine that helped provide Sumgayit with more alumina. At its height during this period, the aluminum complex at Sumgayit was capable of producing 5 million evaporators a year—enough to provide for some 70 percent of all the refrigerators being made annually in the Soviet Union.

In addition to power generation, Sumgayit was already important for sulfuric pyrite and borax waters, helping to produce sulfuric acid and acetylene. There was also a plant for the manufacture of carbon black used in the rubber industry and for printing, and a synthetic rubber plant. In 1959, work started on the construction of an aluminum oxide facility. There was also a plant for making chlorine. It was not long before the city became heavily polluted in spite of its being described in the Soviet Union's literature as a "new town" with some of the best (seemingly) urban facilities in the world.

The work at Sumgayit attracted many people from other parts of Azerbaijan, as well as Armenians, in addition to engineering experts and their families from Siberia and other parts of the Soviet Union. There had long been tensions before the 1920s between the Armenians, who were Christian, and the Azerbaijanis, who were Shiite Muslim, but there were few problems, religious or racial, until the 1980s. Indeed, there were no churches or mosques in the city. By 1990, there were between 200,000 and 265,000 people living there.

On February 29, 1988, before the breakup of the Soviet Union, ethnic tensions in Sumgayit resulted in attacks on ethnic Armenians that resulted in 26 being killed and several thousand injured; six Azerbaijanis also were killed. The rioting lasted for three days, and by the time it was over, some 2,000 of the estimated 10,000 Armenians in the city had fled. The Azerbaijani nationalists blamed the Armenians for inciting the rioting in their effort to discredit them. They also claimed that the fighting had started after reports of Azerbaijani homes being destroyed in Armenia. At the same time, there were immense tensions in the region of Nagorno-Karabakh, which was a mainly Armenian enclave within the Azerbaijani Soviet Socialist Republic. Disturbances within the Armenian Soviet Socialist Republic had seen as many as 3,5000 Azerbaijanis flee and be resettled in Sumgayit. Foreign visitors were barred from the region until August 1988, by which time tensions in Sumgayit had subsided. In January 1990, there were further tensions, causing the Soviet Union's central government, under direction from Mikhail Gorbachev, to send soldiers into Baku and Sumgayit.

The collapse of the Soviet Union, and Azerbaijan proclaiming its independence in 1990, led to the Nagorno-Karabakh War, and some refugees from this conflict were also accommodated in Sumgayit. About this time, Azerbaijan experienced a major breakdown in healthcare, and as had been the case since the 1960s, pollution took its toll, with high death rates and a cancer rate that was 51 percent higher than the national average. Still, the population continued to grow, and there are now (2009 estimate), 280,000 living in the city, which has its own university, the Sumgayit State University (founded in 1961 as a branch of the Azerbaijan State Oil Academy and renamed in 2000).

During the 1980s Sumgayit had already become nicknamed "The Dead Zone" on account of the large number of people succumbing to illnesses associated with pollution. By this time, 32 chemical and metals plants emitted an estimated 120,000 tons of waste each year. Some 36 of every 1,000 babies born annually died before they reached the age of one year, and the infant mortality rate grew to more than three times that of the United States. At its height, the environmental problems were so great that an official from the International Monetary Fund actually suggested that a plan should be drawn up to evacuate the population, which had grown to 330,000.

The growth of Sumgayit from a population of 4,000 to 200,000 had been hailed as an economic miracle. However, the amount of power needed to run the aluminum and other plants was colossal, which contributed heavily to pollution. As the region was viewed as unstable, especially after the 1988 and 1990 disturbances, few Western companies were interested in investing the hundreds of millions of dollars needed to raise the existing factories to the standards of those in Western Europe, let alone to deal with the pollution. In 2007, *Time* magazine listed Sumgayit as one of the world's most polluted places, and *Scientific American* also placed it in the list of the 10 most polluted places in the world. By this time, there was also a good deal of publicity surrounding the city's "Baby Cemetery," which contains the graves of many infants, a large number of whom were born with deformities or mental retardation.

See Also: Carbon Footprint; Ecological Footprint; Vapi, India; Water Pollution.

Further Readings

de Waal, Thomas. *Black Garden: Armenia and Azerbaijan Through Peace and War*. New York: New York University Press, 2003.

Keller, Bill. "Riot's Legacy of Distrust Quietly Stalks a Soviet City." *The New York Times,* (August 31, 1988).

Kostianoy, Andrey and Aleksey N. Kosarev. *The Caspian Sea Environment (Handbook of Environmental Chemistry)*. New York: Springer, 2005.

LeVine, Steve. "Sumgait Journal: A City at Death's Door Turns the Doctors Away." *The New York Times* (May 9, 1997).

Swietochowski, Tadeusz. *Azerbaijan: Legacies of the Past and the Trials of Independence (Postcommunist States and Nations)*. London: Routledge, 2009.

Justin Corfield
Geelong Grammar School, Australia

SUSTAINABILITY INDICATORS

Implementing sustainable development entails large shifts in defining, configuring, and prioritizing public policies. A singular requirement is to create the tools to assess the current situation, as well as the induced societal, economic, and environmental change. Sustainability indicators are in the first instance meant to meet this necessity. "Public choice" has always been a matter of monitoring, and thus valuing, societal evolution and development against criteria defined by individual or group preferences. In this sense, sustainability indicators have ancestors; that is, the System of National Accounts developed since the 1940s, the social indicators' movement of the 1970s, and the formalization of environmental policy performance indicators since the 1980s.

Nearly as many definitions for indicators exist as for sustainability (or sustainable development). Generically, sustainability indicators provide an interpretation of the evolution of stocks and/or flows to account for human–environment interactions. Sustainability indicators are meant to participate in the (self-)generation of sustainable development by enhancing communication, triggering debate, enabling planning, and facilitating strategizing. Because sustainability indicators strive to render the invisible visible, they are enshrined in societal, technical, methodological, and scientific conventions, which entails that the definition, selection, configuration, and interpretation of a sustainability indicator implies articulating scientific and societal values at various levels.

Since the Brundtland Report in 1987 and the Rio Summit in 1992, there have been shifts in the interpretations of sustainability. Over the years, two parallel approaches have emerged to assist the implementation of sustainability into public policy processes. On the one hand, one observed the emergence of specific sustainability policies, and notably of sustainability strategies, which meant providing a top-down reference point for policy makers. On the other hand, a handful of specific instruments, processes, and tools were initiated that should assist in mainstreaming the translation of sustainability criteria and principles into policy processes by influencing the configuration of policy moments, such as public policy evaluation and assessment, policy communication, and policy formulation. Within this second, nonprogrammatic approach to translating sustainability into policies, the debate on the instruments, processes, and tools used generated a wave of processes at national, global, local, and urban levels, concerned with the construction of sustainability indicators. Before the turn of the millennium, every country in the developed world had its initiative on sustainability indicators either accomplished or under development. Hundreds of local and urban communities also had stepped into their indicator processes, often with a direct linkage to their Local Agenda 21 processes, sometimes with considerably interesting results, as has been the case, for instance, of the front-running Sustainable Seattle indicator initiative since 1993.

The reasons for the great number of initiatives are linked to the range of possible objectives pursued with indicator development. Indicators are developed to help define sustainability strategies, whereas other initiatives assess the success of their strategy with indicators. Indicators are used to evaluate and communicate about the performance of buildings and construction sites, as well as about the level of urbanism and land use planning. Indicators are initialized for small-scale evaluations of urban space management or the allocation and use of local development funds. Sustainability indexes are developed to rank stock portfolios and pension funds. Academia, as well as national and international institutions, are striving to configure composite indicators such as the ecological footprint, which is meant to replace or complement gross domestic product, whereas adaptations of the same gross domestic product to integrate environmental and social variables, such as the Index of Sustainable Economic Welfare or Indicator of Genuine Savings, are meant to keep our economic status stable. Sometimes indicators seem to be mere by-products of data treatments, such as maps extruded from geographic information systems software. At times indicators represent the emerged part of an empirical calculus, as is the case with communicating the outcome of an extensive life-cycle analysis. On other occasions, indicators condense the results of complex and time-consuming data collection and structuration efforts, as with the attempts to green the national accounts by constructing satellite environmental accounts (e.g., the National Accounting Matrix including Environmental Accounts). Lately, indicators are outputs of sustainability impact assessments, as well as of processes using multicriteria decision analysis.

Methodologically, sustainability indicators are either statistical aggregates (i.e., composite indicators or indexes) or large-scale collections of individual domain- and sector-specific measurements presented in interlinked scoreboards (such as the international effort within the United Nations Commission for Sustainable Development). In the case of the latter, a prominent organization for indicators is the Organisation for Economic Co-operation and Development–inspired DPSIR framework of indicators, which interlinks indicators representing driving forces (D), pressures (P), states (S), impacts (I), and responses (R).

The above-defined indirect link between (sustainability) indicators and "reality," and the importance of values in indicator definition and interpretation should not be taken as a weakness of sustainability indicators, just as the inherent complexity of the sustainability paradigm itself should not be regarded as undermining the viability of the concept or its assessment by indicators. In every instance, the representativeness of indicators is not only influenced by the extent of the measurable (e.g., advances in scientific knowledge, or data quality) and the measured (e.g., advances in data availability), but more so by the views of the indicators' developers and users on the issues under assessment. For instance, the existing measurement of economic welfare is influenced by our understanding of the dynamic components of welfare (i.e., the measurable), by institutional arrangements between providers of data (i.e., the measured), and by the societal interpretation of what constitutes welfare. The difficult link between indicators and reality leaves much room for interpretation (and thus for dialogue) to those who construct, select, and use indicators. Value judgments are inextricably linked to sustainability indicators at many levels, from the conceptualization to the utilization of indicators, and the societal translation of reality into indicators that will largely be determined by the value referents of the implied actors. As a consequence, sustainability indicators are mostly developed, configured, and selected within societal processes and are often integrated into a participatory procedure.

Indicators are seen as prime policy tools to simplify the communication of facts and evolutions, enabling actors to participate in achieving sustainability via public awareness raising and/or institutional capacity building. Scientific evidence on the role of information for policy making shows that policy actors use information sparsely as a direct input to their decisions. Similar patterns of (non)use are observed with sustainability indicators. Their strength lies more indirectly in inducing change in the observer's way of seeing and interpreting the world, and consequentially, in his or her way of prioritizing problems and implementing solutions. Such a function, again, is not singular to sustainability indicators, as it has been identified with many policy and decision-aiding tools.

See Also: City Politics; Ecological Footprint; Environmental Planning; Sustainable Development.

Further Readings

Bell, Simon and Stephen Morse. *Measuring Sustainability: Learning From Doing.* London: Earthscan, 2003.
Bell, Simon and Stephen Morse. *Sustainability Indicators: Measuring the Immeasurable.* London: Earthscan, 2008.
Lawn, Philippe. *Sustainable Development Indicators in Ecological Economics.* Cheltenham, UK: Edward Elgar, 2006.
Moldan, Bedrich, et al., eds. *Sustainability Indicators: A Scientific Assessment.* SCOPE Publication Series. Washington, D.C.: Island Press, 2007.

Tom Bauler
Free University of Brussels

SUSTAINABLE DEVELOPMENT

Sustainable development is a pattern of resource use that aims to meet human needs while preserving the environment, so that these needs can be met not only in the present but also for future generations to come. It means different things to different people, but the most frequently quoted definition is from the report *Our Common Future* (also known as the Brundtland Report), stating that "sustainable development is development that meets the needs of the present without compromising the ability of future generations to meet their own needs."

Sustainable development is not a new idea. Many cultures over the course of human history have recognized the need for harmony between the environment, society, and economy. What is new is an articulation of these ideas in the context of a global industrial and information society. Sustainable development focuses on improving the quality of life for all of the Earth's citizens without increasing the use of natural resources beyond the capacity of the environment to supply them indefinitely. It requires an understanding that inaction has consequences and that we must find innovative ways to change institutional structures and influence individual behavior. It is about taking action and changing policy and practice at all levels, from the individual to the international.

Discussions about the limits and implications of economic growth have been recurrent in economic history. In the early 1970s, the debate spurred by D. H. Meadows and colleagues (1972) mainly focused on the prospects of shortages in material stocks of nonrenewable natural resources and on whether economic growth would inevitably lead to environmental degradation and social collapse. At that time, economic growth and environmental quality were largely perceived as opposing each other. In the 1980s, the importance of reconciling economic growth with the environment had come to be generally recognized, providing an intellectual underpinning to efforts to elevate the importance of environmental issues in policy making. The origin of sustainable development dates back to 1982, when the World Commission on Environment and Development was initiated by the General Assembly of the United Nations, and its report, *Our Common Future*, was published in 1987. The prime minister of Norway, Gro Harlem Brundtland, was the chair of the commission.

The commission's membership was shared between developed and developing countries. Its roots were in the 1972 Stockholm Conference on the Human Environment, at which the conflicts between environment and development were first acknowledged, and in the 1980 World Conservation Strategy of the International Union for the Conservation of Nature, which argued for conservation as a means to assist development, and specifically for the sustainable development and utilization of species, ecosystems, and resources. Drawing on these, the Brundtland Commission began its work, committed to engage environment and development.

As with previous efforts, the report was followed by major international meetings. The United Nations Conference on Environment and Development in Rio de Janeiro in 1992 (the so-called Earth Summit) issued a declaration of principles, a detailed Agenda 21 of desired actions, international agreements on climate change and biodiversity, and a statement of principles on forests. Ten years later, in 2002, at the World Summit on Sustainable Development in Johannesburg, South Africa, the commitment to sustainable development was reaffirmed.

The Johannesburg Declaration created a collective responsibility to advance and strengthen the interdependent and mutually reinforcing pillars of sustainable development: economic development, social development, and environmental protection at local, national, regional, and global levels. In so doing, the World Summit addressed a running concern over the limits of environment and development, wherein development was widely viewed solely as economic development.

The sustainable development debate is based on the assumption that societies need to manage three types of capital (economic, social, and natural), which may be nonsubstitutable and whose consumption might be irreversible. Some authors point to the fact that natural capital cannot necessarily be substituted by economic capital. Although it is possible that we can find ways to replace some natural resources, it is much more unlikely that they will ever be able to replace ecosystem services, such as the protection provided by the ozone layer or the climate stabilizing function of the rainforests. In fact, natural capital, social capital, and economic capital are often complementarities.

One of the successes of sustainable development has been its ability to serve as a common ground among those who are principally concerned with nature and environment, those who value economic development, and those who are dedicated to improving the human condition. At the global scale, this concept has engaged the developed and developing countries in a common endeavor.

Before this conciliation was formally adopted by the United Nations Conference on Environment and Development, developing countries often viewed demands for greater environmental protection as a threat to their ability to develop, whereas the rich countries viewed some of the development in poor countries as a threat to valued environmental resources. The concept of sustainable development attempts to couple development aspirations with the need to preserve the basic life support systems of the planet.

Sustainable development, as a concept, is an open, dynamic, and evolving idea that can be adapted to fit very different situations and contexts across places, scales, and time. It opens spaces for participation at multiple levels, from local to global, within and across activity sectors, and in institutions of governance, business, and civil society to redefine and reinterpret its meaning to fit their own activities, such as the concept of sustainability that has been adapted to address different challenges, ranging from urban planning to sustainable livelihoods.

Sustainable development requires the participation of different stakeholders and perspectives seeking to reconcile diverse values and goals toward mutual actions to achieve sustainable practices worldwide. Increasingly, goals and targets for sustainable development are being adopted by global and local consensus. Sustainable development can be interpreted as a path along which the maximization of human well-being for today's generations does not lead to declines in future well-being.

Interactions between the economy, the environment, and society must be taken into account in formulating different strategies. All too often, measures targeted to specific dimensions of development do not consider effects on other dimensions, leading to unforeseen effects and costs. Responding to the challenge of sustainable development requires the institutional and technical capacity to assess the economic, environmental, and social implications of development strategies and to formulate and implement appropriate policy responses.

See Also: Agenda 21; Citizen Participation; Sustainability.

Further Readings

The Johannesburg Declaration on Sustainable Development, September 4, 2002. http://www .housing.gov.za/content/legislation_policies/johannesburg.htm (Accessed September 2009).

Kates, R. W., et al. "What is Sustainable Development? Goals, Indicators, Values and Practice." *Environment: Science and Policy for Sustainable Development*, 47/3:8–21 (2005).

Leiserowitz, A. A., et al. *Sustainability Values, Attitudes and Behaviors: A Review of Multi-National and Global Trends*. CID Working Paper No. 112. Cambridge, MA: Science, Environment and Development Group, Center for International Development, Harvard University, 2004.

Meadows, D. H., et al. *The Limits to Growth*. Report for the Club of Rome's Project on Predicament of Mankind. New York: New American Library, 1972.

Organisation for Economic Co-operation and Development (OECD). *Sustainable Development: Critical Issues*. Paris: OECD, 2001.

Wheeler, Stephen M. and Timothy Beatley. *Sustainable Urban Development Reader* (Routledge Urban Reader Series). London: Routledge, 2008.

World Commission on Environment and Development. *Our Common Future*. New York: Oxford University Press, 1987.

Rafael D´Almeida Martins
University of Campinas

SYDNEY, AUSTRALIA

The City of Sydney (CoS) sits on the southeast coast of Australia and serves as the capital of the state of New South Wales (NSW). It is alternatively known as "The Harbour City" or "Australia's Global City." Besides the striking Sydney Opera House and the graceful arch of the Harbour Bridge, another image of Sydney emerged in September 2009, as a dust storm of epic proportions engulfed the city. The sky turned red, as particulate matter pollution reached 1,500 times its normal level and 75,000 tons of dust per hour rained into the Tasman Sea. For Sidneysiders, the event amplified the urgency of implementing their quest to become an environmentally sustainable community, but it was not the only prompt that year. The years from 1991–2000 constituted the hottest decade on record for Sydney, and 2009 was the second hottest year since record keeping began. The annual mean temperature was 0.9 degrees Celsius (1.6 degrees Fahrenheit) above average, and three record-breaking heat waves were responsible for the deaths of 173 people in the metropolitan area.

Red skies aside, the city of Sydney envisions itself as "green, global and connected": a model for the region and the world. The broad environmental agenda for its 176,287 residents (2009 population) outlined in the Sustainable Sydney Vision: 2030 addresses all aspects of urban sustainability. It proposes making Five Big Moves, starting with (1) revitalizing the center city, (2) constructing an integrated transportation system, (3) providing a "liveable green network" of urban villages, (4) creating community hubs with goods and services within walking distance of community members, and (5) undertaking renewal projects to transform Sydney into a model of green urbanism. In drafting the plan, the city enlisted the aid of Jan Gehl, an urban designer renowned for his holistic approach to building livable places. Documents supporting the Sustainable Sydney vision include the CoS Environmental Master Plan, Green Infrastructure Plan, and Decentralized Energy Master Plan. The city has also instituted numerous programs to engage its citizens. Live Green, Watershed, and Zero Waste Partners, are designed to give practical advice and government support to facilitate adoption of green practices by individuals and businesses.

The city of Sydney is an active participant in international organizations focused on the environment. including the United Nations Environment Programme Climate Neutral Network, Project 2°, ICLEI, and C40 Cities. In 2008, Sydney became the first local government in Australia to become carbon neutral.

In order to become carbon neutral, the city adopted a two-pronged approach. First, as a carbon neutral city, it purchases all of the electrical energy used in governmental operations from sources accredited as renewable by GreenPower, a national program for independent auditing. For other facets of governmental operations, such as employees' work-related travel, the city purchases verifiable emissions offsets. To monitor its compliance, Sydney uses the Project 2° emissions tracker and has the results independently verified to ensure its operations are indeed carbon neutral. In becoming a carbon neutral city, Sydney has vowed that its 2012 greenhouse gas (GHG) emissions from city government operations will be 20 percent below 2006 levels. Its GHG emissions in fiscal year 2006–2007 amounted to 48,577 tons CO^2 equivalent (CO^2e), 86 percent of which were attributable to electricity use. While the city of Sydney emissions in 2006–2007 represented less than one percent of total emissions for the metropolitan area (estimated at 5.45 million tons CO^2e), the city's commitment to carbon neutrality sets a sterling example for both citizens and other governments in the metropolitan area.

The city of Sydney has endorsed the Kyoto Protocol targets of limiting total 2020 GHG emissions within its jurisdiction to 70 percent of the 1990 levels and 2050 emissions to 30 percent of the 1990 benchmark. In order to meet the 2050 target, the city will have to cut current emissions in half within the next 12 years. The majority of these cuts will have to come from changes in the provision of electricity, which generates approximately 78 percent of current GHG emissions. Under the business-as-usual scenario (BAU), GHG emissions are forecast to increase by 20 percent by 2030.

Energy: A Green Transformation

Residents and businesses have a choice of using a number of power companies operating in New South Wales (NSW). Of the 71,900 gigawatt hours distributed in NSW in 2008, 43 percent went to industrial uses, 31 percent to households, 24 percent to commercial uses, and 1.6 percent to transportation. While consumption per household has been stabile since 2002–2003, the number of households has increased, as has demand from industry and the service sector. Ninety-five percent of the energy demanded in NSW is supplied from fossil fuels. The source shares of energy in 2008 were coal (39.7 percent), oil (33.6 percent), gas (21.6 percent), and renewables (5 percent). The latter figure reflects a 300 percent increase in renewable generating capacity between 2003–2008, due to the priority placed by NSW on the development of renewable sources. Many new bioenergy, solar, and wind generators came online or were expanded at a time when production of hydropower decreased due to a decade-long drought. All other renewable sources together only provided 1 percent of total energy consumed. Between 2005–2009, the number of households using GreenPower more than tripled (282,000 to 941,000), while the number of business concerns scarcely doubled (14,676 to 32,276).

Sustainable Sydney 2030 proposes to change the way energy is generated and distributed within the city boundaries via a program called Green Transformers. As an element of the city's urban renewal program, the Green Transformers system would comprise cogeneration facilities distributed throughout the city's 10 districts. These facilities would provide low-carbon electricity and low-GHG heating to buildings through a distribution network to be placed under city streets. Benefits expected by the city include a per capita reduction in annual GHG emissions of 66 percent, an overall reduction in demand for electricity, and increased energy efficiency and security for the city, its residents, and businesses. Complementary programs to encourage the transition to energy efficient lighting and appliances are expected to further decrease per capita demand for electricity. Benefits from the Green Transformers system would also extend to waste management and water conservation by diverting waste from landfills and producing recycled water for use in industry and irrigation.

Waste

The average city household generated 500 kilograms of general waste in 2007, approximately 29 percent of which was recycled. Sustainable Sydney 2030 sets goals for recycling or reclamation of 66 percent for general waste and 76 percent of construction and demolition waste by 2014. To advance these goals, the city has already undertaken improvements in collection, including dedicated bins for different waste types and special collection events for discarded appliances, electronic, and hazardous wastes. It promotes recycling of green waste through its urban composting program and engages the business community with its Zero Waste Partners program.

Currently CoS participates in the Eastern Creek facility, a public–private joint venture that has been recognized by C40 Cities as a Best Practices for Recycling site. Eastern Creek uses mechanical and biological processes rather than incineration to convert methane into energy. More than 70 percent of the waste arriving at the facility is diverted from the landfill. Eastern Creek currently has the capacity to handle 260,000 tons of waste a year. The facility is self-sustaining in electricity and water, and generates enough electricity to offload a portion to the grid. In the future, the city will continue to cooperate with suburban governments to expand metropolitan integrated waste management programs. Green Transformers would add a major component to the city's own waste management strategy by converting 50 percent of the waste stream generated within its boundaries to electricity or biogas by 2030.

Water and Air Quality

Changing weather patterns make water a major issue for the city of Sydney. Since 1997 Australia has struggled with drought conditions so severe that they threatened to force farmers out of production. Though the southeast benefited from heavy rainfall in December 2009, the area still faces serious deficiencies in water supply. Water usage decreased by 5 percent in metropolitan Sydney between 2001–2006, mainly as a result of drought-induced restrictions, but the prevailing trend indicates increasing consumption with a 22 percent increase for 2030 forecast under BAU.

The CoS Environmental Management Plan (EMP) establishes goals of cutting water usage in the city back to 2006 levels by 2015. It also proposes to recycle 25 percent of its water by that date. Additionally, the city proposes to establish "Green Links" in the catchment area to allow for storm water capture and treatment.

As the 2009 dust storm so vividly illustrated, air quality is a major concern in metropolitan Sydney. The primary issues are photochemical smog and particulate matter pollution. Photochemical smog arises from the interaction of ozone, nitrous oxides, and sunlight. Eighty percent of nitrous oxide emissions in Sydney come from vehicles and, as in other cities, vehicle use is increasing. Similarly, there has been no apparent decrease in ground level ozone measurements, so the smog problem continues. Particulate matter (PM) pollution is especially problematic, as much of it is from natural sources—sea salt, dust, and pollen. Bush fires and wood smoke created in winter months are also common sources. On normal days, the PM level is 10 mg/m3, but during bush fire incidents the level rises to 500 mg/m3. In comparison, the level during the 2009 dust storm was 15,400 mg/m3. The State of NSW has the authority for air pollution control, and has published Action for Air, which details changes that must occur in order to improve air quality in general and reduce GHG emissions in particular.

Transportation and a Livable City

As its contribution to Sydney's smog indicates, transportation poses major problems for the region. The metropolitan area comprises low density suburbs surrounding the city. With insufficient transit lines, bus routes operating at capacity, and too few bicycle and pedestrian facilities, the region is highly dependent on automobiles. Area roads have a reputation for being clogged, so much so that between 1996 and 2001, commuters shifted from using taxis and cars to traveling by trains, buses, bikes, and foot. In 2009 and early 2010 that trend reversed slightly in response to schedule changes that decreased train frequencies and increased travel times. Train ridership decreased by 2.2 percent, while traffic on roadways increased by 4 percent. Within the city of Sydney, use of alternative modes is much higher than in the suburbs. The EMP set a goal that 20 percent of trips between 2 and 20 kilometers are traveled by bicycle by 2016, but a Sustainable Sydney update indicates that goal may have already been surpassed, with 50 percent of in-city trips now conducted by walking or bicycling.

The second of Sydney's Five Big Moves proposes building an integrated transportation network to effectively and efficiently link the city's 10 districts. It would employ bus routes, dedicated and separate bicycle lanes, light rail loops, jitney services, parking availability, and pricing strategies. The plan is not solely within the purview of the city of Sydney, however. Implementation will require the cooperation of the State (NSW), which has responsibility for public transportation in the region, and the federal government, which controls finances.

Sidneysiders take pride in the high rankings their city receives in quality of life indexes produced by researchers such as Mercer (ranking Sydney number 10 in 2009) and *The Economist* magazine (ranking Sydney number 8 in 2009), but they worry that current rankings are lower than in those awarded in the past. The recognition that livability is key to consolidating the city's position as Global Sydney led to Jan Gehl's involvement in shaping the urban environment envisioned by Sustainable Sydney. Gehl's emphasis is on reintroducing human scale where automobiles now prevail, and three of the Five Big Moves either directly or indirectly focus on making the city attractive to pedestrians and cyclists. Furthermore, much of Sustainable Sydney focuses on tourists and the global corporations Sydney hopes to retain and attract.

A key component of any attractive and sustainable city is its greenspace. Sydney streets are lined with 26,000 trees, while its parks contain 8,000 trees (8.5 per acre), and the city is encouraging the installation of green roofs. The city's Environmental Management Plan commits to providing 24 square meters of public open space for every resident. Currently it is just shy of that mark, at 23 square meters. There are 248 parks and open spaces within the city encompassing 377 hectares (931.6 acres). The two largest parks, Centennial Park and Bicentennial Park, account for 80 percent of this total.

Measured by its population, the city of Sydney is a small city set among populous suburban communities. In the effort to become a sustainable city, being small is both an asset and a liability. On the one hand, dealing with problems on a limited scale makes the tasks more manageable, but on the other hand many aspects of change are not under the city's control. Sydney appears to be using the advantage of its small scale to become a sustainable urban center. While many factors are outside of its jurisdiction, it is forging cooperative relationships to address them. And whether by accident or intent, Sydney's larger than life image encourages other communities to move toward sustainability.

See Also: Carbon Neutral; Distributed Generation; Green Procurement and Purchasing; Recycling in Cities.

Further Readings

City of Sydney. "Environmental Management Plan." http://www.cityofsydney.nsw.gov.au/environment/Overview/EnvironmentalManagementPlan.asp (Accessed January 2010).

City of Sydney. "The Sustainable Sydney Vision: 2030." http://www.cityofsydney.nsw.gov.au/2030/documents/2030Vision/2030VisionBook.pdf (Accessed January 2010).

Gehl, Jan. *Life Between Buildings: Using Public Space*. Copenhagen, Denmark: Danish Architectural Press, 2008.

New South Wales Government, Department of Environment, Climate Change and Water. "Action for Air: 2009 Update." http://www.environment.nsw.gov.au/air/actionforair/index.htm (Accessed January 2010).

Susan H. Weaver
Independent Scholar

Tianying, China

Tianying is a township of around 30,000 people, located in Anhui Province on the outskirts of Jieshou City, China. Not to be confused with Tianjin (some 750 kilometers north), Tianying has been plagued by chronic lead contamination from smelters and battery recycling centers. Half of China's lead comes from the Tianying area. According to some sources, it is one of the world's top 10 most polluted places. Although the Chinese government has taken some action in recent years to improve environmental quality in Tianying, the situation remains poor—mostly because battery businesses drive the local economy. This article considers the various types of pollution that affect Tianying's residents, proposed remediation, and the prospects for a healthier future for Tianying's citizens.

Sources of Pollution and Impacts

Pollution levels in many Chinese cities frequently exceed World Health Organization standards. The World Bank has also reported that 20 of the world's 30 most polluted cities are found in China. Tianying's residents have experienced this firsthand, as they are burdened by chronic air pollution largely derived from poorly regulated lead refineries that use primitive battery recycling and lead smelting technologies. Typically, peddlers collect old batteries and sell them to "backyard" operators, who then render down the batteries to retrieve the lead, which is then refined with low-grade technologies and eventually used to make new batteries.

However, unregulated rendering and refining operations have resulted in chronic lead pollution. Recent studies have found that lead levels in Tianying's air and soils exceed national health standards by as much as 8.5 to 10 times the acceptable limits. Researchers have also found that agricultural land is badly contaminated—local wheat contains up to 24 times the acceptable levels of lead prescribed by Chinese food safety authorities. Standards for suspended lead particulates are exceeded about 85 percent of the time. In the absence of pollution abatement measures, up to 140,000 people have been directly affected by heavy metals and other toxic substances. Other air pollution sources include sulfur dioxide and various particulates (not only iron oxides, but also oxides of mercury, cadmium, arsenic, antimony, and copper). Some sources suggest that chlorides, fluorides, and other chemical contaminants may be present too. In addition, minimal waste treatment practices mean local waterways are contaminated with high levels of lead and arsenic, suspended solids, and hydrocarbons.

Pollution has taken a serious toll on the health of Tianying's residents. Kidney damage, anemia, and brain damage affect both children and adults. Miscarriages, premature babies, birth defects, and lead poisoning of children are common. Children under 6 have suffered most: The mean blood lead levels of children living in polluted areas have been recorded at 496 µg/L (the norm in China is 100 µg/L). The insidious effects of lead poisoning include lower IQs, learning disabilities, attention deficit problems, hyperactivity, hearing and vision disabilities, stomach and colon irritation, kidney damage, anemia, and in severe cases, brain damage.

Government Action

In 2003, China's Xinhua News Agency reported that the local environmental protection administration had ordered a halt to all lead production in Jieshou City and Taihe County. Processing plants were to be relocated to a specialized industrial park, and new technology was to replace crude processing plants. Agricultural land in proximity to the old industries was to be abandoned, and farmers were to be required to have regular health examinations. However, a more recent 2006 report from the same news agency suggested that despite the official order, Jieshou City has continued to process 160,000 tons of lead each year in unregulated plants—mostly from recycled lead-acid batteries.

Some reports state that the Tianying Recycling Economic Park was eventually constructed, and more than 40 companies (mainly related to processing regenerated lead) are said to now operate within the park. The industrial park is reputedly one of the "national pilot recycling economy parks" and is China's largest industrial base for battery recycling and processing. In 2007, Xinhua News Agency praised the economic performance of Tianying for recycling 350,000 tons of batteries and producing 170,000 tons of lead, contributing 26 percent of the local fiscal revenue. Compounding this problem of seemingly contradictory intentions (i.e., growth versus environmental protection), the disposal of Tinaying's waste is still mostly unregulated.

Environmental Injustice in China?

Though government and the media have applauded local business efforts to "protect the environment," data are unavailable to substantiate claims of environmental remediation in Tianying. Despite the rhetoric, little progress appears to have been made in redressing chronic pollution problems. Some scholars have recently observed that disparities in the distribution of pollution throughout China—based on location, ethnicity, and income levels—suggest that environmental injustice appears to be a significant problem in China.

See Also: Air Quality; Cities for Climate Protection; Environmental Planning; Green Communities and Neighborhood Planning; Healthy Cities; LEED (Leadership in Energy and Environmental Design); Linfen, China; Renewable Energy; Sustainable Development.

Further Readings

Blacksmith Institute. *The World's Most Polluted Places: The Top Ten.* New York: Blacksmith Institute, 2006.

Chen, H. Y., et al. "The Lead and Lead-Acid Battery Industries During 2002 and 2007 in China." *Journal of Power Sources*, 191 (2009).

Economy, E. C. "The Great Leap Backward?" *Foreign Affairs*, 86 (2007).

Liu, L. "Urban Environmental Performance in China: A Sustainability Divide?" *Sustainable Development*, 17 (2009).

Ma, L. "From China's Urban Social Space to Social and Environmental Justice." *Eurasian Geography and Economics*, 48 (2007).

Palmer, M. "Towards a Greener China? Accessing Environmental Justice in the People's Republic of China." In *Access to Environmental Justice: A Comparative Study*, A. Harding, ed. Leiden, Netherlands: Brill, 2007.

World Bank. "China Quick Facts." http://web.worldbank.org/WBSITE/EXTERNAL/ COUNTRIES/EASTASIAPACIFICEXT/CHINAEXTN/0,,contentMDK:20680895~ menuPK:318976~pagePK:141137~piPK:141127~theSitePK:318950,00.html (Accessed September 2009).

Zhang, Z. X. "China Is Moving Away From the Pattern of 'Develop First and Then Treat the Pollution.'" *Energy Policy*, 35 (2007).

Zhu, Z. P. "Children's Blood Lead Contamination and the Relationship Between Gender and Age." *Modern Journal of Integrated Traditional Chinese and Western Medicine*, 18 (2009).

Jason Byrne
Griffith University

Liu Li Can
International Centre of Communication Development

Li Xiujuan
Beihai College, Beihang University

TRAFFIC CALMING

Traffic calming reduces the speed and volume of motorized traffic by means of physical alterations to the street layout, street closures, speed limits, and enforcement of traffic laws. Reductions in traffic speed and volume diminish the negative effects of motor vehicles on neighborhoods and the environment. This can improve neighborhood quality of life, promote neighborhood social activity, and enhance mobility options for vulnerable groups. Moreover, having slower and fewer motor vehicles increases traffic safety—especially for pedestrians and cyclists. In addition, less traffic accidents and more walking and cycling can increase public health. Some studies also show increases in the value of adjacent property and a reduction in the need for police enforcement. Common traffic calming measures are street closures or semiclosures,

Traffic calming measures include street closures or partial-closures, speed bumps, raised crosswalks, zigzag road layouts, bulb-outs, and road narrowings, as seen here in Worcestershire, England.

Source: Benkid77/Wikipedia

speed humps, speed tables, and raised crosswalks. Road narrowings, zig-zag road layouts, and bulb-outs—forcing motorists to turn at low speeds—are also common. Most traffic calming measures make car traffic slower and less attractive and, combined with the right policies, can increase the attractiveness of walking and cycling.

Traffic calming has its origins in the Netherlands. In the late 1960s and 1970s, residents of Dutch municipalities responded to increasing cut-through traffic in their neighborhoods. They used tables, benches, sand boxes, and physical streetscape alterations to create an obstacle course for automobiles and to claim the street as living space. This newly designed type of public space was called *Woonerf*—which literally translates into English as "residential"' or "living yard"—today more commonly known as "home zone." Indeed, home zones turn streets from traffic thoroughfares into outdoor living spaces for all. Physical alterations to the streetscape give priority to pedestrians, children at play, and bicyclists over car traffic. Moreover, maximum allowable automobile travel speeds are sharply reduced—often as low as walking speed, or roughly 4 mph. In the 1970s, the home zone concept spread throughout many Dutch municipalities, leading to official adoption by the Dutch national government in the mid-1970s.

In the late 1970s, the concept also spread to other European countries and to North America. In each country the concept was adapted differently. For example, German municipalities considered the extensive alterations to the street surface in home zones as too expensive to be applied to larger areas. Thus, traffic-calmed neighborhoods in Germany include only minor alterations of the streetscape, but restrict car travel speeds to 19 mph (so-called Zone 30 areas). Since the late 1970s, most German municipalities have traffic calmed nearly all of their neighborhood streets. In large cities, such as Berlin or Munich, between 70 and 80 percent of roads have been traffic calmed. Home zones also exist in Germany but are more limited in scope than Zone 30 traffic-calming schemes.

Today, traffic calming in Northern Europe is usually area-wide and not for isolated streets. This ensures that through traffic is displaced to arterial roads and not simply shifted from one residential street to another. Denmark, Norway, and Germany (among others) also successfully applied traffic calming to nonresidential arterials and other major roads. Related to traffic calming, almost every northern European city has created extensive car-free zones in their centers—the ultimate form of traffic calming.

In the United States, Seattle, Washington; Berkeley, California; and Eugene, Oregon, were among the first adopters of citywide traffic calming concepts in the 1970s. Don Appleyard's book *Livable Streets*—first published in 1981—further popularized the idea of urban streets as living spaces instead of automobile thoroughfares. In comparison with its success in Europe, traffic calming in the United States spread slowly. The Institute of Transportation Engineers, under leadership of planning professor Reid Ewing, surveyed the state of traffic calming in the United States in the late 1990s. The Institute of Transportation Engineers provides an extensive overview of the state of practice, affects of traffic calming, and legal considerations.

Physical Measures

Physical traffic calming measures can be divided into volume and speed controls. Volume controls commonly include full and half closures of streets, diagonal diverters, median barriers, and turn islands forcing motorists to turn and restricting access to neighborhood streets. Speed control measures can be further categorized into vertical and horizontal applications. Speed humps, speed tables, raised crosswalks, and raised intersections require automobiles to slow down before traversing these physical obstacles. Traffic circles, chicanes, chokers, narrowings, and realigned intersections require motorists to drive a zigzag course, and thus reduce travel speeds. Parked cars can also be used to narrow streets and reduce traffic speeds. Moreover, changes to the texture of the pavement, such as cobblestones, help to slow traffic.

Evidence of the effect of traffic calming is generally more positive in Europe than in the United States. This might be related to the more widespread application of traffic calming in European countries. Moreover, in contrast to Europe, traffic calming in the United States is rarely area-wide and is often limited to isolated blocks. Nonetheless, U.S. studies also show that traffic calming measures can reduce travel speeds by up to 23 percent and lower traffic volumes by more than 30 percent. According to a survey of traffic-calming impact studies, speed humps seem to have the largest effect on reducing traffic volume and speed. Full street closures, such as pedestrian zones, obviously eliminate all traffic.

European evidence suggests that traffic calming can significantly reduce traffic collisions, injuries, and fatalities—particularly for pedestrians, but also for cyclists. Evidence collected within the United States shows a weaker effect. Researchers speculate that traffic calming in the United States has mainly been applied to streets with low traffic volumes and relatively few traffic collisions. Thus, the measured improvements in traffic safety were small compared with European traffic calming schemes on higher-volume streets with more collisions.

Virtually all European studies report significant increases in overall levels of walking and cycling after the implementation of traffic calming. The evidence is overwhelming that traffic calming enhances both pedestrian and cyclist safety by reducing speeds on secondary roads. A recent study finds that the United States is not following Europe's lead in combining pedestrian- and cycling-enhancing schemes with traffic calming. For example, many European traffic calming schemes include cut-throughs for pedestrians and bicycles at the end of cul-de-sacs, or special speed humps that allow bicycles to pass on either side. The main focus of traffic calming in the United States is still on reducing and slowing car traffic. Only a few cities have attempted to explicitly combine traffic calming with programs that increase walking and cycling.

According to the National Household Travel Survey, roughly 30 percent of all trips made in the United States are shorter than one mile. However, Americans choose to drive for two-thirds of these short trips, compared with less than 30 percent in European countries. Thus, traffic calming schemes, combined with programs that enhance walking and cycling, have the potential to increase the share of short trips made by these nonmotorized modes. Traffic calming can reduce car travel speeds or restrict access by automobile and improve the safety, perceived safety, and attractiveness of walking and cycling. Both would make car use slower, more difficult, and less attractive, while improving conditions for walking and cycling—which virtually do not cause any greenhouse gas emissions and provide important physical exercise. Thus, traffic calming also has the potential to improve public health via increased levels of physical activity.

See Also: Bicycling; Ecological Footprint; Walkability (Pedestrian-Friendly Streets).

Further Readings

Appleyard, Donald. *Livable Streets*. Berkeley: University of California Press, 1981.

Ewing, Reid. *Traffic Calming: State of the Practice*. Washington, D.C.: Institute of Transportation Engineers/Federal Highway Administration, 1999.

Ewing, Reid. "Traffic Calming in the United States: Are We Following Europe's Lead?" *Urban Design International,* 13/2 (2008).

Hass-Klau, Carmen. *The Pedestrian and City Traffic.* Hoboken, NJ: John Wiley & Sons, 1993.

Herrstedt, Lene. "Traffic Calming Design—A Speed Management Method: Danish Experiences on Environmentally Adapted Through Roads." *Accident Analysis and Prevention,* 24 (1992).

Pharoah, Tim. "Traffic Calming in West Europe." *Planning Practice & Research,* 8/1 (1993).

Tolley, Rodney. *Sustainable Transport: Planning for Walking and Cycling in Urban Environments.* Cambridge, UK: Woodhead Publishing, 2003.

Transport for London. *Impacts of 20 mph Zones in London Boroughs.* London: Transport for London, 2003. http://www.tfl.gov.uk/assets/downloads/ResearchSummaryNo2_20mph Zones.pdf (Accessed December 2009).

Ralph Buehler
Virginia Tech University

Transit

As the first decade of the 21st century comes to a close, demand for urban mobility continues to grow worldwide. The economic, environmental and social costs of transportation are immense. The transport sector is responsible for more than 60 percent of world oil consumption. Mounting negative externalities associated with traditional transportation systems are particularly driven by growing dependence on private vehicles, among the most inefficient and costly ways of getting around. However, there are plenty of alternatives for meeting consumers' growing mobility needs. Public transit and transit-oriented development provide a viable path toward greener cities.

A host of factors, including population growth, urbanization, fast economic expansion, and changing lifestyles all contribute to the growing demand for mobility. Virtually all types of motorized transportation produce some type of pollution, but in most countries fossil-fuel based private automobiles are the biggest contributors to emissions growth. The three most dangerous pollutants produced by gasoline-powered vehicles include carbon monoxide, hydrocarbons (HCS), and nitrogen oxides (NO_X); all augment the formation of dangerous photochemical smog. Growing private vehicle use is offsetting recent progress made in reducing other major air pollutants, such as particulates (TSPs) and sulfur dioxide (SO_2).

The health costs of air pollution are immense. A recent World Health Organization (2002) study estimates that more than 800,000 deaths, 6.4 million years of lost life, and tens of millions of adverse health outcomes are directly attributable to urban air pollution. The growth of transport related air pollution is especially acute in developing countries like China and India, where private vehicle use is skyrocketing. In 2009, China's private vehicle sales grew almost 50 percent, making that country's auto market the biggest in the world.

Well-designed public transit systems offer a healthier and more economically and socially viable approach to satisfying growing mobility needs. High capacity public transit options, such as trolley buses, rail (including subways, light rail, monorails, heavy rail, and commuter rail), and ferries all produce less pollution and use less energy per person-km traveled than private automobiles. In many cases the savings are dramatic. For instance, regularly riding the subway to work in New York City or London instead of driving can reduce fossil fuel energy use, greenhouse gas, and smog forming air emissions by up to 90 percent. Public transit infrastructure requires less space and financial resources than building roads, parking lots, and gas stations. Also, public transit is cheaper for end users.

Given the numerous negative externalities associated with widespread adoption of the car culture development paradigm, many planners are no longer content to view urban public transit as an "add-on" to development planning. Development agencies, elected officials, citizen groups, and even private firms increasingly employ "triple bottom line" accounting to evaluate the broad effects of transit alternatives on community development. These methodologies integrate financial, social, and environmental objectives into the overall planning process, and transit-oriented development is almost invariably favored as the best approach to balance competing human needs for mobility, economic growth and sustainable development.

Numerous studies have found that transit-oriented development stimulates economic growth. Transit-oriented development favors creation of higher density communities, which promotes more efficient use of land, labor, and capital resources. Land normally dedicated to accommodating roads, bridges and parking lots can be used to expand public spaces; for example, parks, town squares, and community centers. Consumers increasingly choose to live in transit-oriented communities, which positively impacts property values and business activity. Higher density communities provide more amenities within close walking distance to one another. Also, public transit commuters spend less money on mobility.

Most public transit options generate fewer environmental externalities than private automobiles. Public transit technologies typically use less energy per person-km traveled and often allow switching to cleaner energy inputs. While many auto enthusiasts tout the rise of electric, biofuel and hydrogen cars, less than 5 percent of private vehicles on the road in 2010 employed alternative fuels as a primary propulsion source. In contrast, most modes of urban mass transit allow use of electricity (which can be produced by a variety of fuel sources) as a primary energy input. Emissions from dangerous particulates, carbon monoxide, ozone, and volatile organic compounds are also greatly reduced. Transit-oriented development also permits preservation of natural ecosystems and agricultural land, and increasing the availability of farmland encourages local food production and reduces the environmental footprint of food supply chains.

Transit-oriented development can play a critical role in building healthy, cohesive, civically active and socially equitable communities. Numerous studies have shown that people who regularly use transit also walk and bike more. In most cases, public transit is safer to use than private mobility options. Transit users, no longer isolated in their cars, are more likely to engage in the civic life of their neighborhoods. Trains, buses and subways provide more convenient access to recreational and shopping opportunities. Transit-oriented development promotes social justice and equality by providing greater accessibility for physically, financially, or socially disadvantaged groups to basic government, health, and educational services.

Public transit and transit-oriented development stimulate economic growth, preserve scarce resources, reduce environmental damage, and promote the overall health and well-being of society. Moving forward, it is critical that urban planners employ transit-oriented development and other sustainable practices to serve our current needs, as well as those of future generations.

See Also: Compact Development (New Urbanism); Density; Intermodal Transportation; Sustainable Development; Transit-Oriented Development.

Further Readings

American Public Transportation Association. *Public Transportation: Benefits for the 21st Century*. Washington, D.C.: American Public Transportation Association. http://www.atpa.com (Accessed January 2009).

Dittmar, Hank and Gloria Ohland, eds. *The New Transit Town: Best Practices in Transit- Oriented Development*. Washington, D.C.: Island Press, 2003.

Goilias, Konstadinos, ed. *Transportation Systems Planning: Methods and Applications*. Boca Raton, FL: CRC Press, 2002.

Grava, Sigurd. *Urban Transportation Systems: Choices for Communities*. New York: McGraw-Hill, 2003.

Newman, Peter and Jeffrey Kenworthy. *Sustainability and Cities: Overcoming Automobile Dependence*. Washington, D.C.: Island Press, 2009.

Jonathan Harrington
Troy University

TRANSIT-ORIENTED DEVELOPMENT

Transit-oriented development (TOD) is an urban development strategy that is based on developing urban areas that are close to mass transit facilities. TOD usually occurs in relatively high-density, mixed commercial and residential neighborhoods with rail or bus stations located conveniently nearby. The concept has been embraced as an aspect of regional urban planning that has the potential to revitalize urban areas by making them more walkable and livable. New Urbanists, Smart Growth proponents, and many other urbanists have adopted the idea and see the development of "new transit towns" as a powerful form of urban development. Those scholars are also conscious that it is an evolving approach with many problems. The possibility of gentrification is perhaps the most serious of those problems.

Hank Dittmar and Gloria Ohland edited an invaluable volume on best practices in the first generation of TOD. In introducing the work,

Portland, Oregon, has long been an exemplar of responsible urban development because it employs a strategy based on developing areas close to mass transit facilities.

Source: iStockphoto

Dittmar, Dena Belzer, and Gerald Autler argue that three trends enhance the promise and appeal of TOD. First is the resurgence of investment in downtowns. Underlying this development is an urban rebirth that led more people to move into cities as those cities became increasingly attractive as places to live, thus spurring further demand for homes and amenities. The second trend is the continuing growth of suburbs, many of which have become small cities. Such suburbs often house those who work in larger nearby cities. Residents of these suburban cities often wish for urban amenities nearby. Finally, there is renewed interest in rail travel and investment and a realization of the environmental costs of overreliance on cars. The trends converge to create the "potential for a substantial market for a new form of walkable, mixed-used urban development around new and existing rail or rapid bus stations."

Dittmar and his coauthors suggest that although TOD can improve the quality of life, reduce household transportation costs, and provide stable mixed-income neighborhoods that reduce environmental impacts and have the potential to lessen traffic congestion, that potential is not being met. Overhyped promises often outstrip the reality of TOD projects. This has led to much criticism of the approach. In fact, TOD is not a panacea that will by itself solve problems of urban decline, sustainability, gentrification, and urban sprawl. When combined with other approaches in a broader context of urban development and participatory decision making, however, TOD can be an important part of strategies to improve urban life and to foster urban development.

Peter Calthorpe, one of the most influential urbanists in the development of TOD ideas, early on set out seven principles of TOD-oriented urban design:

1. Regional basis for growth

2. Important destinations within easy walking distance of transit stops

3. Pedestrian-friendly street networks

4. Mix of housing types, densities, and costs

5. Preserve sensitive habitats

6. Orient buildings and neighborhood activities toward public spaces

7. Encourage redevelopment along transit corridors within existing neighborhoods

Although Calthorpe did not complete many of the projects he planned, Portland, Oregon, stands as an impressive achievement by him and his collaborators.

Another major figure in the development of TOD is Peter Cervero, a professor of urban planning. Cervero has emphasized that the types of transit chosen for particular areas should fit the existing urban form. Unfortunately, Dittmar and his collaborators suggest that most actual TOD projects do not live up to the ideas of Calthorpe, Cervero, and other influential figures. Reflecting this situation, the Brookings Center on Urban and Metropolitan Policy and the Great American Station Foundation issued a report in 2002 suggesting that TOD needed to move from "rhetoric to reality" and that most so-called TOD projects were actually just "transit-related." Nonetheless, there are promising examples of the approach, internationally and in the United States.

Curitiba, Brazil, which grew rapidly from 1950 on, used innovative urban planning techniques centered on new transportation policies and choices to enhance the quality of urban life. Unlike most similar rapidly growing urban areas, the city was not designed around the automobile but, rather, on a bus-based public transport system. Although there are many other aspects to urban planning in Curitiba, the emphasis on public transport is a key feature of its development. Some critics suggest that its accomplishments are over-blown and that it relies too heavily on buses, rather than on light rail transport. Nonetheless, Curitiba is an important case, and one worth careful study.

In addition to Portland, long seen as an exemplar of responsible urban development and TOD, there are numerous other examples of TOD projects in the United States. In Arlington County in northern Virginia, across the Potomac River from Washington, D.C., is the Rosslyn–Ballston Corridor. This is a long-term project on a heavily developed, three-mile-long corridor that is tied to the Washington Metropolitan Area Transit Agency. In San Diego, the Barrio Logan's Mercado Project is one of the "City of Villages" developments in that metropolitan area. The Barrio Logan is an older neighborhood in San Diego's urban city that is close to trolley and bus facilities. Located in the Latino center of the city, the project is a hopeful sign that TOD can help revitalize urban-core areas, although it took years to obtain funding for it. More recently, Sacramento's TOD Plan has been hailed as a "model for the nation." Characterized by walkable urban, transit-oriented villages, the plan holds out promise for more areas to move away from automobile-dependent, sprawl types of suburban development.

See Also: Bus Rapid Transit; Citizen Participation; City Politics; Commuting; Curitiba, Brazil; Environmental Justice; Environmental Planning; Green Communities and Neighborhood Planning; Parks, Greenways, and Open Space; Smart Growth.

Further Readings

Belzer, Dena and Gerard Autler. "Transit Oriented Development: Moving From Rhetoric to Reality." Brookings Institution Center on Urban and Metropolitan Policy and the Great American Station Foundation. June 2002. http://www.brookings.edureports/2002/06cities_dena-belzer-and-gerald-autler.aspx (Accessed August 2009).

Calthorpe, Peter. *The Next Urban Metropolis,* 3rd Ed. Princeton, NJ: Princeton Architectural Press, 1995.

Dittmar, Hank and Gloria Ohland, eds. *The New Transit Town: Best Practices in Transit-Oriented Development.* Washington, D.C.: Island Press, 2004.

Leinberger, Christopher B. "Sacramento's Transit-Oriented Development Plan a Model for the Nation." *Sacramento Bee* (March 18, 2009).

Rabinovitch, Jonas and Josef Leitmann. "Urban Planning in Curitiba." *Scientific American,* 27/3 (March 1996).

Walter F. Carroll
Bridgewater State College

Transport Demand Management

Transport demand management (TDM), otherwise known as mobility management (especially in Europe), is an umbrella term that refers to a range of actions (e.g., policies, programs, and projects) that modify the demand for transport infrastructure and services in either space or time to improve the efficiency of the transport system. The term comes from the contradistinction with actions that respond to increases in the demand for transport by adding to transport infrastructure or capacity (i.e., supply management), such as extending highways, expanding airports, or increasing parking space.

The origins of TDM can be traced back to the 1970s, when the oil crises of 1973 and 1979 prompted various attempts in the United States to increase the fuel efficiency of transport and reduce oil dependence. TDM was also seen as a way of addressing other economic and environmental policy issues of the time, such as reducing traffic congestion (and thereby helping to increase economic productivity) and improving air quality, which had become increasingly important in the United States in the 1960s. Addressing air pollution was of particular concern because of growing emissions from vehicles and the incidence of some serious pollution episodes in U.S. cities in the 1960s. In New York City, for example, pollution inversions in 1963 and 1966 were held responsible for several hundred deaths.

TDM encompasses a wide range of measures to influence whether, when, where, and how often people travel. TDM measures are mostly related to urban passenger travel, rather than to intercity travel or freight transport, although the concept is also applicable to these policy areas. Although building additional transport infrastructure or capacity can technically be categorized as supply management, rather than demand management, increasing the capacity of public transport infrastructure or infrastructure for "green modes" (e.g., walking, cycling) is generally considered to be a form of TDM.

There are various ways of categorizing the main types of TDM measures. The grouping of TDM measures here represents just one example of these categorizations:

Education, Promotion, and Outreach Measures: These measures can be used to raise awareness, improve understanding, and build positive attitudes about alternative transport choices. Examples include personalized travel assistance, cycle training, advertising (including sponsorship) and education campaigns to promote specific transport modes, and promotion events for public transport or green modes (e.g., car-free day, cycle to work week).

Regulatory Measures: This group of measures can be used to influence the relative accessibility of different modes of transport. They can be employed either to increase the accessibility of certain types of transport (e.g., bus lanes; bus-only access restrictions; high-occupancy-vehicle lanes) or to decrease the accessibility of other modes (e.g., access restrictions for cars). These measures can be applied either indefinitely or just at certain times of the day (e.g., peak periods), week (e.g., weekdays), or year (e.g., summer). Other examples include priority signaling for cyclists, pedestrians, or public transport, which can help to shorten journey times and increase the incentives for traveling by these modes.

Economic Measures: These measures can be used to make particular transport choices more financially attractive than others. A range of different measures, both push and pull measures (i.e., disincentives and incentives, respectively), are available for this purpose. Examples include fuel taxes, road user charges, parking charges, tax incentives (e.g., for commuting by public transport), and subsidies (e.g., bicycle purchase subsidies for employees).

Land Use Measures: This group of measures can be used to influence the density, diversity, or design of urban development. These characteristics of urban development are important in shaping local levels of accessibility and, as a consequence, can influence travel distance and modal choice. Specific examples of land use measures include street layout patterns (which influence an area's "permeability" for different modes of transport), park-and-ride facilities (which provide an alternative to car journeys), car parking location and capacity (which influences the number of vehicles with access to an area).

Traveler Information Measures: These measures provide travelers with information about how to reach their destination by different modes (and combinations of modes), the travel time involved, and the costs and, sometimes, environmental impacts of these options (e.g., energy consumption, carbon dioxide emissions). These measures can also be linked to real-time travel information to provide travelers with suggestions for alternative routes, modes, or travel times to avoid things like congestion, road closures, or tolls. These measures include online journey planning, real-time roadside travel information, traffic condition broadcasts (by radio, SMS, or e-mail), and "countdown" systems for public transport, indicating waiting time.

Information and Communications Technology Substitution Measures: This group of measures includes a range of electronic applications (mainly Internet-based) that allow communication or transactions to take place remotely, which thereby reduces or eliminates the need to travel. Examples include teleworking, Internet banking, videoconferencing, and e-government. It should be noted that the contribution of these measures to transport demand reduction is controversial. There are ongoing debates about whether some of these measures (e.g., teleworking or videoconferencing) really do lead to less travel or, in fact, actually increase overall demand for travel.

Modal Integration Measures: These measures provide ways of reducing the barriers (both physical and economic) to multimodal journeys (i.e., journeys involving combinations of modes or transport operators) to provide viable alternative options to car journeys. Examples of measures include integrated public transport tickets, coordinated public transport timetables (coordinated between different operators and different modes such as rail, bus, tram, and ferry), cycle carriage facilities on public transport, and multimodal travel planning tools (e.g., online journey planning).

Organizational Measures: These include measures to create or support organizations, organizational structures, or organizational procedures that promote more efficient transport operation. One example of these measures is a car sharing or carpooling scheme (a commonly encountered measure in TDM strategies), which provides a way of increasing the occupancy of vehicles, and hence fuel efficiency, per passenger. A second example is flexible working time arrangements, which provides employees with a way of avoiding peak congestion and/or being able to combine their work journey with other journeys (e.g., doing the shopping on the way to or from work), thereby saving fuel and/or reducing travel distance.

TDM measures can be applied at the national, regional, or local administrative level. In addition, TDM measures can be site specific. In the latter case, TDM measures focus on managing travel demand at a certain location (e.g., office development, school, concert venue, sports arena, or hospital). In some countries (e.g., Germany, Ireland, Sweden, the United Kingdom), approval of planning permission for certain types of development is contingent on preparing a mobility plan that often contains a package of TDM measures.

It is widely accepted that managing travel demand is most effectively achieved by means of combinations (or packages) of several different types of TDM measures, rather than single measures, to create new incentives and disincentives to change travel behavior. A mixture of "push," "pull," "hard," and "soft" measures are usually required. Push measures are disincentives to travel; pull measures are incentives. Hard measures are associated with compulsion, such as traffic regulations and road taxes; soft measures involve making voluntary choices, such as choosing an alternative route or mode of transport. Although pull measures are generally considered to be more acceptable than push measures, more effect on travel demand can often be achieved by means of push measures, which also have the advantage of often being less costly. The example of London illustrates how push, pull, hard, and soft measures have been combined to create a more acceptable but also effective package of TDM measures. Alongside the introduction of the congestion charge (economic measure; push; hard), parking regulations were tightened (regulatory measure; push; hard), public transport capacity was increased (pull; soft measure), traffic signals were coordinated (regulatory measure; pull; hard), and travel information systems at main bus stops were provided (traveler information measure; pull; soft).

TDM gained prominence because it was seen as a cheaper and more effective way of dealing with increases in motorization than building additional transport infrastructure (e.g., roads, parking space). The expansion of existing transport infrastructure, especially in urban areas, was facing increasing difficulties as a result of the high price and limited availability of land for new infrastructure, the financial constraints of public authorities responsible for providing the infrastructure, and the additional "external costs" on society and on the environment associated with the provision of new infrastructure (e.g., noise, disturbance, air pollution, community severance). TDM measures gained appeal as supply-based responses to the growth in transport demand seemed to get more problematic and/or expensive. However, TDM was not without its opponents. There were (and still are) objections that these measures restrict individual freedoms (primarily the freedoms of individuals with a car) and that exaggerated claims were being made about what sort of impacts TDM can really deliver. Others suggested that politicians adopted TDM as a political smokescreen: claiming that action was being taken to reduce transport demand without actually expending much political or financial capital.

As well as promoting efficiency in the transport system, a variety of claims are made about the wider social, economic, and environmental benefits of TDM, such as:

- Promoting transport choice, particularly for nondrivers
- Reducing the environmental impacts of transport
- Increasing the energy efficiency of transport
- Improving the efficiency of land use
- Reducing public spending demands for transport
- Improving public health by increasing walking and cycling
- Reducing congestion and increasing economic productivity

Ultimately, the effectiveness of TDM measures very much depends on the relationship between "travel changing" incentives and disincentives and the propensity of travelers in a particular market to respond. Although there is clear awareness among the majority of travelers about the environmental impact of their actions, their willingness to change behavior is often low as a result of a complex mixture of individual and social interests. Different groups or segments of the population have different needs, and their readiness to change their travel behavior can differ substantially. Although travelers do generally consider distance, time, and cost in making travel decisions, these variables are often measured on a perceptual basis, rather than an absolute basis. Thus, travel decisions are not perfectly rational. Moreover, car drivers are often strongly motivated by their desire for independence and freedom when making travel decisions. The more travelers choose to use the car, the more habitual their car use becomes, and the less susceptible they are to incentives to change mode. As a result, trends in increasing car ownership and use may be very difficult to break without an increasingly effective set of TDM measures.

Interest in TDM continued to grow in the 1990s, particularly in Europe, when the underlying philosophy of transport and infrastructure policy experienced shifts in many countries, moving from a predominantly supply-based "predict and provide" approach toward a "new realism," in which there was greater recognition that, even if new infrastructure is built to accommodate the expected increases in transport demand, levels of traffic, congestion, and pollution can still worsen because there is a substantial latent demand for transport. In other words, new transport infrastructure can lead to further increases in the demand for transport.

See Also: Air Quality; Carpooling; Commuting; Compact Development (New Urbanism); Congestion Pricing; Density; Energy Efficiency; Infrastructure; Intermodal Transportation; London, England; Personal Rapid Transit; Traffic Calming; Transit; Transit-Oriented Development; Walkability (Pedestrian-Friendly Streets).

Further Readings

Ferguson, Erik. "The Evolution of Travel Demand Management." *Transportation Quarterly,* 53/2:57–78 (1999).

Gifford, Jonathan L. and Odd J. Stalebrink. "Transportation Demand Management." In *Handbooks in Transport—Handbook 3: Transport Systems and Traffic Control,* Kenneth Button and David Hensher, eds. Amsterdam: Elsevier Science, 1999.

Meyer, Michael D. "Demand Management as an Element of Transportation Policy: Using Carrots and Sticks to Influence Travel Behavior." *Transportation Research Part A: Policy and Practice,* 33/7–8:575–99 (1999).

Victoria Transport Policy Institute. "Online TDM Encyclopedia." http://www.vtpi.org/tdm (Accessed October 2009).

Dominic Stead
Delft University of Technology

Yusak Octavius Susilo
University of the West of England

Universal Design

Universal design, a term coined by architect and designer Ron Mace, is the design of products and services that are accessible and functional for as many persons as possible, regardless of their age or ability, without the need for adaptation, or if needed, easily adaptable according to different individual requirements. Universal design has a positive effect on society as a whole because everyone—not only persons with a disability—can benefit from it. It meets the needs of persons with different abilities, which means it offers better conditions to all users of a product, service, building, transit system, or urban space. This will be even more important in the future as a result of the increase in life expectancy, which allows more people to live longer, increasing the number of persons with some form of impairment who will benefit directly from universal design.

The concept of universal design or "design for all" was initially associated with the disability rights movement. It is now a more comprehensive perspective referring not only to products and services but also to buildings, infrastructures, transport systems, or built environment. It is based on a set of principles defined and developed over the years by different institutions and international organizations that have adopted guidelines and standards to address the specific needs of children, the elderly, and persons with a disability (e.g., the UN Convention on the rights of persons with a disability, Council of Europe declarations, European Union directives, World Bank guidelines for projects financed by the institution, etc.). At the national level there are also examples of a growing body of legislation, technical guidelines, and action plans that aim to implement universal design principles in different fields (e.g., public buildings and infrastructures, transportation systems, housing, etc.).

A group of experts in the field of universal design, based at North Carolina State University's Center for Universal Design, defined a set of seven principles that should be used in the development of products, services, and environments. These principles are equitable use (must be useful to people with diverse abilities), flexibility in use (should accommodate a wide range of individual preferences and abilities), simple and intuitive use (must be easy to understand), perceptible information (have to communicate necessary information effectively to the user), tolerance for error (should minimize hazards and adverse consequences), low physical effort (can be used with a minimum of fatigue), and size and space for approach and use (must provide appropriate size and space). These principles have been applied in numerous situations including in urban planning and urban design. They are intended to be a guide or a reference framework, leaving it for designers in each discipline and urban planners to decide how exactly they will be included in each particular situation. In some fields, different authors and research groups have developed design guidelines for each of these seven principles, as well as performance measures in the form of surveys to determine how well products, services, and environments designs meet these principles.

In urban planning, universal design—or design for all—is increasingly adopted as a reference by planners and planning authorities to improve accessibility for all, which is a positive development compared with the more limited concept of accessibility for persons with a disability, adopted years before. When applied to the built environment, it means that urban public space, buildings, public transport, and pedestrian infrastructures are designed and built to be accessible and usable by as many people as possible in all periods of their lives. It means to plan and design flexible and inclusive buildings, public spaces, and urban infrastructures for people with different preferences and abilities. In this process, the participation of people with disabilities, the elderly, and children should be stimulated, as they can see barriers where other persons will not see them.

New Urbanism, a relatively recent new paradigm in urban planning, incorporates the principles of universal design, although some of the neighborhoods developed under its label do not always fully apply these values. New Urbanism favors higher densities, mixed uses, pedestrian-friendly neighborhoods, non-car-dependent communities, public transit and other nonmotorized transports, and public spaces more favorable to social interaction and more secure, in contrast with what was common practice under the principles of Modern Urbanism and its Charter of Athens. New Urbanism has been responsible for the creation of neighborhoods with public space that is safer for children, for the elderly, for persons with a disability, and for all those temporarily or permanently unable to walk or to drive. Good practice in this field also means building transport systems that meet the needs of wheelchair users and other mobility-impaired persons, including proper signage to tackle the needs of persons who are deaf or blind. An urban environment that is accessible and safe benefits everyone, regardless of her or his ability.

In sum, the end of the 20th century and the beginning of the 21st century saw a move from a situation in which national governments and municipalities were trying to enhance accessibility for persons with disabilities within the broad framework of social policy to a new perspective that is unequivocally committed to applying universal design principles to all facets of human life, including in urban planning. With an increasingly aging population, it makes economic and social sense to exceed now the minimum levels of accessibility set by current national legislation, and therefore planning authorities and planners should go beyond the minimum requirements set by law in each country and innovate with new and more accessible solutions for as many people as possible.

See Also: Agenda 21; Compact Development (New Urbanism.); Smart Growth; Sustainable Development; Transit-Oriented Development; Walkability (Pedestrian-Friendly Streets).

Further Readings

Afacan, Yasemin and Cigdem Erbug. "An Interdisciplinary Heuristic Evaluation Method for Universal Building Design." *Applied Ergonomics*, 40:731–44 (2009).

Beecher, Valerie and Victor Paquet. "Survey Instrument for the Universal Design of Consumer Products." *Applied Ergonomics*, 36:363–72 (2005).

Bittermana, Alex and Daniel Baldwin Hess. "Bus Rapid Transit Identity Meets Universal Design." *Disability & Society*, 23/5:445–59 (2008).

Danford, Gary Scott. "Universal Design: People With Vision, Hearing, and Mobility Impairments Evaluate a Model Building." *Generations*, 27/1:91–94 (2003).

Demirbilek, Oya and Halime Demirkan. "Universal Product Design Involving Elderly Users: A Participatory Design Model." *Applied Ergonomics*, 35:361–70 (2004).

Grant, Jill. *Planning the Good Community: New Urbanism in Theory and Practice*. London: Routledge, 2006.

Iwarsson, S. and A. Stahl. "Accessibility, Usability and Universal Design: Positioning and Definition of Concepts Describing Person-Environment Relationships." *Disability and Rehabilitation*, 25/2:57–66 (2003).

Saito, Yoko. "Awareness of Universal Design Among Facility Managers in Japan and the United States." *Automation in Construction*, 15: 462–78 (2006).

Sandhu, Jim S. "Citizenship and Universal Design." *Aging International* (March 22, 2000).

Skinner, Jon. "Public Places, Universal Spaces: Taking Accessible Design to the Next Level." *Planning* (July 2008).

Story, Molly F., et al. "Completion of Universal Design Performance Measures." *Proceedings of the RESNA 2001 Annual Conference: The AT Odyssey Continues*. Arlington, VA: Rehabilitation Engineering and Assistive Technology Society of North America (RESNA), 2001.

Talen, Emily. *New Urbanism and American Planning: The Conflict of Cultures*. London: Routledge, 2005.

Carlos Nunes Silva
University of Lisbon

Urban Agriculture

Urban agriculture may be found in backyards, like these raised garden beds in Houston, Texas, which also serve as rotating compost bins.

Source: Ron Francis/Natural Resources Conservation Service

Urban agriculture is any agricultural production—including the cultivation of crops and animal husbandry—both within and on the fringe (peri-urban) of a metropolitan area. Urban agriculture is distinguished from rural agriculture in that it is embedded in an urban political ecology and draws on resources from the city to produce resources for the city. The integration of agricultural production into urban life draws on urbanites for labor-power, provides direct food distribution to urban consumers, reduces urban ecological impacts, and uses urban resources such as organic waste, wastewater, and vacant land. Urban agriculture may be found in a variety of locations, including backyard, patio, and rooftop gardens; commercial operations of all sizes; vacant lot cultivation; institutional gardens (e.g., schools, hospitals, and prisons); and community gardens. Popular perceptions often view urban agriculture as a transitory redress for specific social, political, and economic problems, but advocates of urban agriculture insist on its permanence. Site-specific resources (e.g., land and water availability), combined with varying levels of formal support from urban policy makers, give urban agriculture its distinct character. Diverse agricultural practices and purposes are key features in the development, resiliency, and sustainability of urban agriculture. Urban agriculture helps to build food justice through subsistence production, provides income-generating opportunities for the urban poor, and fosters urban sustainability.

Characteristics of Urban Agriculture

Scholars debate the precise definition of urban and peri-urban, but the definitions are always relational, depending on such things as population density and infrastructure development (particularly transit). As a result, "urban agriculture" itself is loosely defined. Within dense city centers, urban agriculture is readily apparent. Urban agriculture is often defined as any agricultural production that occurs on land with other viable land use options, whereas rural agriculture occupies land that is not threatened by other land uses. Peri-urban agriculture takes advantage of land availability on the urban fringe, particularly on the edges of expanding cities, enjoying close proximity to both labor and markets. Urban agriculture also includes various processing, marketing, and service activities and may be conducted by individuals or by groups of people as a for-profit activity, by state officials, and increasingly, by not-for-profit and nongovernmental organizations.

An examination of the defining characteristics of urban agriculture facilitates an understanding of its place-specific formations within individual cities. Urban farms vary by the actors involved, the types of goods produced, the location of projects, the technological scale of the project, and the form of distribution and economic integration.

The urban poor are the primary actors in urban agricultural activities. Urban farmers are often rural immigrants, people of color, and/or women. In some cases, particularly in the global North, professionals from nongovernmental organizations, lower-level state officials, and schoolteachers are also involved. Throughout the world, women constitute an important segment of urban farmers, as women are most often responsible for household reproduction and combine food provisioning, farm work, and related activities (i.e., preparing, processing, and marketing) with other household labor. Urban agriculture does not readily combine with traditional wage labor, which requires long commutes and long hours, but as more people (particularly women) enter the wage relation, urban farmers are beginning to face this difficult reality.

Food production is the focus of much urban agriculture, but nonfood items are also produced for use as clothing, fuel, and medicine, or simply for ornamental purposes. The different items produced reflect the local (cultural and physical) geography to a greater extent than rural agriculture reflects its context. Urban processes are fully evident in urban agriculture; thus, urban agriculture differs from rural agriculture, just as the city differs from the country. For example, urban agricultural products are normally high-valued perishables that directly reflect cultural preferences, moving away from monocrop commodity production (e.g., cereal crops) unsuitable for urban areas.

Urban agriculture takes place on private, public, and quasi-public (e.g., schoolyards) land, on residential property, and on proper commercial farms. The sites of urban agriculture are categorized according to location of farm relative to residence (on- or off-plot), density of urban development (built-up or open-space), land tenure status (owned or leased, authorized or unauthorized), and land use zoning (residential, industrial, agricultural).

The scale and the technology employed in production distinguish urban agriculture from rural agriculture. Urban agriculture may be individual or family run, may be coordinated by groups as cooperatives, and can be commercial projects varying in size from small scale (as is often the case) to (relatively) large scale. For the most part, urban agriculture employs low levels of heavy machinery and new technologies (e.g., solar-powered irrigation). This is not to say that good small-scale urban production is not technological—it takes a great deal of effort and know-how to make urban agriculture productive—but until recently production techniques steered clear of expensive and oversized high technology. In some places in the global north, the tendency is toward more technologically advanced agriculture, including aquaculture, vertical farming, and intensive hydroponic techniques.

Finally, urban agriculture products are distributed in a variety of ways. Produce may be destined for market and/or for self-consumption. Agricultural production for self-consumption is an important function of urban farming, as subsistence production often supplements wages. Self-consumption has historically been a key driver of urban agriculture, particularly in the global south and in poor areas of the north, but market-oriented urban agriculture is an important economic activity and an increasing part of urban agriculture. Products are sold on-farm, through farmers markets, in local stores, in supermarkets, and by farm carts throughout the city. Thanks to geographical proximity, urban agricultural production enjoys closer time–space relations between the field and markets.

Benefits of Urban Agriculture

Urban farming has many noted benefits. Chief among them are subsistence production aimed at achieving food security, reducing urban hunger, and building food justice; income generation and the development of employment opportunities; social benefits, including social integration, recreation, and community health; and overall contributions to a city's ecology and to urban sustainability through the creation of green space, waste reduction, and resource conservation.

Urban agriculture reduces hunger and fosters food justice through an improvement in both the quantity and quality of food available to urbanities. Hunger reduction and food justice are both urban farming's most important benefits and the most intended goal. Urban food production increases city dwellers' access to affordable, fresh produce, thereby eliminating dependency on highly processed food-like substances. With urban agriculture, production of food occurs where it is most needed—in cities. Urban agriculture also facilitates self-sufficiency, as people directly address the problem of food availability through subsistence production. Food is not produced just for self-consumption but also for other segments of the population, made available through direct markets. Often, farmers markets can better supermarket prices through the elimination of intermediaries and the reduction of transportation and storage costs.

In fact, urban agriculture often emerges as a response to urban hunger and poverty. Urban hunger is the result of spatial restructuring of urban foodscapes (the landscape of food) that create built environments lacking adequate access to quality food. The geography of hunger closely correlates to the geography of urban poverty. Within cities, poverty is more directly linked to hunger than in rural areas, precisely because food is normally available only as a commodity. Urban agriculture resists and struggles against the development of these "food deserts."

Urban agriculture generates income and employment opportunities and alleviates poverty. Subsistence food production reduces expenditures on food, which eat up a substantial portion of household budgets. Growing food, particularly expensive fresh produce, saves money; selling food earns much-needed money. Small-scale production (relative to rural agriculture), characteristic of urban agriculture, is labor-intensive. Urban agriculture's small scale increases labor needs, as animal power and mechanization are uncommon. Urban agriculture can thus encourage widespread participation. As such, economic benefits of urban agriculture extend beyond urban farmers to a whole host of auxiliary activities. Agricultural inputs, processing, packaging, and distribution grow alongside urban agriculture, providing economic development opportunities. Everything from the veterinarian needs of urban livestock to the tools used by urban gardeners create employment and generates economic development.

Urban agriculture provides opportunities for social integration and recreation while improving community health. Often, nongovernmental organizations and/or municipal governments initiate urban agriculture projects targeting specific underserved and isolated communities such as women, recent immigrants, hospital patients, prisoners, and the elderly. These projects are designed to socialize people to cultural norms, provide recreation and rehabilitation, and help improve overall community health. Moreover, in the global North, urban agriculture is often a form of recreation and/or used as a tool for education.

Last but not least, urban agriculture is vital to healthy and sustainable urban ecology. Cities exact tremendous environmental harm, both in terms of resource use and waste creation. Urban agriculture tackles both ends of urban metabolism by converting urban waste to productive purposes. Organic waste is a vital resource for agriculture, and urban farmers can take urban waste streams and put them to good use. In addition, urban agriculture can use wastewater for irrigation purposes and can harvest unused rainwater that normally runs off the impervious urban environment. Urban agriculture is also credited with "greening" cities through the cultivation of vacant lots and marginal land. Urban agriculture reduces petrochemical use through changes in production, transportation, packaging, and distribution—all gas-thirsty activities that contribute to climate change.

As urban gardens are located in areas of high population density, urban agriculture works best when local resources—animals, plants, soil, and water—are properly managed. Hence, thoughtful resource management, through agroecological and permaculture approaches, reduces the use of toxic petrochemicals and conserves water. In addition, the small scale and nonspecialized character of urban agriculture facilitates nutrient recycling. Small-scale agriculture is typically diverse in what it produces, providing natural protection against disease and pests. Crop diversity makes possible a cycle of harvests throughout the year, thereby maximizing overall yields.

Challenges to Urban Agriculture

Some general challenges remain to the long-term sustainability of urban agriculture. Urban agriculture everywhere faces similar limitations and concerns, particularly if practiced widely or if in the process of expanding. Challenges include ecological constraints, environmental health concerns, and popular perceptions of urban agriculture.

Ecological constraints are the most pressing challenges facing urban agriculture and include land availability, water shortages, a lack of adequate soil for production, and pests and diseases. Water shortages are particularly problematic for urban agriculture, as quality water is an increasingly scarce (and expensive) resource, and therefore a major concern for municipal governments. Concerns about land availability and the adequacy of soil to maintain crops fall close behind water as a major concern for urban food production.

Urban environmental problems such as contaminated soil and urban air and water pollution pose great problems to urban food production. Lead and other heavy metals are particularly dangerous contaminants within urban environments. Many types of produce, especially green, leafy vegetables, easily transport such heavy metals. Urban pollution problems threaten to undermine urban agriculture and its central goal of food justice. Notably, these environmental problems can be managed. For example, soil remediation practices can address contamination issues, as can simple solutions such as raised beds for crops.

Popular perceptions of urban farming continue to undermine urban agriculture. Urban food production is often viewed as a sign of poverty and evokes the repression of slavery and colonialism in the global south. The historical legacy of colonialism casts a negative view on urban food production. Colonizers historically viewed urban agriculture as "primitive," and such practices were thus shunned. Food production in former colonies tends to remain modeled on the colonial divisions between town and country established during early periods of colonization. Proponents of urban agriculture must convince the public that the practice is both desirable and viable.

Popular perceptions throughout the world are beginning to change as economic crises proliferate and as successful examples of urban agriculture emerge alongside food movements and growing environmental concern. Beyond public perception, a lack of educated workers in agricultural production and the population's overall lack of experience with diversified agriculture remain as obstacles to urban agriculture. However, urban agricultural initiatives have struggled against this challenge, and both popular perceptions and public knowledge regarding urban agriculture are moving steadily forward.

Despite these advances, weak coordination, planning, and policy for urban agriculture remain problematic. Urban agriculture is practiced under a wide variety of urban planning and policy environments that both support and undermine urban farming practices. Planners and policy makers often view urban agriculture as temporary, at best, and, at times, as an unwanted activity. Although planners are beginning to recognize the importance of urban food production, land use planning impedes the development of urban farming, as agriculture is normally excluded from comprehensive, long-term urban planning.

These problems are specifically socioecological, rather than environmental and/or technical obstacles. However, it is precisely the handling of urban agriculture's socioecological challenges that determine the manageability of other obstacles. The compatibilities and tensions between what central planners and local experts can or want to do with ecological limitations and various environmental and technical inputs influences the degree to which urban agriculture succeeds or fails. Although the challenges appear daunting, advocates press on, and urban agriculture continues to grow in scope and scale.

Examples of Urban Agriculture

Although urban agriculture is practiced in most, if not all, cities, throughout the world, some notable examples highlight the diversity of its practice and indicate the resilience and benefits of urban food production. These cities illustrate the importance of urban agriculture in an increasingly urban world and point toward a promising future for the expansion of urban food production.

The most widely known examples of urban agriculture (at least to Anglophile audiences) emerged in the wake of World War II. Urban agriculture was vital throughout the world during the war. Japanese cities experienced exponential growth of urban agriculture both during and after the war as a necessary survival strategy. England's Allotment Gardens and the U.S. Victory Gardens served as patriotic symbols, freed valuable resources for the war effort, and provided relief to urban hunger.

In North America, many cities have extensive urban agriculture projects. These urban agricultural projects can be well-coordinated state-centered efforts or ad hoc guerrilla efforts working to assert rights to the city and fight for food justice. Major cities such as Chicago, Detroit, Milwaukee, New York City, Philadelphia, San Francisco, Seattle, Montreal, Toronto, and Vancouver are some of the most prominent examples. In Detroit, vast swaths of vacant land created by economic crises have given rise to an informal network of urban agriculture. In New York City and San Francisco, professional not-for-profit organizations team with city officials to coordinate urban agriculture. And in Canadian cities, established food policy councils govern urban agriculture as part and parcel of broader urban food systems.

In Latin America, peri-urban farmers make a living from meat, milk, and corn production around Mexico City. In Havana, Cuba, urban agriculture is credited as a key component in Cuba's response to the economic-induced food crisis brought on by the collapse of the Soviet bloc and violent U.S. embargo. Urban agriculture is expanding in Caracas, Venezuela, with the aid of Cuban agricultural experts.

Throughout urbanizing Asia, urban agriculture plays a significant role in providing food security for the urban poor. In Hanoi, Vietnam, researchers conclude that almost 80 percent of fresh produce and 50 percent of protein consumed by poor people is produced through urban subsistence farming. Throughout China, peri-urban agriculture accounts for well over half of the food consumed by urbanites, poor and rich alike.

Urban agriculture is a vital survival strategy throughout Africa as well. In Accra, Ghana, research indicates that almost all fresh produce consumed in the city is produced within city limits. In Senegal, roughly 60 percent of the nation's food supply is produced in and around Dakar. Urban agriculture is the second-largest employment sector in Dar es Salaam, Tanzania, and provides the third-highest income opportunity in Nairobi, Kenya. In Egypt, university professors have recently taken the lead in expanding urban agriculture through the cultivation of rooftops.

Urban farmers everywhere have been found to eat better than their nonfarming counterparts. In each case, urban agriculture takes a place-specific formation colored by its urban political ecological context, highlighting the importance of urban agriculture as a survival tool throughout the world's cities. The diversity of urban agriculture indicates its resilience and exemplifies the benefits of urban food production.

Looking Ahead

Although it is impossible to predict the future of urban agriculture, some key features hint at future possibilities for urban farming. In part, the expansion of urban agriculture is attributed to increasing urbanization throughout much of the world. For the first time in human history, the world's urban population now surpasses the rural population. Rapid urbanization is accompanied by a rapid increase in urban poverty and food insecurity.

The urban poor experience great difficulties in coping with these trends, as sufficient formal employment opportunities do not exist. Rapidly growing cities face a variety of environmental health issues as infrastructure growth lags behind population shifts, resulting in massive waste problems and worsening air and water quality. Against this background, the benefits of urban agriculture are more readily apparent.

National governments, city officials, and international organizations are all increasingly recognizing the importance of urban agriculture. States are establishing programs to support urban agriculture, and cities are using urban planning to expand urban food production. International organizations like the Food and Agriculture Organization have urban agriculture programs, and the United Nations Division for Sustainable Development pays significant attention to the promise of urban agriculture in Agenda 21.

These benefits notwithstanding, urban agriculture necessarily compels a radical change in town–country relations. Urban agriculture is sometimes viewed as a relict rural holdover or as indicative of arrested urban development—a stopgap that will disappear with economic development. These hegemonic discourses stand in the way of the full development of urban agriculture's potential. To fulfill its promise, urban agriculture needs to be understood not as a relic of the past or as a temporary practice simply brought to the city by rural immigrants. Rather, to be truly transformative, urban agriculture must be viewed as an integral and permanent part of the city. From Baghdad to Havana, New York to Moscow, urban farmers are struggling to make the protest call ring true—another world is indeed possible.

See Also: Agenda 21; Community Gardens; Environmental Planning; Food Deserts; Food Security; Parks, Greenways, and Open Space.

Further Readings

Allen, Patricia. *Together at the Table: Sustainability and Sustenance in the American Agrifood System*. University Park: Pennsylvania State University Press, 2004.

Davis, Mike. *Planet of Slums*. London: Verso, 2006.

Lawson, Laura. *City Bountiful: A Century of Community Gardening in America*. Berkeley: University of California Press, 2005.

Mougeot, Luc J. A., ed. *Agropolis: The Social, Political, and Environmental Dimensions of Urban Agriculture*. London: Earthscan, 2005.

Mougeot, Luc J. A. *Growing Better Cities: Urban Agriculture for Sustainable Development*. Ottawa: International Development Research Centre, 2006.

Evan Weissman
Syracuse University

Urban Forests

Urban forests play a major role in the green city movement, as rapid and global urbanization raises concerns about the sustainability of cities. As valuable components of the urban environment, urban forests enhance local communities by providing the following ecological, social, and economic benefits:

- Climate protection
- Air quality improvement
- Water quality improvement
- Ecological stability (habitat and biodiversity)

- Aesthetically pleasing recreational and educational spaces
- Opportunities for community involvement
- "Green-collar" employment opportunities
- Improvement of land values and the local tax base

More than half of the world's population lives in urban areas, and urban forests help to protect the climate from the problems associated with the resource demands of this burgeoning population. Healthy urban forests mitigate the problems associated with the urban heat island, and urban trees and forests reduce the amount of energy associated with the heating and cooling of urban structures. As windbreaks, urban trees and forests help reduce home heating demands in winter; urban forests positively affect the summer interior temperatures of urban buildings through shading, making non-air-conditioned buildings more comfortable and reducing the energy demands of air-conditioned buildings. As part a long record of urban forest research, E. Gregory McPherson estimated that urban trees in California saved approximately 6,400 GWh per year in electricity use for air conditioning. In addition, as natural carbon sinks, urban trees sequester atmospheric carbon dioxide. In these ways, urban forests help reduce the carbon footprint of the city and protect the global climate.

What Urban Forests Do

Urban forests improve air quality through the absorption of pollutants by the canopy cover. Among the most serious urban pollutants are ozone, nitrogen oxides, sulfuric oxides, and particulate matter. Emissions from motor vehicles and industrial facilities are the main sources of nitrous oxides, sulfuric oxides, and volatile organic compounds. Ground-level ozone, or smog, is formed by chemical reactions between nitrous oxides and volatile organic compounds in the presence of sunlight. High temperatures increase the rate of ozone formation, which aggravates asthma and other respiratory illnesses such as emphysema and bronchitis. Urban trees and forests help to reduce air temperatures, reducing ozone formation and providing cleaner air for urban residents.

Urban forests help to protect watersheds. As cities grow larger, the demand for clean drinking water increases and cities become dependent on peri-urban forests for municipal drinking water. These forests have a great effect on the movement and flow of water. Falling rain is intercepted and absorbed by the tree canopy and, through evapotranspiration, is returned to the atmosphere. Water that does reach the ground is absorbed by the forest floor, where it percolates through the soil and becomes ground or surface water. In addition, healthy forests stabilize slopes and streambanks, minimizing erosion and reducing sediment inputs to urban streams, rivers, and ponds. Unlike impervious surfaces, urban forests act like a sponge and store water, which helps to reduce the load on drainage systems. In storing water, the soil, leaf litter, and vegetation of an urban forest act to remove stormwater pollutants acquired from roadways, sidewalks, yards, and buildings, thereby lessening the amount of potentially harmful substances leaching into surface water. By providing shade, urban forests reduce the temperature of urban waterways, thereby increasing dissolved oxygen and reducing stress for aquatic life.

The expansion of urban areas results in urban sprawl, habitat fragmentation, and the loss of biodiversity. Urban forests, in the form of greenways, parks, and riparian areas, alleviate the problems associated with habitat fragmentation by acting as wildlife corridors. Preserving and expanding urban forests impedes the spread of alien plant and animal species and provides prime habitat for native species. Species richness, and the presence of sensitive native species, is greater in wooded habitats than in formerly wooded spaces converted to lawned environments (i.e., park open spaces and ball fields). In addition to providing ecological benefits, urban forests also provide important social benefits.

The aesthetically pleasing environments of urban forests provide a variety of recreation and education opportunities. Urban forests reflect cultural attitudes and preferences for wilderness, and they provide relief from the built-up urban environment. The scenery associated with urban forests—deciduous trees with their brightly colored autumn leaves contrasting with the green from evergreens, scenic pond views, wild rapids and peaceful pools in streams and rivers, an abundance of wildflowers—act as a restorative environment by offering residents a visual break from the concrete jungle of the city. Depending on the naturalness and level of development of the forest, these spaces provide a variety of recreational activities. Less-developed forests offer serenity and peacefulness for passive activities such as walking, bird watching, and nature appreciation. More-developed forests with paved sidewalks offer spaces for active pastimes such as bicycling, jogging, or rollerblading. Forests with ponds or lakes may provide spaces for fishing or boating in the summer and ice skating in the winter.

In addition to spaces for recreation, urban forests provide spaces for education. Urban forests are the perfect laboratory for kindergarten through grade 12 outdoor environmental education. Experiential and interdisciplinary by nature, environmental education bridges the divide between physical sciences, social sciences, and the humanities. Urban forests provide spaces for a place-based, experiential environmental education curriculum that connects students with the historical, ecological, social, economic, and political attributes of their local community. Service learning, such as tree planting, restoration projects, or cleanup projects, present students with opportunities to foster stewardship, focus on community involvement, and develop civic responsibility.

Caring for Urban Forests

The maintenance of urban forests provides unique opportunities for community involvement. Caring for urban trees fosters a sense of responsibility and ownership in community members. In recognizing the role of urban trees and forests in mitigating the effects of the urban heat island, protecting the global climate, improving water quality, and reducing air pollution, soil erosion, and energy consumption, communities "think globally, act locally" by using urban forests to improve their local communities for future generations. The act of volunteering to plant new trees, as well as to care for and maintain existing forests, can improve the vitality of a community by uniting people and promoting multigenerational and multiethnic community involvement.

Although volunteering improves the local social life and environment of a community, green jobs in urban forestry help to improve the local economy. Green-collar employment opportunities help both the local economy and the urban environment as they train unemployed workers in environmentally friendly jobs. These opportunities can be either in the public or the private sector, but increasingly these types of jobs involve working in public–private partnerships because of ever-shrinking municipal budgets. Green-collar job training prepares unemployed workers to be stewards of the urban ecosystem. Many of these job opportunities are in the high-tech sector, with a focus on renewable energy. However, urban forestry provides unique opportunities for green employment through urban natural resources management and the planting of new trees, as well as the restoration and preservation of existing urban forests. Reminiscent of Aldo Leopold's land ethic, green employment and urban forestry connect people to nature in the community and build the green infrastructure necessary for urban sustainability.

Urban forests enhance nearby property values, which provide an increase in property tax revenues for municipalities. As environmental amenities, urban forests improve the quality of life of an increasingly urbanized society by improving air quality, protecting watersheds, and providing visual access to nature, as well as physical access to a variety of recreation, educational, and community activities. The capitalization of the urban forest into higher property values is known as the proximate principle. The proximate principle rests on the assumption that some people prefer to pay higher premiums for homes near urban environmental amenities. Property values increase from 5 percent to 20 percent or more, depending on the aesthetics of and proximity to an environmental amenity. This increase in property values results in a significant increase in the local tax base. Although urban forests improve urban livability, it is important to note that there is an uneven geography associated with urban forests. The benefits of healthy, well-maintained urban forests and other environmental amenities, such as parks, improve the property values and livability of only those neighborhoods closest to the amenity. In this way, Nikolas Heynen emphasized that urban trees and forests act as "billboards" highlighting elite urban spaces, whereas a lack of trees exposes marginalized ones.

As the global urban population continues to grow, the sustainability of cities increasingly becomes a matter of urgency. Urban areas are consumers of vast resources and therefore must find ways to promote sustainable practices that maintain the ecological integrity, cultural vitality, and economic health of the city. Urban forests help to achieve this in three interrelated ways. First, they improve the urban ecosystem by protecting climate, improving air and water quality, and protecting habitat and biodiversity. Second, urban forests help improve local communities by providing visually appealing spaces for recreation, education, and community involvement. Last, urban forests enhance the local economy by providing environmentally friendly job opportunities and improving the local tax base.

See Also: Air Quality; Biodiversity; Green Belt; Green Jobs; Heat Island Effect; Parks, Greenways, and Open Space; Water Pollution.

Further Readings

Crompton, John L. "The Impact of Parks on Property Values: A Review of the Empirical Evidence." *Journal of Leisure Research,* 33/First Quarter (2001).

Dwyer, John F., et al. "The Deep Significance of Urban Trees and Forests." In *The Ecological City: Preserving and Restoring Urban Biodiversity*, Rutherford H. Platt, et al., eds. Amherst: University of Massachusetts Press, 1996.

Heynen, Nikolas. "The Scalar Production of Injustice Within the Urban Forest." *Antipode,* 35 (November 2003).

McPherson, E. Gregory. "Urban Forestry in North America." *Renewable Resources Journal,* 24 (Autumn 2006).

Miller, Robert W. *Urban Forestry: Planning and Managing Urban Greenspaces*, 2nd Ed. Upper Saddle River, NJ: Prentice Hall, 1997.

Moll, Gary and Sara Ebenreck. *Shading Our Cities: A Resource Guide for Urban and Community Forests.* Washington, D.C.: Island Press, 1989.

Robbins, Paul. *Political Ecology: A Critical Introduction.* Hoboken, NJ: Wiley-Blackwell, 2004.

Phil Birge-Liberman
Syracuse University

Vancouver, Canada

North American cities are often characterized by sprawled, low-density development patterns that result in congestion and the loss of agricultural land, among other environmental impacts. The evolution of low-density urban form is commonly associated with the ascendancy of the automobile, facilitated by government investment in transport infrastructure.

The public transit system of Vancouver consists of over 200 bus routes, a sea bus, a regional commuter rail, and four elevated light-rail transit lines called SkyTrain, shown here.

Source: iStockphoto

In contrast, the Canadian west coast city of Vancouver is widely portrayed as having overcome the dominant paradigm of automobile-based planning, instead promoting more green or sustainable development patterns that emphasize denser, walkable, and transit-oriented urban form. Vancouver is celebrated in the planning community as a model green city because of its large population living downtown, the absence of highways in the central city, and strong emphasis on investment in public transit. Despite the positive reputation, Vancouver faces many of the urban problems afflicting other major cities, such as homelessness and housing affordability, and most growth still occurs in low-density suburbs, where the automobile is the dominant mode of transport.

Vancouver is located in the lower mainland of southwestern British Columbia at the mouth of the Fraser River, an area long inhabited by the Coast Salish aboriginal peoples before colonization. The Vancouver urban area is surrounded by the Cascade mountains and the Pacific ocean to the east and west, the coast mountains to the north, and the U.S. border immediately to the south. According to the 2006 census, the City of

Vancouver has a population of over 570,000, and metro Vancouver, a regional partnership of adjacent municipalities, has 2.1 million inhabitants, making it the largest metropolitan area in the province and the third-largest in Canada. International migration, primarily from Asia, is the largest source of population growth. Vancouver is an important regional service center, in the past mainly for the province's natural resource sector. Tourism, transport, trade, and service industries, such as education, architecture, motion picture, software design, and information technology, are vital to Vancouver's present-day economy.

Vancouver's success in promoting a more walkable, transit-friendly, compact, and mixed-use urban form has been attributed to planning policies that emphasize accessibility by colocating jobs, housing, and amenities to reduce the need for travel by car. Vancouver's transit system provides opportunity for mobility without necessitating car ownership, which is somewhat unique in a North American context. The public transit system consists of over 200 bus routes, a sea bus, four elevated light rail transit lines (SkyTrain), and a regional commuter rail. The system connects most areas of metro Vancouver, resulting in over 170 million passenger trips annually. The 1986 World's Fair was an important turning point in Vancouver's history, as it facilitated investment in public transport infrastructure, such as the SkyTrain, and inner city development. Previously, largely industrial lands were rezoned by the city for residential development, specifically at False Creek, which is often cited today as a successful high-density, mixed-use neighborhood. The provincial government also played a role in the restructuring of Vancouver's downtown through the sale of inner city land to a prominent Hong Kong property developer, contributing to the internationalization of the city's housing market. Declining household size—partly because of the feminization of the labor force and an aging population—and the growth of a well-educated population working in service and professional occupations provided a local market for inner city residential development at higher densities, but also sparked concerns over displacement of low-income populations through gentrification. The term *Vancouverism* is now used by architects and urban planners to refer to Vancouver's development patterns, characterized by narrow residential apartment towers on a three-to-four-story base containing commercial and retail uses, most visible in the Coal Harbour and Yaletown neighborhoods.

Six policy developments dating back to the 1960s and 1970s were essential in facilitating more compact and less-auto-dependent development: the establishment of the Agricultural Land Reserve, which protects agricultural land and acts as an urban growth boundary; the creation of a regional transit agency (TransLink) that coordinates public transport across Metro Vancouver; the rejection of an urban freeway system, largely in part as a result of local activism; density bonusing, which requires private developers to provide amenities and community facilities in exchange for relaxation of zoning restrictions; a regional governance apparatus instrumented by the provincial government to provide municipal governments with powers to stringently regulate development and permit regional service delivery; and the implementation of the Livable Regional Strategic Plan. The goals of the Livable Regional Strategic Plan are the intensification of development within the central city and regional town centers; the implementation of a transit system of private and public modes that connect these centers; the preservation of two-thirds of Metro Vancouver land for green space, agriculture, and watersheds; and the adherence to environmentally acceptable policies of waste removal and treatment, water provision, and pollution control.

Despite a positive environmental reputation, Vancouver is not sustainable: Because of the growth of per capita land, transport, and goods consumption, as well as absolute population growth, Metro Vancouver has an ecological footprint—the hypothetical area required to sustain a city's resource use and waste absorption—almost 300 times its geographical area. Although over half of workers commute by transit, bike, or on foot in the downtown, and that number exceeds 65 percent in the West End, where residential densities are among the highest in North America, up to 80 percent of workers drive to work in suburban municipalities, according to the 2006 census. Single-family dwellings remain the dominant form of housing. One common argument against increasing compact development is the potential for inflationary impacts on housing costs. Although the factors determining housing market outcomes are diverse, housing affordability issues in Vancouver are frequently portrayed as fueling suburban growth and automobile-dependence by displacing households away from high-density employment centers where land values tend to be highest. Vancouver has one of the lowest rental vacancy rates in Canada, and average rents and housing prices are among the highest in the country. The issue of housing affordability is most dramatically visible in Vancouver's Downtown Eastside, Canada's poorest urban neighborhood, known for high incidence of homelessness, drug addiction, and prostitution. The causes of these issues are various, but numerous commentators blame growing concentration of low-income populations in the Downtown Eastside on redevelopment of the inner city, which has reduced the affordable housing stock.

Altogether, Vancouver is attributed with successfully laying the groundwork for less-automobile-oriented urban development patterns by promoting denser, mixed-use development and investing in public transit. Although Vancouver is often seen as a model green city for its urban development patterns, challenges remain in reducing automobile use, lowering the ecological footprint of overall consumption patterns, and addressing housing affordability and social issues. The 2010 Olympics, as well as new regional highway infrastructure intended to increase trade and port functions, are examples of recent challenges as the metropolitan area struggles to promote environmental protection in the face of global growth pressures.

See Also: Density; Ecological Footprint; Transit; Walkability (Pedestrian-Friendly Streets).

Further Readings

Berelowitz, Lance. *Dream City: Vancouver and the Global Imagination*. Vancouver, Canada: Douglas & McIntyre, 2005.

Blomley, Nick. *Unsettling the City: Urban Land and the Politics of Property*. London: Routledge, 2004.

City of Vancouver. "Vancover Green Buildings." http://vancouver.ca/commsvcs/southeast/greenbuildings/index.htm (Accessed December 2009).

Newman, Peter and Jeff R. Kenworthy. *Sustainability and Cities: Overcoming Automobile Dependence*. Washington, D.C.: Island Press, 1999.

Markus Moos
University of British Columbia

Vapi, India

Located in the Valasad District in India, in the state of Gujarat and close to the border with the state of Maharashtra, the town of Vapi is dominated by the chemical industry. It is also the largest industrial area in all of Asia in terms of the number of small-scale industries operating in a single place. The reason for locating this industry in a small area was that it would be easier to control pollution and effluent. However, this has not been the case, and because of the pollution from the factories and processing plants located there, Vapi became recognized as one of the 10 most polluted cities in the world. This shocked many of the locals, and it was not long before the local authorities started to regard this designation with such concern that they immediately sought to end the notoriety and clean up the city. This has resulted in many advances made in the city's environment and its speedy removal from the list of the 10 most polluted cities, unlike most of the other cities that appeared on the initial list.

Most of the businesses that established factories at Vapi were chemical industries, and today over two-thirds of all large factories there remain connected with the chemical industry. These are involved in various different types of chemical distillation, and in the production of dyes, dye intermediaries, paints, and many kinds of pesticides. Vapi also has factories concerned with manufacturing pharmaceuticals, plastics, textiles, paper and packaging, and glass, as well as processing rubber, wood, and timber products. The land area of some 11.4 square kilometers (1,140 hectares) was divided into 1,382 industrial units, of which some 800 were operational by August 2009.

For many people, especially economists, Vapi was a major success story, as people from many regions in India found work there, leading to the establishment of residential suburbs around Vapi. The city had helped provide fertilizers and chemicals for the local economy and exports for India, and marked a major stage in the industrialization of the country. The population of Vapi consisted of 71,395 people in 2001, with a literacy rate of 73 percent, higher than the Indian national average of 59.5 percent. The male literacy rate was 79 percent (compared to 75 percent for the national average), and the female literacy rate was 64 percent (compared to 54.2 percent for the national average). Approximately 14 percent of the population is younger than 6 years. For many, Vapi was a major success for the emergence of India as a major industrial power.

Although Gujarat has much fertile land, it has to import food each year, and much of the state has seen widespread deforestation, with much of the land officially declared barren or unsuitable for agriculture. Gradually, more and more of the state finances tended to rely on industry, which in Vapi increased voraciously.

Soon after the creation of the Vapi Industrial Estate, criticism of the Central Effluent Treatment Plant in Vapi was heard, which was said to have been inadequately funded and performed poorly in pollution management. The plant cost $9.2 million to construct, and the hope had been that a central plant would be able to deal with the toxic waste from all over Vapi. The Blacksmith Institute, using the pollution levels published in 1998 by the Gujarat Pollution Control Board, first highlighted Vapi as one of the 10 most polluted places in the world. This was because chemical waste was regularly dumped near the factories where it was created, and during the seasonal monsoons each year, some of the waste washed into groundwater reservoirs and polluted the drinking water. Indeed, some environmental activists argued that it was the laxity of the Insecticides Act of 1968 that permitted the use of organochlorine pesticides, which were responsible for the pollution of groundwater and reservoirs.

The city has improved itself between 1998—when it was listed as one of the most polluted cities in the world—and 2009, when it was removed from the list. A large part of this was the result of the Central Effluent Treatment Plant being used for all the waste, rather than 90 percent of it, which had been the case before 1998.

Illegal Dumping

The Blacksmith Institute pointed out that the Central Effluent Treatment Plant that was failing the community. The idea of establishing it had been commendable, and the vast majority of the industries at Vapi were using it, but a small minority was not. These industries were essentially responsible for much of the pollution and toxicity in the water and air in Vapi. Although the Central Effluent Treatment Plant had the capability of treating up to 55 million liters per day, and production in Vapi was only at 50 million liters per day, only 45 million liters was being collected through the Gujarat Industrial Development Corporation drainage system for treatment at the plant. This meant that 10 percent of the waste was being illegally dumped—often openly. The Blacksmith Institute urged that controls be enforced to prevent this from happening. It also recommended that the size of the Central Effluent Treatment Plant be increased so that it would be capable of dealing with as much as 70 million liters per day. This upgrade would come at a cost of $2 million, but would, in return, dramatically increase its performance. At the same time, major changes were needed to ensure that all effluent was sent for treatment and filtering at the plant.

The Vapi Industries Association and the Vapi Waste & Effluent Management Company Limited, operated by the Vapi Industries Association, both condemned the Blacksmith Institute. However, it led to the enlargement of the Common Effluent Treatment Plant, aiming to make it far more effective. From 2000, local politicians and government officials managed to obtain substantial capital investment, as well as improved enforcement of government health regulations, which saw a dramatic change in Vapi. The result was that in a report issued in August 2009, the Blacksmith Institute removed Vapi from its list of the top 10 worst-polluted cities in the world, highlighting the major efforts made in Vapi in terms of reducing water pollution. Tests of groundwater in the region have shown an improvement in quality. The Blacksmith Institute also recognized that some of the data regarding the level of mercury in the groundwater, which had contributed to Vapi gaining a bad report for pollution, were actually false. In addition, air pollution in the city is now well within the government standards and is monitored regularly to ensure that this remains the case.

In addition to improved water and air pollution levels, the Common Effluent Treatment Plant has been upgraded and is undergoing further upgrades. The Vapi Industrial Association has also gone to great lengths to prevent the illegal dumping of waste, especially toxic waste, in the industrial estate, and has also tried to prevent toxic waste from the industrial estate from being dumped outside it.

Furthermore, the roads and footpaths have been upgraded, with some $6 million invested in their improvement and an additional $10 million earmarked to continue the process. This money has seen open gutters covered and water from them taken by an underground pipeline to the Common Effluent Treatment Plant, where it can be treated. The Blacksmith Institute noted that in August 2009, the underground pipeline was about to become operational, and that work was proceeding at improving what had already been completed. In addition to improving the environment, tens of thousands of trees have been planted around Vapi, which has considerably improved the appearance of the city. A Center of Excellence was established by the Industries Association, which has ensured that levels of pollution are monitored and that action is taken against anybody or any company involved in pollution.

The Blacksmith Institute commended the authorities in Vapi, but also urged continued vigilance to ensure that all waste is routed to the Common Effluent Treatment Plant, and that groundwater and surfacewater continue to be monitored to prevent contamination. The prevention of dumping waste from the industrial area was important in helping the overall environment. Although it could have had access to Indian hospital data, the Blacksmith Institute chose to embark on a long process of interviews with selected locals. They sought to assess the health status of people in the Vapi industrial area to compare them with inhabitants from the nearby city of Dharampur, also in the state of Gujarat, which was far less polluted than Vapi had been in 1998, but which also relied on heavy industry. As well as checking the key indicators such as health, reproductive cycle, and longevity, they also sought to compare aspects such as the prevalence of skin diseases and mental retardation. As a result, the Blacksmith Institute was able to report that the health standards for men, women, and children were similar to those elsewhere in India, and that abortions, stillbirths, and infant deaths were minimal. When compared with control groups elsewhere in India, the figures were the same, and the institute was unable to observe any higher cases of birth abnormalities or other abnormalities in both the physical and mental growth of children between the ages of 1 and 5 years.

See Also: Ecological Footprint; Sukinda, India; Sumgayit, Azerbaijan; Water Pollution; Water Sources and Delivery.

Further Readings

Dabke, Suneet V. "Improvements in Pollution Management —Vapi Industrial Estate." Blacksmith Institute, August 2009. http://www.worstpolluted.org/files/FileUpload/files/VIA%20Pollution%20Management%20Improvement%20August%202009%20RF.pdf (Accessed October 2009).

Davidar, David J. "Beyond Bhopal: The Toxic Waste Hazard in India." *Ambio*, 14/2:112–16 (1985).

Davidar, David J. "India: Every River Polluted, and Few Effective Controls." *Ambio*, 11/1:63–4 (1982).

Justin Corfield
Geelong Grammar School, Australia

Walkability (Pedestrian-Friendly Streets)

Walkability is a term applied to the built environment to describe a confluence of factors that facilitate human-centered activities such as strolling, shopping, commuting, and socializing. Walkability has an objective, measurable aspect; however, it manifests as a subjective experience with a strong aesthetic component. The essence of walkability speaks to basic human values such as scale, community, and safety and is found in both indigenous villages and the finest examples of intentional urban design.

Walkability celebrates the pedestrian. The texture of the pavement, the shop windows, eye contact, and the open air are all among the experiences that cannot be enjoyed from a speeding automobile. There is a psychological value to this engagement with our surroundings—an intimacy that stimulates all of the senses—but it is the health effects of walking over riding, the aerobic oxygenation and muscular stimulation, that are most significant for our overall well-being. Despite an increased emphasis on fitness, walking is for many people the only means of exercise available on a daily basis. If walking is not a part of daily life, health risks increase and quality of life decreases. Walkability, therefore, is a necessity and a prerequisite for our welfare and fulfillment as human beings.

Walkability has many aspects. Clearly a space or route intentionally designed for pedestrian use is a fundamental starting point. This requires a modicum of segregation from vehicular transportation routes, a scale that is inviting, landscaping and/or retail for interest and diversion, and a sense of safety, preferably derived from the presence of other human beings. A "walkable" neighborhood will have access to amenities such as supermarkets, schools, restaurants, stores, and public transportation within a one-quarter-mile radius for all residents. This distance is generally accepted as equaling a 15-minute walk, which studies show to be the outer time limit for most people's strolls. Some protection from environmental extremes is desirable, such as shade and/or canopies.

Perhaps the most significant aspect of walkability, however, is one that is difficult to quantify or manufacture: aesthetic delight. The steep and windy streets of a medieval hill town are walkable, despite the exertion required, because of the endless interest provided by changing views and the subtlety of the materials and colors of the architecture. Hill town street patterns were formed over hundreds, and sometimes thousands, of years, based solely on pedestrian patterns. A boulevard, by contrast, is an imposed pattern with wheeled vehicles and movement of military forces in mind. Nevertheless, when wide sidewalks are lined with shops and restaurants and buffered from vehicular traffic by trees and other plantings, boulevards attract crowds of strollers. Squares, piazzas, parks, and fountains all contribute to the aesthetic quotient, creating pleasurable punctuation for the syntax of the street.

The presence of people enjoying themselves is itself an aesthetic pleasure. It is human nature to be affected by the moods of others. However, activity that may be appropriate on an athletic field can be disruptive to the atmosphere of a promenade. Our ability to enjoy requires that we let go of the "fight or flight" response and appreciate the senses. For this reason walkability requires that certain rules of behavior be consensually adhered to. Loud noises, regardless of the source, are counterproductive to walkability. Skateboarding and even bicycles can be disruptive to the cultivation of a meditative stroll. Graffiti and litter are advertisements of disorder.

The pace of walking allows a full appreciation of details: Shop windows and signs, the curb at a crosswalk, building design, and the clothes and expressions on the people we pass are all of interest, and when we tire of walking, an outdoor table at a café or restaurant provides the optimum perch from which to continue surveying our surroundings. The food or beverages are of interest, of course, but the real draw is the street, the boulevard, the avenue, the piazza, or the square. To take pleasure in being in public surrounded by strangers is a uniquely urban experience.

The Kansas City Plan

Among other jurisdictions, Kansas City, Missouri, has developed a walkability plan, adopted by its city council on March 20, 2003. The plan is expected to provide economic and health benefits to the city and to reduce automobile use. Increasing the number of individuals who walk to work, for instance, enhances several spheres of the commonwealth, extending from the health of the individuals themselves to the use of streets where traffic is reduced, and on to the wider region and the atmosphere in terms of the effect of reduced greenhouse gas emissions. The Kansas City plan also intends to facilitate pedestrian access for children to local schools. Subdivision planning since the 1950s has virtually eliminated any pedestrian circulation beyond the immediate neighborhood, resulting in increased dependence on automobiles for all members of a family. Providing a safe and walkable connection between homes and schools frees both the child and the parent from a need to drive. And, of course, walking is far more economical for the household, as well as for the planet, in terms of efficient use of resources.

The Kansas City plan sees walkability as a necessary aspect of a larger transportation network, with pedestrian routes radiating from public transportation stops. This encourages the use of buses and light rail while further reducing automobile dependency.

The Kansas City Plan identifies five components of walkability, described below:

1. *Directness:* Does the network provide the shortest possible route?

2. *Continuity:* Is the network free from gaps and barriers?

3. *Street Crossings:* Can the pedestrian safely cross streets?

4. *Visual Interest and Amenities:* Is the environment attractive and comfortable?

5. *Security:* Is the environment secure and well lighted with good line of sight to see the pedestrian?

Using these criteria, "pedestrian level of service standards" were developed and a citywide assessment was conducted to measure walkability and to identify potential areas for improvement. As a result, 10 "Pedestrian Zones" were established, with guidelines for future development created for these zones based on the pedestrian level of service standards.

Creating an effective pedestrian environment is not a simple task. In the 1970s, based on successful examples from Europe, several American cities banned automobiles from portions of downtown shopping areas. The intent was to create vibrant "streets for people" that would energize the center city and stem the exodus of economic activity to the suburbs. Experience has shown that not all of the pedestrian streets created over the past 30 years have resulted in improved economic activity. This can be ascribed to a number of factors, including a lack of proximate mixed-use development (residential use always benefits walkability and vice versa) and sheer cultural unfamiliarity. When it works, walkability is of benefit to all. As experience grows and the culture changes, this pedestrian amenity will become an ingredient in all good design.

See Also: Compact Development (New Urbanism); Traffic Calming; Transit-Oriented Development.

Further Readings

Holmgren, David. *Future Scenarios: How Communities Can Adapt to Peak Oil and Climate Change*. White River Junction, VT: Chelsea Green, 2009.

Hopkins, Rob and Richard Heinberg. *The Transition Handbook: From Oil Dependency to Local Resilience*. White River Junction, VT: Chelsea Green, 2008.

Rudofsky, Bernard. *Streets for People: A Primer for Americans*. New York: Van Nostrand Reinhold, 1969.

A. Vernon Woodworth
Boston Architectural College

Waste Disposal

Solid waste is generally known as "trash" or "garbage" in layman's terms. Generally it refers to both domestic waste, generated from households, as well as commercial waste, and includes among other things food scraps, paper, newspaper, clothes, packaging, cans, bottles, grass clippings, furniture, paints, and batteries. Municipal waste disposal practices have changed over period of time.

Early Urban Waste Disposal

Dumping on land was a common method in urban communities because it was simple and inexpensive to haul solid wastes to the edge of town and dump them there. Burning this waste also was a common practice. Because open dumps attracted flies and rodents that spread diseases, haphazard disposal became a matter of great concern to public health authorities responsible for the control of infectious diseases. Over the years, in developed countries this problem has been controlled via engineered landfills.

Dumping waste in water was generally used in some coastal cities, although it was not favored because the pollution consequences were well recognized. The disfigurement of Coney Island beach in New York City became a case in point. Nevertheless, the practice has continued even in developed countries until relatively recently.

Plowing refuse into the soil was a method of disposal used mainly for food wastes and street sweepings. Because of the large land requirements and the fact that the food wastes had to be separated from other wastes, this method was not used extensively, but interest in it was rekindled in the 1970s.

Food waste was often separately collected to serve as animal feed, usually to pigs, on farms close to urban areas. This practice led to a number of health problems. For example, as much as 16 percent of the U.S. population was infected by eating uncooked pork from hogs fed on food waste around the 1930s.

Historically, raw food wastes have been treated to separate them into solid and liquid portions and to recover the grease to make pomades and cheaper grades of perfumery, as well as wagon grease.

Open burning was one of the easiest methods of disposal, either at common open dumping grounds or outdoors, in front of the houses. Although this practice causes a lot of health problems and air quality issues resulting from burning, it is still carried out in some countries even now.

Current Methods of Waste Disposal

Landfills are disposal facilities in which solid wastes are openly dumped on a designated piece of land. Uncontrolled dumping has been carried out for years, but in most developed countries, "sanitary" landfill sites still exist. Sanitary landfilling is controlled disposal of waste on land, and involves deliberate engineering of the landfill sites, which includes lining the landfill to prevent wastewater seeping into groundwater, compacting waste, daily covering of waste, collection of leachate, and on-site treatment ponds.

Recycling is a "process by which materials that would otherwise become solid waste are collected, separated, or processed and returned to the economic mainstream to be reused in the form of raw materials or finished goods." In a sense, recycling can turn discarded materials into valuable resources. In addition, recycling generates environmental, economical, and social benefits to the society. Resource recovery in its totality includes material recovery through recycling, where certain components are used in the production of "new" materials. Through incineration of wastes or through biogas production, energy is recovered for heating, and recovery of organic materials and plant nutrients is accomplished by composting. In many Western countries, campaigns offering incentives and subsidies are organized by government departments, industry, and municipalities for the collection and separation of wastes. In developing countries, these activities are self-organized through a chain of self-employed individuals or groups of dealers and agents for whom this work is a good source of income.

Composting has been practiced by farmers and gardeners for centuries. In earlier times, "night soil," vegetable matter, animal manure, household garbage, and so on, were placed in piles or in a pit and allowed to decompose until ready to be used to fertilize crops. Composting is a controlled aerobic process carried out by successive mesophilic and thermophilic activities, leading to the production of carbon dioxide, water, minerals, and stabilized organic matter. The main development of modern composting has been in India, China, southeast Asia, and east and south Africa. In recent years, there has been a lot of interest in composting in developed nations for treatment of waste, principally to mechanize the process of composting. Composting can be done either by windrow (where long heaps of materials, called windrows, can be either statically or periodically turned) or by in-vessel composting (done inside reactors to minimize odor and rodent spread). Some parts of the world also use "vermicomposting," in which worms are used to facilitate the process of composting.

In the past (to some extent this still occurs in some developing countries), the easiest way of disposing of waste was simply to burn it in the open air. Incineration of raw waste has been practiced throughout the world for decades in the most crude form of incineration—indiscriminate burning of waste. Large-scale indiscriminate open burning of waste, however, causes air, water, and soil pollution. Technology has now progressed from open burning to sophisticated incinerators, and the purpose has changed from simply getting rid of the waste to recovering energy from waste. The process occuring inside of an incinerator is called pyrolysis, which is a thermal decomposition of waste at high temperatures in an absence or near absence of oxygen. The by-products of incineration exist in all three forms—solid, liquid, and gaseous. It is important to regulate the temperature of the furnace, depending on the quality of waste, so that more hazardous products are not produced upon decomposition and released into the environment.

See Also: Composting; Garbage; Landfills; Recycling in Cities.

Further Readings

Grover V. I., et al. *Solid Waste Management*. Oxford: IBH Ltd., 2000.

Holmes, J. R., ed. *Managing Solid Wastes in Developing Countries*. Hoboken, NJ: John Wiley & Sons, 1984.

Pichtel, John. *Waste Management Practices: Municipal, Hazardous, and Industrial*. Boca Raton, FL: CRC Press, 2005.

Porter, Richard C. *The Economics of Waste*. London: RFF, 2002.

Rathje, William and Cullen Murphy. *Rubbish! The Archaeology of Garbage*. Tucson: University of Arizona Press, 2001.

Tammemagi, Hans Y. *The Waste Crisis: Landfills, Incinerators, and the Search for a Sustainable Future*. New York: Oxford University Press, 1999.

Williams, Paul T. *Waste Treatment and Disposal*. Hoboken, NJ: John Wiley & Sons, 2005.

Velma I. Grover
United Nations University–Institute for Water,
Environment and Health (UNU–IWEH)

WATER CONSERVATION

Water conservation refers to a set of ideas and strategies orientated toward making less water do more. Thus, water conservation is about both increasing the efficiency of every unit of water used in all spheres—domestic, industrial, and agricultural—and also using less water overall. After decades of blind faith in "supply-side" solutions (ever-larger dams, etc.), policy makers now prefer to focus their energies on reducing demand across the entire range of users. The British government's strategy for water provision published in 2008, *Future Water*, talks about revising policies to encourage conservation behavior, including altering architectural and planning codes, changing the way water services are priced and regulated, and investing in consumer education. Water conservation policy in the United States is more complex, with some states (e.g., Arizona) working hard to promote conservation and others (e.g., Alaska, South Carolina) doing relatively little. Federal agencies such as the Natural Resources Conservation Service of the U.S. Department of Agriculture act more in an advisory than a regulatory capacity.

One traditional method of water conservation that is being revisited is a system of rain barrels that catch roof runoff through a downspout.

Source: iStockphoto

The exact nature of conservation measures undertaken will, of course, also vary, with the structure of demand prevailing in any given country or region. In the United Kingdom, approximately two-thirds of water consumption is related to domestic use, so this is the relevant point of application. The Environment Agency for England and Wales estimates that the introduction of water metering, low-flow showers, low-flush toilets, and a variety of other water-saving technologies could save approximately 65 megaliters per day in the southeast region and London. Parliament has established a cross-party "Water Saving Group," and the private water companies have all written conservation and demand management into their 25-year "water resources plans" published in summer 2008. Across Europe the situation is broadly the same—though some governments are pursuing completion of national water grids long under development (e.g., Spain), most are now concentrating their energies on disciplining demand. In countries (such as France) in which agriculture consumes a greater share of available water resources, more effort has gone into promoting efficient irrigation through cooperative water management schemes and through the metered pricing of irrigation water. In the Midwest and Great Plains of the United States, water conservation in agriculture has become both pressing (because surface and groundwater resources are overdrafted) and intensely political (because the economies of entire counties and regions depend on continued overdraft).

Over the course of the 1990s, new development standards in Britain, such as "Eco Homes," PPG 12, and the Water Supply Regulations (1999) provided various forms of regulatory encouragement in the United Kingdom. The 1999 Water Supply Regulations Act, for example, did make it mandatory for low-flush toilets to be fitted to all new developments, with the incentive of an enhanced capital allowance to encourage developers to take a more sustainable approach to water management in their developments. These gains, however, seem timid and a long time coming when compared with the more proactive and energetic approach taken by water managers in some U.S. states. For example, in the area served by Tucson Water in southeastern Arizona, water-intensive "xerophytic" landscaping has been strongly encouraged as water managers seek to push water conservation outside as well as inside the home. With respect to water fixtures inside the home, Tucson Water has instituted a rebate program that will pay up to half the cost of installing low-flush toilet fixtures and other water-saving devices. Tucson Water has also instituted tiered water-usage rates that encourage conservation by charging progressively more for higher water users. It is important to note that these sorts of strategies work in Tucson because they operate within the context of a complex network of local water-user organizations that give key stakeholders (homeowners associations, landscaping/gardening interests, etc.) "voice" within the water-regulatory apparatus.

Growing Use of Greywater

In both Europe and the United States there has been some attention to rainwater harvesting and greywater reuse, but these technologies tend not to survive cost-benefit analysis, as well as having specific technical problems. Greywater reuse systems, for example, raise water flow imbalance problems that can only be rectified by the installation of on-site storage tanks—easy at the point of initial construction, difficult to retrofit. Moreover, repeated studies have shown that the greater the storage capacity, the greater the problem of bacterial formation and degradation within the storage tank. Most jurisdictions either have already or are currently developing strict regulations for the installation and operation of greywater systems to protect public health.

An alternative to such systems are the "reclaimed water" or "purple pipe" systems that are now growing in popularity, especially in East Asia and the desert southwest of the United States. Such systems recognize a distinction between the types of water needed for safe drinking water provision and that needed for many other uses, especially landscaping and environmental gain. For example, in Maricopa County, Arizona, water authorities are increasingly providing dual water inputs to households—one input for drinking and personal hygiene and the other for everything else.

Another area targeted for special attention in many water management districts is leakage—most modern distribution systems have leakage rates of between 15 percent and 25 percent, meaning that as much as one-quarter of all input water is lost through the pipes before it ever reaches a user. Although such statistics provide great copy for national and local press, the realities of significantly reducing leakage rates are more complex. It is common in the industry to think in terms of an "economic level of leakage," which is that point at which the costs of further leakage reduction become higher than the benefits obtained. Remember that water (unlike other utilities) is a low marginal-cost good, meaning that the cost of each additional unit of water is very little, especially relative to the high costs of updating infrastructure. As a consequence, the costs of capital works (e.g., mains replacement) must be apportioned across all users and can add significantly to the costs of water services. Water companies therefore tend to try to strike a balance between active leakage detection programs (through network-wide pressure monitoring and metering systems) and mains replacement programs that will provide value for money over 5- or 10-year asset management cycles.

Economic Value of Water

The economic valuation of water is a key element in demand-wide management. This argument is based on the suggestion that as a low marginal-value (but high fixed-cost) good, there has been little financial incentive for either water companies or water users to economize. The U.K. government has recently signaled that it is considering legislating for universal water metering in England, Scotland, and Wales. Echoing statements from other government ministers and senior civil servants, the climate change minister, who is drawing up the government's water strategy, recently said, "The case for universal metering is now overwhelming—provided there is protection for low income and large families." Government officials contend that paying only for what you use (as measured by a meter) is both the fairest way to pay and also offers the potential for significant conservation savings, an important issue, as the balance between supply and population continues to shift in favor of the latter. The overall context for this renewed enthusiasm for water metering includes the proliferation of larger numbers of smaller household units, the adoption of new more water-intensive appliances (power showers, etc.), and the changing U.K. hydrological balance as a result of climate change. Because only 28 percent of U.K. households are at present metered for domestic water use, such a policy change would have significant effects on U.K. households and the water sector in general.

Water-metering trials have been undertaken around the world over the preceding 30 years. D. Gadbury and M. J. Hall report on the establishment of the original U.K. metering trials of 53,000 households in the Isle of Wight and in 11 other areas of the United Kingdom. The "National Water Metering Trials" ran from April 1989 to March 1993 and were set up to assess the practicalities of large scale. By 1993, the early results of this trial had shown that the average reduction in domestic consumption associated with compulsory metering was 11 percent, though Department for Environment, Food and Rural Affairs (2006) itself claims savings of between 10 and 15 percent (and ministers have been known to claim as much as 20 percent). However, perhaps as much as 40 percent of this reduction was actually the product of better leak detection, rather than reduced/disciplined household consumption. The trials also suggested that although compulsory metering had marked effects on peak demand, with a 30 percent reduction recorded in peak monthly, weekly, daily, and hourly demand, they had relatively little effect on average demand. Finally, there was evidence, even in this limited trial, that the conservation effects wore off after a short time, suggesting that consumers quickly become inured to the existence of the water meter under the kitchen sink (or in the road outside). Thus, based on the U.K. evidence, the true effect of metering needs to be seen in terms of better leak detection, modest (though likely temporary) reductions in domestic consumption, and reduced peak consumption, but higher cost and complexity in customer billing and management.

These findings have been replicated in studies of metering trials in other parts of the globe. Virtually all such studies have shown that there is no easy relationship between water metering and conservation. Instead, these studies have found that for metering to achieve real conservation benefits, a strong regulator is required to manage the transfer of rights to meter from consumer to provider, that the links between metering and tariffs must be clarified, and that distinctions ought to be made between universal metering, optimal metering, and demand metering.

Rising urban populations, and especially smaller household size, are increasing the demand for water at the same time that climate change is changing the absolute and relative availabilities of water through the year and across geographic space. As a consequence, water managers around the world are required to think quickly and creatively about how to make less water go further.

See Also: Greywater; Water Pollution; Water Sources and Delivery.

Further Readings

Alcamo, J., et al. "A New Assessment of Climate Change Impacts on Food Production Shortfalls and Water Availability in Russia." *Global Environmental Change* (2007; Corrected Proof in Press).

Appleby, P. "Water Use." *Regeneration Magazine* (April 2006).

Brouwer, R. and D. Pearce. *Cost Benefit Analysis and Water Resources*. Cheltenham, UK: Edward Elgar, 2005.

Chambouleyroun, A. "An Incentive Mechanism for Decentralized Water Metering Decisions." *Water Resources Management*, 17/2:89–111 (2003).

Chambouleyroun, A. "Optimal Water Metering and Pricing." *Water Resources Management*, 18/4:305–19 (2004).

Department for Environment, Food and Rural Affairs. *Water Saving Group*. 2006. http://www.defra.gov.uk/environment/water/conserve/wsg/pdf/wsg-meteringpaper.pdf (Accessed November 2006).

Dovey, W. J. and D. V. Rogers. "The Effect of Leakage Control and Domestic Metering on Water Consumption in the Isle of Wight." *Water and Environment Journal*, 7/2:156–60 (1993).

Drury, I. "Meters Could Add £200 to the Family Water Bill." *Daily Mail* (October 15, 2007).

Gadbury, D. and M. J. Hall. "Metering Trials for Water Supply." *Journal–Institution of Water & Environmental Management*, 3/2:182–87 (1989).

Glennon, R. *Water Follies: Groundwater Pumping and the Fate of America's Fresh Waters*. Washington, D.C.: Island Press, 2002.

Hall, M. J., et al. "Domestic Per Capita Water Consumption in South West England." *Journal–Institution of Water & Environmental Management*, 2/6:626–31 (1988).

Jenerette, G. and L. Larsen. "A Global Perspective on Changing Sustainable Urban Water Supplies." *Global and Planetary Change*, 50/3-4:202–11 (2006).

King, D. "Managing Water Resources in England and Wales." London: Lecture at the Foundation for Science and Technology, June 2006.

King, D. "A Measure of Change." *Utility Week* (October 26, 2007).

Mercer, D., L. Christensen and M. Buxton. "Squandering the Future—Climate Change, Policy Failure and the Water Crisis in Australia." *Futures*, 39/2–3:272–87 (2007).

Organisation for Economic Co-operation and Development (OECD). *The Price of Water: Trends in OECD Countries*. Paris: OECD, 1999.

Organisation for Economic Co-operation and Development (OECD). *Social Issues in the Provision and Pricing of Water Services*. Paris: OECD, 2003.

Organisation for Economic Co-operation and Development (OECD). *Water Resources Management: Integrated Policies*. Paris: OECD, 1989.

Smith, A. L. and D. V. Rogers. "Isle of Wight Water Metering Trial." *Journal—Institution of Water & Environmental Management*, 4/5:403–09 (1990).

Staddon, C. *Managing Europe's Water: 21st Century Challenges*. London: Ashgate, 2009.

Chad Staddon
University of the West of England, Bristol

WATER POLLUTION

The contamination of point sources like rivers, lakes, seashores, and coastal waters is an example of the impact of city-generated wastes on wider regions. Here a scientist samples pollution levels in a creek.

Source: iStockphoto

Water that falls on the Earth's surface moves through runoff (surface flow), cutoff (groundwater), and fly-off (atmospheric water). In each of these three levels (surface, ground, and atmosphere), water is being polluted by several agents. Water pollution means unfavorable alteration of water, largely by human actions through modification of energy and/or matter to such an extent that degradation disturbs the natural balance. Heat and chemicals alter the physical and chemical properties of water and thus contaminate surface water, groundwater, and atmospheric water adversely. Most water pollutants fall into one of these three categories: liquid organic wastes, liquid inorganic wastes, and waterborne or water-based pathogens.

Surface water includes both ocean water and water that stands or flows on land. Ocean water pollution, which mainly occurs as a result of leakage of crude oils from oil tankers and offshore oilfields, needs special mention. The world's largest oil spill, estimated to be around 8 million barrels, happened during the Gulf War in 1991. Drains and rivers also add new pollutants to oceans. However, nonpoint sources of pollution (fertilizers, garbage, plastic bags, and chemicals) do more harm to oceans. Lake water, which gets polluted through varying agencies, has an added disadvantage. As it is stagnant water, unlike rivers and oceans, the pollutants are concentrated in lake water over time. Organic wastes from hilly slopes, toxic effluents from urban areas, and chemical nutrients from agricultural fields also increase the toxicity of lake water. Lake water eutrophication is a common phenomenon.

Such wastes generally stretch the biological capacities of aquatic system, resulting in a foul-smelling, waste-filled body of water. This has already occurred in such places as Lake Erie and the Baltic Sea, and is a growing problem in freshwater lakes all over the world. The U.S. Department of Health, Education, and Welfare has classified surface water pollutants in the following eight categories: sewerage and waste, infectious agents (germs and viruses), plant nutrients and dissolved substances (fertilizers, detergents), particulate matters (soil and mineral particles), radioactive substances, mineral and chemical substances, heat, and organic chemical exotics (e.g., pesticides, insecticides, herbicides, and rodenticides).

Compared with surface water, groundwater is generally less susceptible to contamination and pollution. Seawater intrusion is the biggest threat to groundwater pollution, in addition to chemical contamination through fertilizers, pesticides, septic tanks, landfills, storage tanks, road salts, and so on. Indiscriminate use and overexploitation also cause aquifer contamination. For instance, arsenic contamination of the groundwater is now alarming in many Asian countries, and globally over 137 million people are exposed to arsenic through drinking water. Arsenic is derived from the oxidation of arsenic-rich pyrite in the shallow aquifers as a result of lowering of the water table because of overextraction of groundwater for irrigation. Similarly, incidences of fluoride, nitrate, and chlorine contamination above permissible levels can make water unfit for human consumption. Pollution of groundwater resulting from industrial effluentsand municipal wastes in water bodies is another major concern in many cities and industrial clusters.

Acidification of surface water—mainly lakes and reservoirs—is one of the major environmental effects of transport over long distances of air pollutants, such as sulfur dioxide (from power plants and heavy industries like steel plants) and motor vehicles. This problem is more severe in the United States and in parts of Europe. Excessive ejection of sulfur dioxide into the air through the combustion of fossil fuels from factories and automobiles leads to formation of sulfuric acid. Presence of acid in rainwater is termed *acid rain*. Such rain, which basically pollutes the atmospheric water, ultimately contaminates the surface water, destroys crops and forests, and damages buildings and monuments.

Urban Water and Pollution

The scale of water pollution has affected urban areas more than rural locales. There are usually four main sources of water pollution in cities: sewage, industrial effluents, storm and urban runoff, and agricultural runoff. The relative importance of the different pollutants and the relative contribution of the different sources vary greatly from city to city. This not only damages the water bodies (and aquatic life within them including fisheries), but also contaminates freshwater sources that cause major health problems for those who subsequently use them.

In most cities in low- and middle-income nations, the problems are not so easily addressed—they have much more serious nonpoint sources of water pollution because of the lack of sewers and drains and inadequate services to collect solid wastes. A lack of solid-waste collection adds to water pollution problems because many of the uncollected wastes are washed into streams, rivers, or lakes, adding to the pollution load. Although no city or smaller urban center can exist without reliable sources of freshwater, very few urban authorities have paid sufficient attention to safeguarding sources and preventing water pollution. Existing evidence suggests that a large proportion of the population in many cities has no access to protected piped water supplies. This often means an increasing number of people drawing on polluted sources (e.g., using polluted rivers, streams, lakes, and groundwater).

The contamination of point sources, namely, rivers, lakes, seashores, and coastal waters, is an example of both the effect of city-generated wastes on the wider region and of governments' negligent attitude toward protecting open areas. Untreated waste-water and sewage from households and commercial and industrial enterprises are usually discharged into nearby lakes, rivers, or seas. This often leads to serious health problems for large numbers of people, whose water supply is drawn from these sources. Liquid wastes from city activities have environmental impacts stretching beyond the immediate hinterland. It is common for fisheries to be damaged or destroyed by toxic chemicals, as a result of which hundreds or even thousands of people lose their livelihood. Among the places where major declines in fish catches have been documented are many rivers, estuaries, or coastal waters in China, India, Malaysia, the Gulf of Paria between Venezuela and Trinidad, Manila Bay, the Bay of Dakar, and the Indus Delta near Karachi.

Conclusion

Water pollution causes irreparable damage to all living beings, including humans. A majority of infectious diseases, namely typhoid, cholera, paratyphoid fever, bacillary dysentery (bacterial infections), infectious hepatitis, poliomyelitis (viral infection), and amoebic dysentery (protozal infection), are all spread by polluted water. The organophosphates, carbonates, and chlorides present in pesticides affect and damage the nervous system and may also cause reproductive and endocrinal damage. Pesticides containing carcinogens that exceed the recommended levels may even cause cancer.

See Also: Greywater; Nonpoint Source Pollution; Stormwater Management; Water Conservation; Water Sources and Delivery; Water Treatment.

Further Readings

Dash, M. C. *Ecology, Chemistry and Management of Environmental Pollution.* New Delhi, India: Macmillan, 2004.

The Energy and Research Institute. "Water Pollution." http://edugreen.teri.res.in/EXPLORE/water/pollu.htm (Accessed April 2009).

Hardoy, Jorge E., et al. *Environmental Problems in an Urbanizing World.* London: Earthscan, 2004.

Kolsky, Peter J. "Water, Health and Cities: Concepts and Examples." Paper Presented at the International Workshop on Planning for Sustainable Urban Development, University of Wales, July 1992.

Sharma, B. K. *Environmental Chemistry.* Meerut, India: Goel, 2005.

Viessman, Warren, Jr., et al. *Water Supply and Pollution Control*, 8th Ed. Upper Saddle River, NJ: Prentice Hall, 2008.

Mohua Guha
Aparajita Chattopadhyay
International Institute for Population Sciences, Mumbai

WATERSHED PROTECTION

Watersheds are topographically defined regions that are drained by a given stream or river system. They occur along a range of scale—anywhere from small headwater streams with small catchments to large rivers with watersheds that cover continental regions. The qual-

Watershed protection includes maintaining adequate flows in streams and rivers, like these terraces, buffer strips, and grass plantings at a lake in Iowa.

Source: Lynn Betts/Natural Resources Conservation Service

ity and quantity of water entering streams, rivers, lakes, and bays is often reflected in the land uses and land cover within the watershed. Watershed protection is considered any of the myriad of human measures (laws, policies, programs, and practices) intended to protect water quality and quantity in a watershed. These measures are numerous and often specific to land uses.

The 1972 Clean Water Act (CWA) and subsequent amendments represent the primary federal law intended to maintain and improve water quality in the United States. Although water pollution continues to be problem, substantial improvements have been achieved through the CWA. Some of the most important advances have resulted in better regulation of point source pollution (such as discharge from a factory pipe). However, to meet the goals of the CWA, it will be necessary to decrease pollution from nonpoint sources as well. These include the many diffuse sources that occur throughout the landscape.

To address nonpoint source pollution, the CWA includes provisions to maintain or improve water quality in watersheds. For instance, states are responsible for designating total maximum daily loads (TMDLs) to streams that are not meeting water quality standards. TMDLs are the maximum pollutant load that the stream can assimilate while still meeting water quality standards. Once a stream is designated, meeting the TMDL often is dependent on a variety of watershed partners (both public and private) to work collectively and meet water quality goals. Because stream water quality and quantity are a function of all the contributing land factors in a watershed, the challenge of managing watersheds is gaining the cooperation of multiple owners, users, and stakeholders. As a consequence, programs and measures intended to protect watersheds are often a combination of regulatory and incentive-based.

Incentive programs have been used extensively to implement watershed protection measures. According to the most recent National Water Quality Inventory by the Environmental Protection Agency, excess sediment from agricultural lands remains one of the main sources of pollution in streams and rivers of the United States. In response to this, the U.S. Department of Agriculture administers several incentive programs, including the

Conservation Reserve Program. The Conservation Reserve Program was initiated in 1985 as a voluntary program for agricultural landowners. It uses rental payments and cost-shares to encourage farmers to stop production on lands that are susceptible to erosion or sensitive in other ways. Once land is enrolled, it is usually planted with a perennial grass and retired from production for 10 years. Often croplands on historical floodplains are enrolled because they flood regularly and are susceptible to soil loss. From the farmer's perspective, these fields may only be productive during drier years, and thus have only marginal value. In this case, there is usually an incentive for the farmer to enroll these lands as well, so they can receive a predictable payment for them.

Best management practices (BMPs) are a wide variety of preventive measures developed to protect water quality from a variety of land uses. In the timber industry, BMPs are used to reduce soil erosion and disconnect sediment runoff from streams. Sediment runoff can originate from road construction, skid trails, harvesting operations, and postharvest site preparation. Other water quality problems can involve the improper use of pesticides and fertilizers. Nevertheless, water quality from managed forests generally remains high because of the high rate of forestry BMP implementation. For example, streamside management zones are forested areas alongside streams that are preserved to buffer streams from nearby operations. These zones can intercept runoff and protect the land immediately next to a stream from soil disturbance and erosion.

Roads are common features in watersheds and have been identified as a major source of pollutants to streams and rivers. The cumulative effects of road construction, use, and maintenance can contribute pollutants such as sediments and chemicals. Roads that are graded but unimproved (not paved or graveled) usually leave exposed mineral soils. When precipitation occurs, the energy from falling raindrops can dislodge soil particles and move them with storm runoff into streams and rivers. Paving or graveling roads reduces potential erosion; however, runoff can contain high levels of hydrocarbons and other chemicals related to automobiles. Other pollutants may come from the breakdown of road surface material. In northern latitudes, roadway deicing chemicals have been shown to affect streams and other aquatic habitats. Increasingly, stormwater basins and detention areas are being incorporated into roadway plans to collect runoff and remove pollutants before they discharge into streams.

Stormwater and land use planning have evolved to protect watersheds from urban growth. There is normally a limit to the amount of land that can be urbanized before water quality declines. As lands become urban, they also become increasingly impervious. Roads, roof tops, and other features associated with urbanization reduce the amount of precipitation that percolates into the ground and increases rapid drainage via storm sewers into streams and waterways. This can drastically affect streams by altering hydrology, increasing pollutant loads, and damaging habitats. However, these effects can be mitigated by careful stormwater planning that may include stormwater ponds or basins to attenuate storm flows. There is also increasing emphasis on the preservation of urban green space to reduce stormwater volumes.

As the demand for water increases throughout the world, there is less available in rivers and streams to support ecosystem requirements. Watershed protection also includes measures designed to maintain adequate flows in streams and rivers. *Environmental flow* is the term used to describe the quantity of water needed to maintain a healthy aquatic ecosystem. Fish and wildlife populations are dependent on a certain amount of water being available. Other communities such as riparian wetlands and bottomland forests require certain flows to function properly. Humans also benefit from adequate river flows—we rely on these waters to dilute our wastewater, provide recreation, support power generation, and maintain viable fisheries. Unfortunately, determining an adequate environmental flow requires considerable investigation. Rivers and streams are dynamic systems and do not maintain steady flows. Conditions naturally change within and between years. Often, seasonal variability (high and low) of river flow is a key condition for aquatic organisms. Few water agencies around the world have implemented guidelines to protect environmental flows, but with greater societal demand for water, it has become increasingly necessary. Determining how much water can be removed from a river system without impairing ecosystems will be a major challenge for watershed protection this century.

See Also: Coastal Zone Management; Green Infrastructure; Stormwater Management; Water Pollution; Water Sources and Delivery.

Further Readings

Brooks, K. N., et al. *Hydrology and the Management of Watersheds*, 3rd Ed. Ames, IA: Blackwell, 2003.
Richter, B. D., et al. "Ecologically Sustainable Water Management: Managing River Flows for Ecological Integrity." *Ecological Applications,* 13/1 (2003).
U.S. Environmental Protection Agency. "National Water Quality Inventory: Report to Congress, 2002 Reporting Cycle." EPA 841-R-07-001. Washington, D.C.: U.S. Environmental Protection Agency, 2007.

Christopher Anderson
Auburn University

WATER SOURCES AND DELIVERY

Access to sufficient supplies of clean water is absolutely central to maintaining human life. Biologically, we are more than two-thirds composed of water, and even a minor deficiency in water—say 5 percent of our biophysical need—can seriously debilitate a human being. We can survive weeks without food, but only days without water (less in hotter, more arid places). Indeed, water is so central, and so contested, that there is now a movement to establish a legally enforceable global right to water.

Because the need for water is so universal it can hardly be a surprise that the history of water provision is rich, varied, and reaches back to the earliest human conurbations. Only slightly less obvious, water has a complex geography in at least two senses—it exists in some places in abundance, and not at all in others. What is more, water moves. Thus, a river that services the needs of an upstream city may also serve one or more towns and cities farther downstream. Both the geographies of water as a "stock" resource, as well as the geographies of water as "flow," generate complex "hydropolitics," which, many commentators agree, may do much to shape the 21st century.

A Short History of Water Sources and Delivery Systems

Possibly inspired by this deep cultural groundwater, various urban historians have pointed to the centrality of control over water in the urbanization process. Virtually all urban places established before the 20th century were located on or very near sources of clean water, for both water supply and for sewerage. To take just one early example, the water system that supplied Imperial Rome from the 4th century B.C.E. eventually deployed 11 major aqueducts to bring water more than 40 kilometers to the city through underground tunnels made of stone and terra cotta. Other cities, especially ones associated with Islam, raised the functional manipulation of water and watercourses into philosophical and aesthetic principles. At the Alhambra in Granada, Spain, one can still see an entire palace complex organized around the skillful manipulation of water.

By the 19th century, increasingly rapid urbanization made the reorganization of water supply and sewerage networks a priority for local and national governments. Advances in public health science attended famous instances such as Dr. John Snow's dramatic demonstration of the link between contact with water contaminated with human fecal waste and diseases like cholera, dysentery, and typhus. Simultaneously, urban growth boosters realized that it would be necessary to bring water from farther afield to supply the growing industrial cities of the United Kingdom, Western Europe, and the northeastern seaboard of the United States, inspiring the golden age of hydroengineering that lasted at least until the late 20th century. In the United Kingdom, evidence of this golden age abounds in the 21st-century landscape, from the neoromantic crenellated structures associated with the reservoirs of central and north Wales, to the Great Central Interceptor constructed by Joseph Bazalgette under Victorian London, with the sole purpose of collecting together the burgeoning metropolis's sewage so that it could be pumped and dumped farther away up the Thames Estuary. Associated with these engineering achievements was the development of water-quality treatment systems.

Of course, technical know-how could not have enacted such remarkable feats of hydroengineering (whatever your opinion of large dams from social or environmental perspectives) as China's Three Gorges project (one of the largest hydroengineering schemes ever built), or even the Roman aqueducts of southern Europe, in the absence of strong political will. Indeed, the story of hydroengineering is also the story of the convergence of political and technical programs around a modernist ideal of "progress" and "development" that made such projects not only conceivable but imperative.

A Short Geography of Water Sources and Delivery

Although almost 100 percent of North Americans and Europeans have access to abundant clean drinking water in the home, only 28 percent of Kenyans, only 38 percent of Congans, and only 69 percent of Mexicans have such access. Moreover, approximately 40 percent of the world's population does not have access to adequate sanitation. In water resources planning, it is common to classify countries with a water balance in excess of 2,000 m³/capita as "water rich," whereas those with less than 1,000 m³/capita are considered "water short." Middle-range countries, such as India and China, have 1,000–2,000 m³/capita and are considered "water scarce." It is immediately obvious from readily available data that not only are there countries of clear abundance (Canada or Iceland) and countries of worrying shortage (Egypt, Libya, and Kuwait), but also regions of abundance (North America) and regions of shortage (North Africa and the Middle East). Yet these data present, at best, a partial picture, as even physical hydrological abundance does not guarantee that everyone has equal access to the water they need. Nor do these statistics take account of the likely effects of climate change, particularly in terms of making much of even the relative abundance enjoyed by the United Kingdom unavailable by concentrating it into higher-frequency storm events. In late 2007, Sir Nicholas Stern published a report on the projected economic effects of climate change in which he pointed out that reduced water availability in low-latitude countries could result in widespread crop failures and displacement of human populations.

So-called hydropolitics have emerged onto the international diplomatic scene as a key threat to the stability of the global political economic order. Water resources in certain regions of the globe are under increasing pressure, and this pressure is resulting in a new and dangerously belligerent attitude toward the allocation of water resources, especially in watersheds that are shared by a number of states. Because most watersheds are shared, the potential for conflict over water is great indeed. Speaking about the countries of the Nile Basin in 1985, former UN Secretary-General Boutros Boutros-Ghali said, "The next war in our region will be over the waters of the Nile." Since that time, the Nile Basin Initiative has worked hard to bring together the riparian nations of the Nile basin to work together to achieve collaboration, consensus, and equitable division of increasingly scarce water resources. Periodic threats by the Egyptian government against upstream users, however, shows how fragile the situation remains.

Contemporary Challenges

In recognition of the histories and geographies of water sources and delivery given earlier, the United Nations has twice attempted to focus international attention and investment on water and sanitation through the proclamation of "International Decades of Water and Sanitation" (1981–1990 and 2005–2015). Although more than 1 billion people were reached over the course of the first International Water Decade, global inequalities in access grew even faster. By the end of the decade, more than a billion people were still without access to clean water—a situation arguably worse than a decade before. Several years into the second International Water Decade, the situation remains much the same. An interim "report card" for water services published in 2008 presented a mixed picture: though great strides have been made, more than a billion people remain without access to clean water and sanitation facilities.

In developed countries the challenge is less about absolute levels of provision and more about doing more with less water. For example, although much of the world's population gets by with less than 50 liters of water per person per day, Americans use on average upward of 300 liters per person, per day, and Europeans 150 liters per person, per day. To date, efforts to encourage water conservation have largely concentrated on the promotion of lower-water-use fixtures and fittings, such as dual-flush toilets, reduced-flow appliances and shower fittings, and decreased landscape water use. The latter generally comprises about 30 percent of a municipality's total treated water use, with the residential sector being the largest user of urban water. Trials are also under way linking universal water metering with so-called rising block tariffs to provide strong economic imperatives to conserve.

Notwithstanding the above, however, a key challenge for the 21st century will be to achieve some sort of global consensus on the status of water (and water services) as either private goods or common goods. Although the logic of treating at least water services as a matter for private enterprise seems compelling, many commentators are not so sure, essentially arguing for a "Blue Covenant" that would treat access to water and sanitation as an enforceable human right. Because the past generation has witnessed the extensive rolling out of the logic of services privatization across the globe, challenging this established movement will be a major and a contentious undertaking with potentially far-reaching consequences for us all, requiring what R. Varady, B. Karkkainen, and others have called a new "ontology of global water initiatives."

See Also: Millennium Development Goals; Sustainable Development; Water Conservation; Water Treatment.

Further Readings

Barlow, M. Blue Covenant: *The Global Water Crisis and the Coming Battle for the Right to Water*. New York: The New Press, 2008.

Gardner-Outlaw, T. and R. Engelman. *Sustaining Water, Easing Scarcity: A Second Update*. Washington, D.C.: Population Action International, 1997.

Gleick, P *The World's Water: The Biennial Report on Freshwater Resources*. Washington, D.C.: Island, 1999.

Karkkainen, B. "Post-Sovereign Environmental Governance." *Global Environmental Politics*, 4/1:72–96 (2004).

Lowi, M. *Water and Power: The Politics of a Scarce Resource in the Jordan River Basin*. Cambridge: Cambridge University Press, 1995.

Reisner, M. *Cadillac Desert: The American West and Its Disappearing Water*. New York: Penguin, 1993.

Staddon, C. *Managing Europe's Water Resources: 21st Century Challenges*. London: Ashgate, 2009.

Varady, R., et al."Charting the Emergence of 'Global Water Initiatives' in World Water Governance." *Physics and Chemistry of the Earth*, 34:150–55 (2009).

Chad Staddon
University of the West of England, Bristol

WATER TREATMENT

There are at least three ways to protect water quality in water supply systems:

- Use a relatively pristine natural source
- Manufacture a pristine source through storage reservoirs
- Filter or treat nonpristine water sources

Historically, water providers have tended to take a supply-side perspective, always seeking to solve quantitative or qualitative shortage issues by extending their supply systems deeper into remaining pristine sources (first strategy above), but over time they have been forced to enact the second and third strategies. As explored by other articles in this volume, strategy two dominated water management practices until the last quarter of the 20th century. Some commentators now believe that the time has long since passed when such supply-side strategies were appropriate and that demand-side management (forcing/encouraging water users to use water more efficiently) is required.

Water pollution sources may be split into "point" and "nonpoint" (or diffuse) sources. Comparison of the share of nonpoint and point sources in water pollution in the United States demonstrates the high degree of variability of pollution by source—some areas being dominated by one or the other, or perhaps both. Heavy metals such as cadmium and mercury can often enter the environment primarily through diffuse mechanisms, whereas organic pollution is usually localizable; that is, traceable to a source. Nonpoint source pollution is, by definition, harder to treat at source, and so treatment options tend toward reliance on the natural self-purification processes attendant on surface water and groundwater and on industrial wastewater collection, treatment, and release back into the natural environment. Urbanization also results in a certain amount of surface and groundwater contamination. For example, groundwater quality in the Marlborough and Berkshire Downs and the Kennet Valley in southern England is monitored using a network of public supply and private abstraction boreholes. In all cases there remains a heavy (over)reliance on the age-old water engineering mantra: "dilution is the solution to pollution."

The management of phosphate pollution provides a good example of problems attendant on a pollutant with both point and nonpoint sources. At this time, the European Union and its member states are looking closely at strategies for managing phosphates in surface and subsurface waters. Although the two primary sources of phosphates are well known (i.e., detergents and fertilizers), removing or reducing them from water can be much more difficult than for other pollutants. Several member states have banned phosphates in detergents, but others argue that the cost of the measure is disproportional to the potential benefit, whereas in other states, the contribution from agriculture or even from sewage effluent is more significant and more difficult to manage. Some member states, including Germany, Italy, Hungary, and the Netherlands, have phased out phosphorus-based detergent activators and have seen gradual recovery of surface water quality.

It has already been mentioned that natural self-purification processes occur in groundwater and surface water, which can lead to improvements of its quality. Self-purification of surface water is the complete set of biological, physical, chemical, and hydrological processes leading to reduction of water pollution in water courses and reservoirs through biodegradation, sedimentatation, reaeration, dilution, and adsorption. The microorganisms that participate in the self-purification process consist primarily of bacteria: aerobes (which require the presence of oxygen dissolved in water) and anaerobes (which do not occur in the presence of oxygen, but take oxygen from chemical compounds). In aerobic conditions, decomposable organic compounds (carbohydrates, fats, and proteins) give the following final products: carbon dioxide, water, nitrates, and phosphates. In anaerobic conditions, the following are produced: methane, ammonia, hydrogen sulfite, alcohols, and organic acids. The aerobic decomposition of pollution is a much more intensive and beneficially productive process that does not produce as many toxic decomposition products (though too much of anything can be toxic in certain circumstances). Self-purification of water sources is usually described through the well-known Streeter-Phelps formula:

$$C_t = C_0 * e^{-k_1 * t}$$

in which C_t stands for the pollutant concentration after a period t (in milligrams per liter), C_0 stands for the initial pollutant concentration (in milligrams per liter), t stands for the duration of the decomposition process (in days), and k_1 stands for the constant decomposition rate (1 per day).

Wastewater treatment plants are hydrotechnical facilities focused on the reduction of pollution loads found in wastewater to a level that ensures safe discharge into recipient water bodies. These plants tend to attempt to mimic, and intensify, the self-purification processes outlined earlier. Depending on the exact technology applied, we may distinguish the following types of treatment method:

- Mechanical techniques use hydrodynamics (including the creation of vortices), as well as screening and deposition processes, to provide primary treatment of wastewaters
- Biological techniques use various sorts of aeration techniques to speed the otherwise natural processes of aerobic digestion of pollutants to provide secondary treatment of wastewaters
- Chemical techniques add specific reagents or apply specific processes to provide tertiary treatment of wastewater. Exposure to ultraviolet light is an increasingly common method of tertiary chemical treatment of wastewater

Municipal wastewater (which most often consists of domestic wastewater with the addition of stormwater and pretreated industrial wastes) is usually treated with mechanical and biological, or biological with improved removal of specified nutrients, wastewater treatment plants. Urban wastewater is collected using a system of household drains, pipes, sewer systems, sewer catch basins, and combined sewer overflows (CSOs), and is transported by wastewater pumping stations to treatment plants.

The most common nutrients requiring specific tertiary treatment (often with expensive flocculation or ultraviolet treatments) are nitrates and phosphates. Mechanical treatment is insufficient—reduction of pollutant load is not high enough. In some cases biological treatment may also be insufficient because of the amounts of phosphorus and nitrogen left in the effluent; in such cases, it is necessary to extend conventional biological treatment onto a phase for removal of nitrogen compounds (denitrification) and/or chemical precipitation of phosphorus compounds. In the case of relatively small sources of domestic wastewater, specially prepared facilities such as filtration beds, sprinkled slopes, algae, and oxidation ponds and bogs are used. Technical solutions differ, but the pollutant reduction principle is always the same: They are based on biological decomposition. Such wastewater treatment methods are called natural or seminatural. Industrial wastes are basically treated by mechanical and chemical treatment, and the treatment technology applied depends on the type of pollution.

In Europe, water treatment standards are codified in the Water Framework Directive (2000), the Urban Waste Water Directive of 1991, and its many "daughter" directives, and in the United States, water standards are detailed in the Clean Water Act (1977) and the Water Quality Act (1987), and by their associated federal and state regulations. Most regulatory systems use maximum permitted concentration limits to control an increasingly wide variety of pollutants.

See Also: Denitrification; Nonpoint Source Pollution; Water Conservation; Water Pollution; Water Sources and Delivery.

Further Readings

Bartlett, R. E. "Developments in Sewerage." *Applied Science*, 1 (1979).
Barty-King, H. *Water: An Illustrated History of Water Supply and Wastewater in the United Kingdom.* London: Quiller, 1992.

Cech, T. *Principles of Water Resources: History, Development, Management and Policy.* Hoboken, NJ: John Wiley & Sons, 2005.

Downing, R. A. *Groundwater: Our Hidden Asset.* Keyworth, UK: British Geological Survey, 1998.

European Commission. "Urban Waste Water Directive Overview." http://ec.europa.eu/ environment/water/water-urbanwaste/index_en.html (Accessed December 2009).

Glennie, E. B., et al. *Phosphates and Alternative Detergent Builders: Final Report.* Copenhagen, Denmark: EU Environment Directorate, 2002.

Gray, N. F. *Drinking Water Quality: Problems and Solutions.* Cambridge, UK: Cambridge University Press, 2008.

McDonald, A. and D. Kay. *Water Resources.* Harlow: Longmans, 1988.

Tebbutt, T. H. Y. *Principles of Water Quality Control,* 5th Ed. London: Butterworth Heinemann, 1998.

Chad Staddon
University of the West of England, Bristol

WETLANDS

The term *wetlands* encompasses a wide variety of topographic conditions, including swamps, bogs, marshes, fens, pocosins, and vernal pools. The defining characteristics of all of these locations is frequent saturation (whether by saltwater or fresh), a high level of biodiversity and energy transfer, and extreme sensitivity to human interference. The Environmental Protection Act (40 CFR 230.3(t)) defines wetlands as "those areas that are inundated or saturated by surface or groundwater at a frequency and duration sufficient to support, and that under normal circumstances do support, a prevalence of vegetation typically adapted for life in saturated soil conditions."

Some wetland plants adapt to specific salinity-tolerance ranges in estuarine locations where fresh water meets coastal marine systems. In a Louisiana bayou, a crane installs a self-regulating tide gate that will restrict the inflow and retention time of saline water.

Source: Erik Zobrist/NOAA Restoration Center

Wetlands are significant because of their contribution to the health and stability of both the terrestrial and the aquatic habitats that they straddle. Wetlands have been called *ecotones*—a term that describes a transitional ecosystem between two adjacent and distinct communities. It is this transitional position that gives wetlands both their fertility and their vulnerability. Shallower and occasionally saturated locations support species that cannot survive on uplands or in marine ecosystems. Solar energy is available at all depths in a wetland and cyclical saturation, whether tidal or seasonal, and promotes a resilient adaptation to changing conditions. Some wetland plants, for instance, adapt to very specific tolerance ranges of salinity in estuarine locations, where freshwater rivers and streams meet the tidal conditions of coastal marine systems. Disturbances that vary these salinity conditions, such as either excessive salt- or freshwater or alterations to the elevation of the shoreline, will allow other plant species to invade, crowding out the native wetland plants and interrupting the delicate food chain dynamics.

The variety of wetlands types and the fact that each of these may not in fact seem "wet" at any point in time has created ambiguity and confusion. Vegetation tolerant of saturated soil conditions has become the defining characteristic, as these plants cannot live anywhere but in a wetland. However, identifying such vegetation requires knowledge and experience, and therefore typically involves the services of an expert. In most cases it is the transitional nature of a wetland that makes it readily identifiable.

An estuary may provide one of the more dramatic examples of such a transitional wetland. Most estuaries occur where a river meets the sea, as in a bay, fjord, sound, and so on. The point of transition between the two water bodies is marked by tidal fluctuations, a mixture of fresh- and saltwater, and a high degree of biodiversity and ecological productivity. Because such areas provide abundant food sources and navigability, estuaries are favored locations for human settlements. In fact, of the 32 largest cities in the world, 22 are located on estuaries.

Marshes are characterized by the presence of grasses and reeds (cattails being emblematic), as well as occasional shrubs or woody plants. Marshes are often found in estuaries and constitute critical habitat for migrating birds, as well as productive generators of nutrients for marine food chains. A marsh that borders a body of saltwater is saline, referred to as a *saltwater marsh*. A marsh in an estuary where saltwater predominates over freshwater has specific characteristics and is known as a brackish marsh. There are also freshwater marshes that occur inland, near rivers or lakes. The largest freshwater marsh in the United States is the Everglades in southern Florida.

A swamp is distinguished from a marsh by the presence of woody vegetation and acidic soils. The distinction can be a fine line (some sources refer to the Everglades as a swamp). However, a "true" (forest) swamp is easily distinguished from the flat grassy contours of a marsh. Trees adapted to growth in standing water such as bald cypress and Water Tupelo are defining features. Swamps do not occur next to bodies of saltwater, although rivers often flow into and out of swamps.

A bog is a specific wetland type characterized by the presence of acidic peat. Bogs are typically found in colder northern locations, including Finland, Estonia, Canada, and Scotland, where lower average temperatures retard the decomposition of organic material. A bog can be a stage in succession from a body of water to land (a process known as "hydrosere"). The typical bog types include valley, raised, blanket, and quaking bogs, each having distinct characteristics. The common thread is peat, an assortment of partially decayed plant material as well as fungi and other organic remains. Similar to all wetlands, bogs are fertile, vulnerable places. The use of peat for burning as a heat source is common in settlements located near bogs.

Marshes, swamps, and bogs all qualify as *palustrine wetlands*. The term stems from the Latin word for marsh (*palus*) and denotes a wetland without flowing water, tidal influence, or salt content. A pocosin is a palustrine wetland with many overlapping characteristics of a bog, although a pocosin is more likely to support shrubs and other wooded vegetation. The term *pocosin* is thought to derive from the Algonquin term for "swamp on a hill," a reference to the characteristic elevated nature of this wetland type. The accumulation of organic material over millennia creates a wet deposit of acidic and nutrient-poor soil, which supports only very hearty vegetation. Pocosins are found in the mid-Atlantic states, especially North Carolina. The ecotone condition in which a pocosin meets a long-leaf pine savannah is productive of plants with peculiar adaptations such as the Venus fly-trap (*Dionaea muscipula*) and rough-leaf loosestrife (*Lysimachia asperulifolia*).

A fen is distinguished from a bog by its soil, which is alkaline or neutral more often than it is acidic, and by the fact that it is fed by surface and subsurface groundwater, rather than just by rainwater. A fen will support a more varied plant community than a bog because of lesser acidity and higher levels of nutrients. A vernal pool is a seasonal body of water that is dry for some period each year, typically recharging through snowmelt or seasonal rain events. This cycle ensures that fish are unable to live in a vernal pool, consequently making these areas an ideal breeding habitat for certain insect and amphibian species. Unlike other wetlands, a vernal pool is more typically defined by the presence of certain amphibians (wood frogs, spadefoot toads, and salamanders) or branchiopods (freshwater crustaceans), such as fairy shrimp.

Vernal pools have gained attention recently because of their critical importance in the life cycles of endangered amphibian species and their fragility and elusiveness. As such, they represent the vulnerable nature of wetlands in general. Our understanding of the ecological importance of wetlands is less than a century old. It is still common to destroy wetlands for development, and the current existing wetlands in North America cover a fraction of the area of the wetlands present when the first European settlers arrived. Because of their capacity to absorb large volumes of water, wetlands have a storage dimension that acts as a flood control mechanism. Areas where settlement has compromised this capacity are now more vulnerable to flooding. Rising sea levels resulting from the melting of polar ice caps are threatening millions of acres of coastal wetlands around the globe. Protection of wetlands is therefore a critical policy priority with implications for numerous species and habitats, including our own. This lesson was made amply clear in New Orleans as a result of the devastation wreaked by Hurricane Katrina. New York City, located at the mouth of the Hudson River, is also a prime example of a metropolis where the loss of extensive wetlands has resulted in a vulnerability to flooding due to major storm events or sea level rise. Sea level rise, which some estimates suggest could be as much as 26 inches along U.S. coasts by 2050, will cause significant areas of remaining coastal wetlands to disappear, further depleting this natural resource for flood control and habitat diversity. Whereas the coastal cities of the world have developed, based on the many advantages provided by their locations, along estuaries and wetlands, this development pattern is now subject to acute vulnerability. While massive flood control engineering projects, similar to the Thames flood barrier, may work in certain circumstances, addressing the critical importance of wetland preservation and restoration is certainly an essential ingredient in a comprehensive response to the threats posed by this aspect of global climate change.

See Also: Environmental Impact Assessment; Habitat Conservation and Restoration; Parks, Greenways, and Open Space; Sea Level Rise.

Further Readings

Biebighauser, Thomas R. *Wetland Drainage, Restoration, and Repair*. Lexington: University Press of Kentucky, 2007.

Mitsch, William J., et al. *Wetland Ecosystems*. Hoboken, NJ: Wiley, 2009.

U.S. Environmental Protection Agency. "Wetlands." http://www.epa.gov/wetlands (Accessed November 2009).

U.S. Geological Survey. "About Wetlands." http://www.nwrc.usgs.gov/wetlands.htm (Accessed November 2009).

A. Vernon Woodworth
Boston Architectural College

XERISCAPING

Xeriscaping means landscaping designed to require little to no supplemental water above and beyond natural precipitation in the location being landscaped. The Environmental Protection Agency's Office of Water predicts that 36 states will experience water shortages by 2013. The Colorado Water Wise Council estimates that more than 50 percent of residential water use is applied to landscape and lawns and that 40 to 50 percent of total urban water use is for landscaping. Encouraging urban residents to use water more efficiently, then, is extremely important to ensure that water supplies are available for the indefinite future. These concerns are especially important in the water-scarce cities of the western and southwestern United States where the term xeriscape was first coined.

The 1977 drought in the western United States created a scare that led to the creation of a task force of water, horticultural, and landscaping experts from the Denver Water Department, Colorado State University, and the Associated Landscape Contractors of Colorado. The focus of the conference was water supply planning, which attempts to extend existing water supplies to meet future needs. One of the greatest challenges was to demonstrate that it was indeed possible have both beautiful gardens and use significantly less water. The term *xeriscape* was coined in 1981 by this task force. The term is trademarked by Denver Water and refers

Xeriscaping promotes landscapes that need little supplemental water and the use of mulches to suppress weeds. The Hartman Prehistoric Garden in Zilker Park in Austin, Texas, is an example of xeriscaping.

Source: iStockphoto

to water-efficient and bioregionally appropriate landscaping. The name is derived from the Greek *xeros* for "dry" and the English term *landscape*. In 1982, Denver Water donated land and 55 landscape businesses contributed money for a successful demonstration garden.

There are seven principles that serve as the basis for the concept of xeriscaping:

- Planning and design
- Limiting turf areas
- Selecting and zoning plants appropriately
- Improving the soil
- Using mulches
- Irrigating efficiently
- Doing appropriate maintenance

The term *xeriscape* is flexible and allows for various levels of effort that can be tailored to meet specific local situations. Xeriscaping promotes landscapes that need little supplemental water. Acknowledging that droughts occur, it focuses on planting drought-tolerant landscapes, yet does not refer to a dry, barren landscape or a no-maintenance landscape. Instead it promotes the use of plants that are well adapted, in tandem with mulches that suppress weeds and conserve water. Xeriscaping does not imply a rock-strewn landscape devoid of vegetation and, in fact, can be lush, colorful, and easy to care for.

In addition, xeriscaping does not mean creating a landscape that requires no water. In fact, a xeriscaped lawn may require watering on a regular basis. However, it does encourage smart design that decreases the need for water. Efficient irrigation, such as watering at night instead of during the heat of the day and using properly designed hoses and sprinklers, is crucial to xeriscaping. Proper weeding, mowing, pruning, and fertilization can also help decrease water use. Zoning plants with similar needs in the same area can also greatly reduce the amount of water wasted. Ideally, the principles of xeriscaping benefit homeowners, businesses, and landscapers by encouraging them to plant and maintain shrubs, flowers, and lawns that are more suited to local soil and climate conditions.

The benefits of xeriscaping include the following:

- Sustainable water supply
- Reduced energy and site-maintenance costs
- Increased property values
- Lower water bills
- Improved landscape aesthetics
- Protection of native habitat including estuaries, streams, ponds, and lakes
- Reduced desertification

By using plants that are already adapted to the local climate, precipitation, and soil type, xeriscaping can also result in fewer disease and insect problems and lower maintenance needs. Successfully implemented, xeriscaping can raise property values by up to 15 percent.

Challenges to Xeriscaping

The most obvious challenge to the promotion of xeriscaping is fear of change. Many plants commonly used in landscaping come from the cool and moist climate of England. These plants are not necessarily well adapted to their new homes, and thus can require more water than the climate generally offers.

City governments can and do play a key role in promoting xeriscaping. In fact, it is city governments and water departments that have surfaced as the greatest promoters of xeriscaping. Many cities provide public information relating to climate-appropriate landscaping on their websites, and some even have xeriscape pages. For example, the city of Atlanta, Georgia, provides homeowners with water conservation literature. In addition, public areas can be xeriscaped to provided a how-to example for residents and to display the aesthetic beauty possible with xeriscaping. Fort Collins, Colorado, for example, xeriscaped the front lawn of city hall to increase the acceptability of such landscaping. Furthermore, educational demonstration gardens, workshops, and public information materials provided by municial governments can provide bioregionally specific information on xeriscaping. City landscape ordinances promoting xeriscaping have also been put into effect. The Lawn Permit Ordinance in Aurora, Colorado, for example, limits the amount of turf that a resident may install and also regulates soil composition to encourage proper water drainage.

Many municipal governments have discovered that financial incentives encourage change faster than arguments for environmental health, and many have instituted appropriate policies. The Southern Nevada Water Authority, for example, offers a Water Smart Landscapes rebate, which helps property owners convert grass to a more desert-friendly landscape. The Southern Nevada Water Authority will rebate customers $1.50 per square foot of grass removed and replaced with desert landscaping up to the first 5,000 square feet converted per property, per year. Beyond the first 5,000 feet, the authority provides a rebate of $1 per square foot. The maximum award for any property in a fiscal year is $300,000. The East Bay Municipal Utility District of California offers customers a free, five-spray, water-saving hose nozzle that allows the user to select any one of five sprays, from full force to a water-saving mist, that ultimately promotes more efficient water use. Denver Water offers similar rebates as part of its Water Conservation Plan.

The concept of xeriscaping has spread far and wide from its origins in Colorado. Even the wetter states along the eastern United States have embraced the concept as a way to save water and prevent yard waste and pesticides from contaminating watersheds. The practice has even spread abroad to cities in Saskatchewan, Canada, and in Spain.

This water-prudent method of landscaping has great potential to encourage native and regionally appropriate landscaping that is particularly aware of available water sources. As cities adopt sustainable landscaping principles, xeriscaping is likely to play a key role in defining standards, making laws, and providing landscape guidance for urban residents.

See Also: Ecological Footprint; Green Design, Construction, and Operations; Green Landscaping; Greywater.

Further Readings

Denver Water. "Xeriscape Principles." http://www.denverwater.org (Accessed April 2009).

Lockhart, Connie, et al. *Xeriscape Gardening: Water Conservation for the American Landscape.* New York: Macmillian, 1992.

Weinstein, Gail. *Xeriscape Handbook: A How-To Guide to Natural Resource–Wise Gardening.* Golden, CO: Fulcrum, 1999.

Shannon Tyman
University of Oregon

Green Cities Glossary

A

Abatement: Reducing the degree or intensity of, or eliminating, pollution.

Accident Site: The location of an unexpected occurrence, failure, or loss, either at a plant or along a transportation route, resulting in a release of hazardous materials.

Activated Sludge: Product that results when primary effluent is mixed with bacteria-laden sludge and then agitated and aerated to promote biological treatment, speeding the breakdown of organic matter in raw sewage undergoing secondary waste treatment.

Activity Plans: Written procedures in a school's asbestos-management plan that detail the steps a Local Education Agency (LEA) will follow in performing the initial and additional cleaning, operation, and maintenance-program tasks; periodic surveillance; and reinspection required by the Asbestos Hazard Emergency Response Act (AHERA).

Affected Landfill: Under the Clean Air Act, landfills that meet criteria for capacity, age, and emissions rates set by the EPA. They are required to collect and combust their gas emissions.

Affected Public: 1. The people who live and/or work near a hazardous waste site. 2. The human population adversely impacted following exposure to a toxic pollutant in food, water, air, or soil.

Air Quality Standards: The level of pollutants prescribed by regulations that are not to be exceeded during a given time in a defined area.

B

Backflow/Back Siphonage: A reverse flow condition created by a difference in water pressures that causes water to flow back into the distribution pipes of a drinking water supply from any source other than the intended one.

Back Pressure: A pressure that can cause water to backflow into the water supply when a user's wastewater system is at a higher pressure than the public system.

Biological Integrity: The ability to support and maintain balanced, integrated, functionality in the natural habitat of a given region. Concept is applied primarily in drinking water management.

Biome: Entire community of living organisms in a single major ecological area.

Broadcast Application: The spreading of pesticides over an entire area.

C

Chemical Treatment: Any one of a variety of technologies that use chemicals or a variety of chemical processes to treat waste.

Chlorination: The application of chlorine to drinking water, sewage, or industrial waste to disinfect or to oxidize undesirable compounds.

Clean Coal Technology: Any technology not in widespread use prior to the Clean Air Act Amendments of 1990. This act achieved significant reductions in pollutants associated with the burning of coal.

Climate Change: The term *climate change* is sometimes used to refer to all forms of climatic inconsistency, but because the Earth's climate is never static, the term is more properly used to imply a significant change from one climatic condition to another. In some cases, climate change has been used synonymously with the term global warming; scientists, however, tend to use the term in the wider sense to also include natural changes in climate.

Closure: The procedure a landfill operator must follow when a landfill reaches its legal capacity for solid waste, ceasing acceptance of solid waste, and placing a cap on the landfill site.

Coal Cleaning Technology: A precombustion process by which coal is physically or chemically treated to remove some of its sulfur so as to reduce sulfur dioxide emissions.

Collector: A public or private hauler that collects nonhazardous waste and recyclable materials from residential, commercial, institutional, and industrial sources.

Combined Sewer Overflows: Discharge of a mixture of storm water and domestic waste when the flow capacity of a sewer system is exceeded during rainstorms.

Community Water System: A public water system that serves at least 15 service connections used by year-round residents, or regularly serves at least 25 year-round residents.

Complete Treatment: A method of treating water that consists of the addition of coagulant chemicals, flash mixing, coagulation-flocculation, sedimentation, and filtration. Also called conventional filtration.

Contingency Plan: A document setting out an organized, planned, and coordinated course of action to be followed in case of a fire, explosion, or other accident that releases toxic chemicals, hazardous waste, or radioactive materials that threaten human health or the environment.

Curbside Collection: Method of collecting recyclable materials at homes, community districts, or businesses.

D

Dead End: The end of a water main that is not connected to other parts of the distribution system.

Decontamination: Removal of harmful substances such as noxious chemicals, harmful bacteria or other organisms, or radioactive material from exposed individuals, rooms, furnishings in buildings, or the exterior environment.

Drinking Water State Revolving Fund: The fund provides capitalization grants to states to develop drinking water revolving loan funds to help finance system infrastructure improvements, assure source-water protection, enhance operation and management of drinking-water systems, and otherwise promote local water-system compliance and protection of public health.

Dump: A site used to dispose of solid waste without environmental controls.

E

Ecosystem: The interacting system of a biological community and its nonliving environmental surroundings.

Endangered Species: Animals, birds, fish, plants, or other living organisms threatened with extinction by anthropogenic (human-caused) or other natural changes in their environment. Requirements for declaring a species endangered are contained in the Endangered Species Act.

Environment: The sum of all external conditions affecting the life, development, and survival of an organism.

Environmental Medium: A major environmental category that surrounds or contacts humans, animals, plants, and other organisms (such as surface water, ground water, soil, or air) and through which chemicals or pollutants move.

Episode (Pollution): An air pollution incident in a given area caused by a concentration of atmospheric pollutants under meteorological conditions that may result in a significant increase in illnesses or deaths. May also describe water pollution events or hazardous material spills.

Erosion: The wearing away of land surface by wind or water, intensified by land-clearing practices related to farming, residential or industrial development, road building, or logging.

F

Federal Implementation Plan: Under current law, a federally implemented plan to achieve attainment of air quality standards, used when a state is unable to develop an adequate plan.

Finished Water: Water is "finished" when it has passed through all the processes in a water treatment plant and is ready to be delivered to consumers.

Fluoridation: The addition of a chemical to increase the concentration of fluoride ions in drinking water to reduce the incidence of tooth decay.

Food Waste: Uneaten food and food preparation wastes from residences and commercial establishments such as grocery stores, restaurants, produce stands, institutional cafeterias and kitchens, and industrial sources like employee lunchrooms.

Fuel Economy Standard: The Corporate Average Fuel Economy Standard (CAFE) effective in 1978. It enhanced the national fuel conservation effort imposing a miles-per-gallon floor for motor vehicles.

G

Game Fish: Species like trout, salmon, or bass caught for sport. Many of them show more sensitivity to environmental change than "rough" fish.

Gasohol: Mixture of gasoline and ethanol derived from fermented agricultural products containing at least 9 percent ethanol. Gasohol emissions contain less carbon monoxide than those from gasoline.

Global Warming: An increase in the near surface temperature of the Earth. Global warming has occurred in the distant past as the result of natural influences, but the term is most often used to refer to the warming predicted to occur as a result of increased emissions of greenhouse gases. Scientists generally agree that the Earth's surface has warmed by about one degree Fahrenheit in the past 140 years. The Intergovernmental Panel on Climate Change (IPCC) recently concluded that increased concentrations of greenhouse gases are causing an increase in the Earth's surface temperature, and that increased concentrations of sulfate aerosols have led to relative cooling in some regions, generally over and downwind of heavily industrialized areas.

Greenhouse Gas: A gas, such as carbon dioxide or methane, which contributes to potential climate change.

Ground-Water Disinfection Rule: A 1996 amendment of the Safe Drinking Water Act requiring the EPA to promulgate national primary drinking water regulations requiring disinfection for all public water systems, including surface waters and ground water systems.

H

Hazardous Waste Landfill: An excavated or engineered site where hazardous waste is deposited and covered.

Heavy Metals: Metallic elements with high atomic weights, such as mercury, chromium, cadmium, arsenic, and lead, which can damage living things at low concentrations and tend to accumulate in the food chain.

High-Level Nuclear Waste Facility: Plant designed to handle disposal of used nuclear fuel, high-level radioactive waste, and plutonium waste.

High-Line Jumpers: Pipes or hoses connected to fire hydrants and laid on top of the ground to provide emergency water service for an isolated portion of a distribution system.

High-Risk Community: A community located within the vicinity of numerous sites of facilities or other potential sources of environmental exposure/health hazards that may result in high levels of exposure to contaminants or pollutants.

I

Incident Command Post: A facility located at a safe distance from an emergency site, where the incident commander, key staff, and technical representatives can make decisions and deploy emergency manpower and equipment.

Indirect Discharge: Introduction of pollutants from a nondomestic source into a publicly owned waste-treatment system. Indirect dischargers can be commercial or industrial facilities that discharge wastes into local sewers.

Infiltration Gallery: A sub-surface groundwater collection system, typically shallow in depth, constructed with open-jointed or perforated pipes that discharge collected water into a watertight chamber from which the water is pumped to treatment facilities and into the distribution system. Usually located close to streams or ponds.

J

Joint and Several Liability: Under CERCLA, this legal concept relates to the liability for Superfund site cleanup and other costs on the part of more than one potentially responsible party. If there were several owners or users of a site that became contaminated over the years, they could all be considered potentially liable for cleaning up the site.

L

Landfills: 1. Sanitary landfills are disposal sites for nonhazardous solid wastes spread in layers, compacted to the smallest practical volume, and covered by material applied at the end of each operating day. 2. Secure chemical landfills are disposal sites for hazardous waste, selected and designed to minimize the chance of release of hazardous substances into the environment.

Large Water System: A water system that services more than 50,000 customers.

Litter: The highly visible portion of solid waste carelessly discarded outside the regular garbage and trash collection and disposal system.

M

Majors: Larger publicly owned treatment works (POTWs) with flows equal to at least one million gallons per day, or servicing a population equivalent to 10,000 persons; certain other POTWs having significant water quality impacts.

Materials Recovery Facility (MRF): A facility that processes residentially collected mixed recyclables into new products available for market.

Measure of Exposure: A measurable characteristic of a stressor (such as the specific amount of mercury in a body of water) used to help quantify the exposure of an ecological entity or individual organism.

Medium-Size Water System: A water system that serves 3,300 to 50,000 customers.

Methoxychlor: Pesticide that causes adverse health effects in domestic water supplies and is toxic to freshwater and marine aquatic life.

Minors: Publicly owned treatment works with flows less than one million gallons per day.

N

National Municipal Plan: A policy created in 1984 by the EPA and the states in 1984 to bring all publicly owned treatment works into compliance with Clean Water Act requirements.

National Secondary Drinking Water Regulations: Commonly referred to as NSDWRs.

Noncommunity Water System: A public water system that is not a community water system, such as the water supply at a campsite or national park.

Nuclear Reactors and Support Facilities: Uranium mills, commercial power reactors, fuel reprocessing plants, and uranium enrichment facilities.

O

Oil Spill: An accidental or intentional discharge of oil that reaches bodies of water. Can be controlled by chemical dispersion, combustion, mechanical containment, and/or adsorption. Spills from tanks and pipelines can also occur away from water bodies, contaminating the soil, getting into sewer systems, and threatening underground water sources.

Ozone Depletion: Destruction of the stratospheric ozone layer that shields the Earth from ultraviolet radiation harmful to life. This destruction of ozone is caused by the breakdown of certain chlorine and/or bromine containing compounds (chlorofluorocarbons or halons), which break down when they reach the stratosphere and then catalytically destroy ozone molecules.

Ozone Hole: A thinning break in the stratospheric ozone layer. There is an "ozone hole" when the detected amount of depletion exceeds 50 percent. Seasonal ozone holes have been observed over the Antarctic and Arctic regions, part of Canada, and the extreme northeastern United States.

Ozone Layer: The protective layer in the atmosphere, about 15 miles above the ground, that absorbs some of the sun's ultraviolet rays, thereby reducing the amount of potentially harmful radiation that reaches the Earth's surface.

P

Pandemic: A widespread epidemic throughout an area, nation, or the world.

Performance Bond: Cash or securities deposited before a landfill operating permit is issued, which are held to ensure that all requirements for operating and subsequently closing the landfill are faithfully performed. The money is returned to the owner after proper closure of the landfill is completed. If contamination or other problems appear at any time during operation, or upon closure, and are not addressed, the owner must forfeit all or part of the bond, which is then used to cover clean-up costs.

Permit: An authorization, license, or equivalent control document issued by EPA or an approved state agency to implement the requirements of an environmental regulation; such as a permit to operate a wastewater treatment plant, or to operate a facility that may generate harmful emissions.

Pollutant: Generally, any substance introduced into the environment that adversely affects the usefulness of a resource or the health of humans, animals, or ecosystems.

Pollution: Generally, the presence of a substance in the environment that because of its chemical composition or quantity prevents the functioning of natural processes and produces undesirable environmental and health effects. Under the Clean Water Act, for example, the term has been defined as the man-made or man-induced alteration of the physical, biological, chemical, and radiological integrity of water and other media.

Potable Water: Water that is safe for drinking and cooking.

Pretreatment: Processes used to reduce, eliminate, or alter the nature of wastewater pollutants from nondomestic sources before they are discharged into publicly owned treatment works.

Primary Drinking Water Regulation: Applies to public water systems and specifies a contaminant level, which, in the judgment of the EPA Administrator, will not adversely affect human health.

Proposed Plan: A plan for a site cleanup that is available to the public for comment.

Public Health Context: The incidence, prevalence, and severity of diseases in communities or populations and the factors that account for them, including infections, exposure to pollutants, and other exposures or activities.

Publicly Owned Treatment Works (POTWs): A waste-treatment works owned by a state, unit of local government, or Indian tribe, usually designed to treat domestic wastewaters.

Public Water System: A system that provides piped water for human consumption to at least 15 service connections, or regularly serves 25 individuals.

R

Raw Sewage: Untreated wastewater and its contents.

Recycle/Reuse: Minimizing waste generation by recovering and reprocessing usable products that might otherwise become waste (like the recycling of aluminum cans, paper, and bottles).

Refuse Reclamation: Conversion of solid waste into useful products; such as composting organic wastes to make soil conditioners, or separating aluminum and other metals for recycling.

Reserve Capacity: Extra treatment capacity built into solid waste and wastewater treatment plants and interceptor sewers to accommodate flow increases due to future population growth.

Reservoir: Any natural or artificial holding area used to store, regulate, or control water.

Reverse Osmosis: A treatment process used in water systems by adding pressure to force water through a semi-permeable membrane. Reverse osmosis removes most drinking water contaminants. Also used in wastewater treatment. Large-scale reverse osmosis plants are being developed.

S

Safe Water: Water that does not contain harmful bacteria, toxic materials, or chemicals, and is considered safe for drinking even if it may have taste, odor, color, and certain mineral problems.

Salinity: The percentage of salt in water.

Secondary Drinking Water Regulations: Non-enforceable regulations applying to public water systems and specifying the maximum contamination levels that, in the judgment of EPA, are required to protect the public welfare. These regulations apply to any contaminants that may adversely affect the odor or appearance of such water and consequently may cause people served by the system to discontinue its use.

Secondary Treatment: The second step in most publicly owned waste treatment systems in which bacteria consume the organic parts of the waste. It is accomplished by bringing together waste, bacteria, and oxygen in trickling filters or in the activated sludge process. This treatment removes floating and settleable solids, and about 90 percent of the oxygen-demanding substances and suspended solids. Disinfection is the final stage of secondary treatment.

Sewer: A channel or conduit that carries wastewater and storm-water runoff from the source to a treatment plant or receiving stream. Sanitary sewers carry household, industrial, and commercial waste. Storm sewers carry runoff from rain or snow. Combined sewers handle both.

Significant Municipal Facilities: Those publicly owned sewage treatment plants that discharge a million gallons per day or more and are therefore considered by states to have the potential to substantially affect the quality of receiving waters.

Sole-Source Aquifer: An aquifer that supplies 50 percent or more of the drinking water of an area.

Storm Sewer: A system of pipes (separate from sanitary sewers) that carries water runoff from buildings and land surfaces.

Superfund: The program operated under the legislative authority of CERCLA and SARA that funds and carries out EPA solid waste emergency and long-term removal and remedial activities. These activities include establishing the National Priorities List, investigating sites for inclusion on the list, determining their priority, and conducting and/or supervising cleanup and other remedial actions.

Susceptibility Analysis: An analysis to determine whether a public water supply is subject to significant pollution from known potential sources.

T

Teratogen: A substance capable of causing birth defects.

Tertiary Treatment: Advanced cleaning of wastewater that goes beyond the secondary or biological stage, removing nutrients such as phosphorus, nitrogen, and most biological oxygen demand and suspended solids.

Total Maximum Daily Load (TMDL): A calculation of the highest amount of a pollutant that a water body can receive and safely meet water quality standards set by the state, territory, or authorized tribe.

Toxaphene: Chemical that causes adverse health effects in domestic water supplies and is toxic to fresh water and marine aquatic life.

Transporter: Hauling firm that picks up properly packaged and labeled hazardous waste from generators and transports it to designated facilities for treatment, storage, or disposal. Transporters are subject to EPA and DOT hazardous waste regulations.

U

Ultra Clean Coal (UCC): Coal that is washed, ground into fine particles, then chemically treated to remove sulfur, ash, silicone, and other substances; usually briquetted and coated with a sealant made from coal.

Underground Sources of Drinking Water: Aquifers currently being used as a source of drinking water, or those capable of supplying a public water system. They have a total dissolved solids content of 10,000 milligrams per liter or less, and are not "exempted aquifers."

Urban Runoff: Storm water from city streets and adjacent domestic or commercial properties that carries pollutants of various kinds into the sewer systems and receiving waters.

W

Waste: 1. Unwanted materials left over from a manufacturing process. 2. Refuse from places of human or animal habitation.

Waste Treatment Stream: The continuous movement of waste from generator to treater and disposer.

Wastewater: The spent or used water from a home, community, farm, or industry that contains dissolved or suspended matter.

Water Pollution: The presence in water of enough harmful or objectionable material to damage the water's quality.

Water Supplier: One who owns or operates a public water system.

Water Well: An excavation where the intended use is for location, acquisition, development, or artificial recharge of ground water.

Source: U.S. Environmental Protection Agency (http://www.epa.gov/OCEPAterms)

Green Cities Resource Guide

Books

Abbott, Carl. *Greater Portland: Urban Life and Landscape in the Pacific Northwest.* Philadelphia: University of Pennsylvania Press, 2001.

Allen, Patricia. *Together at the Table: Sustainability and Sustenance in the American Agrifood System.* University Park: Pennsylvania State University Press, 2004.

Appleyard, Donald. *Livable Streets.* Berkeley: University of California Press, 1981.

Barlow, M. *Blue Covenant: The Global Water Crisis and the Coming Battle for the Right to Water.* New York: The New Press, 2008.

Barty-King H. *Water: An Illustrated History of Water Supply and Wastewater in the United Kingdom.* Shrewsbury, UK: Quiller Press, 1992.

Bell, Simon and Stephen Morse. *Measuring Sustainability. Learning From Doing.* London: Earthscan, 2003.

Berelowitz, Lance. *Dream City: Vancouver and the Global Imagination.* Vancouver, Canada: Douglas & McIntyre, 2005.

Blomley, Nick. *Unsettling the City: Urban Land and the Politics of Property.* New York: Routledge, 2004.

Brooks, K. N., P. F. Ffolliott, H. M. Gregersen and L. F. DeBano. *Hydrology and the Management of Watersheds.* Ames, IA: Blackwell Publishing, 2003.

Calthorpe, Peter. *The Next Urban Metropolis.* Princeton, NJ: Princeton Architectural Press, 1995.

Dash, M. C. *Ecology, Chemistry and Management of Environmental Pollution.* New York: Macmillan, 2004.

Dittmar, Hank and Gloria Ohland, eds. *The New Transit Town: Best Practices in Transit-Oriented Development.* Washington, DC: Island Press, 2004.

Energy Information Administration (EIA). *International Energy Outlook 2008.* Washington, DC: EIA Publications, 2008.

Ewing, Reid. *Traffic Calming: State of the Practice.* Washington, DC: Institute of Transportation Engineers/Federal Highway Administration, 1999.

Foster, K., A. Stelmack and D. Hindman. *Sustainable Residential Interiors.* Hoboken, NJ: John Wiley & Sons, 2007.

Gleick, P. *The World's Water: The Biennial Report on Freshwater Resources*. Washington, DC: Island Press, 1999.

Global Wind Energy Council. *Global Wind 2008 Report*. Brussels, Belgium: Global Wind Energy Council, 2008.

Grant, Jill. *Planning the Good Community: New Urbanism in Theory and Practice*. London: Routledge, 2006.

Grava, Sigurd. *Urban Transportation Systems: Choices for Communities*. New York: McGraw-Hill, 2003.

Gray, N. F. *Drinking Water Quality: Problems and Solutions*. New York: Cambridge University Press, 2008.

Hardoy, Jorge E., Diana Miltin and David Satterthwaite. *Environmental Problems in an Urbanizing World*. London: Earthscan, 2004.

Hatch, Alex. *Cracks in the Asphalt: Community Gardens of San Francisco*. San Francisco, CA: Pasha Press, 2008.

Hellmund, Paul Cawood and Daniel Smith. *Designing Greenways: Sustainable Landscapes for Nature and People*. Washington, DC: Island Press, 2006.

Hills, Richard L. *Power From Wind: A History of Windmill Technology*. New York: Cambridge University Press, 1994.

Ighacimuthu, S. *Environmental Awareness and Protection*. New Delhi, India: Phoenix Publishing House Pvt. Ltd., 1998.

International Energy Agency (IEA). *World Energy Outlook 2008*. Paris: IEA Publications, 2008.

Kellogg, Scott and Stacy Pettigrew. *Toolbox for Sustainable City Living*. Cambridge, MA: South End Press, 2008.

Kidd, J. S. and Renee A. Kidd. *Air Pollution Problems and Solutions*. New York: Chelsea House, 2006.

Klingle, Matthew. *Emerald City: An Environmental History of Seattle*. New Haven, CT: Yale University Press, 2007.

Kopec, D. *Health, Sustainability and the Built Environment*. New York: Fairchild Books, Inc., 2009.

Lansing, Jewel. *Portland: People, Politics, and Power, 1851–2001*. Corvallis: Oregon State University Press, 2003.

Lawn, Philippe. *Sustainable Development Indicators In Ecological Economics*. Cheltenham, UK: Edward Elgar, 2006.

Lawson, Laura. *City Bountiful: A Century of Community Gardening in America*. Berkeley, CA: University of California Press, 2005.

Leeds, Rob, et al. *Non-Point Source Pollution: Water Primer*. Columbus: Ohio State University Extension, 1996.

Lowi, M. *Water and Power: The Politics of a Scarce Resource in the Jordan River Basin*. New York: Cambridge University Press, 1995.

Lundkvist, Lennart J. *The Hare and the Tortoise: Clean Air Policies in the United States and Sweden*. Ann Arbor: University of Michigan Press, 1980.

MacColl, E. Kimbark. *The Growth of a City: Power and Politics in Portland, Oregon 1915 to 1950*. Portland, OR: Georgian Press, 1979.

MacColl, E. Kimbark. *The Shaping of a City: Business and Politics in Portland, Oregon 1885 to 1915*. Portland, OR: Georgian Press, 1976.

Mallon, Karl. *Renewable Energy Policy and Politics: A Handbook for Decision-Making*. London: Earthscan, 2006.

Miller, Robert W. *Urban Forestry: Planning and Managing Urban Greenspaces*. Upper Sadlle River, NJ: Prentice Hall, 1997.

Moll, Gary and Sara Ebenreck. *Shading Our Cities: A Resource Guide for Urban and Community Forests*. Washington, DC: Island Press, 1989.

Morris, D. *Self-Reliant Cities: Energy and Transformation in Urban America*. San Francisco, CA: Sierra Club Books, 1982.

Mougeot, Luc J. A. *Growing Better Cities: Urban Agriculture for Sustainable Development*. Ottawa, Canada: International Development Research Centre, 2006.

National Research Council. *Compensating for Wetland Losses Under the Clean Water Act*. Washington, DC: The National Academies Press, 2001.

Newman, Peter and Jeffrey Kenworthy. *Sustainability and Cities: Overcoming Automobile Dependence*. Washington, DC: Island Press, 2009.

Ohlsson, L. *Hydropolitics: Conflicts Over Water as a Development Constraint*. London: Zed Books, 1995.

Palahniuk, Chuck. *Fugitives and Refugees: A Walk in Portland, Oregon*. New York: Crown, 2003.

Perlin, John. *From Space to Earth: The Story of Solar Electricity*. Cambridge, MA: Harvard University Press, 2002.

Reisner, M. *Cadillac Desert: the American West and Its Disappearing Water*. New York: Penguin Books, 1993.

Roseland, M. *Toward Sustainable Communities; Resources for Citizens and Their Governments*. Gabriola Island, British Columbia, Canada: New Society Publishers, 1998.

Rosillo-Calle, Frank, Sarah Hemstock, Peter de Groot, and Jeremy Woods. *The Biomass Assessment Handbook: Bioenergy for a Sustainable Environment*. London: Earthscan, 2008.

Sharma, B. K. *Environmental Chemistry*. Meerut, IN: Goel Publishing House, 2005.

Staddon, C. *Managing Europe's Water Resources: 21st Century Challenges*. Surrey, UK: Ashgate Books, 2009.

Talen, Emily. *New Urbanism and American Planning: The Conflict of Cultures*. London: Routledge, 2005.

Tchobanoglous, G., H. Theisen and S. A. Vigil. *Integrated Solid Waste Management*. New York: McGraw-Hill, 1993.

Tolley, Rodney. *Sustainable Transport: Planning for Walking and Cycling in Urban Environments*. Cambridge, UK: Woodhead Publishing, 2003.

Weinberg, Adam S., David N. Pellow and Allan Schnaiberg. *Urban Recycling and the Search for Sustainable Community Development*. Princeton, NJ: Princeton University Press, 2000.

Westra, L. *Environmental Justice and the Rights of Ecological Refugees*. London: Earthscan Press, 2009.

World Commission on Environment and Development (WCED). *Our Common Future*. New York: Oxford University Press, 1987.

Journals

American Journal of Public Health (American Public Health Association)
American School & University (Penton Media)

Cities (Elsevier)
Conservation Biology (John Wiley & Sons)

Ecological Applications (Ecological Society of America)
Energy & Buildings (Elsevier)
Environment: Science and Policy for Sustainable Development (MetaPress)
Environment and Urbanization (SAGE Publications)

Geoforum (Elsevier)
Global Environmental Change (Elsevier)
Global Environmental Politics (MIT Press)

Journal of Engineering, Design & Technology (Emerald)
Journal of Planning Literature (SAGE Publications)

Landscape Journal (University of Wisconsin Press)

Physics and Chemistry of the Earth (Elsevier)
Planning (American Planning Association)
Planning Practice & Research (Taylor & Francis Group)

Renewable Resources Journal (Renewable Natural Resources Foundation)

The Science of the Total Environment (Elsevier)

Urban Design International (Palgrave Macmillan)
Urban Studies (SAGE Publications)

Water, Science and Technology (IWA Publishing)
Water and Environment Journal (John Wiley & Sons)

Websites

American Council for an Energy Efficient Economy (ACEEE)
    www.aceee.org

American Public Transportation Association
    www.atpa.com

Encyclopedia of Life
    www.eol.org

The Energy and Research Institute
    edugreen.teri.res.in

Global Biodiversity Information Facility
    www.gbif.org

Global Footprint Network
    www.footprintnetwork.org

The Green Guide
www.thegreenguide.com

Greenhouse Gas Protocol
www.ghgprotocol.org

Our Green Cities – The Top 12 Most Sustainable Cities
www.ourgreencities.com

Sustainable Seattle
www.sustainableseattle.org

United Nations Millennium Goals
www.un.org/millenniumgoals

U.S. Department of Energy – Energy Efficiency and Renewable Energy (EERE)
www.eere.energy.gov

U.S. Environmental Protection Agency
www.epa.gov

U.S. Environmental Protection Agency – ENERGY STAR
www.energystar.gov

Green Cities Appendix

American Community Garden Association

http://www.communitygarden.org/about-acga

The American Community Gardening Association (ACGA) is a nonprofit membership organization of individuals in the United States and Canada interested in promoting community gardening through means such as encouraging research, conducting educational programs, developing resources about community gardening, and supporting state and regional community gardening networks. The website explains what community gardens are, and reasons for supporting their formation (which include improving nutrition, saving money, preserving green space, stimulating social interaction, and conserving the environment). It provides information about starting a community garden, and a searchable interface to find a community garden by geographic area or name. The website also includes resources for children and young people, information about the ACGA annual conference and other training opportunities, downloadable draft documents, how-to manuals and start-up guides, and information about political activism related to community gardens.

Built Environment/Healthy Community Design

http://www.cdc.gov/nceh/ehs/Topics/BuiltEnvironment.htm

This website, created by the Centers for Disease Control and Prevention (CDC) of the U.S. government, brings together a number of resources about the relationship between the built environment and human health. Topics covered include methods to deal with blighted properties in small towns, information about improving the air quality in buildings, community design and the built environment, health impact assessments of construction projects and developments, and land use and planning. Most of the documents come from within agencies of the U.S. government, but the Rollins School of Public Health at Emory University (in Atlanta, Georgia) provides several documents. There are also extensive links to other relevant agencies within federal and state governments, as well as universities and private associations. Topics include watershed protection, sustainable buildings and cityscapes, reduction of environmental hazards in schools and childcare settings, promotion of walking and bicycling for transportation, accessibility and universal design, and urban sprawl.

5 Amazing Green Cities

http://science.howstuffworks.com/five-amazing-green-cities.htm

This website, part of the HowStuffWorks site run by Discovery Communications, provides a history and overview of the concept of green cities before looking at five specific cities, each notable for at least one aspect of protecting the environment. Malmö, Sweden, is featured for its use of renewable resources, including generation and use of green electricity, and for transforming many of its neighborhoods into ecologically friendly enclaves favoring bicycle and pedestrian traffic. Copenhagen, Denmark, is noted for its clean harbors, use of windmills to generate electricity, and wide use of bikes and the metro system for transportation. Portland, Oregon, is noted for preserving green space within urban boundaries, being the first U.S. city to enact a plan to reduce greenhouse gas emissions, and for its plan to supply 100% of its energy in 2010 from renewable sources. Vancouver, Canada, is noted for leading the world in the use of hydroelectricity, having a 100-year plan for green living, and for the use of solar-powered trash compactors. Reykjavik, Iceland, is noted for its plan to eliminate use of fossil fuels by 2050, and for its use of hydropower and geothermal resources. Each article is sourced and includes links to relevant topics within the HowStuffWorks site (such as hydroelectric energy and urban heat island), as well to exterior sites.

Global Ecovillage Network

http://www.ecovillage.org

Global Ecovillage Network is a confederation of people and communities that promotes the development and use of technologies for sustainable living, including the creation of ecovillages, which may incorporate elements such as permaculture, green building and production, alternative energy sources, ecological design, and social and community building. The Global Ecovillage Network website includes information about ecovillages and many other topics related to sustainable living, as well as tools such as the Community Sustainability Assessment to allow a community to evaluate its current condition on dimensions such as sense of place, consumption, waste and pollution management, health care, education, and cultural sustainability. It also includes a searchable database of ecovillages and events, resources, and news relating to ecovillage concerns.

ICLEI–Local Governments for Sustainability

http://www.iclei.org

This is the website of the ICLEI–Local Governments for Sustainability, with a membership including over 1,000 cities, towns, and counties, as well as national and regional government organizations interested in promoting sustainable development. A brochure explaining the organization's activities (specific initiatives include creating local action plans to reduce greenhouse gases, standards for the measurement and reduction of municipal carbon emissions, and coordination of the representation of local governments in the UN Framework Convention on Climate Change) is downloadable from the website in several languages. Information is also available on the organization, its history, members, supporters, and programs. The website has links to resources (including publications and case studies) about local sustainable development, training opportunities, conferences, news about sustainable development, prizes and awards, links to ICLEI–Local Governments for Sustainability research, and information about technical and consulting services available for sustainable development projects.

Traffic Calming

http://www.trafficcalming.org/index.html

This website, created by the transportation consultant firm Fehr & Peers, collects a number of resources related to traffic calming. Traffic calming includes procedures intended to reduce vehicle speeds, increase pedestrian safety, and improve quality of life. Examples of such procedures include changing street alignment and traffic routes, installing devices such as speed bumps and raised crosswalks, and increasing police enforcement of existing traffic laws. The website provides a history of traffic calming (beginning in Europe in the 1960s and in the United States in the 1970s) and provides discussion of the advantages and disadvantages, including estimated costs for many different means of traffic calming. Different means of traffic calming are evaluated in terms of speed reduction, volume of traffic reduction, and reduction in collisions. Links are provided to the websites of many traffic-calming programs in the United States and Canada. The website also includes an extensive bibliography of literature on traffic calming.

World Alliance for Decentralized Energy

http://www.localpower.org

World Alliance for Decentralized Energy (WADE), founded in 1997 during the UN Framework Convention on Climate Change, is an international organization that promotes the development and use of cogeneration, on-site power, and decentralized energy (DE) systems as a means to increase efficiency, decrease costs, and reduce harm to the environment. The website includes basic information about DE technologies, case studies of successful DE projects in Europe and China, benefits of DE (including improving access to energy in developing countries, greater efficiency, reduced harm to human health and the environment, reliability, and energy security), barriers to DE, and policies affecting DE. It also includes a searchable database of news items and press releases, calendar of upcoming events, list of chapters, multimedia files, and many downloadable publications from WADE and other organizations. The website also has an interface to look for resources and information by geographic region.

Sarah Boslaugh
Washington University in St. Louis

Index

Article titles and their page numbers are in **bold**.